Lecture Notes in Physics

Springer
Berlin
Heidelberg
New York
Barcelona
Hong Kong
London
Milan
Paris
Tokyo

Physics and Astronomy ONLINE LIBRARY

http://www.springer.de/phys/

Editorial Policy

The series *Lecture Notes in Physics* (LNP), founded in 1969, reports new developments in physics research and teaching -- quickly, informally but with a high quality. Manuscripts to be considered for publication are topical volumes consisting of a limited number of contributions, carefully edited and closely related to each other. Each contribution should contain at least partly original and previously unpublished material, be written in a clear, pedagogical style and aimed at a broader readership, especially graduate students and nonspecialist researchers wishing to familiarize themselves with the topic concerned. For this reason, traditional proceedings cannot be considered for this series though volumes to appear in this series are often based on material presented at conferences, workshops and schools (in exceptional cases the original papers and/or those not included in the printed book may be added on an accompanying CD ROM, together with the abstracts of posters and other material suitable for publication, e.g. large tables, colour pictures, program codes, etc.).

Acceptance

A project can only be accepted tentatively for publication, by both the editorial board and the publisher, following thorough examination of the material submitted. The book proposal sent to the publisher should consist at least of a preliminary table of contents outlining the structure of the book together with abstracts of all contributions to be included.
Final acceptance is issued by the series editor in charge, in consultation with the publisher, only after receiving the complete manuscript. Final acceptance, possibly requiring minor corrections, usually follows the tentative acceptance unless the final manuscript differs significantly from expectations (project outline). In particular, the series editors are entitled to reject individual contributions if they do not meet the high quality standards of this series. The final manuscript must be camera-ready, and should include both an informative introduction and a sufficiently detailed subject index.

Contractual Aspects

Publication in LNP is free of charge. There is no formal contract, no royalties are paid, and no bulk orders are required, although special discounts are offered in this case. The volume editors receive jointly 30 free copies for their personal use and are entitled, as are the contributing authors, to purchase Springer books at a reduced rate. The publisher secures the copyright for each volume. As a rule, no reprints of individual contributions can be supplied.

Manuscript Submission

The manuscript in its final and approved version must be submitted in camera-ready form. The corresponding electronic source files are also required for the production process, in particular the online version. Technical assistance in compiling the final manuscript can be provided by the publisher's production editor(s), especially with regard to the publisher's own Latex macro package which has been specially designed for this series.

Online Version/ LNP Homepage

LNP homepage (list of available titles, aims and scope, editorial contacts etc.):
http://www.springer.de/phys/books/lnpp/

LNP online (abstracts, full-texts, subscriptions etc.):
http://link.springer.de/series/lnpp/

K. Mecke D. Stoyan (Eds.)

Morphology of Condensed Matter

Physics and Geometry of Spatially Complex Systems

Springer

Editors

Klaus Mecke
Max-Planck-Institut für Metallforschung
Heisenbergstrasse 1
70569 Stuttgart, Germany

Dietrich Stoyan
TU Bergakademie Freiberg
Fakultät für Mathematik und Informatik
Institut für Stochastik
09596 Freiberg, Germany

Cover Picture: 3D Image of a block copolymer (SBS) microdomain structure reconstructed from a series of TappingMode scanning force microscopy phase and height images. The enclosed volume can be interpreted as PS cylinders within a $200 \times 160 \times 45$ nm^3 large portion of the SBS film. For size comparison and illustration of the molecular structure of the material, a SBS molecule bridging two PS domains is sketched (Reprinted from: R.Magerle, 'Nanotomography', *Physical Review Letters* **85**, 2749 (2000); ©2000 by The American Physical Society)

Cataloging-in-Publication Data applied for

A catalog record for this book is available from the Library of Congress.

Bibliographic information published by Die Deutsche Bibliothek

Die Deutsche Bibliothek lists this publication in the Deutsche Nationalbibliografie; detailed bibliographic data is available in the Internet at http://dnb.ddb.de

ISSN 0075-8450
ISBN 3-540-44203-0 Springer-Verlag Berlin Heidelberg New York

Springer-Verlag Berlin Heidelberg New York
a member of BertelsmannSpringer Science+Business Media GmbH

http://www.springer.de

Typesetting: Camera-ready by the authors/editor
Camera-data conversion by Steingraeber Satztechnik GmbH Heidelberg
Cover design: *design & production*, Heidelberg

Printed on acid-free paper
SPIN: 10891631 54/3141/du - 5 4 3 2 1 0

Preface

What does a typical foam cell look like? How do we measure spatially structured phases of composite materials? What patterns can be observed in thin films, emulsions, or polymer blends? Is it possible to describe and compare structures of complex, often disordered, materials quantitatively and relate the observed structure to macroscopic physical properties? Does the liquid flow through porous rocks, for instance, depend on the shape of the pores? These are only a few questions which are of enormous interest to both the scientific and the industrial communities.

The rapid growth of high resolution three-dimensional imaging technology is likely to impact on numerous experimental areas in physics, chemistry, geology and biology as well as on technical disciplines, including materials science, chemical engineering, or diagnostic medicine. Nowadays, experimental techniques such as high-resolution X-ray computed tomography or confocal microscopy are available to measure the morphology of complex materials and visualise, for instance, multiple fluid phases in porous materials. Such experimental facilities will provide researchers in the near future with a rich source of experimental data to further the goals of their research. The development of facilities which can acquire images, perform geometric analysis, visualise and calculate physical properties will have a major impact on the goal to catalogue material structure and infer physical properties from structural characteristics. Therefore, the development of expertise in computational image analysis, spatial statistics and materials modelling using typical three-dimensional data sets will become an essential requirement in these fields. Morphological and spatial statistics techniques are therefore certain to be more generally useful for a range of researchers.

The objective behind this volume is to bring the tools of physics, spatial statistics and mathematics to bear on these problems, which concern not only workers in those fields but also, for instance, engineers, geologists and forestry researchers. While the previous volume 'Statistical Physics and Spatial Statistics – The Art of Analysing and Modelling Spatial Structures and Pattern Formation', edited by Klaus Mecke and Dietrich Stoyan (Lecture Notes in Physics, Vol. 554, Springer, July 2000) focuses on the statistical characterisation of random spatial configurations, in particular point processes and Gibbsian particle models, we try to emphasise here applications of morphology on real physical systems ranging from porous and composite materials to microemulsions and foams . Additional techniques of spatial statistics and integral geometry, such as shape analysis and mark correlations, as well as measures such as contact distributions, topological Betti numbers and tensor valuations are also considered.

In an attempt to guide the reader and categorise the diverse contributions we have divided the articles into two parts: the first focusing on physical systems such as spatially structured solids and complex fluids. The second part comprises contributions with a background in spatial statistics and morphology. Of course, a strict distinction is not possible since each article provides some insight in each category but nevertheless the book tries to group together main features, such as solids, fluids, shape measures and spatial distributions.

Part I. Complex Structured Condensed Matter

Porous media, foams , microemulsions and liquid crystals are examples of spatially structured materials. Since they are nowadays experimentally accessible, these systems are particularly interesting for spatial statisticians and mathematicians in order to apply their techniques to characterise and model random spatial configurations.

Porous materials such as sandstones or papers are considered as standard examples in nature for a complex spatial structure. Worldwide, the petroleum industry spends an excess of a billion dollars annually on the characterisation of reservoirs and on laboratory measurements of core materials. These measurements are critical in the development of oil and gas fields because they play a major role in predicting hydrocarbon recovery. Despite the large sums spent on measurements and the ability to image and visualise, there is little basic science to support the interpretation of data and to improve the quantitative morphological description and modelling of laboratory core measurements. Modest improvements in relating physical properties to geometric shapes of pores would significantly reduce the economic risk associated with new oil and gas developments and have a major impact on the petroleum industry. Such a structure–property relationship is also essential in crustal geophysics. Within the Earth's crust, fluid flow in fractured rock masses influences a range of important geological processes, particularly the genesis of many types of precious metal and base metal ore deposits and hydrocarbon migration in some reservoir types. A rigorous understanding of fracture-controlled fluid migration based on statistical measures of structure and distribution of fractures is also required to develop more effective strategies for the sustainable development of geothermal energy resources and for the analysis of risk associated with toxic waste - both nuclear and chemical - containment in deep underground repositories. Another important application of morphological techniques to describe the shape of solids is in paper manufacture: the development of improved printing mediums and technologies depends on an understanding of how a complex material made up of cellulose fibre matrix, often coated with a consolidated mass of pigment and binder, can be modified to improve printability, durability and appearance. To understand the flow, optical and mechanical properties of paper products, one must develop a realistic structural description of the pore space coupled with an ability to simulate flow and mechanical properties.

A major shortcoming in the understanding of processes involving complex porous and composite materials has been the inability to accurately characterise the statistical distribution of shape and structure of disordered microstructures. Successful predictive modelling of the properties of 'real world' materials is reliant on accurate statistical characterisation, such as for the generation of equivalent network structures required for multiphase flow studies or fracture prediction. Dominique Jeulin shows in 'Spatial Statistics and Micromechanics of Materials' how useful bounds of elastic properties can be determined from the three-point statistics of random media and how fracture statistics models based on random functions allow us to predict the probability of fracture of materials.

The availability of direct measurements of the three-dimensional microstructure of sedimentary rocks, for example, requires the development of computational tools to gauge the stochastic nature of the void space and to construct realistic model representations of the complex space. We need to generate stochastic model networks with topological properties representative of real sedimentary rocks and measure and quantify pore geometry and correlations that occur at the pore scale. This task is approached in the contribution 'Characterising the Morphology of Disordered Materials' by Christoph Arns, Mark Knackstedt and Klaus Mecke, where advanced methods of image analysis and integral geometry (see Part II) are applied to reconstruct and predict physical properties of porous media based on the measurement of pore space. A major aim is to understand the structure–property correlation since the relationship between microstructure and macroscopic properties is a central issue in materials science. To date, engineers have relied on simple empirical models that ignore actual microstructural information, to correlate macroscopic properties of disordered materials to their microstructure. The bulk of experimental and theoretical work has been devoted to establishing the empirical coefficients for each class of material. Imaging materials via high-resolution X-ray CT, for instance, and subsequent laboratory measurement of material properties may help us form a more accurate and comprehensive picture of the role of the shape of disordered materials in governing the mechanical and transport properties. The development of theoretical models describing the relationships between disordered media and fluid transport properties is of specific interest to geologists, both in the laboratory and the field. Hans-Jörg Vogel demonstrates in 'Topological Characterisation of Porous Media' the significance of the Euler characteristic for the prediction of transport in natural soil where it is used as a function of pore diameter to generate network models of the porous structure. Other applications of the Euler characteristic are studied in Part II of this volume, for instance, in the contributions by Vanessa Robins on Betti numbers, Joachim Ohser et al. on discretized sets, and Claus Beisbart et al. on vector and tensor-valued descriptors, as well as in Part II of the previous Lecture Notes in Physics, Vol. 554.

Most modern materials have complex three-dimensional structures on the micron and nanometer scale. Unfortunately, most microscopy techniques available today are limited to two-dimensional imaging and the choices for obtaining nanometer-resolution three-dimensional volume images are rather limited, at least in materials science. This considerably hinders a detailed study of the structure–property relationship and renders a direct comparison between simulation results, the materials microstructure and its physical properties difficult, if not impossible. In particular, quantitative experimental data is needed as input for reliable computer simulations and for benchmarking of models. To this end, Robert Magerle presents nanotomography as a general procedure for high-resolution three-dimensional volume imaging of composite materials based on scanning probe microscopy. The specimen under study is eroded gradually and the remaining material is imaged with scanning probe microscopy at each freshly exposed surface.

From the resulting series of images, the specimen's three-dimensional microstructure can be reconstructed with methods adopted from computed tomography. This approach is expected to be a simple and versatile means for the imaging of various materials and physical properties with micron, nanometer and even atomic resolution.This effort is of immediate benefit to modelling scientists, spatial statisticians and physicists to interprete and compare model assumptions to reliable data of real-space volume images in three dimensions.

In contrast to solid structures, soft condensed matter, such as foams and fluids, have a typical energy scale which is comparable with thermal energies at room temperature. Thus, thermally activated movement of its molecular components are sufficient to induce configurational changes, which may eventually lead to complex spatial structure at room temperatures. Such thermally induced patterns are of particular importance for biological systems, such as living organisms, which are based on self-organisation of molecules in membranes and cells. However, the understanding of the underlying mechanisms is also essential for the application of many technologies in everyday life, including colloidal dispersions (paints, inks, food, creams, lotions), foams (beverages, see the contribution by Francois Graner), liquid crystals (displays, see the contribution by Friederike Schmid and Nguyen Phuong), polyelectrolyte gels used in diapers, as well as soaps for washing and cleaning. Most biomaterials, for example blood or cartilage, belong to the material class of soft condensed matter. One particularly important biomaterial is the biomembrane, that is, the protein carrying lipid bilayer which surrounds each cell and its organelles and which form spontaneously in mixtures of water and lipids. A good model system for the structural properties of biomembranes are self-assembling amphiphilic systems studied by Ulrich Schwarz and Gerhard Gompper in this volume, but also Langmuir monolayers studied by Mathias Lösche und Peter Krüger.

Soft condensed matter has received much attention in the past mainly because of its spatial structure and relevance for biological studies. Bicontinuous phases in microemulsions (see U. Schwarz and G. Gompper), complex domain patterns in fluid layers on substrates (see M. Lösche und P. Krüger), orientational order in liquid crystals (see F. Schmid and N. H. Phuong), or cellular shapes in foams (see F. Graner) are just a few spatial features observable in soft condensed matter. As for solid structures, the shape and connectivity of the constituents are essential for physical properties and biological function, so that techniques from spatial statistics and morphology become useful.

Ulrich Schwarz and Gerhard Gompper demonstrate in 'Bicontinuous Surfaces in Selfassembling Amphiphilic Systems' the richness of emerging structures in the case of self-organising amphiphiles. These are molecules with both hydrophilic and hydrophobic parts and which are particularly relevant for biological studies. In water, they self-assemble into extended sheet-like structures due to the hydrophobic effect. Many different interfacial structures emerge such as random surfaces, bicontinuous labyrinths, triply periodic minimal surfaces, their constant mean curvature and parallel surface companions. The free energy of an amphiphilic system can be written as a functional of its geometry and phase diagrams can be calculated by comparing the free energies following from different geometries. Due to the richness of its spatial structures, self-organising amphiphiles may serve as model systems for the study of the relationship of shape and energy in soft condensed matter.

Related systems are Langmuir monolayers which are commonly used as model systems for biological membranes. Biomembranes are complex, quasi-two-dimensional lipid-protein structures that self-assemble in aqueous media. For a quantitative understanding of the physics that underlies the functioning of biological cells, an assessment of the structural aspects of the molecular interactions within the membrane is necessary. These interactions may lead to complex two-dimensional domain patterns which are studied and characterised by Mathias Lösche and Peter Krüger in 'Morphology of Langmuir Monolayer Phases' using geometric tools such as Minkowski functionals (see Part II in this volume as well as in the previous Lecture Notes in Physics, Vol. 554).

Orientational order in liquid crystals is another important example of complex structures in condensed matter. Randomly distributed hard spheres in three dimensions form two types of structures, depending on their density: a disordered fluid phase and an ordered crystalline phase. For randomly distributed anisotropic particles, the situation is different: several so-called 'mesophases' exist at intermediate densities between the fluid and the solid state. For instance, the particles may be oriented in one common preferred direction, but do not have crystalline translational order (nematic phase). Alternatively, the particles may be arranged in layers of two dimensional fluids (smectic phases). Since these mesophases are neither crystalline nor truly liquid, they are commonly referred to as 'liquid crystal phases'. Friederike Schmid and Nguyen Phuong present computer simulations of simple model systems with special emphasis on systems of hard ellipsoids. The occurring orientational order of the molecules can be described by mark correlation functions which are introduced and studied in detail in Part II.

Whereas in liquid crystals geometric order is induced by molecular interactions, in foams the reverse effect occurs: geometrical constraints determine the energy landscape of the configurations. Liquid foams consist of gas bubbles separated by a continuous film of liquid which exhibit a complex spatial structure. The morphology of the vapour cells is a fascinating topic as shown by Graner in 'Two-Dimensional Fluid Foams at Equilibrium'. Fluid foams try to minimise their surface energy by decreasing the area of the bubbles. This mathematical 'minimum perimeter' problem has previously been approached by Gompper and Schwarz dealing with amphiphilic systems, but the minimal surfaces in foams are subjected to complex topological constraints which are very common in nature. Such foam-like structures occur also as grain boundaries in crystals, as magnetic domains in solids, as honeycombs, as biological epithelia, or as retina patterns.

In all these examples of structured condensed matter, the use of techniques stemming from spatial statistics and morphology are indispensable. The methods are described in detail in Part II, although the tools of mathematical morphology will have already been introduced and applied in the previous sections: for instance, mark correlations by F. Schmid and N. H. Phuong, Minkowski functionals and parallel surfaces by C. Arns et al., by U. Schwarz and G. Gompper, and by M. Lösche and P. Krüger, or the topology of minimal surfaces and perimeters by F. Graner. These techniques and other methods of spatial statistics to characterise the morphology of structures are considered in detail in the second part of this volume.

Part II. Spatial Statistics and Morphology

The imaging of complex topologies and geometries, typical of all real world materials, is now feasible, due to advances in three-dimensional imaging technology and computer processing. These advances are of importance to a variety of scientists, both academic and industrial, leading to an increasing interest not only to capture images but also to model and characterise them, that is, to quantify disordered morphologies. The ability to generate non-destructively three-dimensional images will allow the cataloguing of the detailed microstructure of a range of complex materials. The question remains: How can one describe disordered forms of arbitrary shape? We need to quantify random morphologies obtained experimentally utilising tools from integral, statistical and differential geometry and topology, and generate a complete inventory of forms using these techniques. This is of fundamental interest to scientists grappling with the quantification and comparison of disorder in general. Therefore, in Part II, stochastic techniques and geometric measures are introduced to characterise spatial structures occurring, for instance, in the physical systems introduced in Part I. Its first part focuses on the morphology, i.e., on the shape and connectivity of spatial domains, whereas in the second part the spatial arrangement, i.e., correlation functions and distance distributions, play a more important role.

Various statistical methods for the characterisation of textures, shapes and topological properties of geometrical structures are described. In contrast to the papers in Part I, the texts here are method-oriented. Pierre Soille describes in 'Morphological Texture Analysis: An Introduction' texture analysis methods based on ideas from mathematical morphology. Texture is usually defined as 'the characteristic physical structure given to an object by the size, shape, arrangement, and proportions of its parts'. It is shown that granulometries, certain families of transformations of images constructed by means of structuring elements, are very helpful for describing textures. A particular case of such a transformation is the construction of the parallel set for various radii. Soille shows that it is useful to evaluate the transformed images by geometrical measures such as area or volume. A project for the near future is to use other geometrical measures as well and, in particular, to use Minkowski functionals of dimensions smaller than d if the object of interest lives in R^d. Soille studies in detail two of the possible structuring elements: discrete lines and discs, and then describes the application of granulometries for the characterisation of anisotropy.

Claus Beisbart, Robert Dahlke, Klaus Mecke and Herbert Wagner demonstrate in 'Vector- and Tensor-Valued Descriptors for Spatial Patterns' the successful application of an advanced concept of integral geometry: additive tensor valuation. This is an important generalisation of the well-known Minkowski functionals and enable novel characterisations of orientations in structures of physical interest. Applications to density functional theory for fluids (see Part I) are possible, as are applications to the galaxy distribution in the universe and the geometry of the electric charge distribution in molecules, which may be relevant in chemistry.

The next two papers describe methods of computational topology. As already emphasised in Part I, one important problem in the analysis of three-dimensional image data is the estimation of topological characteristics such as the Euler characteristic of a porous medium (cf. Arns et al. and Vogel). The Euler characteristic is an appropriate parameter for the quantification of the connectivity of a phase and of its complement. If the structure is on the scale of the resolution of the grid of observation points, then the crucial point for the study of topological properties of a discretized set is the choice of an adjacency system in a point lattice. The theoretical foundation of the concepts of adjacency is still under development. In the paper 'The Euler Number of Discretized Sets – On the Choice of Adjacency in Homogeneous Lattices' by Joachim Ohser, Werner Nagel, and Katja Schladitz, the commonly used graph theoretic approach is related to convex and integral geometry. This allows an improved mathematical approach and a proof of the consistency of certain estimators of the Euler characteristics of a set and of its complement ('foreground' and 'background' of an image, respectively). Sufficient conditions for the unbiasedness of the estimators are also formulated.

Vanessa Robins extends in 'Computational Topology for Point Data: Betti Numbers of α-Shapes' the topic of Minkowski functionals by considering Betti numbers. These numbers, which were formulated in algebraic topology, yield more detailed information on the topology of structures in physics than the Euler characteristic. In fact, the Euler characteristic is simply a linear combination of Betti numbers. In this paper, the Betti numbers are determined for α-shapes, which are planar sets of the form $S \oplus b(o, \alpha)$, where S is the data set, a finite point set approximating some target set X, and $b(o, \alpha)$ is the disc of radius α centred at the origin o. In other words, the α-shape is the parallel set of S with radius α. The aim of these calculations is to obtain topological information on X by using various values of α.

Dietrich Stoyan, Ashot Davtyan and Daulet Turetayev present in 'Shape Statistics for Random Domains and Particles' a short survey on the methods developed by statisticians for the description of the shape of planar sets such as images of particles, grains or vesicles. This includes the simple shape indices like area/perimeter ratios, but offers more sophisticated approaches using various functions and models. Additionally, a family of models of planar 'lattice animals' is introduced. These models have some similarity to the smoother objects studied in Part I by Schwarz et al. and Lösche et al.

The rest of Part II is closer still to classical methods of spatial statistics. Daniel Hug, Günter Last and Wolfgang Weil discuss in their long paper 'A Survey on Contact Distributions' contact distribution functions systematically. The classical spherical contact distribution $H_s(r)$ is the probability distribution function of the random distance from a random test point outside a structure X to the boundary of X. This function is closely related to the d-dimensional Minkowski measure intensity of $X \oplus b(o, r)$, which is the simplest of Mecke's morphological functions. For the Boolean model with isotropic grains there are beautiful and well-known formulae for $H_s(r)$. This contribution gives formulae for contact distribution functions corresponding to structuring elements different to the sphere or disc used in $H_s(r)$, also for anisotropic Boolean models. Furthermore, Poisson cluster point processes and grain models constructed by means of such point processes are considered. Finally, contact distribution functions of inhomogeneous cluster processes and Boolean models are presented.

Claus Beisbart, Martin Kerscher and Klaus Mecke apply in 'Mark Correlations: Relating Physical Properties to Spatial Distributions' ideas of the second-order theory of marked point processes to physical data. In one of their applications, the 'points' are positions of galaxies and the 'marks' are characteristics of the galaxies such as luminosity.

They use known types of mark correlation functions and introduce some new concepts which are particularly powerful for the case where the marks are directions. Their aim is to interpret physically the information obtained.

Kasper Berthelsen and Jesper Møller study in 'Spatial Jump Processes and Perfect Simulation' finite spatial jump processes. These are Markov processes with piecewise constant sample paths which play an important role in the simulation of Gibbs point processes. Known particular cases are spatial birth-and-death processes. It is shown how perfect simulation enables an efficient form of simulation of these processes. As a particular case, their application to the case of planar hard core Gibbs processes is demonstrated.

Gunter Döge and Dietrich Stoyan describe in 'Statistics for Non-Sparse Spatially Homogeneous Gibbs Point Processes' the application of the Takacs–Fiksel method to point data. The idea is to interpret a given pattern as a sample of a homogeneous Gibbs point process with a pair potential and to estimate parameters of the potential function by means of a least squares technique. The new problem here is that the patterns analysed are dense, while in the statistical literature typically sparse patterns have been analysed.

Turbulence is ubiquitous in nature and is not only observed in fluid flow. Although responsible for many interesting disordered structure formation, its detailed mechanisms are still not understood. Martin Greiner introduces in 'Spatial Statistics of a Turbulent Random Multiplicative Branching Process' random multiplicative branching processes as empirical models for fully developed turbulence. In order to compare these geometric cascade models with experimental data, it is necessary to discuss spatial statistics in terms of a generating functional. This application of a statistical standard technique closes the second part of this volume.

It is a pleasure to thank all participants of the second international Wuppertal workshop on 'Statistical Physics and Spatial Statistics' for their valuable contributions, their openness to share their experience and knowledge, and for the numerous discussions which made the workshop so lively and fruitful. The workshop took place at the University of Wuppertal from 5 to 9 March 2001. Its aim was to provide a forum for the exchange of fundamental ideas between physicists and spatial statisticians, both working in a wide spectrum of science related to stochastic geometry. This volume comprises the majority of the papers presented orally at the workshop as plenary lectures. The editors are in particular grateful to all authors of this volume for their additional work.

Stuttgart and Freiberg,
September 2002

Klaus Mecke
Dietrich Stoyan

Table of Contents

List of Contributors

Dr. Claus Beisbart
Ludwig-Maximilians-Universität München
Sektion Physik, Theoretische Physik
Theresienstr. 37
80333 München
Germany
Tel: +49-89-21804559
Fax: +49-89-21804517
claus.beisbart@physik.uni-muenchen.de

Kasper Klitgaard Berthelsen
Department of Mathematical Sciences
Fredrik Bajers Vej 7 G
DK-9220 Aalborg
Denmark
Tel: +45-9635-9981
kkb@math.auc.dk

Ashot Davtian
TU Bergakademie Freiberg
Fakultät für Mathematik und Informatik
Institut für Stochastik
09596 Freiberg
Germany
Tel: +49-3731-39-3394
Fax: +49-3731-39-3598
davtian@merkur.hrz.tu-freiberg.de

Robert Dahlke
Ludwig-Maximilians-Universität München
Sektion Physik, Theoretische Physik
Theresienstr. 37
80333 München
Germany
Tel: +49-89-23944548
Fax: +49-89-23944517
dahlke@theorie.physik.uni-muenchen.de

Dr. Gunter Döge
TU Bergakademie Freiberg
Fakultät für Mathematik und Informatik
Institut für Stochastik
09596 Freiberg
Germany
Tel: +49-3731-39-2282
doege@math.tu-freiberg.de

Prof. Dr. Gerhard Gompper
Forschungszentrum Jülich
Institut für Festkörperforschung
D-52425 Jülich
Germany
Tel: +49-2461-61-4012
Fax: +49-2461-61-2850
http://www.fz-juelich.de/iff/Institute/it2/
g.gompper@fz-juelich.de

Dr. Francois Graner
Laboratoire de Spectrométrie Physique
Université Grenoble I BP 87
38402 Saint Martin d'Héres Cedex
France
Tel: +33-476-514774
Fax: +33-476-635495
Francois.Graner@ujf-grenoble.fr

Dr. Martin Greiner
Corporate Technology Dpt. CT IC4, Siemens AG
D-81730 Muenchen
Tel: +49-89-636-49500
Fax: +49-89-636-49767
martin.greiner@mchp.siemens.de

PD Dr. Daniel Hug
Universität Freiburg
Mathematisches Institut
Eckerstr. 1
D-79104 Freiburg
Germany
daniel.hug@math.uni-freiburg.de

Prof. Dominique Jeulin
CMM, Centre de Morphologie Mathématique
ENSMP, Ecole des Mines de Paris
35 rue Saint Honoré
F77300 Fontainebleau
FRANCE
Tel: +33-164-694795
Fax: +33-164-694707
http://cmm.ensmp.fr/ jeulin/
jeulin@cmm.ensmp.fr

Dr. Martin Kerscher
Ludwig-Maximilians-Universität München
Sektion Physik, Theoretische Physik
Theresienstr. 37
80333 München
Germany
kerscher@theorie.physik.uni-muenchen.de

Dr. Mark Knackstedt
Department of Applied Mathematics
Research School of Physical Sciences and Engineering
Australian National University
Canberra ACT 0200
Australia
Tel: +61-2-6125-2495
Fax: +61-2-6125-0732
mak110@rsphy1.anu.edu.au

Christoph Arns
Department of Applied Mathematics
Research School of Physical Sciences and Engineering
Australian National University
Canberra ACT 0200
Australia
cha110@rsphysse.anu.edu.au

Prof. Dr. Günter Last
Universität (TH) Karlsruhe
Fakultät für Mathematik
Institut für Mathematische Stochastik
76128 Karlsruhe
Germany
g.last@math.unikarlsruhe.de

Dr. P. Krüger
Institute of Experimental Physics I
Leipzig University D-04103 Leipzig
Germany

Prof. Dr. Mathias Lösche
Institute of Experimental Physics I
Leipzig University D-04103 Leipzig
Germany
phone: +49-341-97-32488
fax: +49-341-97-32479
http://hypatia.exphysik.uni-leipzig.de
loesche@physik.uni-leipzig.de

Dr. Robert Magerle
Universität Bayreuth
Physikalische Chemie II
D-95440 Bayreuth
Germany
Tel: +49-921-55-2641
Fax: +49-921-55-2059
robert.magerle@uni-bayreuth.de

Dr. Klaus Mecke
Max-Planck-Institut
fuer Metallforschung,
Heisenbergstr. 1,
D-70569 Stuttgart,
Germany
Tel: +49-711-689-1936
Fax: +49-711-689-1922
mecke@mf.mpg.de

Prof. Dr. Jesper Møller
Department of Mathematical Sciences,
Aalborg University,
Fredrik Bajers Vej 7E, DK-9220 Aalborg,
Denmark
Tel: +45-96358863 (direct call)
Tel: +45-96358080 (switchboard)
Fax: +45-98158129
http://www.math.auc.dk/~jm
jm@math.auc.dk

Dr. Werner Nagel
Friedrich-Schiller-Universität Jena
Fakultät für Mathematik und Informatik
Institut für Stochastik
07740 Jena
Germany
Tel: +49-3641-946261
Fax: +49-3641-946252
nagel@minet.uni-jena.de

Priv.-Doz. Dr. Joachim Ohser
Universität Kaiserslautern
Fraunhofer Institut für
Techno- und Wirtschaftsmathematik
Gottlieb-Daimler-Str.
67653 Kaiserslautern
Germany
Tel: +49-631-205-4426
Fax: +49-631-205-4139
ohser@itwm.fhg.de

Dr. Nguyen H. Phuong
Theoretische Physik
Fakultät für Physik
Universiät Bielefeld
Postfach 100131
D-33501 Bielefeld
Germany

Dr. Vanessa Robins
Department of Applied Mathematics
Research School of Physical Sciences
and Engineering
Australian National University
Canberra ACT 0200
Australia
Vanessa.Robins@anu.edu.au

Dr. Katja Schladitz
Universität Kaiserslautern
Fraunhofer Institut für
Techno- und Wirtschaftsmathematik
Gottlieb-Daimler-Str.
67653 Kaiserslautern
Germany

Prof. Friederike Schmid
Theoretische Physik
Fakultät für Physik
Universiät Bielefeld
Postfach 100131
D-33501 Bielefeld
Germany
Tel: +49-521-106-6901 (secretary)
Tel: +49-521-106-6191
Fax: +49-521-106-6455
schmid@physik.uni-bielefeld.de

Dr. Ulrich Schwarz
Max-Planck-Institute of Colloids and Interfaces
14424 Potsdam (Golm), Germany
Tel: +49-331-567-9610
Fax: +49-331-567-9602
http://www.mpikg-golm.mpg.de/theory/people/schwarz
Ulrich.Schwarz@mpikg-golm.mpg.de

Dr. Pierre Soille
Institute for Environment
and Sustainability
EC Joint Research Centre, T.P. 262
I-21020 Ispra (VA)
Italy
Tel: +39-0332-785-068
Fax: +39-0332-789-803
http://ams.egeo.sai.jrc.it/soille
`Pierre.Soille@jrc.it`

Prof. Dr. Dietrich Stoyan
TU Bergakademie Freiberg
Fakultät für Mathematik und Informatik
Institut für Stochastik
09596 Freiberg
Germany
Tel: +49-3731-39-2118
`stoyan@orion.hrz.tu-freiberg.de`

Dr. D. Turetayev
TU Bergakademie Freiberg
Fakultät für Mathematik und Informatik
Institut für Stochastik
09596 Freiberg
Germany

Dr. Hans-Jörg Vogel
University of Heidelberg
Institute of Environmental Physics
Im Neuenheimer Feld 229
D-69120 Heidelberg
Germany
Tel: +49-6221-54-5481
Fax: +49-6221-54-6405
http://www.iup.uni-heidelberg.de
`hjvogel@iup.uni-heidelberg.de`

Prof. Dr. H. Wagner
Ludwig-Maximilians-Universität
München
Sektion Physik,
Theoretische Physik
Theresienstr. 37
80333 München
Tel: +49-89-21804537
Fax: +49-89-21804517
`herbert.wagner@physik.uni-muenchen.de`

Prof. Dr. Wolfgang Weil
Universität (TH) Karlsruhe
Fakultät für Mathematik
Mathematisches Institut II
76128 Karlsruhe
Germany
Tel: +49-721-6083942
`weil@math.uni-karlsruhe.de`

Part I

Complex Structured Condensed Matter

Spatial Statistics and Micromechanics of Materials

Dominique Jeulin

Centre de Morphologie Mathématique Ecole des Mines de Paris,
35 rue Saint Honoré, F77300 Fontainebleau, France

Abstract. The prediction of elastic properties and of fracture strength of materials from their microstructure is an old but still active problem, implying many theoretical questions and important practical consequences. In both cases, statistical correlation functions deduced from the geometry by image analysis measurements, or obtained for specific models of random media, enter into the estimation of overall properties of materials. Concerning elastic properties of random media, useful bounds can be worked out from their three-points statistics. This is illustrated for various models of random media, including multiscale textures producing so-called "optimal microstructures". This method also provides estimators (upper bounds) of the elastic moduli of porous media. Fracture statistics models based on random functions allow us to predict the probability of fracture of materials, depending on the scale, the applied stress field, and the microstructure. Recent developments are reviewed in brittle fracture (using the weakest link model, or Griffith crack propagation in random media) and in the case of materials submitted to an increasing damage during the fracture process.

1 Introduction

The presence of heterogeneities at different scales in materials requires the implementation of probabilistic tools. For instance, the prediction of elastic properties and of fracture strength of materials from their microstructure is an old but still active problem, implying many theoretical questions and important practical consequences. For these two types of micromechanical problems (more results are given in [44]), statistical moments involving a finite number of points (for the estimation of effective elastic moduli), as well as more involved probability functions (for fracture statistics) are required. They are deduced from the geometry by image analysis measurements, or obtained for specific models of random media.

Concerning the prediction of overall elastic properties (or more generally "linear" properties such as thermal conductivity or permeability) of random media, convenient and useful bounds can be worked out from their three-points statistics. This is illustrated for various models of random media, including multiscale textures producing so-called "optimal microstructures". This method also provides estimators (upper bounds) of the elastic moduli of porous media. Other contributions on the prediction of properties of porous media, based on a geometric and on a topological characterization, are found in Sects. 1.2 and 1.3 of this volume [1, 74].

Fracture statistics models based on random functions allow us to predict the probability of fracture of materials, depending on the scale, the applied stress field, and the microstructure. Some models are based on probabilistic properties of extreme values.

We review recent developments in brittle fracture (using the weakest link model, or Griffith crack propagation in random media) and in the case of materials submitted to an increasing damage during the fracture process.

2 Probabilistic Models of Structures

2.1 The Choquet Capacity $T(K)$

When deciding to use a probabilistic approach, it is important to be able to characterize a random medium by appropriate tools. A generalization of the notion of random variable can be proposed for random sets, through the Choquet capacity [52], which can be physically interpreted according to its morphological meaning as illustrated below. Examples of models of random sets are detailed in many references, such as, among others [32, 33, 38, 45, 66, 69].

To characterize a set A in a space E (for instance the grains of a porous medium seen as a subset of the euclidean space R^3), we can select a set $K \subset E$, and examine the mutual locations of K and of A, by means of the answer to the two following exclusive questions:

$$\begin{aligned} &\text{i) is } K \text{ disjoint from } A\text{? } (K \cap A = \emptyset)\text{?} \\ &\text{ii) } K \text{ hits } A \ (K \cap A \neq \emptyset)\text{?} \end{aligned} \tag{1}$$

For instance if $E = R^3$ and if K is the point x, the test 1 enables us to test whether $x \notin A$ (i) or $x \in A$ (ii), i.e. to which component of the two-phase medium (A, A^c) the point x belongs. The same process can be repeated for K made of the pair of points $\{x, x+h\}$, or more generally of a denumerable set of points $\{x_1, x_2, ..., x_n, ...\}$. By increasing the number of points, the number of positive answers to questions i) decreases and to ii) increases, and a richer information is obtained about the structure of A. Note that limiting the set of points K to a denumerable one, as in the standard definition of random functions (RF) by Kolmogorov, is not pertinent to discriminate between random structures, despite a common belief in the physical literature: for instance, for two stationary random sets differing by a stationary point process (or more generally by a stationary random set with zero measure), the probability of events of the types i) or ii) would be the same. Let us assume now that $x \notin A$ and that we want to estimate the distance separating the point x from the set A, denoting $\vee$ by and $\wedge$ the supremum and the infimum:

$$R(x, A) = \wedge_{y \in A}\{d(x, y)\} \tag{2}$$

Let $B_x(r)$ be the closed ball with radius r and with center x: $y \in B_x(r) \Leftrightarrow d(x,y) \leq r$.

$$R(x, A) = \vee\{r; B_x(r) \subset A^c\} \tag{3}$$

Equation (3) involves sets $B_x(r)$ containing a non-denumerable set of points. This type of information is used in Sect. 4.1 on generalized contact distributions in this volume [27].

Theorem 1. *Models derived from the theory of Random Closed Sets (RACS) by G. Matheron [49, 52] are characterized by their Choquet capacity $T(K)$ defined on the compact sets K. For a RACS noted A,*

$$T(K) = P\{K \cap A \neq \emptyset\} = 1 - P\{K \subset A^c\} = 1 - Q(K) \tag{4}$$

In practice, various geometrical figures are used for the compact set K, in order to test various morphological properties of a structure.

For upper semi-continuous random functions (usc RF) $Z(x)$, a generalization of the Choquet capacity, $T(g)$ can be defined [33, 45] on the lower semi continuous functions (lsc) g with a compact support K

$$T(g) = P\{x \in D_Z(g)\} \text{ where } D_Z(g)^c = \{x : Z(y) < g(y - x), \forall y \in K\} \tag{5}$$

A particular and usual case is the spatial law defined on a finite number of points $x_1, x_2, ..., x_n$,

$$F(x_1, x_2, ..., x_n, z_1, z_2, ..., z_n) = P\{Z(x_1) < z_1, Z(x_2) < z_2, ..., Z(x_n) < z_n\} \tag{6}$$

Other particular cases give the changes of supports by $\vee$ (supremum) or by $\wedge$ (infimum), which generate new RFs when the compact set K is moved by translations:

$$Z_\vee(K) = \vee_{x \in K}\{Z(x)\}$$

$$Z_\wedge(K) = \wedge_{x \in K}\{Z(x)\}$$

We must stress the following points:

- The Choquet capacity of a RACS is equivalent to the distribution function of a random variable.
- Two models (RACS, usc RF) with the same functional $T(K)$, $T(g)$, cannot be distinguished (theoretically as well as experimentally). This gives a definite theoretical answer to the question: Which information is required in order to be able to reconstruct a random set from statistical information?
- The functional $T(K)$ (or $T(g)$) connects theory and experiments; it is used to estimate the parameters of a model and to test its validity.

2.2 Calculation and Estimation of the Functional $T(K)$

The functional $T(K)$ (or $T(g)$) has to be determined as a function of the assumptions of the model, of its parameters, and of the compact set K or the function g. For a given model, the functional T is obtained:

- by theoretical calculation
- by estimation on simulations of the random structure, or on samples of the real structures. In the latter case, it is possible to estimate the parameters from the "experimental" T, and to test the validity of assumptions [60, 66]. The estimation of the functional $T(K)$ is easily performed from the implementation of the basic operations of mathematical morphology, namely erosions or dilations.

The functions $T(K)$ (obtained for different types of compacts sets K) for a given model are compatible with each other (which is not the case of any prior analytical model, as often used in the literature).

After specification and validation of the model from available data, it is possible to use a predictive implementation of its properties (such as $T(K)$ for compact sets K not used during the identification step). Practical examples are the following: 3D properties deduced from 2D observations (stereology); change of support by $\vee$ or $\wedge$ in the case of a change of scale in fracture statistics (as in Sect. 4).

2.3 Morphological Content of the Choquet Capacity

In the Euclidean space R^n, with translations by vector x, the Choquet capacity of A for K at the origin of coordinates O can be expressed by:

$$T(K) = P\{K \cap A \neq \emptyset\} = P\{O \in A \oplus \check{K}\} \tag{7}$$

where $A \oplus \check{K}$ is the result of the dilation of A by K:

$$A \oplus \check{K} = \{x, K_x \cap A \neq \emptyset\} = \cup_{-y \in K} A'_y = \cup_{x \in A, y \in K} \{x - y\} \tag{8}$$

For K translated at x :

$$T(K_x) = P\{K_x \cap A \neq \emptyset\} = P\{x \in A \oplus \check{K}\} \tag{9}$$

The functional $Q(K) = 1 - T(K)$ is connected to the erosion by:

$$Q(K_x) = P\{K_x \subset A^c\} = P\{x \in A^c \ominus \check{K}\} \tag{10}$$

with

$$A \ominus \check{K} = \{x, K_x \subset A\} = \cap_{y \in K} A_{-y} \tag{11}$$

The dilation and erosion (as illustrated in Fig. 1) are the basic operations of Mathematical Morphology, introduced by G. Matheron and J. Serra in 1965 [25, 49, 66, 67].

A statistical estimation of $T(K)$ can be obtained by an image analysis performed on realizations of A, involving the dilation operation, in relation with (9). In the general case, several realizations of A are required, and $T(K_x)$ is estimated from a frequency for every point x. For a stationary random set, $T(K_x) = T(K)$ ($T(K_x)$ is invariant by translations of K). If in addition the RACS A is ergodic, $T(K)$ can be estimated from a single realization of A with:

$$T(K)^* = P\{x \in A \oplus \check{K}\}^* = V_V(A \oplus \check{K})^* \tag{12}$$

where V_V is the volume fraction (for a RACS in R^3). Equation (12) provides a practical process to estimate $T(K)$ from image analysis on samples of a microstructure. For an isotropic random set, $T(K)$ is invariant by the rotations of K.

Every compact set K brings its own information on the set A, considered as a realization of a random set [32, 49, 52, 66, 69]. For instance, we consider the following types of compact sets:

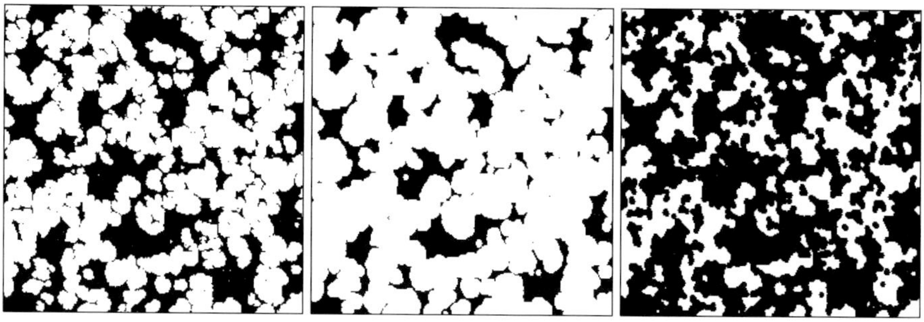

(a) (b) (c)

Fig. 1. Fe (white phase) in a Fe-Ag composite (a) [12]; Dilation of Fe by an hexagon with size 2 (b); Erosion of Fe by an hexagon with size 2 (c).

- $K = \{x\}$

$$T(x) = P\{x \in A\} \tag{13}$$

For a stationary RACS in R^3, we get the volume fraction V_V from $T(x) = p = V_V(A)$

- $K = \{x, x+h\}$

$$T(x, x+h) = P\{x \in A \cup A_{-h}\}$$
$$Q(x, x+h) = P\{x \in A^c \cap A^c_{-h}\} \tag{14}$$

$Q(x, x+h)$ is the covariance of A^c. It depends only on h for a stationary random set. The covariance $C(h)$ of the set A is deduced from $Q(h)$ by

$$C(h) = 1 - 2Q(0) + Q(h)$$

- $K = B_O(r)$ (closed ball with radius r centred at 0)

$$T(B_x(r)) = P\{x \in A \oplus B(r)\} \tag{15}$$

$T(B_x(r))$ enables us to estimate the distribution of the random variable $R(x, A)$ defined in (2,3):

$$F_x(r) = P\{R(x, A) < r \mid x \in A^c\} = \frac{T(B_x(r)) - T(x)}{1 - T(x)} \tag{16}$$

The distribution function $F_x(r)$ (also called spherical contact distribution function and noted $H_S(r)$) does not depend on x for a stationary RACS.

The covariance and the distribution $F(r)$ are useful for studying the spatial distribution of A. The covariance enables us to measure the notion of scale by its range, defined as the length a for which the correlation between $1_A(x)$ and $1_A(x+a)$ vanishes, $1_A(x)$ being the indicator function of the set A at point x ($1_A(x) = 1$ when $x \in A$ and $1_A(x) = 0$ when $x \in A^c$). The covariance is a convenient tool to study the same materials at different scales [66]. In addition, as it depends on the orientation in space of the vector with origin x and extremity $x + h$, it is sensitive to anisotropy. This is the most convenient way to characterize anisotropic textures.

2.4 Minkowski Functionals of Dilated Sets

The Choquet's capacity in (12) involves the measurement of the volume V (after dilation of the random set A), which is the Minkowski functional of degree $n = 3$ in R^3. The Minkowski functionals, denoted by W_k for $k = 0, 1, ..., n$, with W_k homogeneous and of degree k: $W_k(\lambda A) = \lambda^k W_k(A)$, were first defined for convex bodies A, and then extended to unions of convex sets in integral geometry [52, 66, 69].

From Theorem 1 and using (7) and (12) in the case of stationary random closed sets, this type of information, involving the volume V and the dilation, obtained for all compact sets, is necessary and sufficient to completely characterize a random set. However, useful data are obtained by means of the measurements $W_k(A \oplus B(r))$ (and by the intensities of Minkowski measures for stationary A), for all functionals W_k, as systematically investigated in [54, 55].

In a particular case, this idea already appeared in a study on 2D sections of sintered materials [28, 29], from which is borrowed Fig. 2: the curves $W_2(A \ominus \ell)$ obtained for segments with length ℓ, show a maximum (denoted by N_{Am} on Fig. 2) when A is the solid phase. This maximum was inversely correlated to the strength of iron ore sinters.

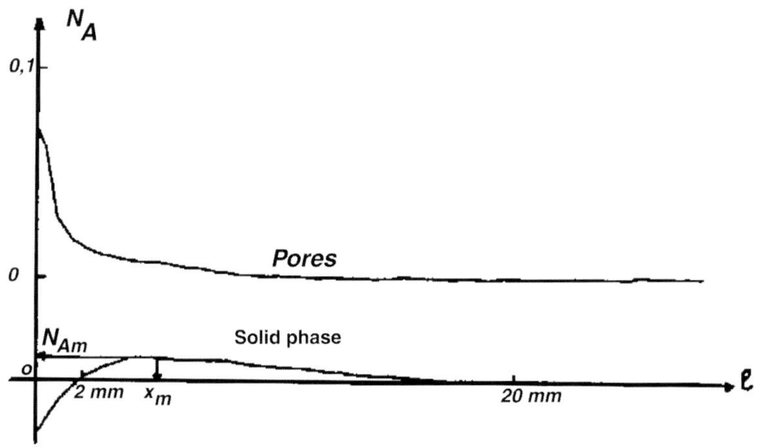

Fig. 2. Connectivity number (N_A in 2D) of pores and of the solid phase of iron ore sinter after erosion by segments of increasing length ℓ.

For multi-component systems (involving the random sets A_i, with $A_i \cap A_j = \emptyset$ for $i \neq j$), there is a possible generalization of this approach and of the standard cross-covariances $C_{ij}(h) = P\{x \in A_i, x + h \in A_j\}$ to degenerate components (with dimension $d < n$), such as cracks, boundaries, or points, for which covariances are equal to zero [31]. This generalization compares the microstructure of a given component A_i in neighborhoods $B(r)$ with increasing sizes r to the overall microstructure, by means of Minkowski functionals measured after dilation by balls with increasing radius, $W_k(A_i \oplus B(r) \cap A_j)$.

3 Change of Scale in Random Media

3.1 Introduction

An important field of application of models of random media concerns the prediction of the macroscopic behavior of a physical system from its microscopic behavior. This so-called "homogenization" process goes back to the early preoccupations of statistical physics when was developed the theory of gas kinetics from the mechanical interaction of particles with different velocities. Two types of approaches can be followed: we can first look for estimations of the effective properties (namely the overall properties of an equivalent homogeneous medium) of random heterogeneous media from their microstructure. Using variational principles, bounds of the effective properties for linear constitutive equations can be established from a limited amount of statistical information, as will be illustrated from third order bounds. Another technique of estimation of the effective behavior involves numerical solutions of partial differential equations on simulations of random media, as will be briefly described. These two types of approach can be followed for all physical properties involving a conservation law, as for instance the dielectric permittivity [5, 6, 9, 37, 39, 56, 58, 59], the magnetic permeability of materials [23], the permeability of porous media [49, 50], the elastic moduli of composites [10, 11, 24, 47, 53, 57, 59, 75]. In what follows, general results concerning bounds are introduced; they are illustrated by examples of applications to some models of random sets. Most results presented in this section are detailed in [43].

3.2 Third Order Bounds of the Dielectric Permittivity

Introduction. We first give some results on the dielectric permittivity, as an introduction to the process of homogenization of random media. For electrostatics of a linear dielectric medium, there is a proportionality between the electric displacement D and the electric field E, ϵ being the dielectric permittivity tensor, and ϕ the potential:

$$D = \epsilon E = -\epsilon \text{grad}\phi \tag{17}$$

$E(x)$ is obtained from the applied boundary conditions and from the partial differential equation expressing the Gauss' law:

$$\text{div}(D) = \sum_i \frac{\partial}{\partial x_i}(D_i(x)) = 0 \tag{18}$$

The effective dielectric tensor ϵ^* of the equivalent homogeneous medium contained in B is defined for (17) to be satisfied on a macroscopic scale, that is after taking a spatial average of the fields (denoted by $\langle D \rangle$ and $\langle E \rangle$ respectively) over the domain B:

$$\langle D \rangle = \epsilon^* \langle E \rangle \tag{19}$$

In general, it is not possible to know exactly the effective (or macroscopic) permittivity ϵ^*, except for some specific geometries (some examples are given below). In fact, the exact estimation of ϵ^* requires a very large amount of information (the set of all n

point correlation functions, which naturally appear in a perturbation expansion of the solution of (18) combined to (17) for infinite domains B containing realizations of ergodic random media [6, 47, 49, 50]). Using a limited amount of statistical information, bounds (upper bound ϵ_+ and lower bound ϵ_-) are derived from a perturbation expansion of the electric field and a variational principle on the stored electrostatic energy [5, 6]. Consider now the case of an isotropic multiphase material in R^d, where for every component i (with the volume fraction p_i), ϵ_i is a scalar. The Hashin and Shtrikman (H-S) bounds [23] are obtained after introduction of a reference homogeneous medium (with the permittivity ϵ_0 satisfying $\epsilon_m = \inf_i(\epsilon_i) \leq \epsilon_0 \leq \epsilon_M = \sup_i(\epsilon_i)$) from

$$K = \sum_{j=1}^{j=N} \frac{p_j}{(\epsilon_j - \epsilon_0)^{-1} + 1/(\epsilon_0 d)} \tag{20}$$

Choosing $\epsilon_0 = \epsilon_m$ or alternatively $\epsilon_0 = \epsilon_M$,

$$\begin{aligned} \epsilon^* \geq \epsilon_- = \epsilon_m + \frac{K}{1 - \frac{K}{d\epsilon_0}} \\ \epsilon^* \leq \epsilon_+ = \epsilon_M + \frac{K}{1 - \frac{K}{d\epsilon_0}} \end{aligned} \tag{21}$$

Third order bounds. We consider now random composites made of two phases A_1 (with fraction p) and A_2 (with fraction $q = 1 - p$) having a scalar real dielectric permittivity ϵ_1 and ϵ_2 (with $\epsilon_2 > \epsilon_1$) (the case of tensorial permittivity is detailed in [26]). The composite is modelled by a stationary and isotropic random set A (with $A = A_2$ and $A^c = A_1$). The third order bounds are expressed in R^d as a function of ϵ_1, ϵ_2 and of the three-point probability $P(h_1, h_2) = P\{x \in A_1, x + h_1 \in A_1, x + h_2 \in A_1\}$. In R^d $(d = 2, 3)$, using G. Milton's [58, 59] and S. Torquato's [73] notations,

$$\epsilon_-/\epsilon_1 = \frac{1 + ((d-1)(1+q) + \zeta_1 - 1)\,\beta_{21} + (d-1)\,(((d-1)q + \zeta_1 - 1))\,\beta_{21}^2}{1 - (q + 1 - \zeta_1 - (d-1))\,\beta_{21} + ((q - (d-1)p)\,(1 - \zeta_1) - (d-1)q)\,\beta_{21}^2} \tag{22}$$

$$\epsilon_+/\epsilon_2 = \frac{1 + ((d-1)(p + \zeta_1) - 1)\,\beta_{12} + (d-1)\,(((d-1)p - q)\,\zeta_1 - p)\,\beta_{12}^2}{1 - (1 + p - (d-1)\zeta_1)\,\beta_{12} + (p - (d-1)\zeta_1)\,\beta_{12}^2} \tag{23}$$

$$\beta_{ij} = \frac{\epsilon_i - \epsilon_j}{\epsilon_i + (d-1)\epsilon_j} \tag{24}$$

where the function $\zeta_1(p)$ introduced by G. Milton [59] is obtained from the probability $P(h_1, h_2) = P(|h_1|, |h_2|, \theta)$, with $u = \cos\theta$, θ being the angle between the vectors h_1 and h_2. We have:

$$\zeta_1(p) = \frac{9}{4pq} \int_0^{+\infty} \frac{dx}{x} \int_0^{+\infty} \frac{dy}{y} \int_{-1}^{+1} (3u^2 - 1) P(x, y, \theta) du \text{ in } R^3 \tag{25}$$

$$\zeta_1(p) = \frac{4}{\pi pq} \int_0^{+\infty} \frac{dx}{x} \int_0^{+\infty} \frac{dy}{y} \int_0^{\pi} P(x,y,\alpha) \cos(2\alpha)\, d\alpha \text{ in } R^2 \qquad (26)$$

The integrals in (25) and (26) can be evaluated analytically, but most often numerically (after replacement of $P(x,y,\theta)$ by $P(x,y,\theta) - P(x)P(y)/p$, with $P(h) = P\{x \in A_1, x+h \in A_1\}$.

After exchanging the phases A_1 and A_2 we define the function $\zeta_2(q)$ with $\zeta_2(q) = 1 - \zeta_1(p)$. The function $\zeta_1(p)$ satisfies $0 \leq \zeta_1(p) \leq 1$. For given p, ϵ_1 and ϵ_2 (with $\epsilon_1 > \epsilon_2$), ϵ_+ and ϵ_- increase with $\zeta_1(p)$, so that higher values of the effective properties are expected. This gives a guideline to compare the properties of random media from the comparison of their corresponding geometrical function $\zeta_1(p)$, as will be illustrated later. If the two phases A_1 and A_2 are symmetric, the case of $p = 0.5$ produces an autodual random set (the two phases having the same probabilistic properties, as for the mosaic model introduced below), for which the third order central correlation function is equal to zero. Therefore in two and three dimensions, the third order bounds of a symmetric medium present a fixed point at $p = 0.5$. In addition, it is known [49, 50], that for an autodual random set in two dimensions the effective permittivity is exactly equal to the geometrical average of the two permittivities ($\epsilon^* = \sqrt{\epsilon_1 \epsilon_2}$).

Third order bounds of the Boolean model. We give now examples of application of third order bounds in the case of the Boolean model introduced by G. Matheron [49, 52], and studied by many authors [32, 33, 66, 69]. It is obtained by implantation of random primary grains A' (with possible overlaps) on Poisson points x_k with the intensity θ: $A = \cup_{x_k} A'_{x_k}$. Any shape (convex or non convex, and even non connected) can be used for the grain A'.

Fig. 3. Realization of a Boolean model of spheres ($V_V = 0.5$).

Fig. 4. Section of a Boolean model from Poisson flats in 3D.

An example is shown in Fig. 3 in the case of spheres, as in [12, 28, 29]. In [64], Poisson polyhedra are used for the grains of WC-Co composites, while in [42], parallelepiped random grains are used for the microstructure of gypsum. Replacing the Poisson points by Poisson varieties [52], enables us to generate random sets models with fiber or strata textures [33, 45] (Fig. 4). This version of the Boolean model is used in Sect. 3.3. In [19], the fiber version is used for the simulation of cellulosic fiber composites. Anisotropic versions of this model can be easily proposed, from anisotropic primary grains, as for instance ellipsoids in R^3.

The Choquet capacity of this model can be deduced from the fact that the number of primary grains hit by K follows a Poisson distribution with average $\theta\overline{\mu}_n(A' \oplus \check{K})$; denoting $q = P\{x \in A^c\}$, we have:

$$T(K) = 1 - Q(K) = 1 - \exp\left(-\theta\overline{\mu}_n(A' \oplus \check{K})\right) = 1 - q^{\frac{\overline{\mu}_n(A' \oplus \check{K})}{\overline{\mu}_n(A')}} \tag{27}$$

The three point probability for the random set A^c is given by

$$\begin{aligned} Q(h_1, h_2) &= P\{x \in A^c, x + h_1 \in A^c, x + h_2 \in A^c\} \\ &= \exp\left(-\theta\overline{\mu}_n(A' \cup A'_{-h_1} \cup A'_{-h_2})\right) \\ &= q^{3-r(h_1)-r(h_2)-r(h_2-h_1)+s(h_1,h_2)} \end{aligned} \tag{28}$$

with the geometrical covariogram

$$K(h) = \overline{\mu}_n(A' \cap A'_{-h}) \text{ and } r(h) = \frac{K(h)}{K(0)} \tag{29}$$

and with

$$s(h_1, h_2) = \frac{\overline{\mu}_n(A' \cap A'_{-h_1} \cap A'_{-h_2})}{K(0)} \tag{30}$$

The function $\zeta_1(p)$ for the Boolean model is obtained by numerical integration of (25) or (26), using (28) and (30). With a good approximation, we get a linear expression with coefficients α and β:

$$\zeta_1(p) \simeq \alpha p + \beta$$

$\zeta_1(0)$ is obtained when $p \to 0$. We have $\zeta_1(0) = a$ with a given by (42) in Sect. 3.3 for the random grain A' (or for any size distribution of A'):

- in R^3 $\zeta_1(0) = 0$ for spheres, and $\zeta_1(0) = \frac{1}{4}$ for Poisson polyhedra
- in R^2, $\zeta_1(0) = 0$ for disks and $\zeta_1(0) = 3 - 4\ln 2 \simeq 0.2274$ for Poisson polygons.

The complementary function $\zeta_2(p) = 1 - \zeta_1(1-p)$ is obtained by exchange of the two phases:

$$\begin{gathered} \zeta_1(p) \simeq 0.5615p \text{ for spheres [10, 71, 72, 73]} \\ \zeta_2(p) \simeq 0.5615p + 0.4385 \text{ for (spheres)}^c \\ \zeta_1(p) \simeq \tfrac{2}{3}p \text{ for discs in } R^2 \text{[37, 48, 73]} \\ \zeta_2(p) \simeq \tfrac{2}{3}p + \tfrac{1}{3} \text{ for (discs)}^c \text{ in } R^2 \\ \zeta_1(p) \simeq 0.5057p + 0.2274 \text{ for Poisson polygons in } R^2\text{[48]} \\ \zeta_2(p) \simeq 0.5057p + 0.2669 \text{ for (Poisson polygons in)}^c\ R^2 \end{gathered} \tag{31}$$

For the Boolean model, different sets of bounds are obtained when exchanging the properties of A and A^c (which is not the case for the second order H-S bounds), as seen in (31). This is illustrated for the case of the thermal conductivity λ of ceramic materials [62], modelled by two Boolean models of spheres with a constant radius (in Figs. 5 (a) and 5 (b), the morphology of the low conducting bright phase is exchanged): here, the third order bounds increase when $\lambda(x) = \lambda_1 > \lambda_2$ for $x \in A^c$ (Fig. 6). This is due to the fact that it is easier for the "matrix" phase A^c to percolate than for the overlapping inclusions building A. We see here that third order bounds are sensitive to the morphology in a more accurate way than the H-S bounds [23].

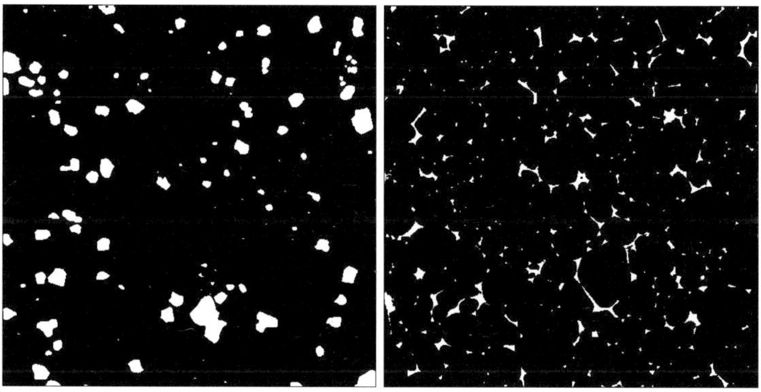

(a) (b)

Fig. 5. AlN textures (AlN in black; Y rich phase in white).

In [37, 48], various morphologies are studied in two and three dimensions. Two-scale hierarchical models accounting for local fluctuations of the volume fraction of a Boolean model can be optimized with respect to the third order bounds. As for the elastic case (see below in Sect. 3.3), an infinite superposition of multiscale Boolean strata generates random media with the effective permittivity exactly equal to the upper (or lower) H-S bound.

The more general approach to bounds considers multiphase and even continuous models (scalar or multivariate) [33, 45], for which the calculation of third order bounds can be made using the general derivation based on the third order correlation function [6, 26, 40, 47, 56]. An extension of bounds to the complex dielectric permittivity was developed by D. Bergman [9] and by G. Milton [58]. It was applied to various types of random textures [39], showing that third order bounds could generate separate domains in the complex plane.

3.3 Third Order Bounds for Elasticity

A similar approach can be made for the third order bounds of the bulk modulus K and of the shear modulus G in the case of isotropic elastic media, with an isotropic geometry. Beran and Molyneux developed bounds (depending only on $\zeta_1(p)$ for two-phase composites) for the bulk modulus K [5]. McCoy worked out bounds for the shear modulus G [53]. These bounds are valid for multiphase media, and more generally for

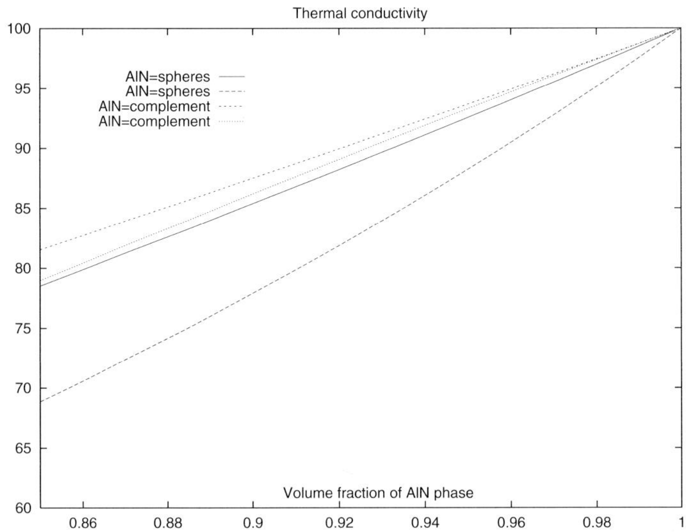

Fig. 6. Bounds of the thermal conductivity of AlN materials ($\lambda_1 = 100$ W/(m.K) for AlN $\lambda_2 = 10$ W/(m.K) for the second phase).

elastic moduli modelled by random functions [6, 32, 47, 49, 50]. In the case of isotropic two-phase composites, they depend on $\zeta_1(p)$ and on a similar function, $\eta_1(p)$ defined later. In what follows, we first recall the expressions of third order bounds, and then introduce some examples for various models of random sets.

Third order bounds for two-phase composites. We consider now two-phase random composites, modelled by a stationary and isotropic random set A ($A = A_1$ and $A^c = A_2$). The same approach can be followed for the bounds of transverse moduli, in the case of materials showing an isotropic transverse morphology [59]. The third order bounds depend on the moduli of the two phases, the volume fraction p, and on two morphological functions $\zeta_1(p)$ and $\eta_1(p)$ ($0 \leq \zeta_1(p) \leq 1$ and $0 \leq \eta_1(p) \leq 1$). We will use the notation of Berryman and Milton [10, 11, 59]. For any composite with properties P_1 (with the probability p) and P_2 (with the complementary probability $q = 1 - p$), we define the following averages (with $\zeta_2(q) = 1 - \zeta_1(p)$ and $\eta_2(q) = 1 - \eta_1(p)$):

$$\begin{aligned} \langle P \rangle &= pP_1 + qP_2 \\ \langle P \rangle_\zeta &= \zeta_1(p)P_1 + \zeta_2(q)P_2 \\ \langle P \rangle_\eta &= \eta_1(p)P_1 + \eta_2(q)P_2 \end{aligned}$$

Bounds on the bulk modulus K. The upper K_+ and lower bounds K_- are expressed by means of the auxiliary function $\Lambda(x)$:

$$\Lambda(x) = \left\langle \frac{1}{K + \frac{4}{3}x} \right\rangle^{-1} - \frac{4}{3}x$$

$$K_+ = \Lambda(\langle G \rangle_\zeta) \tag{32}$$

$$K_- = \Lambda\left(\left\langle \frac{1}{G} \right\rangle_\zeta^{-1}\right) \tag{33}$$

The third order bounds of K dependent only on the function $\zeta_1(p)$. If $\zeta_1(p) = 1$ or 0, $K_+ = K_-$ $(= K^{HS-}$ $(\zeta_1(p) = 1)$ or K^{HS+} $(\zeta_1(p) = 0))$. The Hashin and Shtrikman bounds are therefore a limiting case of third order bounds of K.

Bounds on the shear modulus G. The upper G_+ and lower bounds G_- are expressed by means of the auxiliary function $\Gamma(y)$, and of the following expressions Θ and Ξ:

$$\Gamma(y) = \left\langle \frac{1}{G + y} \right\rangle^{-1} - y$$

$$\Theta = \left[10 \langle G \rangle^2 \langle K \rangle_\zeta + 5 \langle G \rangle \langle 3G + 2K \rangle \langle G \rangle_\zeta + \langle 3K + G \rangle^2 \langle G \rangle_\eta\right] / \langle K + 2G \rangle^2$$

$$\Xi = \left[10 \langle K \rangle^2 \langle 1/K \rangle_\zeta + 5 \langle G \rangle \langle 3G + 2K \rangle \langle 1/G \rangle_\zeta + \langle 3K + G \rangle^2 \langle 1/G \rangle_\eta\right] / \langle 9K + 8G \rangle^2$$

$$G_+ = \Gamma\left(\frac{\Theta}{6}\right) \tag{34}$$

$$G_- = \Gamma\left(\frac{1}{6\Xi}\right) \tag{35}$$

When $\eta_1(p) = \zeta_1(p) = 1$ (0) and when $K_1 = K_2 = \infty$, $G_+ - G_- = G^{HS+}(G^{HS-})$

Bounds on the Young's modulus E. Starting from the expression of the elastic modulus E as a function of K and G,

$$\frac{1}{E} = \frac{1}{9K} + \frac{1}{3G}$$

we get bounds of E deduced from bounds of K and G, since E increases with these variables:

$$\frac{1}{E_+} = \frac{1}{9K_+} + \frac{1}{3G_+} \tag{36}$$

$$\frac{1}{E_-} = \frac{1}{9K_-} + \frac{1}{3G_-} \tag{37}$$

Bounds on the Poisson coefficient ν. The Poisson coefficient is expressed as a function of K and G by:

$$\nu = \frac{3K - 2G}{2(3K + G)} = \frac{1}{2} - \frac{3}{2}\frac{G}{3K + G}$$

ν increases with K and decreases with G. We get the following non optimal bounds $\nu_+(G_-, K_+)$ and $\nu_-(G_+, K_-)$:

$$\nu_+ = \frac{1}{2} - \frac{3}{2}\frac{G_-}{3K_+ + G_-}; \nu_- = \frac{1}{2} - \frac{3}{2}\frac{G_+}{3K_- + G_+} \tag{38}$$

We can check that $K_+ - K_-$ and $G_+ - G_-$ are of the order of $(K_1 - K_2)^4$ and $(G_1 - G_2)^4$ when $K_1 \to K_2$ and $G_1 \to G_2$.

Function $\eta_1(p)$. The function $\eta_1(p)$ [59] is obtained from the probability $P(|h_1|, |h_2|, \theta)$, with $u = \cos\theta$, θ being the angle between the vectors h_1 and h_2. We have:

$$\eta_1(p) = \frac{5\zeta_1(p)}{21} + \frac{150}{7pq}\int_0^{+\infty}\frac{dx}{x}\int_0^{+\infty}\frac{dy}{y}\int_{-1}^{+1} P_4(u)P(x, y, \theta)du \tag{39}$$

in R^3

where $P_4(u)$ is the Legendre polynomial of order 4 with

$$P_4(u) = \frac{1}{8}(35u^4 - 30u^2 + 3)$$

After exchanging the phases A_1 and A_2 we define the function $\eta_2(q)$ with $\eta_2(q) = 1 - \eta_1(p)$. The function $\eta_1(p)$ satisfies $0 \leq \eta_1(p) \leq 1$. The integral in (39) can be calculated analytically in some cases, but more generally numerically.

For two symmetrical phases A_1 and A_2, the case of $p = 0.5$ produces an auto-dual random set (the two phases having the same probabilistic properties), for which $\zeta_1(0.5) = \eta_1(0.5) = 0.5$. This case corresponds to a third order central correlation function equal to zero. As a consequence, the third order bounds of a symmetrical medium have a fixed point in $p = 0.5$. Recent developments on composites with a non linear behavior propose bounds using the function $\zeta_1(p)$, after introduction of comparison media with a linear behavior [63, 76]. We now introduce the third order bounds of some models of random structures.

The mosaic model. The mosaic (or cell) model [33, 50, 56] is obtained from a random tessellation of space into classes (each class or grain is denoted by A'). Any class is allocated to the random set A with the probability p and to the set A^c with the probability $q = 1 - p$. The medium is symmetrical in A and A^c (after permutation of p and q). The third order bounds depend on a parameter G introduced by Miller [56, 57], obtained from the function $s(|h_1|, |h_2|, \theta)$ (μ_d being the Lebesgue measure in R^d, $\overline{\mu}_d$ its average value over the realizations of A', and A'_h the set A' after translation by vector $\mathbf{h}$):

$$s(|h_1|, |h_2|, \theta) = \frac{\overline{\mu}_d(A' \cap A'_{h_1} \cap A'_{h_2})}{\overline{\mu}_d(A')} \tag{40}$$

$$G = \frac{1}{9} + \frac{1}{2}\int_0^{+\infty} \frac{dx}{x} \int_0^{+\infty} \frac{dy}{y} \int_{-1}^{+1} (3u^2 - 1)s(x, y, \theta)du \text{ in } R^3 \tag{41}$$

we have $\frac{1}{9} \le G \le \frac{1}{3}$ in R^3. From (25) we obtain [56, 57, 59]:

$$\begin{aligned} \zeta_1(p) &= p + a(q-p) \text{ with } a = \frac{d^2 G - 1}{d - 1} \\ \zeta_2(p) &= \zeta_1(p) \end{aligned} \tag{42}$$

and $\zeta_1(p)$ is a linear function of p, with the slope $1 - 2a$ $(0 \le a \le 1)$.

Similarly, for the mosaic model:

$$\eta_1(p) = p + (q-p)\left[4(5E-1) - 5(9G-1)\right]/6 \tag{43}$$

E being another parameter depending on the random tessellation [59, 68]. These coefficients were obtained for several types of tessellation with spheroidal grains, and are given in the next table, where needles are cylinders with infinite length, and plates are domains between two parallel planes (it is rather difficult to imagine an isotropic random tessellation of space from these two types of grains).

grain	spheres	needles	plates
E	$1/5$	$3/8$	1
G	$1/9$	$1/6$	$1/3$

The parameters E and G are not sensitive to the scale of the grain A', and are therefore invariant for a population of similar grains. Using (43), we get $\eta_1(p) = \zeta_1(p) = p$ for spheres, and $\eta_1(p) = \zeta_1(p) = 1 - p$ for plates. In [48, 50] are given the functions $\zeta_1(p)$ for the Poisson mosaic, built from a tessellation of space by Poisson lines in R^2 and by Poisson planes in R^3.

Bounds for the Boolean model. As opposite to the mosaic model, the Boolean model is not symmetrical: two sets of bounds are obtained by permutation of A and A^c. The functions $\zeta_1(p)$ and $\eta_1(p)$ are obtained by a numerical integration. With a good approximation, they are linear functions of p. We have $\eta_1(0) = \left[4(5E-1) - 5(9G-1)\right]/6$. In R^3 $\zeta_1(0) = \eta_1(0) = 0$ for spheres. The functions $\eta_1(p)$ and $\eta_2(p)$ of a Boolean model of spheres are given by:

$$\begin{aligned} \eta_1(p) &\simeq 0.711p \text{ , see [10, 71, 72, 73]} \\ \eta_2(p) &\simeq 0.711p + 0.289 \text{ (for the complementary of spheres)} \end{aligned} \tag{44}$$

For illustration Fig. 7 shows the third order bounds of elastic moduli of Boolean models of equal spherical grains: for the studied contrast and for a volume fraction of phase 1 lower than 0.3, the macroscopic elastic moduli are higher if this phase is affected to the complementary set of a Boolean model of spheres, this last one percolating while the union of spheres does not percolate (the percolation threshold p_c of a Boolean model of spheres is given by $p_c = 0.2895 \pm 0.0005$ [65]). This result is an illustration of the fact that third order bounds are more sensitive to percolation effects than Hashin and Shtrikman bounds. Other percolation effects are mentioned in Sect. 1.2 of this volume [1].

Bounds for the hard spheres model. A hard spheres model is obtained by a population of spheres in equilibrium under a potential preventing from any overlap [72, 73]. By construction, the fraction of spheres p is limited to 0.64 for this process. The functions $\eta_1(p)$ and $\zeta_1(p)$ were computed numerically, and fit to the following expressions [72, 73]:

$$\eta_1(p) \simeq 0.48274p \text{ for } p \leq 0.64 \qquad (45)$$
$$\zeta_1(p) \simeq 0.21068p - 0.04693p^2 + 0.00247p^3$$

Combination of basic random sets. More complex models can be built from elementary ones. We consider here multi-scale models. An interesting construction is obtained for the union or the intersection of independent random sets [37, 39]. For $A = A_1 \cap A_2$ we have:

$$P(h_1, h_2) = P_1(h_1, h_2)P_2(h_1, h_2)$$

for widely separate scales (in particular for A_2 with a lower scale), we have approximately [37]:

$$pq\zeta_{H1}(p) \simeq p_1p_2q_2\zeta_{12}(p_2) + p_2^3p_1q_1\zeta_{11}(p_1)$$
$$pq\eta_{H1}(p) \simeq p_1p_2q_2\eta_{12}(p_2) + p_2^3p_1q_1\eta_{11}(p_1)$$

If the two structures are built from the same random set with $p_1 = p_2$, we obtain after n iterations of the process involving intersections:

$$\zeta_{H1}^{(n)}(p) = \frac{1+p}{1+p^{1/n}}\zeta_1(p^{1/n}) \qquad (46)$$
$$\eta_{H1}^{(n)}(p) = \frac{1+p}{1+p^{1/n}}\eta_1(p^{1/n})$$

for $n \to \infty$ we get asymptotically $\zeta_{H_1}^{(n)}(p) \to \frac{1+p}{2}\zeta_1(1)$ and $\eta_{H1}^{(n)}(p) \to \frac{1+p}{2}\eta_1(1)$. Starting from a mosaic model with $a = 1$ (plates in R^3), or starting from a Boolean variety of dilated flats [33] (as in Fig. 4), we have $\zeta_1(1) = \eta_1(1) = 0$ and therefore $\zeta_{H_1}^{(\infty)}(p) = \eta_{H1}^{(n)}(p) = 0$. The complementary set is a union of mosaics (or a union of dilated flats), for which $\zeta_{H_1}^{(\infty)}(p) = \eta_{H1}^{(n)}(p) = 1$. The obtained limit structure admits two equal third order bounds for the bulk modulus K. If $K_2 > K_1$ is attributed to the union of mosaics,

$$K^{eff} = K_- = K_+ = K^{HS+}$$

If $K_2 > K_1$ is attributed to the intersection of mosaics,

$$K^{eff} = K_- = K_+ = K^{HS-}$$

We obtain the construction of an optimal structure corresponding to these bounds, based on the mosaic model, or on the Boolean model. We have the same result when considering the case of the dielectric permittivity. On the other hand, by this process are obtained two separate bounds for the shear modulus G. However they are very close if the materials are well ordered, that is if $(K_2 - K_1)(G_2 - G_1) > 0$. Structures with optimal properties are generated, based on multiscale models. The geometrical construction is unknown for the mosaic, but is easily obtained and simulated for the Boolean layered random model.

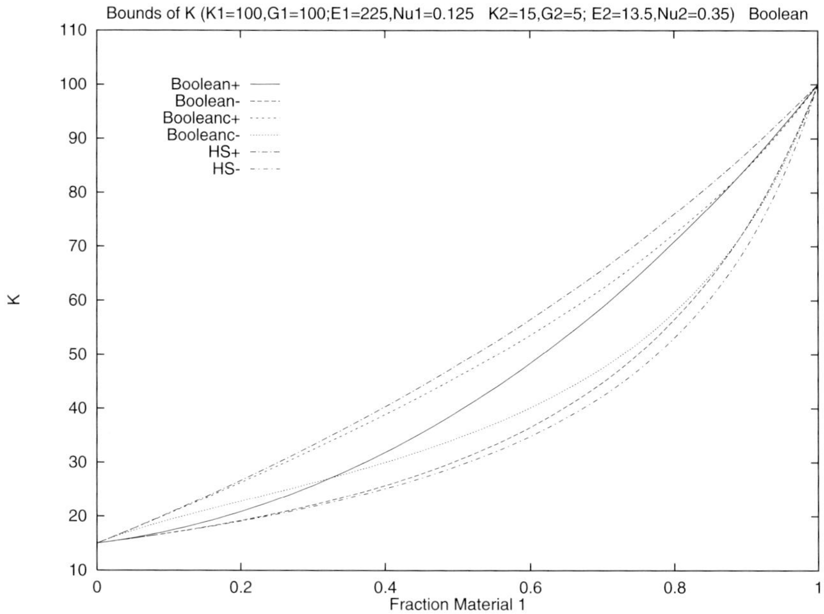

(a)

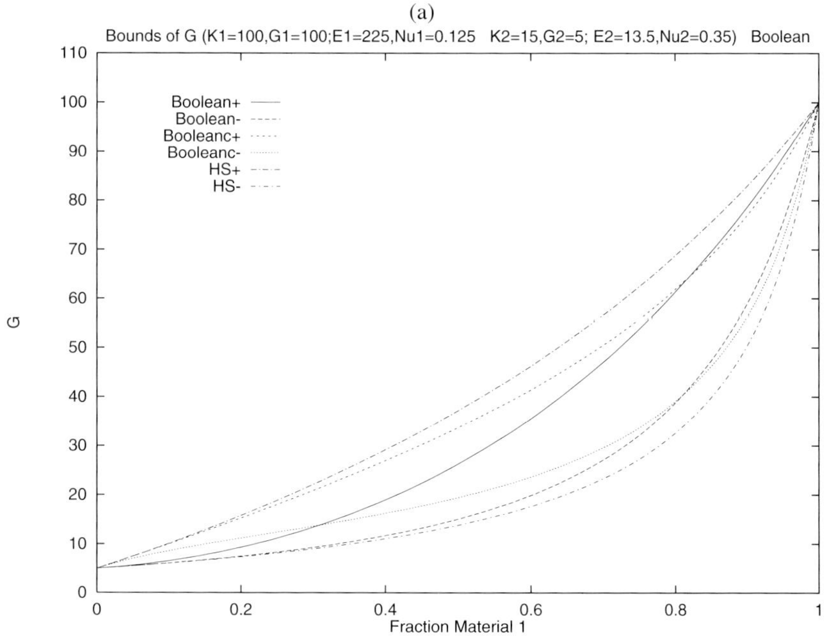

(b)

Fig. 7a-b. Third order bounds on elastic moduli of the Boolean model of spheres (Phase 1: $K_1 = 100$, $G_1 = 100$, $E_1 = 225$, $\nu_1 = 0.125$; Phase 2: $K_2 = 15$, $G_2 = 5$, $E_2 = 13.5$, $\nu_2 = 0.35$). The spheres are attributed to phase 1 (Boolean) or to phase 2 (Boolean^c).

For comparison are given the upper (HS+) and the lower (HS-) Hashin and Shtrikman bound (a) bounds of K. (b) bounds of G. (c) bounds of E. (d) bounds of ν.

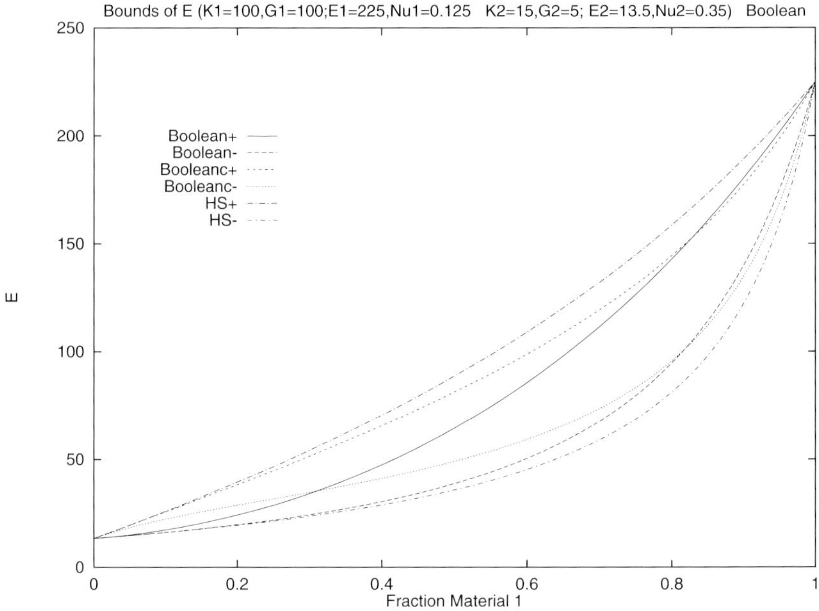

(c)

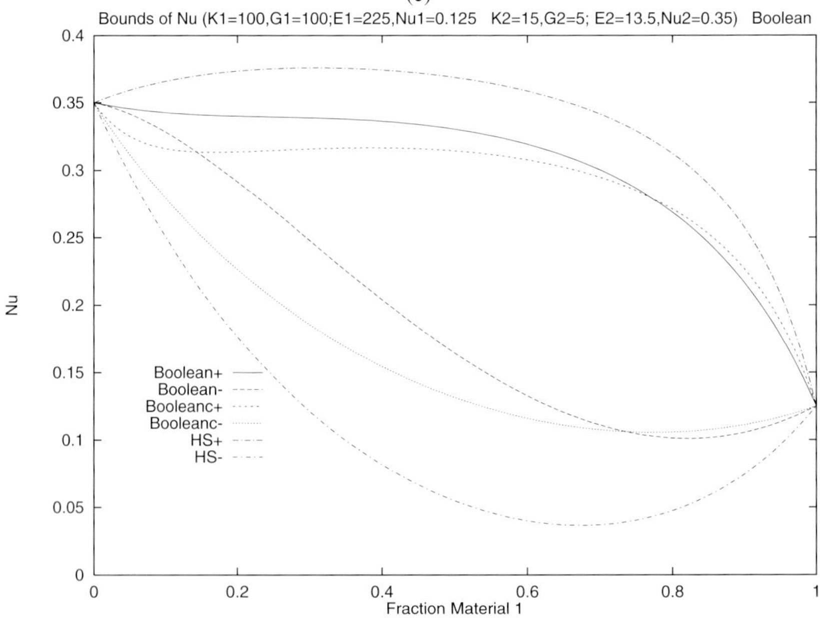

(d)

Fig. 7c-d. Third order bounds of the Boolean model of spheres (continued).

Case of porous media. For a porous medium, we have $K_- = G_- = E_- = 0$, $\nu_- = -\frac{1}{2}$, $\nu_+ = \frac{1}{2}$. The upper third order bounds can be used as an estimator of moduli. They are expressed as a function of the volume fraction of the solid phase p (with A_1 = solid and $A_2 = A_1^c$ = pores), and of its properties K, G, $\zeta_1(p)$, $\eta_1(p)$:

$$K_+ = \frac{4pGK\zeta_1(p)}{3(1-p)K + 4G\zeta_1(p)} \tag{47}$$

$$G_+ = \frac{pG}{1 + \dfrac{6(1-p)(K+2G)^2}{5G(4K+3G)\zeta_1(p) + (3K+G)^2\eta_1(p)}} \tag{48}$$

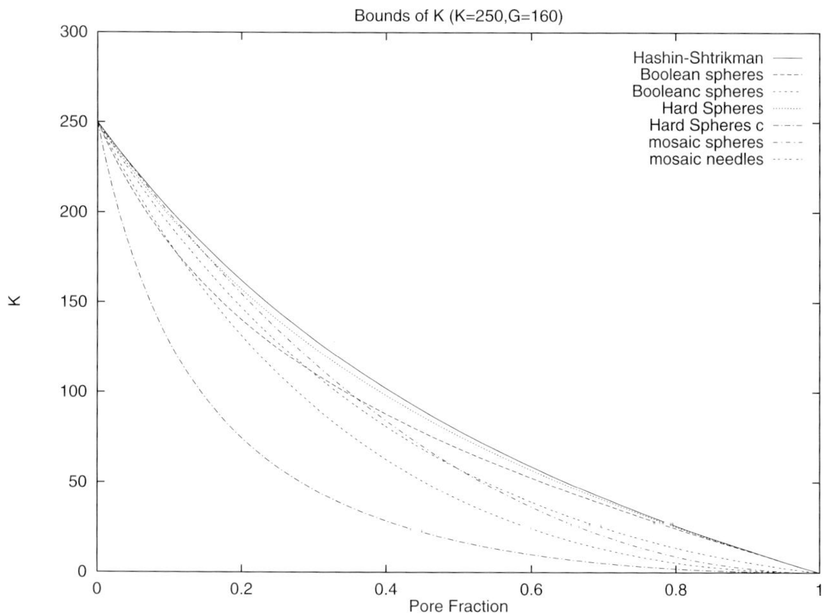

(a)

Fig. 8a. Upper third order bound of various porous random structures (solid phase: $K = 250$, $G = 160$). The pores are spherical for the Boolean model or the hard sphere model (only the part of the curve corresponding to a porosity lower than 0.64 should be considered); the solid grains are spherical for the complementary set of these models (Boolean^c and Hard Spheres^c (only the part of the curve corresponding to a porosity larger than 0.36 is correct)). (a) bounds of K. (b) bounds of G.

When $\eta_1(p) = \zeta_1(p) = 1$, we have , $K_+ = HS_+$ and $G_+ = HS_+$. Therefore, porous media where the solid phase is made of a combination of random Boolean strata with many scales, as in the previous subsection, show optimal elastic properties. The upper bounds of K and G are compared for various models of random structures in Fig. 8. For spherical grains (corresponding for instance to a model describing the beginning of a

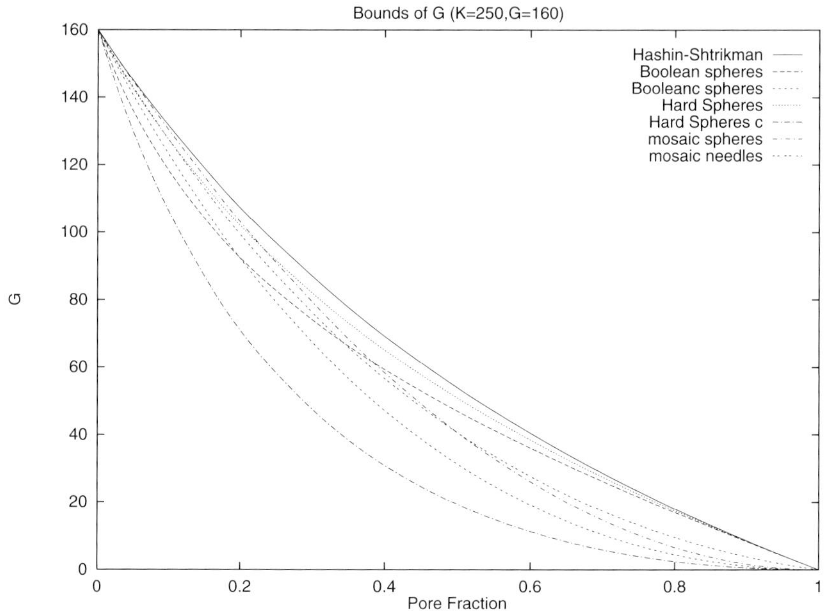

(b)

Fig. 8b. Upper third order bound of various porous random structures (continued).

sinter process), the lower values of bounds of moduli are obtained (the effective moduli remaining close to zero for the hard spheres model). For spherical pores (hard spheres, Boolean model of spheres, spherical mosaic), higher upper bounds are obtained, and they are closer to the upper Hashin and Shtrikman bound. These third order bounds were used to predict the elastic properties of gypsum in [42].

3.4 Estimation of Overall Properties from Numerical Simulations

An alternative way to solve the homogenization problem is based on the numerical resolution of the boundary value problem, using for instance the finite element method, as illustrated in [2, 44]. By numerical techniques, it is possible to solve problems for materials with a non linear behavior (involving for instance some plastic deformations). This requires the use of 3D information on the microstructure, mostly generated by simulations based on a model of random set. A convenient way to solve numerically the homogenization problem is to use periodic boundary conditions applied on a periodic cell. For this purpose, we developed simulations of periodic random media, like the Voronoi 3D polycrystal in Fig. 9 [17]. Beside the finite element method, efficient iterative techniques operating by Fast Fourier Transform on periodic media were recently developed for micromechanics problems [61], and for the dielectric permittivity [18, 21]. A key-point, still under study, concerns the control of the representativeness (from a statistical point of view) of the periodic cell used in the calculations. Recent

Fig. 9. Simulation of a 3D Voronoi mosaic, with periodic boundary conditions [17].

developments use a geostatistical approach based on the experimental determination of the integral range (defined in Sect. 4.4) from simulations [13, 18, 51].

4 Fracture Statistics Models

Fracture statistics appear when measuring the strength of materials from specimens: usually, a wide dispersion of strength is observed on brittle materials like ceramics or like steels at low temperature. In addition, size effects are observed in experiments: there is most often a decrease of the average strength with the size of specimens. This is of practical importance for engineers, who have to design parts with a given reliability. Therefore, models based on random structures and accounting for the dispersion of strength and for scale effects are very useful [33, 35, 45]. The main purpose of this approach is to estimate the probability of fracture of a specimen under given loading conditions. Differently from the case of effective properties, there is in fracture a great sensitivity of the macroscopic fracture behavior to local defects, and mainly to the tail of their probabilistic distribution. One of the main points of fracture statistics approaches deals with scale effects. This point is of major importance when we need to predict the strength of large parts (like in aeronautics, dams or buildings in engineering...) from data available at a laboratory scale; the fact that the average strength may decrease (or sometimes increase) with the size of parts must be known and evaluated in practice. This can be accounted for by models based on the reproduction, at a point scale, of the variations of a fracture criterion (critical stress σ_c, or critical stress intensity factor K_{Ic}, local fracture energy γ, for brittle materials). These models are based on different macroscopic fracture criteria: the weakest link assumption is suitable for the sudden fracture of brittle materials; the models with a damage threshold allow several sites with a crack initiation before fracture; models based on the fracture energy (Griffith's criterion) account for the propagation and arrest of cracks in a random medium; finally, models of random damage mimic the generation of damage zones in quasi-brittle materials. For every family of models the probability of fracture is worked out as a function of the loading conditions and of the parameters of the selected random structure models. Some interesting aspects for applications, such as the prediction of expected scale effects, are derived. The proposed models can be tested at different scales (including the microscopic scale, by use of image analysis). The diversity of the obtained theoretical distributions for fracture statistics offers new possibilities for the microstructure based interpretation and modelling of mechanical data obtained on materials. A more detailed presentation is given in [43].

4.1 Weakest Link Model and Boolean Random Functions

Consider first the case of the weakest link model, where there is a sudden propagation of a crack after its initiation. In this case, the fracture statistics is governed by the most critical defects. The fracture probability is expressed by

$$P\{\text{non fracture}\} = P\{\inf(\sigma_c(x) - \sigma(x)) \geq 0\} \tag{49}$$

where $\sigma_c(x)$ is the random critical stress, $\sigma(x)$ is the applied stress field, and $\inf$ is the minimal value over the sample.

Explicit expressions of the fracture probability are available for uncoupled critical and applied stress fields and for specific models (like Boolean RF). They are more general than the standard Weibull model. The expected scale effect is a decrease of strength (to a constant for large samples) with the size of specimens.

For illustration, consider a medium where defects are introduced in a matrix with an infinite strength. For instance defects with a critical stress σ_c being a random variable Z' inside a closed random set A'_0, implanted in space on Poisson points x_k with the intensity $\theta(\sigma)$. We take for σ_c the minimum of the values of Z' on overlapping grains, building by this process a infimum ($\wedge$) Boolean RF [30, 32, 33, 66]. The general construction of this model, which is a continuous version of the Boolean random set model, is obtained by replacing the $\cup$ operation for overlapping grains (implanted on a Poisson point process or on Poisson varieties) by the supremum ($\vee$) or by the infimum ($\wedge$), as illustrated in Fig. 10. We have:

$$Z(x) = \vee_{x_k} Z'(x - x_k) \text{ for the supremum } (\vee) \text{ Boolean RF}$$
$$Z(x) = \wedge_{x_k} Z'(x - x_k) \text{ for the infimum } (\wedge) \text{ Boolean RF}$$

In the present case, the resulting field $\sigma_c(x)$ is a mosaic with domains where σ_c is constant. In brittle tensile fracture, we can restrict the purpose to the scalar case with $\sigma > 0$ (σ being the maximal main stress). For a uniform applied stress field, the probability distribution of the apparent fracture stress σ_R of a specimen B is obtained from (50), generalizing (27) obtained for the Boolean random set:

$$P\{\sigma_R \geq \sigma\} = P\{\text{non fracture of } B\} = \exp(-\bar{\mu}(A'_0 \oplus \check{B})\Phi(\sigma)) \tag{50}$$

where $\bar{\mu}$ is given by averaging over the realizations of the random set A'_0, and $\Phi(\sigma)$ is the average number of defects per unit volume, with strength lower than σ:

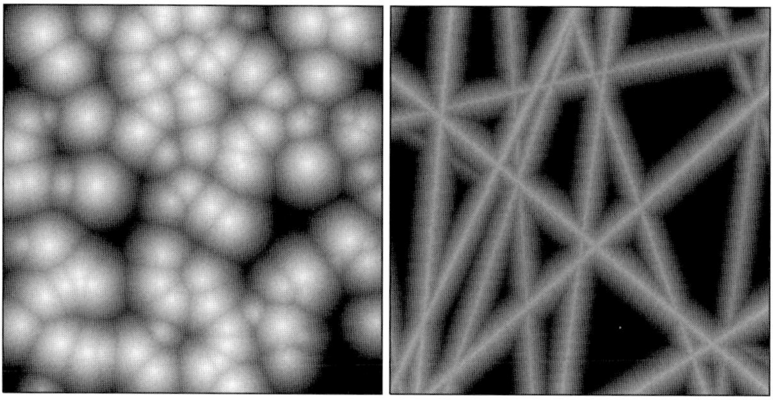

(a) (b)

Fig. 10. Supremum Boolean RF with cone primary grains (a) and Boolean Variety RF (b): the supremum of observed grey values (dark points corresponding to low values, and bright points to high values) is taken in every point of the image.

$$\Phi(\sigma) = \int_0^{\sigma} \theta(t)dt \tag{51}$$

A model of this kind was used for fracture of ceramics, the function $\Phi(\sigma)$ being estimated by image analysis by some transformation of the distribution function of defects (inclusions and porosities ranging from 20 μm to 70 μm) [7].

When applying a non uniform stress field $\sigma(x)$ and for point defects,

$$P\{\text{non fracture of } B\} = \exp(-\int_B \Phi(\sigma(x))dx) = \exp(-\Phi(\sigma_{eq})) \tag{52}$$

with $\Phi(\sigma_{cq}) = \int_B \Phi(\sigma(x))dx$. This formulation with the equivalent stress σ_{eq} makes possible to put together results of different types of fracture tests, for the identification of a model, as made in [8] for the Weibull statistics, σ_{eq} being the so-called Weibull stress.

For defects with $\theta(\sigma) = \theta_1 m(\sigma - \sigma_0)^{m-1}$ and $m > 1$ and $\sigma > \sigma_0$, we have $\Phi(\sigma) = (\sigma - \sigma_0)^m \theta_1$, and in the case of a homogeneous stress field, σ_R follows a Weibull distribution with $\sigma_u^m = 1/\theta$:

$$P\{\sigma_R \geq \sigma\} = \exp(-\bar{\mu}(A'_0 \oplus \check{B}) \left(\frac{\sigma - \sigma_0}{\sigma_u}\right)^m) \tag{53}$$

This distribution is well known in the practice of fracture statistics, in the case of point defects [22, 46], but most users of this model for experimental data are not aware of the underlying Boolean RF model. When $\sigma_0 = 0$, the coefficient of variation of σ_R, $\sqrt{Var[\sigma_R]}/E[\sigma_R]$ does not depend on V, which can be easily checked from data. Scale effects can be seen from the variation of the median strength σ_M, as a function of the specimen volume V:

$$\sigma_M = \sigma_0 + KV^{-1/m} \tag{54}$$

For a mixture of two populations of defects following Weibull distributions with parameters $(\theta_1, \sigma_{01}, m_1)$ and $(\theta_2, \sigma_{02}, m_2)$, resulting into a bimodal Weibull distribution when a uniform stress field is applied:

$$\Phi(\sigma) = \theta_1(\sigma - \sigma_{01})^{m_1} + \theta_2(\sigma - \sigma_{02})^{m_2} \tag{55}$$

Other distributions functions can be derived from other functions $\theta(\sigma)$, like the Pareto distribution, the sigmoidal distribution, and so on [33, 35, 43, 45], where the scale effects induced by these models are studied.

4.2 Critical Defect Fraction

We can relax the weakest link assumption in two different ways, letting the defects (where a potential crack can initiate) reach a critical volume fraction of the domain where $\sigma_c(x) < \sigma(x)$ or a critical density (number of defects per unit volume) [33, 35, 43]. In both cases, no size effect is expected for the median strength, while the dispersion of the observed strength decreases with the size of the specimen.

4.3 Griffith Crack Arrest Criterion

Fracture statistics models were derived for two dimensional media with a random fracture energy $\Gamma(x)$ [14, 35, 43]. According to the Griffith criterion, a possible fracture path $P(s, d)$ connecting a source s to a destination d must satisfy for every point x of the path:

$$2\Gamma(x) \leq G(x) \tag{56}$$

where the energy release rate $G(x)$ depends on the location of the crack front x, on the loading conditions, and on the crack path connecting s to x. From (56) a potential crack path P must satisfy

$$\vee\{2\Gamma(x) - G(x); x \in P\} \leq 0 \tag{57}$$

where $\vee$ is the upper bound. For a random medium, the condition 57 is a random event, and

$$P\{\text{fracture}\} = P\{\vee\{2\Gamma(x) - G(x); x \in P\} \leq 0\}$$

where the fracture path $P(s, d)$ separates the specimen in two parts. In general, $P\{\text{fracture}\}$ is difficult to calculate, due to a very complex crack path geometry, related to the microstructure. Usually $G(x)$ is a realization of a random function correlated to $\Gamma(x)$. For simplification, the following situations were considered: straight crack propagation according to the mode I opening under various loading conditions, resulting in stable or unstable crack propagation; the case of a microcrack initiation on defects, followed by an advance or arrest of the crack, was also examined. For a given model of RF concerning the fracture energy $\Gamma(x)$, the following types of probability distributions can be expressed as a function of the model and of the loading conditions: fracture stress σ_R, toughness (standard G_c, and at crack arrest G_a), critical length of defects (unstable microcrack a_c or crack after arrest l_a). Since these probability distributions are related in a coherent way, it is possible to use them in practice to test models from data at different scales:

– mechanical data $(\sigma_R, G_c, \sigma_a, G_a)$,
– size distribution of cracks observed for given loading conditions (a_c, l_a).

Calculations made on a Poisson mosaic, and on a Boolean random mosaic, with the distribution function $F(\gamma) = P\{\Gamma < \gamma\}$ provide various scale effects. An interesting example concerns a model of random fracture energy $\Gamma(x)$ corresponding to the situation of a polycrystal, every crystal showing its own fracture energy. A convenient model for this case is the Poisson mosaic: we start from a Poisson random tessellation with parameter λ delimiting grain boundaries (Poisson lines in the plane in 2D); independent realizations of the fracture energy Γ are attributed to each grain of the tessellation, with the cumulative distribution function

$$F(\gamma) = P\{\Gamma < \gamma\}$$

As a result, it is known that $F(\gamma)$ is the distribution function of the RF $\Gamma(x)$. We consider now the unstable propagation of a surface crack with the initial length $2a$ in a random medium, until it reaches the length $2(a+b)$. For any loading where G increases with x,

$$P(a+b) = F(G(a)/2) \exp\left[-\lambda \int_{2a}^{2(a+b)} (1 - F(G(u)/2))du\right] \tag{58}$$

From (58), the fracture probability $P(a+b)$ increases (and converges to 1)

– with the crack length $2a$, or the applied stress σ
– for lower λ (corresponding to a coarser microstructure): small grains improve the ability to resist the crack growth, as a result of a higher probability to meet grains with a large fracture energy γ, blocking the crack propagation.

The predicted scale effect depends on the distribution $F(\gamma)$ for specimens with a variable size (constant a/b):

– for a finite range of $F(\gamma)$ ($F(\gamma) = 1$ for $\gamma \geq \gamma_c$), $P\{\text{fracture}\} \to 1$ when $a \geq a_c$ with $2a_c = 4\gamma_c E/(\pi\sigma^2)$
– for a distribution $F(\gamma)$ with a tail for large fracture energies:

$$1 - F(\gamma) \simeq \gamma^{-\alpha} \text{ for } \gamma \to \infty$$

the scale effects depend on the positive coefficient α
  – if $\alpha = 1$, there is asymptotically no scale effect for $P\{\text{fracture }\}$

$$P(a+b) = y^{\lambda c} \text{ with } y = a/(a+b) \text{ and } c = E/(2\pi\sigma^2) \tag{59}$$

  – if $\alpha \neq 1$, the large scale behavior of $P\{\text{fracture }\}$ is given by

$$P(a+b) = \exp\left[-\lambda(2c)^\alpha a^{1-\alpha}(1 - y^{\alpha-1})/(\alpha-1)\right]$$

  Therefore, if $\alpha < 1$, corresponding to a slow growth of $F(\gamma)$ towards 1, the crack is almost surely stopped during its propagation by reinforcing obstacles with a high fracture energy. If $\alpha > 1$, $P(a+b) \to 1$ for increasing sizes, as in the case of a distribution function of a bounded RV γ.

The scale effect for the median strength σ_M depends on α:

- if $\alpha = 1$, there is no scale effect

$$\sigma_M \simeq \sqrt{-\log y}$$

- if $\alpha \neq 1$,

$$\sigma_M \simeq d^{\frac{1-\alpha}{2\alpha}} \left|1 - y^{\alpha-1}\right|^{1/(2\alpha)}$$

where d is the size of the specimen: σ_M increases with d for $\alpha < 1$ and σ_M decreases with d for $\alpha > 1$.

The probability distributions of the other variables (toughness G_c and G_a, critical length l_c and length at arrest l_a) are found in [35, 45]. Similar conclusions, concerning size effects, were obtained for a model where $\Gamma(x)$ is a Boolean random function [34, 45].

4.4 Models of Random Damage

The occurrence of damage in a loaded material is followed by a progressive loss of integrity, ending by its ruin. This damage is the result of a local degradation of weaker parts in the material, that can be generated by the presence of defects. The presence of defects and of their heterogeneity can be accounted for by a probabilistic approach. This is generally made in the case of brittle materials by the weakest link model, as described above. In what follows, we propose probabilistic models of damage based on the presence of random defects. After a presentation of our main assumptions, we consider the case of a random damage generated under a homogeneous loading. More details are given in [41].

We consider a homogeneous elastic material (with the elasticity tensor C) containing point defects. Under the action of a stress field $\sigma(x)$ ($\sigma(x) = \sigma$ in the case of a uniform load), any defect located at x, with a critical stress σ_c lower than $\sigma(x)$, generates a damaged zone with volume v centered in x. The damaged zone can be replaced by a domain v with a constant elasticity tensor C_D. We consider here the extreme case where damage zones behave like pores ($C_D \equiv 0$). This is equivalent to the notion of statistical volume element (containing at least a critical defect) introduced in Sect. 4.5 for the finite elements simulation of damage in composites [3]. One model of this type is obtained as a tessellation of space into cells with a volume v, the cell covering x remaining unbroken with probability $P(\sigma(x))$, and generating an evolving two-phase mosaic model. Another model considers point defects distributed according to a Poisson point process, $\Phi(\sigma)$ being the average number of defects with a critical stress lower than σ, per unit volume of material. The function $\Phi(\sigma) = \left(\frac{\sigma}{\sigma_u}\right)^m$ gives the Weibull distribution for the "weakest link" model. Here, critical defects at Poisson points x_k generate damaged zones A'_k (namely pores in what follows) with centers in x_k. Therefore the damaged part of a specimen builds a Boolean model with primary grain A'. This primary grain can be random, and can be oriented in a given direction, in the case of anisotropic damage. In examples below, we consider isotropic and uniform damaged zones, generating a

Boolean model with spheres having a constant volume v. If overlaps between damage zones at different points x_k are forbidden, a hard-core process is generated.

The next assumption in the model concerns the redistribution of stress in the non damaged zone. As in the case of the damage of bundles of fibers, we assume a uniform load sharing [15, 70], the damage generating a uniform amplification of stresses over the non damaged parts. With this simplification, the damage behaves as if it was totally diffuse, no localization being permitted. This is typical of a quasi brittle material, in opposition to a brittle material submitted to the weakest link assumption.

The progression of damage in the material processes as follows: defects are ranked with an increasing severity $(\sigma_c - \sigma(x))$; the most critical defect is converted into a damaged domain; the load is then redistributed over the matrix, and the process goes on for the remaining defects.

We consider now a continuous model in the case of large homogeneous specimens. The average volume fraction of the specimen with a critical stress larger than σ is given by $P(\sigma)$. For the previous cell model, this is the probability for one cell to remain undamaged. For the Poisson point defects, generating a Boolean model, we have:

$$P(\sigma) = \exp\left(-\Phi(\sigma)v\right) \tag{60}$$

Consequently, when the non damaged parts of the material are loaded with the homogeneous stress field σ, the average macroscopic stress σ_M acting on the material is:

$$\sigma_M = \sigma P(\sigma) \tag{61}$$

From (61), the average macroscopic stress is expected to reach a maximum σ_{ult}, when it exists, since $P(\sigma)$ decreases with σ. For a specimen of volume V, fluctuations of $P(\sigma)$ (and therefore of $P(\sigma_{ult})$) are expected. Its variance $Var(\sigma, V)$ is given, as a function of the covariance $Q(h)$ of the damaged zone (with $Q(0) = q = P(\sigma)$), by

$$Var(\upsilon, V) = \frac{1}{V^2}\int_V \int_V \left(Q(x-y) - q^2\right) dxdy \tag{62}$$

For a large specimen (with $V \gg A_3$), (62) can be expressed as a function of the integral range in the space R^3, A_3, by

$$\begin{aligned} Var(\sigma, V) &= \frac{q(1-q)A_3}{V} \\ \text{with } A_3 &= \frac{1}{q(1-q)}\int_{R^3} \left(Q(h) - q^2\right) dh \end{aligned} \tag{63}$$

For a model with cells having a constant volume v, we get $A_3 = v$. For a Boolean model of spheres, we illustrate this point in Table 1 by the value of two standard deviations (giving the confidence interval of $q \pm 2\sqrt{Var(\sigma, V)}$ for q expected on realizations) when the specimen is a cube with the volume $V = L^3$ and when the diameter a of spheres (damaged zones) is such that $a/L = 0.1$. The same statistical property follows for σ_{ult}, since from (61), the variance of σ_{ult} is obtained from the variance of $P(\sigma_{ult})$, after multiplication by σ^2_{ult}.

Table 1. Variability of the proportion of undamaged material $P(\sigma_{ult})$

$P(\sigma_{ult})$	0.95	0.9	0.8	0.7	0.5
$2\sqrt{Var(\sigma_{ult}, V)}$	0.0054	0.00865	0.0134	0.0167	0.02

The macroscopic average relationship between the stress σ_M and the deformation ε is known, provided the effective elastic tensor $C(\sigma)$ is known as a function of the damage induced in the material at the level of stress σ acting on the matrix. It depends on the geometrical arrangement of the damage (and therefore on its configuration for a given volume V). For simplification, we consider large specimens and assume that $C(\sigma)$ is given by the effective tensor of an infinite porous medium with the solid volume fraction $P(\sigma)$. It is therefore obtained by homogenization of a random medium. We can use the upper third order bound 3.3. The average stress-strain relation is deduced in a parametric way (as a function of σ) from

$$\sigma_M = \sigma P(\sigma), \varepsilon = (C(\sigma))^{-1} \sigma P(\sigma) \tag{64}$$

For an isotropic elastic medium in traction, we get the relation between the elongation ε, the traction σ and the Young's modulus $E(\sigma)$: $\varepsilon = \sigma P(\sigma)/E(\sigma)$. To illustrate the approach, we consider specimens in tension, with defects on Poisson points. For defects obeying to the Weibull distribution, the ultimate stress σ_{ult} and the fraction of non damaged material are given by

$$\begin{aligned} \sigma_{ult} &= \sigma_u \left(\frac{1}{mv}\right)^{1/m} \\ P(\sigma_{ult}) &= \exp(-\frac{1}{m}) \end{aligned} \tag{65}$$

In Table 2 is given the variation, as a function of m, of σ_{ult}/σ_u (taking $v = 1$), of $P(\sigma_{ult})$ and of the reduced coefficient of variation $c = \sqrt{\frac{1-P(\sigma_{ult})}{P(\sigma_{ult})}}$, from which the ratio $\frac{\text{standard deviation}}{\text{average}}$ of $P(\sigma_{ult})$ and of σ_{ult} is deduced by $c\sqrt{\frac{A_3}{V}}$. The ultimate strength increases with m, while the proportion of damaged material $1 - P(\sigma_{ult})$ and the fluctuations decrease. Increasing m results in a more deterministic behavior. For $m \geq 3$, the volume fraction of the damaged material is lower than the percolation threshold of the phase made of spheres for the Boolean model, which means that the physical behavior of the model is correct. Concerning the $\sigma_M - \varepsilon$ curve, high values of m generate a more brittle elastic material, while a low value of m results in a ductile type macroscopic behavior, as illustrated on Fig.11. The curves own a common point obtained for $\sigma = \sigma_u$. Note that it should be possible to estimate the parameters of the statistical model by identification from an experimental $\sigma_M - \varepsilon$ curve.

For defects obeying to a Pareto distribution, we have $\Phi(\sigma) = m \log \frac{\sigma}{\sigma_u}$ and $P(\sigma) = \left(\frac{\sigma_u}{\sigma}\right)^{mv}$ for $\sigma \geq \sigma_u$. When $mv > 1$, we have $\sigma_{ult} = \sigma_u$ and $P(\sigma_{ult}) = 1$: large specimens of the material are brittle with a constant ultimate stress. When $mv < 1$, we have $\sigma_{ult} = +\infty$ and there is no bounded maximum for $\sigma_M = \sigma P(\sigma)$.

Table 2. Effect of m on the ultimate strength, $P(\sigma_{ult})$, and c (Weibull distribution).

m	1	2	3	4	5	10	25	50
σ_{ult}/σ_u	1	0.707	0.693	0.707	0.725	0.794	0.879	0.925
$P(\sigma_{ult})$	0.368	0.606	0.716	0.779	0.819	0.905	0.961	0.980
c	1.311	0.805	0.629	0.533	0.470	0.324	0.202	0.142

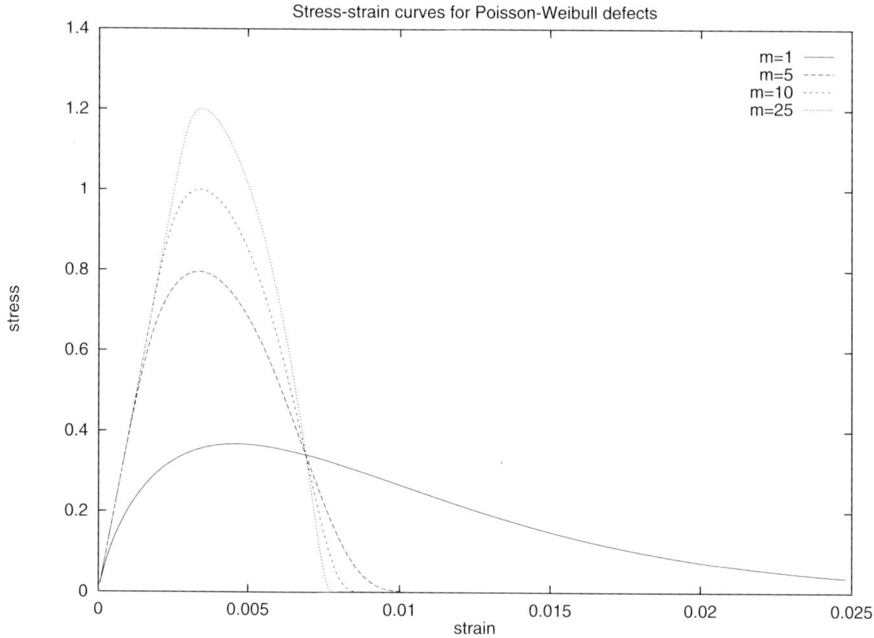

Fig. 11. Average macroscopic stress-strain (σ_M/σ_u) relation for Weibull populations of defects with various m and a constant median $\sigma_u \log 2$.

4.5 Fracture Statistics Models and Simulations

When considering damaging materials at different scales, like fiber composites or metals under a ductile fracture, the fracture process must be studied by means of numerical simulations [3, 16, 20, 36, 4], since it is difficult to account for the complex stress field resulting from the interaction between various damage sites. In the case of composite materials, we used realizations of the random fracture criterion, combined with finite elements (FE) calculations. The following methodology was developed to study the fracture statistics unidirectional composites for fiber fracture [3, 4], for transverse fracture, and finally for the fracture of laminate composites. The first step is based on the experimental identification of the population of defects (point defects are considered here), by appropriate mechanical tests and by calculation of the local stress field seen by the microstructure; in this experimental part, a statistical volume element (SVE) is determined, as the elementary volume element broken during the progression of damage. Then, the statistical information and the fracture criterion are introduced in a FE calculation: to every SVE is attributed a random strength σ_R corresponding to the population of defects; during the calculation, it is broken if the average stress in the SVE, or more rigorously the equivalent stress σ_{eq} defined by (52) is larger than σ_R. Random simulations enable us to study the fracture behavior on a first scale, namely a representative volume element RVE. If necessary, the material can be considered as a set of RVE to study its behavior on the next scale. The main difficulty in running simulations is to determine the appropriate microstructural element converted into a SVE, and to know its fracture statistics. When operating on scales with increasing lengths, the statistical models proposed in the previous sections can be used to generate the necessary random variables. However, the correct corresponding type of assumption (weakest link, critical density,...) has to be introduced on a physical basis corresponding to the behavior of the material, as known from experiments. Similar approaches are developed for ductile fracture statistics in steels (see [16, 20], and the chapter by A. Pineau in [44]).

From these examples of simulations, general guidelines can be derived. Firstly, a damage parameter (density of point defects, microcrack network parameter, cavity volume fraction, cavity growth rate,...), connected to the microstructure, should be selected at a given scale, the SVE. Secondly, statistical properties of this parameter should be obtained by image analysis (or by micromechanical tests); until now this information was limited to the probability distribution function over domains with a given size, but higher order information can be estimated to recover probabilistic information on the damage parameter, considered as the realization of a random function. Finally, simulations of the damage parameter as initial conditions for the prediction of its evolution by means of finite elements have to be performed; for this step, morphological models of random media, as well as change of scale models to generate correct simulations on different scales, can be useful.

5 Conclusions

A wide range of random models of structures is available to simulate the complex morphology of materials at a microscopic level. The experimental approach of random sets, based on measurements obtained by image analysis is fruitful to select appropriate models, to test their validity, and to estimate their parameters. Moreover, models of random media are useful for the prediction of overall properties of materials, and for an attempt to optimize the morphology with respect to some physical properties. From the same approach, it is possible to model the fracture statistics behavior of materials. The various available models are useful for applications, to predict expected scale effects. They can be tested at different scales (including the microscopic scale, by use of image analysis). The diversity of theoretical probability distributions obtained for fracture statistics opens the door to a microstructure based interpretation and modeling of mechanical data obtained on materials. Efforts are in progress to integrate the models presented in this chapter in numerical simulation techniques and in the design of microstructures.

References

1. Arns, C., M. Knackstedt, K. Mecke (2002): 'Characterising the morphology of disordered materials', this volume.
2. Barbe, F., L. Decker., D. Jeulin, G. Cailletaud (2001): 'Intergranular and intragranular behavior of polycristalline aggregates. Part 1: F.E. model', *Int. J. Plasticity* **17** (4), pp. 513-536 .
3. Baxevanakis, C., D. Jeulin, J. Renard (1995): 'Fracture Statistics of a Unidirectional Composite', *Int. J. Fracture* **73**, pp. 149-181.
4. Baxevanakis, C., D. Jeulin, B. Lebon, J. Renard (1998): 'Fracture statistics modeling of laminate composites', *Int. J. Solids Struct.* **35**, N° 19, pp. 2505-2521.
5. Beran, M.J., J. Molyneux (1966): 'Use of classical variational principles to determine bounds for the effective bulk modulus in heterogeneous media', *Q. Appl. Math.* **24**, pp. 107-118.
6. Beran, M. J. (1968): *Statistical Continuum Theories.* (J. Wiley, New York).
7. Berdin, C., G. Cailletaud, D. Jeulin (1993): 'Micro-Macro Identification of Fracture Probabilistic Models'. In: *Proc. of the International Seminar on Micromechanics of Materials, MECAMAT'93*, Fontainebleau, 6-8 July 1993 (Eyrolles, Paris), pp. 499-510.
8. Beremin, F.M. (1983): 'A local criterion for cleavage fracture of a nuclear pressure vessel steel', *Metall. Trans. A.* **14A**, pp. 2277-2287.
9. Bergman, D. (1978): 'The dielectric constant of a composite material: a problem in classical physics', *Phys. Rep. C* **43**, pp. 377-407.
10. Berryman, G.J. (1985): 'Variational bounds on elastic constants for the penetrable sphere model', *J. Phys. D: Appl. Phys.* **18**, pp. 585-597.
11. Berryman, G.J., G.W. Milton (1988): 'Microgeometry of random composites and porous media', *J. Phys. D: Appl. Phys.* **21**, pp. 87-94.
12. Bretheau, T., D. Jeulin (1989): 'Caractéristiques morphologiques des constituants et comportement à la limite élastique d'un matériau biphasé Fe/Ag', *Rev. Phys. Appl.* **24**, pp. 861-869.
13. Cailletaud, G., D. Jeulin, Ph. Rolland (1994): 'Size effect on elastic properties of random composites', *Eng. Comput.* **11** (2), pp. 99-110.
14. Chudnovsky, A., B. Kunin (1987): 'A probabilistic model of brittle crack formation', *J. Appl. Phys.* **62** (10), pp. 4124, 4129.
15. Daniels, H.E. (1945): 'The statistical theory of the Strength of Bundles of Threads', *Proc. Roy. Soc. London* A **183**, pp. 405-435.
16. Decamp, K., L. Bauvineau, J. Besson, A. Pineau (1997): 'Size and geometry effect on ductile fracture of notched bars in a C-Mn steel: experiments and modelling', *Int. J. Fracture* **88**, pp. 1-18.
17. Decker, L., D. Jeulin (2000): 'Simulation 3D de matériaux aléatoires polycristallins', *Revue de Métallurgie - CIT/Science et Génie des Matériaux*, Feb. 2000, pp. 271-275.
18. Delarue, A., D. Jeulin, in preparation.
19. Delisée, Ch., D. Jeulin, F. Michaud (2001): 'Caractérisation morphologique et porosité en 3D de matériaux fibreux cellulosiques', *C.R. Acad. Sci. Paris*, **vol. 329**, Série II b, pp. 179-185.
20. Devillers-Guerville, L., J. Besson, A. Pineau (1997): 'Notch fracture toughness of a cast duplex stainless steel: modeling of experimental scatter and size effect', *Nucl. Eng. Design* **168**, pp. 211-225.
21. Eyre, D.J., G.W. Milton (1999): 'A fast numerical scheme for computing the response of composites using grid refinement', *Eur. Phys. J. Appl. Phys.* **6**, pp. 41-47.
22. Harter, H.L. (1978): 'A bibliography of extreme-value theory', *Int. Stat. Rev.* **46**, pp. 279-306.
23. Hashin, Z., S. Shtrikman (1962): 'A variational approach to the theory of the effective magnetic permeability of multiphase materials', *J. Appl. Phys.* **33**, pp. 3125-3131.

24. Hashin, Z., S. Shtrikman (1963): 'A variational approach to the theory of the elastic behavior of multiphase materials', *J. Mech. Phys. Solids* **11**, pp. 127-140.
25. Heijmans H.J.A.M. (1994): *Morphological Image Operators. Advances in Electronics and Electron Physics.* Vol. Suppl. 24. (Academic Press, Boston).
26. Hori, M. (1973): 'Statistical theory of the effective electrical, thermal, and magnetic properties of random heterogeneous materials. II. Bounds for the effective permittivity of statistically anisotropic materials', *J. Math. Phys.* **14**, pp. 1942-1948.
27. Hug, D., G. Last, W. Weil (2002): 'A Survey on Contact Distributions', this volume.
28. Jeulin, D. (1978): 'Amélioration des performances des hauts fourneaux: recherche des relations entre morphologie et qualité des agglomérés', *IRSID Report RE 578*, Rapport final CECA (convention n°6 220. AA/3/304).
29. Jeulin, D. (1981): 'Improvement of blast furnace performance: research into relations between morphology and quality of sinters', *E.E.C. Publication EUR 6280.*
30. Jeulin, D., P. Jeulin (1981): 'Synthesis of Rough Surfaces by Random Morphological Models'. In: Proc. 3rd European Symposium of Stereology, *Stereol. Iugosl.* **3**, suppl. 1, pp. 239-246.
31. Jeulin, D. (1986): 'Study of spatial distributions in multicomponent structures by image analysis'. In: Proc. 4th European Symposium of Stereology (Göteborg, 26-30 August 1985), *Acta Stereol.* **5/2**, pp. 233-239.
32. Jeulin, D. (1987): 'Random structure analysis and modelling by Mathematical Morphology'. In: *Proc. CMDS5,* ed. by A. J. M. Spencer (Balkema, Rotterdam), pp. 217-226.
33. Jeulin, D. (1991): *Modèles morphologiques de structures aléatoires et de changement d'échelle,* Thèse de Doctorat d'Etat, University of Caen.
34. Jeulin, D. (1992): 'Some Crack Propagation Models in Random Media'. In: *Proc. Symposium on the Macroscopic Behavior of the Heterogeneous Materials from the Microstructure,* ASME, Anaheim, Nov 8-13, 1992. AMD Vo. 147, pp. 161-170.
35. Jeulin, D. (1994): 'Random structure models for composite media and fracture statistics'. In: *Advances in Mathematical Modelling of Composite Materials*, ed. by K.Z. Markov (World Scientific Company, Singapore), pp. 239-289.
36. Jeulin, D., C. Baxevanakis, J. Renard (1995): 'Statistical modelling of the fracture of laminate composites'. In: *Applications of Statistics and Probability*,ed. by M. Lemaire, J.L. Favre, A. Mébarki (Balkema, Rotterdam), pp. 203-208.
37. Jeulin, D., A. Le Coënt (1996): 'Morphological modeling of random composites', *Proceedings of the CMDS8 Conference (Varna, 11-16 June 1995)*, ed. by K.Z. Markov (World Scientific, Singapore), pp. 199-206.
38. Jeulin, D. (ed) (1997) Proceedings of the Symposium on the *Advances in the Theory and Applications of Random Sets* (Fontainebleau, 9-11 October 1996) (World Scientific, Singapore).
39. Jeulin, D., L. Savary (1997): 'Effective Complex Permittivity of Random Composites', *J. Phys. I/ Condens.Matter* **7**, pp. 1123-1142.
40. Jeulin, D. (1998): 'Bounds of physical properties of some random structure'. In: *Proceedings of the CMDS9 Conference* (Istanbul, Turkey, June 29-July 3, 1998), ed. by E. Inan and K.Z. Markov (World Scientific, Singapore), pp. 147-154.
41. Jeulin, D. (2000): 'Models of random damage'. In: *Proc.Euromat 2000 Conference*, Tours, France, ed. by D. Miannay, P. Costa, D. François, A. Pineau, pp. 771-776.
42. Jeulin, D., P. Monnaie, F. Péronnet (2001): 'Gypsum morphological analysis and modeling', *Cement and Concrete Composites* **23 (2-3)**, pp. 299-311.
43. Jeulin, D. (2001): 'Random Structure Models for Homogenization and Fracture Statistics', In: *Mechanics of Random and Multiscale Microstructures*, ed. by D. Jeulin, M. Ostoja-Starzewski (CISM Lecture Notes N° 430, Springer Verlag).
44. Jeulin, D., M. Ostoja-Starzewski (eds) (2001): *Mechanics of Random and Multiscale Microstructures.* (CISM Lecture Notes N° 430, Springer Verlag).

45. Jeulin, D. : *Morphological Models of Random Structures,* in preparation.
46. Kittl, P., G. Diaz (1986): 'Weibull's fracture statistics, or probabilistic strength of materials: state of the art', *Res. Mechanica* **25**, pp. 99-207.
47. Kröner, E. (1971): *Statistical Continuum Mechanics.* (Springer Verlag, Berlin).
48. Le Coënt, A., D. Jeulin (1996): 'Bounds of effective physical properties for random polygons composites', *C.R. Acad. Sci. Paris,* **323**, Série II b, pp. 299-306.
49. Matheron, G. (1967): *Eléments pour une théorie des milieux poreux.* (Masson, Paris).
50. Matheron, G. (1968): 'Composition des perméabilités en milieu poreux hétérogène: critique de la règle de pondération géométrique', *Rev. IFP* **23**, pp. 201-218.
51. Matheron, G. (1971): *The theory of regionalized variables and its applications.* (Paris School of Mines publication).
52. Matheron, G. (1975): *Random sets and integral geometry.* (J. Wiley, New York).
53. McCoy, J.J. (1970): 'On the deplacement field in an elastic medium with random variations of material properties'. In: *Recent Advances in Engineering Sciences*, Vol. 5 (Gordon and Breach, New York), pp. 235-254.
54. Mecke, K., H. Wagner (1991): 'Euler characteristics and related measures for random geometric sets', *J. Statist. Phys.* **64**, pp. 843-850.
55. Mecke, K. (2000): 'Additivity, convexity, and beyond: Applications of Minkowski functionals in statistical physics'. In: *Statistical Physics and Spatial Statistics*, ed. by K.R. Mecke, D. Stoyan, Lecture Notes in Physics 554 (Springer Verlag, Berlin), pp. 111-184.
56. Miller, M.N. (1969): 'Bounds for the effective electrical, thermal and magnetic properties of heterogeneous materials', *J. Math. Phys.* **10**, pp. 1988-2004.
57. Miller, M.N. (1969): 'Bounds for effective bulk modulus of heterogeneous materials', *J. Math. Phys.* **10**, pp. 2005-2013.
58. Milton, G. (1980): 'Bounds on the complex dielectric constant of a composite material', *Appl. Phys. Lett.* **37**, pp. 300-302.
59. Milton, G. (1982): 'Bounds on the elastic and transport properties of two component composites', *J. Mech. Phys. Solids* **30**, pp. 177-191.
60. Molchanov, I. (1997): *Statistics of the Boolean model for practitioners and mathematicians.* (J. Wiley, Chichester).
61. Moulinec H., P. Suquet (1994): 'A fast numerical method for computing the linear and nonlinear mechanical properties of composites', *C.R. Acad. Sci. Paris,* **318**, Série II, pp 1417-1423.
62. Pélissonnier Grosjean, C., D. Jeulin, L. Pottier, D. Fournier, A. Thorel (1997): 'Mesoscopic modeling of the intergranular structure of Y_2O_3 doped aluminium nitride and application to the prediction of the effective thermal conductivity'. In: *Key Engineering Materials*, Volumes 132-136, Part 1 (Transtech Publications, Switzerland), pp. 623-626.
63. Ponte Castaneda, P. (1996): 'Variational methods for estimating the effective behavior of nonlinear composite materials'. In: *Proceedings of the CMDS8 Conference (Varna, 11-16 June 1995)*, ed. by K.Z. Markov (World Scientific, Singapore), pp. 268-279.
64. Quenec'h, J.L., J.L. Chermant, M. Coster, D. Jeulin (1994): 'Liquid phase sintered materials modelling by random closed sets'. In: *Mathematical morphology and its applications to image processing,* ed. by J. Serra, P. Soille (Kluwer, Dordrecht), pp. 225-232.
65. Rintoul, M.D., S. Torquato (1997): 'Precise determination of the critical threshold and exponents in a three-dimensional continuum percolation model', *J. Phys. A: Math. Gen.* **30**, pp. L585-L592.
66. Serra, J. (1982): *Image analysis and mathematical morphology.* (Academic Press, London).
67. Soille, P. (1999): *Morphological Image Analysis: Principles and Applications.* (Springer, Berlin).
68. Silnutzer, N.R. (1972): *Effective constants of statistically homogeneous materials*, PhD thesis, University of Pennsylvania.

69. Stoyan, D., W.S. Kendall, J. Mecke (1987): *Stochastic Geometry and its Applications*. (J. Wiley, New York).
70. Suh, M.W., B.B. Bhattacharyya, A. Grandage (1970): 'On the distribution and moments of the strength of a bundle of filaments', *J. Appl. Prob.* **7**, pp. 712-720.
71. Torquato, S., G. Stell (1983): 'Microstructure of two-phase random media III. The n-point matrix probability functions for fully penetrable spheres', *J. Chem. Phys.* **79**, pp. 1505-1510.
72. Torquato, S., F. Lado (1986): 'Effective properties of two phase disordered composite media: II Evaluation of bounds on the conductivity and bulk modulus of dispersions of impenetrable spheres', *Phys. Rev. B* **33**, pp. 6428-6434.
73. Torquato, S. (1991): 'Random heterogeneous media: microstructure and improved bounds on effective properties', *Appl. Mech. Rev.* **44**, pp. 37-76.
74. Vogel, H.J. (2002): 'Topological characterization of porous media', this volume.
75. Willis, J.R. (1981): 'Variational and related methods for the overall properties of composites'. In: *Advances in Applied Mechanics*, **21**, pp. 1-78.
76. Willis, J.R. (1991): 'On methods for bounding the overall properties of nonlinear composites', *J. Mech. Phys. Solids* **39**, 1, pp. 73-86.

Characterising the Morphology of Disordered Materials

Christoph H. Arns[1], Mark A. Knackstedt[1,2], and Klaus R. Mecke[3,4]

[1] School of Petroleum Engineering, University of New South Wales, Sydney NSW 2052, Australia
[2] Department of Applied Mathematics, Research School of Physical Sciences and Engineering, Australian National University, Canberra ACT 0200
[3] MPI für Metallforschung, Heisenbergstr. 3, 70569 Stuttgart, Germany
[4] ITAP, Fakultät für Physik, Universität Stuttgart, Pfaffenwaldring 57, 70569 Stuttgart, Germany
e-mail: mecke@mf.mpg.de

Abstract. We introduce Minkowski functionals to characterise, reconstruct and discriminate different complex material microstructures, for instance, experimental data sets generated from X-ray computer tomography imaging; samples include a suite of Fontainebleau sandstone, and a heterogeneous cross-bedded sandstone. Three distinct classes of digitised complex microstructure are considered: particle based Boolean models, structures generated by level-cuts through Gaussian fields, and models based on a Voronoi tesselation of space. One can define a set of measures for random composite media from a single image at any phase fraction ϕ which allows one to accurately reconstruct the medium for all other phase fractions and to predict, for instance, the percolation threshold p_c. The evolution of the Minkowski functions during erosion and dilation operations on non-convex morphologies leads to a very accurate discrimination of morphology – better than commonly used techniques such as structure functions or chord length distributions.

1 Spatial Structures: Experimental Data and Stochastic Models

The structure of a disordered material – an oil bearing rock, a piece of paper, or a polymer composite – is a remarkably incoherent concept. Despite this, scientists and engineers are asked to predict the properties of a disordered material based on the "structure" of its constituent components. A major shortcoming in the understanding of processes involving complex materials has been an inability to accurately characterize microstructure. The specification of "structure" requires topological as well as geometric descriptors to characterize the connectivity and the shape of the spatial configuration. In oil recovery from petroleum reservoir rocks, an area of particular interest to the authors, recovery depends crucially on the topology of the pore space and on the mean curvature of the surfaces where immiscible phases meet at a contact angle. To determine accurate flow models and to devise intelligent recovery strategies requires an accurate characterization of reservoir rocks in terms of topology and geometry.

To date, the main toolkit used to quantify complex structures has been primarily that of the statistical physicist and not the advanced techniques developed in spatial statistics [11, 59] and digital image analysis [53] presented partially in this volume. For instance, contact distributions are discussed by Hug et al., tensor valuations by Beisbart et al, texture analysis by Soille, and mark correlation functions by Beisbart, Kerscher and Mecke. Complete characterization of the effective morphology however requires knowledge of an infinite set of n-point statistical correlation functions. In practice only

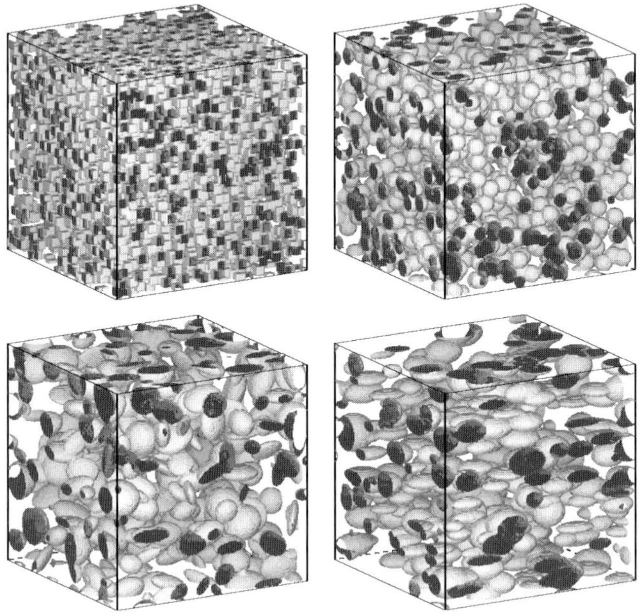

Fig. 1. The interface of various Boolean models where overlapping convex grains (particles) are distributed randomly. Top: the grains are cubes of sidelength $\lambda = 8$ and overlapping spheres of radius $r = 8$. Bottom: the grains are overlapping spheroids with an half-axis range of $r = 4..20$; randomly oriented (left), fully aligned (right). The volume fraction of the particle phase is $\phi = 0.25$ in each case.

lower order morphological information is available; common methods [20] are based on matching the first two moments (volume fraction and two-point correlation function) of the binary phase function to a random model. It is widely recognised that although the two-point correlation function of a reference and a reconstructed system is in good agreement, this does not ensure that the structures of the two systems will match well and attempts to reconstruct materials from experimentally measured two-point information have not been very successful [1, 14]. The same problem has been encountered at the atomic scale in characterizing and comparing amorphous atomic glasses, where two-point correlation functions are recognised to give poor structural signatures. The function is non-unique and does not capture many important features of the microstructure.

Other useful two-point characterizations of microstructure include the chord-length distribution function [47, 53] (and the related lineal-path function [60]), the pore size distribution function [51] and other techniques from spatial statistics and image analysis [25, 38, 53, 59]. However, reconstructions of experimental data sets based on these characterisations have been shown to give a poor representation of the connectivity of the systems [26]. Functions that may provide more complete information about connectivity [16] are unfortunately too complex to incorporate into reconstruction schemes [26]. Incorporation of three- and four-point information may lead to a better estimation of structure, but their measurement is very complex and it is not clear how to incorporate the information within reconstruction algorithms. There is a need for morphological mea-

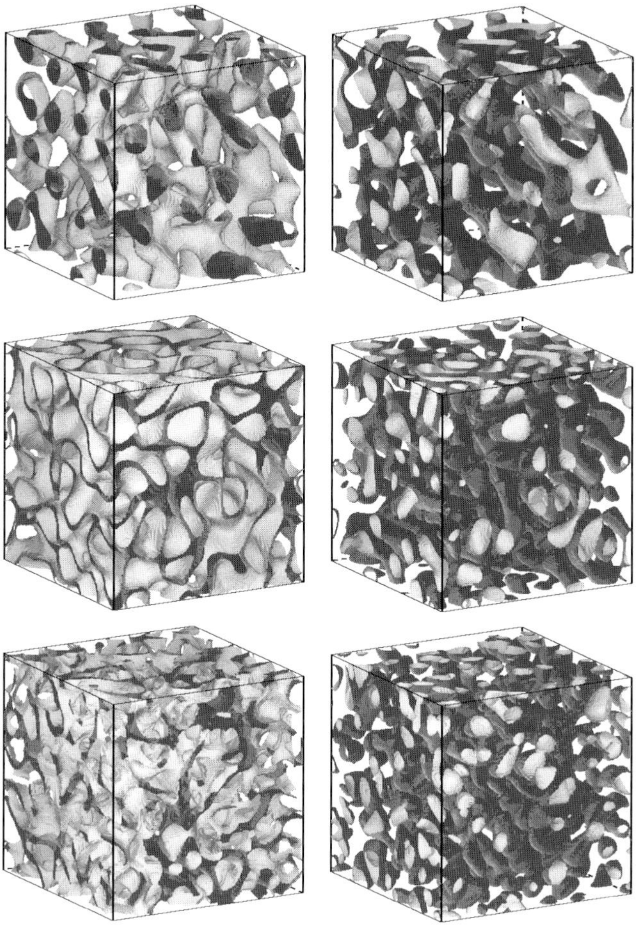

Fig. 2. The interface of Gaussian models of periodicity $t = 10$ for volume fractions of $\phi = 0.25$ (left) and $\phi = 0.75$ (right). Top: 1-level-cut. Middle: 2-level-cut. Bottom: intersection of two 2-level-cuts.

sures which include higher-order correlations, but are fast and reliable for characterizing the morphology of a structure.

Statistical measures which are sensitive to the morphology or shape of structures, i.e., to curvatures and connectivity, have been extensively investigated in other fields such as image analysis and pattern recognition [25, 38, 48, 53, 59]. Integral geometry [30, 32, 34, 50] provides a suitable family of morphological descriptors, the Minkowski functionals (MFs). These measures embody information from every order of the correlation functions in an integral way, are numerically robust even for small samples, and yield global as well as local morphological information. One of the Minkowski functionals is the Euler characteristic discussed also in the contributions to this volume by Ohser et al and Robins as a topological measure for the connectivity of spatial structures. The Minkowski functionals characterize not only the connectivity, but the shape and content

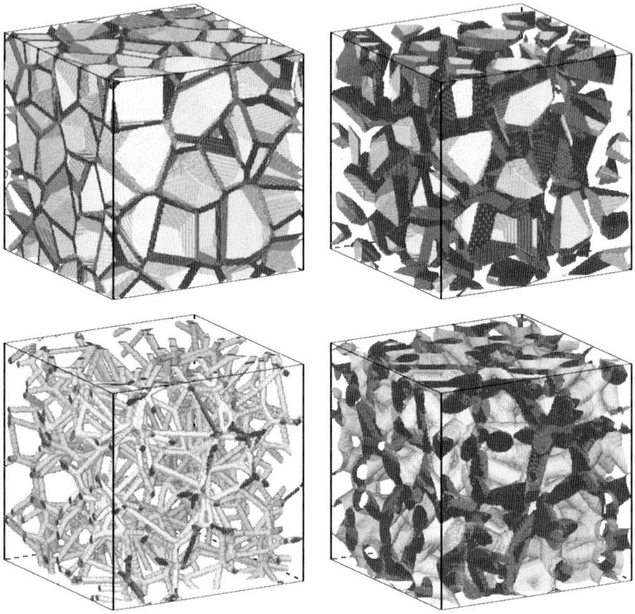

Fig. 3. The interface of Voronoi models with 100 seeds on a 200^3 lattice. First row: facet models for volume fractions of $\phi = 0.26$ (left) and $\phi = 0.74$ (right). Second row: edge models for volume fractions of $\phi = 0.06$ (left) and $\phi = 0.50$ (right).

of spatial figures. In three dimensions the functionals are related to the familiar measures of volume fraction, surface area, integral mean curvature and Euler characteristic. These measures are efficiently calculated at the local scale from digital images [29, 34]. Minkowski functionals have been used previously to distinguish quantitatively between different complex morphologies; to characterize turbulent and regular Turing patterns from chemical reaction-diffusion systems [31], to show that the hole distribution in thin films is inconsistent with the concept of spinodal decomposition, but consistent with a nucleation scenario [18]. and to discriminate between different cosmological models of the early universe [30]. Measurements of the Minkowski functionals for model random materials has to date been made on only a simple random filling of a cubic network [19].

In this paper we use the Minkowski functionals to characterise, reconstruct and discriminate different complex material microstructures. We consider three distinct classes of digitised complex microstructure; particle based composites, structures generated by level-cuts through Gaussian fields, and models based on a Voronoi tesselation of space. The models are chosen to represent a wide range of composite structure. Particle models with different grain shapes and mixtures of grains are also employed. We also consider several experimental data sets generated from X-ray-CT imaging; samples include a suite of Fontainebleau sandstone, and a heterogeneous cross-bedded sandstone. We show first how the Minkowski functionals can discriminate different classes of structure (Sect. 2 Characterising via Minkowski functionals). Next a reconstruction scheme for mixtures of overlapping particles is developed. We show that for a Boolean model of *any* mixture of grains, one can, from a single image of a system at any particle fraction and with no prior knowledge of the ensemble of particle size and shape, define a set of parameters based on Minkowski functionals which allows one to accurately reconstruct the medium for *all other* phase fractions (Sect. 3 Reconstruction via Minkowski functionals). Third, the Minkowski functionals are extended via the application of erosion and dilation operations. The method is used to sensitively discriminate between different morphologies (Sect. 4 Characterising via Minkowski functionals of parallel surfaces).

1.1 Generation of Model Media

The models we considered in this paper can be loosely separated into three different classes. The first class, particle based models, include Poisson distributed cubes and Poisson distributed overlapping oblate and prolate spheroids. Examples of these Boolean models are given in Fig. 1.

A class of material that is not in general well described by particulate models is that of amorphous composites. Recently model random materials have been described by level cuts of a superposition of random plane waves; the levelled-wave model [7, 56]. Originally developed to describe the morphologies associated with spinodal decomposition [10], and later to describe the structure of bicontinuous microemulsions [6], the levelled wave model accounts for many features observed in real disordered materials [7] including polymer blends [21] and foams [46]. Examples of the morphologies generated by level-cuts of Gaussian models are illustrated in Fig. 2.

The third class of morphologies, random cellular solids [15] are constructed using a Voronoi tesselation. In this model we subdivide space randomly into convex polyhedra by scattering Poisson points to a given density and construct the corresponding Voronoi cells, i.e., the bisecting planes between each pair of points. Within each polyhedra every point is closer to the given Poisson point than to any other. The resultant structure is similar to that of a closed-cell foam or random honeycomb. By thickening the walls of the Voronoi cells we obtain a connected system of tubes (starting from edges) or plates (starting from facets), which yields a large range of phase fractions (see similar constructions of stochastic geometries in Refs. [28] and [53]). To do this we generate a Euclidean Distance Map (EDM), a mapping giving the Euclidean distance $(1, \sqrt{2}, \sqrt{3}, 2, ..)$ of each voxel to its nearest surface voxel. By stepping through the different distances a range of phase fractions for the morphology is generated. To form a foam-like network of low coordination number we considered the trisecting planes of the Poisson points. By using the EDM we grow a network of connected cylinders. At fractions below ca. 10% this may be a good model of foams (see i.e. Figs. 2 & 4 in [35]) and of aerogels [43]. Examples of the Voronoi structures are given in Fig. 3.

1.2 Experimental Images from Microtomography

In this paper a number of morphologies derived from experimental images are considered. Direct measurement of a 3D structure is now available via micro X-ray computed microtomography [5, 12, 54]. These techniques provide the opportunity to experimentally measure the complex morphology of a range of materials in three dimensions at

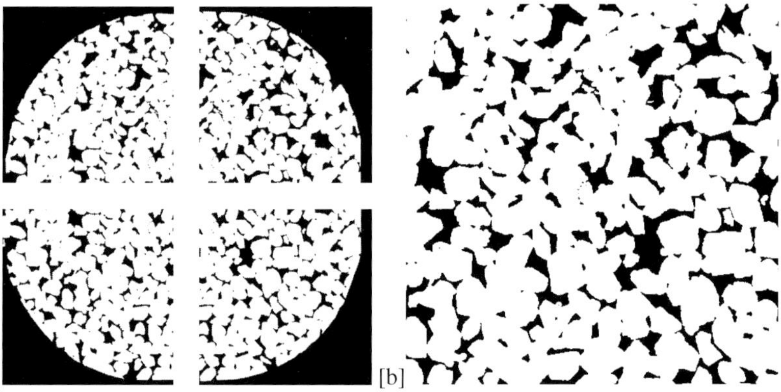

[a] [b]

Fig. 4. Right: Compilation of the four 450^2 slices. Left: Final first 480^2 slice of fb15, after removing overlap, and recombination.

resolutions down to 5 μm and lower. From this technology we obtained data sets for two types of rock; Fontainebleau sandstone and a cross-bedded sandstone. X-ray computed tomographic images of porous media are grey scale images, usually with a bimodal population apparent, one mode corresponding to the signal from the void space, the second to the signal from the grain space. A simple thresholding based on matching a predetermined bulk measurement (phase fraction) is often used to segment a tomographic image. We use a kriging-based thresholding method [37] to give a binary pore-solid image [23, 24]. The images are also corrected for noise by re-identifying the phase type of all isolated grain and void voxel clusters. These isolated voxels would otherwise have a strong effect on many morphological measures.

Fontainebleau sandstone. The tomographic data of the Fontainebleau sandstone was obtained from four 4.52 mm diameter cylindrical core samples with bulk porosities $\phi = 7.5\%$, 13%, 15%, and 22%. In the following the samples are defined by fbϕ, i.e. fb13 for the 13% sample of Fontainebleau sandstone. A 2.91 mm length of each core was imaged [12, 13, 54]. The reconstructed images have a resolution of 5.68 μm resulting in $795 \times 795 \times 512$ imaged sections. Due to limitations in computational hardware – at the time the image was initial processed and segmented [22, 23] – the original image was divided into 4 subsamples of size $450 \times 450 \times 512$ by Lindquist, overlapping by 50 pixels on the inner sides. The kriging was carried out on these subsamples, which then were recombined to a single binary image (see Fig. 4). Finally from each of the recombined binarised cylindrical plugs a (centered) 480^3 cubic subset was extracted for analysis corresponding to a volume of 20.3 mm^3.

From each of these final 480^3 samples a central 120^3 subsample is illustrated in Fig. 5. The samples show the variability observed in the Fontainebleau cores. Moreover, the porous fraction of the central 120^3 section of the core fb15 is 19.6%, which does not match the macroscopic porosity (Fig. 5.[c]), and is higher than 19.0% for the central section of the fb22 sample (Fig. 5.[d]). This illustrates a possible strong variability in the porosity observed at a local scale. Figure 6 illustrates the porosity variation per slice

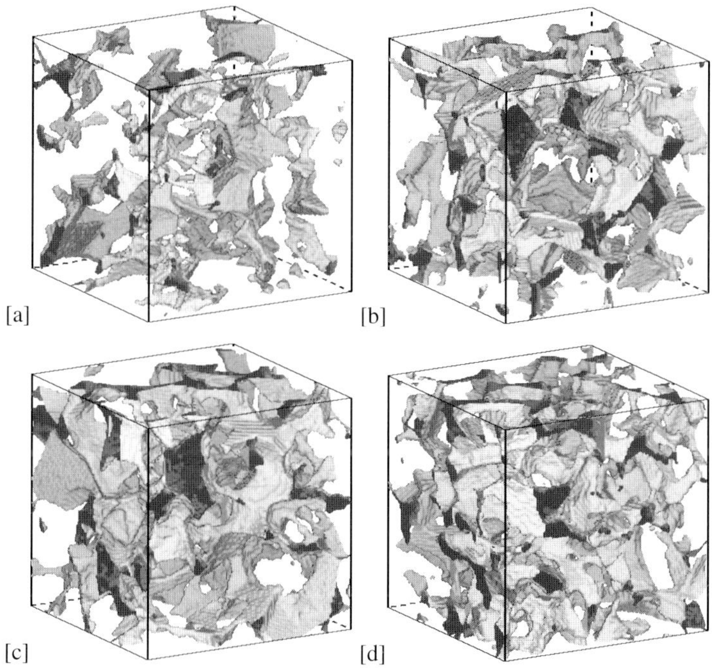

Fig. 5. Pore space images of a central 120^3 piece of each of the four 480^3 samples, each taken from the centre. From top left to bottom right: fb7.5 ($\phi = 7.04\%$), fb13 ($\phi = 12.6\%$), fb15 ($\phi = 19.6\%$), fb22 ($\phi = 19.0\%$). Here ϕ notes the porosity of the 120^3 subsamples.

along the x,y,z directions of the 480^3 samples as well as the 2-point void-void correlation functions $p^{(2)}(r)$. Each of the slices represents a volume of $0.042\ \mathrm{mm}^3$.

Fontainebleau is considered very homogeneous, and considered a "benchmark" of a homogeneous rock in the petroleum industry, but the plot of the porosity variations shows considerable variability on the local grain scale. A statistical summary of average ϕ and standard deviations is compiled in Table 1. The standard deviation $\sigma(\phi_{\text{slice}})$ of the data increases with fraction. Defining a normalised deviation as a measure of hetero-

Table 1. Volume fractions and variability of the Fontainebleau samples. Given are the nominal porosity ϕ_{plug} of the core plugs, from which the samples were extracted, the porosities ϕ_{480} of the 480^3 samples, as well as the absolute and relative standard deviation of the slice porosity ϕ_{sl} along the coordinate axes of the samples extracted from the core plugs ($z \parallel$ cylindrical axis).

Sample	ϕ_{plug}	ϕ_{480}	$\sigma_x(\phi_{\text{sl}})$	$\frac{\sigma_x(\phi_{\text{sl}})}{\langle\phi_{\text{sl}}\rangle}$	$\sigma_y(\phi_{\text{sl}})$	$\frac{\sigma_y(\phi_{\text{sl}})}{\langle\phi_{\text{sl}}\rangle}$	$\sigma_z(\phi_{\text{sl}})$	$\frac{\sigma_z(\phi_{\text{sl}})}{\langle\phi_{\text{sl}}\rangle}$
fb7.5	7.5%	.0829	.0078	9.4%	0.0091	11%	.0054	6.5%
fb13	13%	.129	.0076	5.9%	0.0095	7.4%	.0095	7.4%
fb15	15%	.177	.0116	6.5%	0.0095	5.4%	.0103	5.8%
fb22	22%	.210	.0102	4.8%	0.0113	5.4%	.0109	5.2%

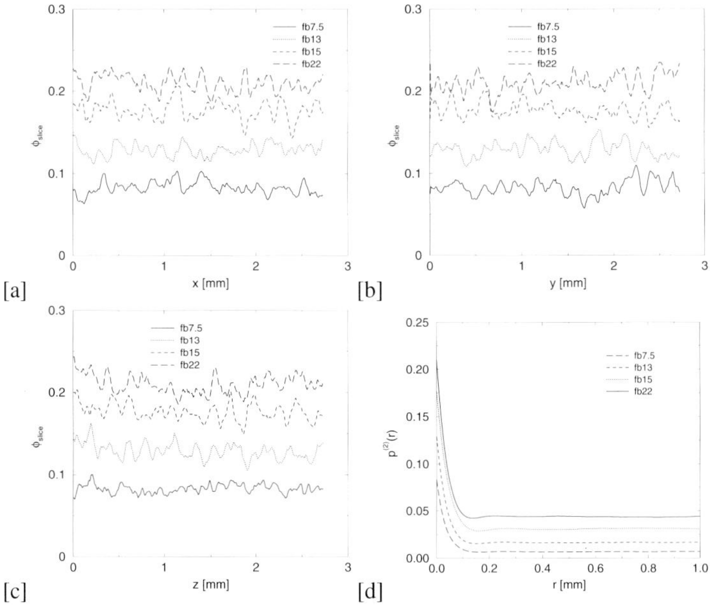

Fig. 6. Variations in porosity and 2-point correlation functions of the 4 Fontainebleau sandstone samples (480^3 subvolumes). [a-c]: porosity variation along the x,y,z axes of the core plugs. [d]: 2-point correlation functions $p^2(r)$, i.e., covariance $C(r) = p^2(r)/\phi$.

Table 2. Volume fractions and variability of the two cross-bedded sandstone 300^3 subsamples. Given are the porosities ϕ_{300} of the 300^3 samples, as well as the absolute and relative standard deviation of the slice porosity ϕ_{sl} along the coordinate axes of the samples extracted from the core plugs ($z \parallel$ cylindrical axis).

Sample	ϕ_{300}	$\sigma_x(\phi_{\mathrm{sl}})$	$\frac{\sigma_x(\phi_{\mathrm{sl}})}{\langle\phi_{\mathrm{sl}}\rangle}$	$\sigma_y(\phi_{\mathrm{sl}})$	$\frac{\sigma_y(\phi_{\mathrm{sl}})}{\langle\phi_{\mathrm{sl}}\rangle}$	$\sigma_z(\phi_{\mathrm{sl}})$	$\frac{\sigma_z(\phi_{\mathrm{sl}})}{\langle\phi_{\mathrm{sl}}\rangle}$
#1	.132	.0151	11.5%	0.0171	13.0%	.0311	23.7%
#2	.143	.0185	12.9%	0.0222	15.5%	.0230	16.1%

geneity, $\sigma(\phi_{\mathrm{slice}})/\langle\phi_{\mathrm{slice}}\rangle$, shows fb22 as the most homogeneous, and fb7.5 as the most heterogeneous sample.

Cross-bedded sandstone. A $512 \times 512 \times 666$ image of a cross-bedded sandstone was obtained at 10 μm resolution via micro-CT imaging (raw data from the University of Texas). The image was converted to a binary image in the same way as the other rock samples. Of the cylindrical sample a $300 \times 300 \times 600$ subset was selected, having a bulk porosity of 13.7% and representing a volume of 54 mm^3. For further analysis this volume was subsampled into 2 disjoint cubes of size 300^3, 16 disjoint cubes of size 150^3, 54 disjoint cubes of size 100^3, and 250 disjoint cubes of size 60^3. Examples of the subvolumes are shown in Fig. 7.

Cross-bedding geologically implies a higher energy depositional environment and therefore stronger heterogeneity of the microstructure. This behaviour is observed in

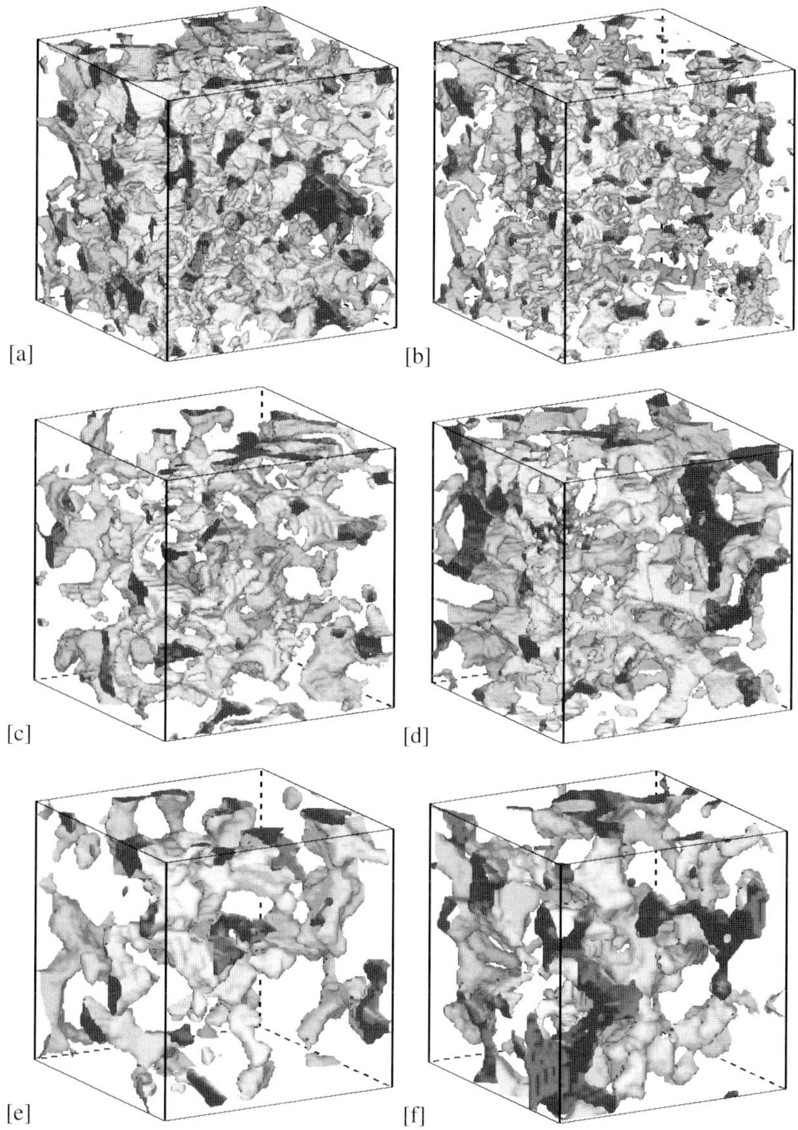

Fig. 7. The interface of sections of the $300 \times 300 \times 600$ sized segmented X-ray-CT scan of a cross-bedded sandstone. [a]: $150^3_{(112)}$, [b]: $150^3_{(213)}$, [c]: $100^3_{(214)}$, [d]: $100^3_{(335)}$, [e]: $60^3_{(211)}$, [f]: $60^3_{(221)}$. The subscripts note the position in the sandstone sample matrix. Top: 2x2x4, middle: 3x3x6, bottom: 5x5x10.

Fig. 8. In particular, strong heterogeneity perpendicular to the primary bedding plane can be seen. The porosity per slice ranges between 5% and 21% for the cross-bedded as compared to the homogeneous Fontainebleau sandstone. The standard deviations of the slice fractions along the cylindrical axis of the core plug are given by $\sigma_z(\phi_{\text{slice}}) =$

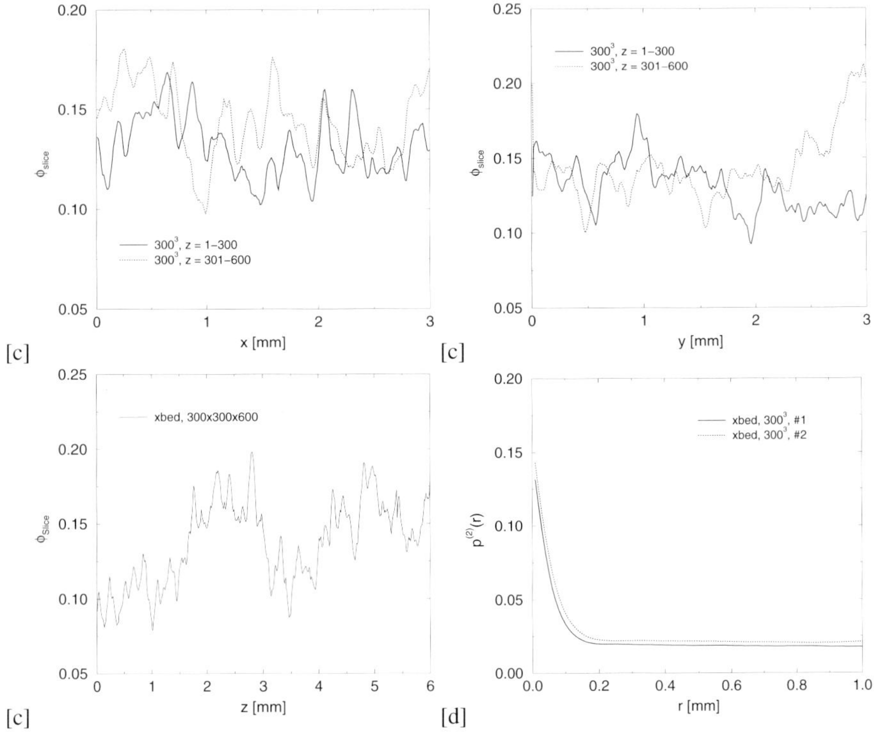

Fig. 8. Variations in porosity and 2-point correlation functions of the cross-bedded sandstone data. [a-c]: porosity variation along the x,y,z axes for the extracted $300 \times 300 \times 600$ subvolume. The cylindrical axis of the core plug coincides with the z-direction and the volume is subdivided into 2 samples of size 300^3 for the x,y directions to compare the same slice volumes. [d]: 2-point correlation functions (covariance $C(r)$) of the two 300^3 subsamples.

0.028, or normalised $\sigma_z(\phi_{\text{slice}})/\langle\phi_{\text{slice}}\rangle = 20.4\%$. A more detailed statistical summary analogous to the Fontainebleau samples, but for two 300^3 subvolumes of the cross-bedded sandstone, is given in Table 2. Cleary this sample is the more heterogeneous sandstone.

2 Characterization of Complex Morphologies via Minkowski Functionals

2.1 Integral Geometric Measures

Integral geometry [50] provides complimentary methods and tools for measuring spatial structure. These tools are commonly used in other disciplines (e.g, digital image analysis [48]) but have only recently been developed in a systematic way as measures of complex media [3]. A family of measures, the Minkowski functionals in particular seem to be promising measures for describing the morphology of complex materials. The functionals characterise not only the connectivity, but the shape and content of spatial figures such as the configurations shown in Fig. 1. As many physical phenomena depend essentially on the geometry of the spatial structure, integral geometry may provide useful tools to study physical systems. In particular, integral geometry provides powerful formulae for the Boolean model. A review of these measures and applications in physics is given in [3, 32, 34].

In this study we consider the integral geometric measures of digitized representations of complex media. We consider a two-component medium filling a cubic volume $N = L^3$. A digitised set of either component can be described by a collection of voxels of compact (closed and bounded) convex sets. The global morphology of a Boolean model, i.e., of a structure $\cup_i K_i$ made up of individual overlapping grains K of number density ρ are [3, 32, 34];

$$\begin{aligned} v_0 &= \phi(\rho) &&= 1 - e^{-\rho V_0} \\ v_1 &= \tfrac{1}{6} S(\rho) &&= e^{-\rho V_0}(1 - e^{-\rho V_1}) \\ v_2 &= \tfrac{1}{3\pi} H(\rho) &&= e^{-\rho V_0}(-1 + 2e^{-\rho V_1} - e^{-\rho(2V_1+V_2)}) \\ v_3 &= \chi(\rho) &&= e^{-\rho V_0}(1 - 3e^{-\rho V_1} + 3e^{-\rho(2V_1+V_2)} - e^{-\rho(3V_1+3V_2+V_3)}), \end{aligned} \tag{1}$$

where ϕ is the particle phase fraction, S is the surface area to volume ratio, H is the specific integral mean curvature and χ is the specific Euler characteristic (genus). The $V_i = V_i(K)$ are the measures of the individual grains whereas v_i denote the intensities of the measures $V_i(\cup_i K_i)$, i.e., the mean values of the Minkowski functionals of the union of all grains normalized by the total number of voxels. Note, that these intensities were derived by Miles (1976) for *non*-discretised configurations, whereas (1) holds for unions of digitised grains K defined as a collection of voxels. In this equation we assume the particulate phase is the continuous phase– each voxel has 26 neighbour connections, while the inverse phase is 6-connected. The dependence of these measures on the definition of the voxel-neighbourhood is discussed in detail in [3] as well as in the contribution by Ohser et al. in this volume. In the examples discussed below, the resolution of the digitisation was always smaller than the size of typical spatial features, so that the differences between neighbourhood definitions can be neglected since they are usually smaller than the symbol size used. It is convenient to normalize the Minkowski functionals for digitised data so that for a single voxel one has $V_i = 1$, in contrast to the normalization $W_i = \omega_i V_i$ usually applied for shapes in the Euclidean space with the volume ω_i of an i-dimensional unit-sphere. Thus, for cubes of sidelength λ voxels in three dimensions we have $V_0 = \lambda^3$, $V_1 = \lambda^2$, $V_2 = \lambda$, $V_3 = 1$, and for rectangular sticks of edge length λ_i: $V_0 = \lambda_1\lambda_2\lambda_3$, $V_1 = (\lambda_1\lambda_2 + \lambda_1\lambda_3 + \lambda_2\lambda_3)/3$, $V_1 = (\lambda_1 + \lambda_3 + \lambda_3)/3$, and $V_3 = 1$. These powerful formulae given in (1) illustrate that it is sufficient to know the morphology of the individual grains $V_i = V_i(K)$ and the density of the grains ρ to derive the mean values v_i of the global morphology, measured in terms of ϕ, S, H, and $\chi(\rho)$.

This result holds also for complex mixtures of grains where V_ν is now replaced by proper averaged values of arbitrary shaped grains, for instance, for mixtures of polyhedra as used later. One can replace the quantities V_ν of a single grain by averages over an ensemble of n grains, weighted by the probability p_j of their occurence

$$\langle V_\nu \rangle = \sum_{j=1}^{n} p_j V_{\nu j} \quad \text{for} \quad \nu = 0, \ldots, d. \tag{2}$$

where $p_j = \rho_j / \rho$ is the ratio of the densities of the Poisson processes.

Calculation of integral geometric measures. The intensities $\phi, H, S, \chi(\rho)$ of the global Minkowski functionals can be measured directly from any image made up of discrete voxels. For example, the volume fraction ϕ of a phase is trivially obtained by dividing the number of voxels of that phase by the total number of voxels. The other functionals are obtained by considering the interface associated with the vertices of each voxel. Global measures $v_i = V_i(\cup_i K_i)/N$ for each configuration are obtained by configuration counts over all vertices on any voxelated structure after normalizing by the total number $N = L^3$ of vertices [3, 17, 33].

In [3] the algorithm used to calculate these measures was validated against (1) for a monodisperse grain ensemble. To further validate the algorithm we measure the global morphology ϕ, S, H, χ for an ensemble made up of mixtures of grains. For each of these mixtures the Minkowski functionals were computed on a minimum lattice size of 200^3 for a minimum of 50 realisations over the full range of the volume fraction ϕ in steps of $\Delta\phi = 0.02$. Figure 9 illustrates the results for a system made up of a mixture of cubes (2^3) and sticks ($8 \times 1 \times 1$) at equal particle density. We observe excellent agreement between the theoretical and computational results.

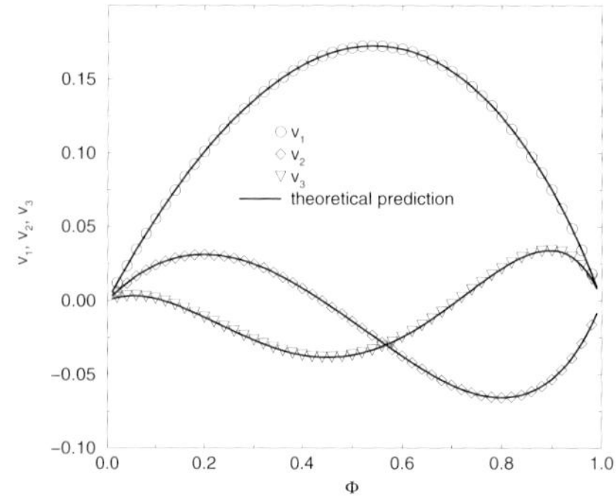

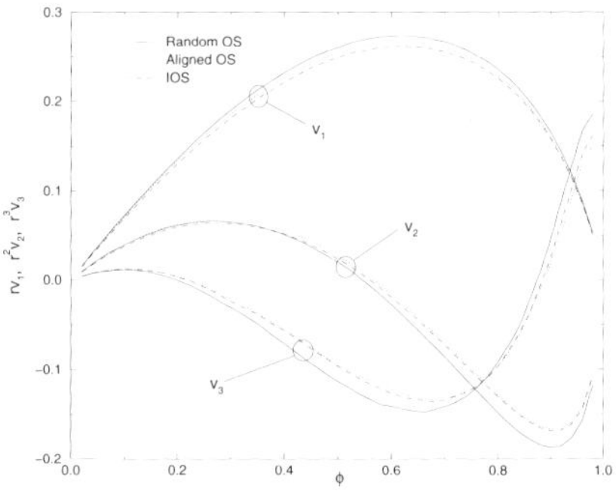

Fig. 9. (a) Minkowski functionals over particle fraction ϕ for a mixture of cubes and sticks ($\lambda_{1i} = 2, \lambda_{21} = 8, \lambda_{22} = \lambda_{23} = 1, p_1 = 1/2$; see (2)). (b) Minkowski functionals over volume fraction ϕ for a Boolean model with Poisson distributed overlapping spheroids (random OS) of half-axes $r = 4..20$, with $\bar{r} = 12$, compared to fully aligned overlapping spheroids (aligned OS) of the same size distribution and overlapping spheres of radius $r = 12$ (IOS). The measures are scaled as $S \to \bar{r} \cdot S$, $H \to \bar{r}^2 \cdot H$, and $\chi \to \bar{r}^3 \cdot \chi$.

2.2 Morphology of Model Systems

Particle based Boolean models: In Fig. 9 we plot the integral geometric measures for a range of particle-based models. In particular, we compare the sensitivity of the Minkowski functionals to deviations in form or alignment. To do this we consider a Boolean model with overlapping spheroids. We generate randomly oriented and fully aligned overlapping spheroids of uniformly distributed half-axes $r = 4, .., 20$ and $\bar{r} = 12$ and compare to the a Boolean model with overlapping spheres of radius $r = 12$. As Fig. 9 shows, the Minkowski functionals for all these Boolean models are quite similar. It may be non-trivial to distinguish the measures for these different systems and to estimate the shape of the grains. However, in Sect. 3 we demonstrate that it is possible to estimate V_i quite accurately if v_i is measured at any porosity ϕ. Furthermore, in Sect. 4 we propose a novel method to go beyond v_i by studying the morphology $v_i(\epsilon)$ of the parallel surface of distance ϵ which is an even more sensitive technique to estimate the local grain shape V_i. In particular, by applying erosion-dilation operations, i.e., by studying the morphology of parallel surfaces it is possible to distinguish spatial systems which exhibit similar global morphological measures v_i.

Field based Gaussian models: The Gaussian 1-level cut model results in symmetric Minkowski functionals around $\phi = 0.5$ (see Fig. 10(a)). The integral mean curvature is much less than for the particle based models. For small and large ϕ isolated elongated (elliptical-shaped) inclusions are present and the Euler characteristic is positive. In the regime $\phi \simeq .20$ to $\phi \simeq .80$ the interface becomes predominantly hyperbolic ($\chi < 0$) and both phases are continuous.

For the symmetric 2-level cut Gaussians of small ϕ the surface to volume ratio is much larger than for the previous systems. Particularly at low volume fractions the morphology exhibits a sheetlike phase, and the surface to volume ratio quickly reaches a maximum (Fig. 10(b)). Further densification is associated with a thickening of the sheets and the surface to volume ratio drops. Also for small ϕ, discretization effects are important and the different neighbourhoods play an important role. The Euler characteristic is very large and becomes negative for small ϕ, once the sheetlike phase connects, implying a strongly bicontinuous structure, as is evident in Fig. 2.

The intersection set of the two 2-level cut Gaussians has a distinct signature (Fig. 10(c)) when compared to the 1- and 2-level cut models. The strong Gaussian curvature feature at small ϕ may be useful as a signature of this structure when compared with the other models.

Cell based Voronoi models: For the Voronoi facet model, the connected polyhedra phase always percolates, and the other phase is made up of disconnected inclusions. Accordingly, the curvature never changes sign – the holes remain convex for all ϕ (Fig. 11(a)). Similarly, the Euler characteristic always remains positive (see [28] for χ of random networks), while the surface area due to the construction of the fields starts at maximum and decreases with thickening facet boundaries. Small ambiguities arise at the intersections of the facets, again giving rise to a separation of the Minkowski functionals based on the choice of continuity.

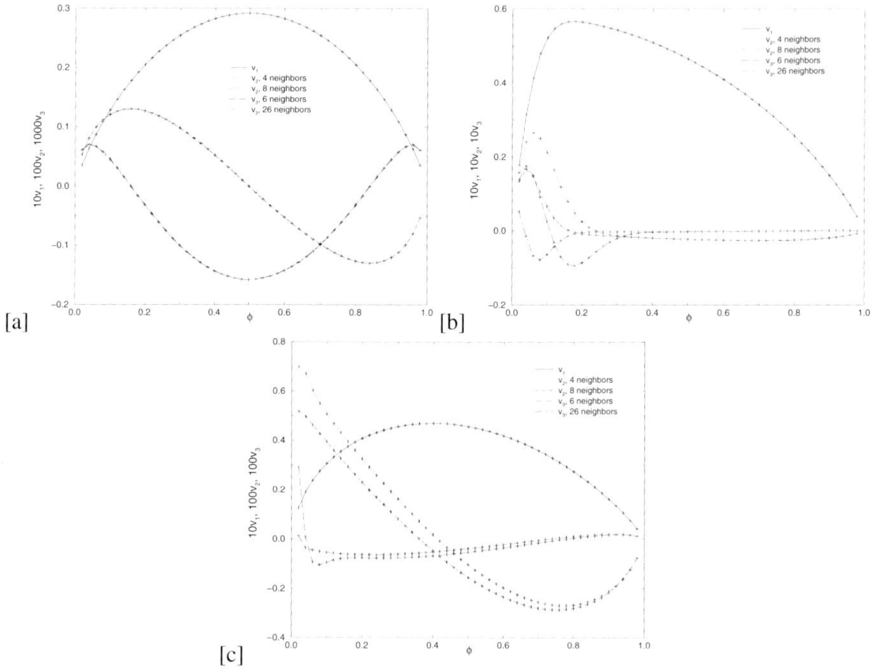

Fig. 10. Minkowski functionals over volume fraction ϕ for (a) Gaussian 1-level-cut models of periodicity $t = 20$, (b) 2-level-cut models of periodicity $t = 20$ and (c) intersection models of periodicity $t = 20$.

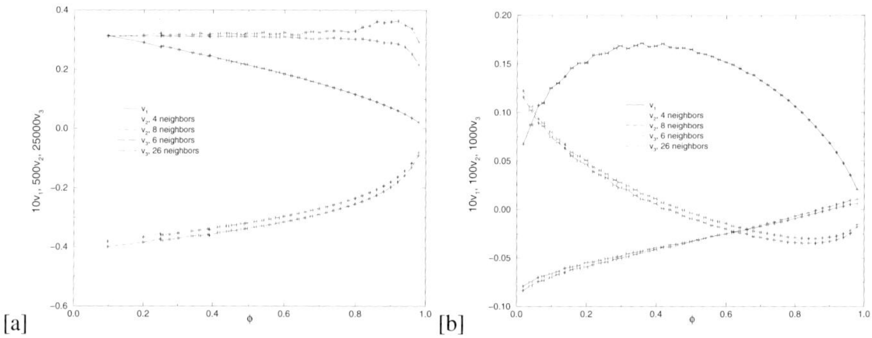

Fig. 11. Minkowski functionals over volume fraction ϕ for (a) Voronoi facet model and (b) Voronoi edge model of 100 seeds on a 200^3 lattice.

For the Voronoi cylinder model, both the network of cylinders and the background phase percolate over a wide range of ϕ. Here the surface area first grows with increasing cylinder radius, and then decreases again, after the cylinders begin merging (Fig. 11(b)). The curvature decreases with increasing dilation, and becomes negative, once individual edges of the Voronoi cells begin to generate isolated inclusions of the background phase.

The Euler characteristic is initially negative, implying that both phases percolate, and then increases almost linearly, finally becoming positive.

2.3 Characteristics of a Sandstone Sample

A number of statistical models have been proposed for reconstructing porous media from statistical information [1, 9, 20, 41, 42, 44, 45]. These methods, based on different underlying model microstructures, are generated in such a manner that they match the observed two-point statistical properties of the rock (see p. 93 in [11]). Of course, one can derive more complex model systems which incorporate other two-point correlation information (e.g., chord distribution functions [47, 53]. Recent work [26] has shown that these measures give a poor representation of connectivity. Preliminary results on Minkowski functionals of these more complex models indicate that this is reflected in a poor match to the v_2 measure. In contrast, we compare the morphological measures v_i for the sandstone to three standard stochastic models used to generate realisations of sedimentary rock microstructures from two-point information. The first is based on a germ-grain-model with spherical grains [41], the second on the 1-level cut model [1, 20], and the third on the intersection set model [45]. The quantities used to characterize the microstructure of these systems are the volume fraction, the surface to volume ratio s/v and $p^{(2)}(r)$, the two-point correlation function. Note that $\phi = p^{(2)}(0)$, and $s/v = -4\frac{dp^{(2)}(0)}{dr}$.

The correlation function for the phase *external* to the spheres of radius r_0 in the IOS model is $p^{(2)}(r) = p^{(\nu(r))}$ for $r < 2r_0$ and $p^{(2)}(r) = \phi^2$ for $r > 2r_0$ where $\nu(r) = 1 + \frac{3r}{4r_0} - \frac{r^3}{16r_0^3}$, and $s/v = -3\phi ln\phi/r_0$. To generate matching Gaussian random field (GRF) models we employ the field-field correlation function [44, 57]

$$g(r) = \frac{e^{-r/\xi} - (r_c/\xi)e^{-r/r_c}}{1 - (r_c/\xi)} \frac{sin 2\pi r/d}{2\pi r/d}, \tag{3}$$

characterized by a correlation length ξ, domain scale d and a cut-off scale r_c. The three length scale parameters are obtained by a best fit procedure to minimize the non-linear least squares error [44],

$$Ep^{(2)} = \frac{\sum_{i=1}^{M}[p_{fit}^{(2)}(r_i) - p_{exp}^{(2)}(r_i)]^2}{\sum_{i=1}^{M}[p_{fit}^{(2)}(r_i) - p_{exp}^2]^2}, \tag{4}$$

and a downhill simplex method is used to find the best parameter set [36]. We generate the stochastic models that best match the two 300x300x300 samples and compare the normalised 2-point correlation functions $S(r) = \frac{p_2(r) - p^2}{p - p^2}$ of these models in Fig. 12.

Estimates of r_0 for the IOS model, and d, ξ and r_c are summarised in Table 3 along with the error estimate $E(p^{(2)})$. The average fit is worst for the sphere model, and more than a factor two better for the Gaussian 1-level-cut and intersection set. The sphere model however has only one free parameter as compared to three for the Gaussian models.

We then calculate the integral geometric measures of the three stochastic models and compare with the sandstone data (Fig. 13). The measures are compared to the values

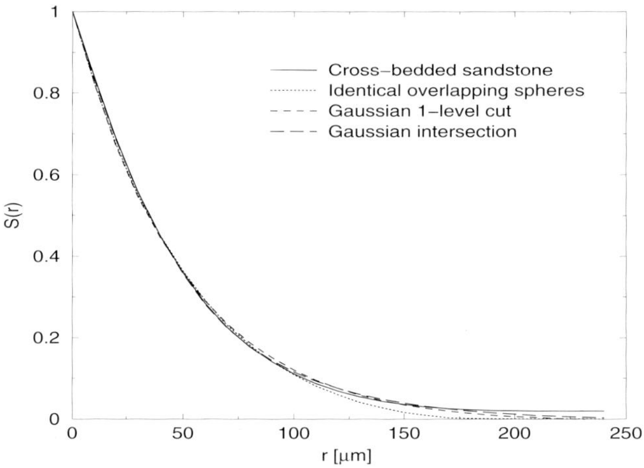

Fig. 12. Normalized 2-point correlation functions $S(r)$ of one of the two 300^3 sandstone samples and the averaged best fits for the model structures.

resulting from the sampling window at 100^3. The values for the samples were binned in steps of porosity $\Delta\phi = 0.02$.

None of the models satisfactorily match the experimental data. The surface area (S) is matched best across all ϕ by the intersection model. Integral mean curvature H is best matched by the IOS models which honours the granular structure of the sedimentary rock. The topology χ is described well by both the IOS and 1-cut models. The intersection model is particularly poor for this model over the full range of ϕ. The model which best captures the characteristics of the sandstone is the IOS model. Even though the other models do reasonably well at the fraction where the 2-point correlation functions were matched, they fail to describe the structure across a range of phase fractions. Different transport and mechanical processes will depend more strongly on the agreement with *specific* morphological measures. For example, single phase flow and conductivity will be most strongly affected by surface-to-volume (related to average constriction size) and the topology. A model which accurately describes *these* two measures may still yield good agreement with experiment. However, multiphase flow properties depend crucially on the curvature of the surfaces where immiscible phases meet. For these processes a model that also accurately matches H will be required. In this case the IOS model is the best of the three candidates as a reconstructed data set. Recently, a model for describing

Table 3. Parameters of the models for the cross-bedded sandstone in microns. The original sandstone data set image has resolution at 10μm per pixel. For the IOS model, the equivalent sphere radius is in the r_c column.

Model	$r_c[\mu m]$	$\xi[\mu m]$	$d[\mu m]$	$Ep^{(2)}$
Spheres	94.2	—	—	$2 \cdot 10^{-3}$
1-level-cut	20.2	22.4	1080	$9 \cdot 10^{-4}$
intersection	54.3	80.5	410	$4 \cdot 10^{-4}$

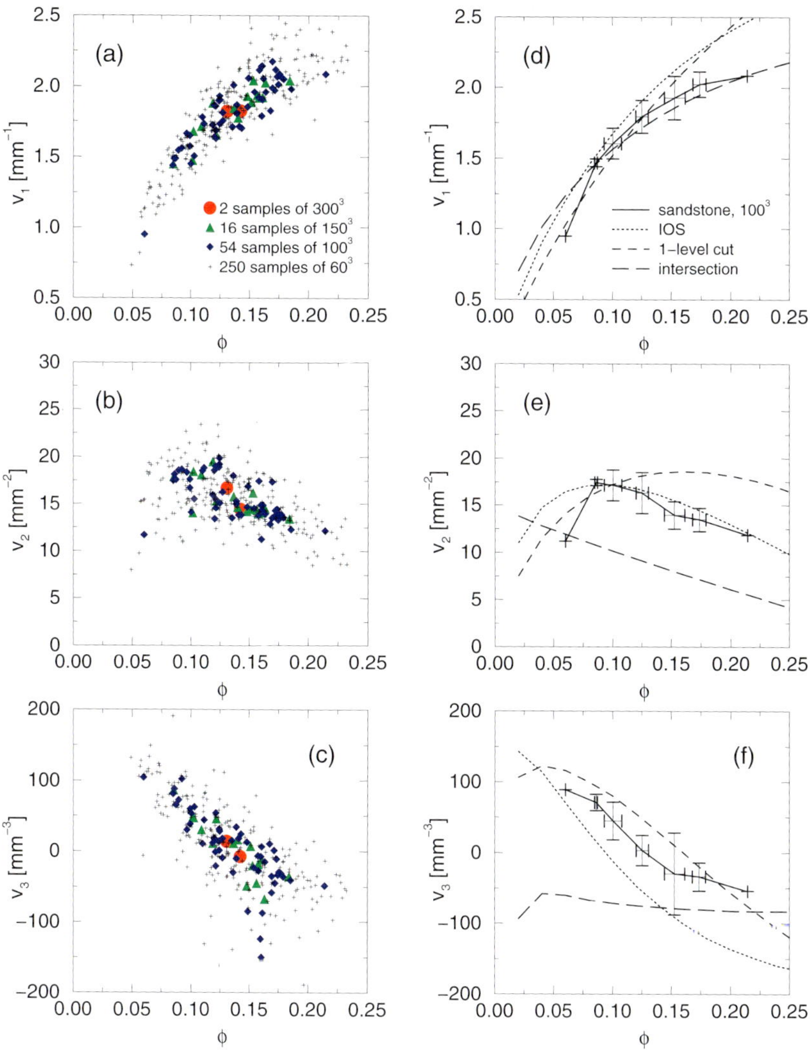

Fig. 13. Comparison of the Minkowski functionals for the sandstone data to the three stochastic models based on a window size of 100^3.

sandstone morphology has been developed based on a full process-based sedimentation, compaction and diagenesis model [40]. This model honours both the shape of the original grains and the geological formation processes and therefore may provide a more accurate description of pore space morphology of sedimentary rock.

3 Reconstruction of Complex Morphologies via Minkowski Functionals

In the previous section we have used knowledge of the local 'grain' morphology to predict global morphology of a Boolean model. Here we consider the inverse process; from a single snapshot of a complex Boolean model derive an equivalent local 'grain' ensemble which will generate for the same complex global morphology. All Boolean models have grains with $\langle V_3 \rangle = 1$. As one can evaluate the four global morphological measures ϕ, S, H, χ from a single 3D image, one can use (1) to obtain estimates of $\langle V_0 \rangle$, $\langle V_1 \rangle$, $\langle V_2 \rangle$, and density ρ. This technique is called 'method of intensities' (see p. 89 in [11]) and goes back to Santalo (1976). Statisticians consider it as the best estimation method. We illustrate this first for Boolean samples made up of monodisperse and polydisperse grains and in Sect. 3.2 for sandstone data.

3.1 Reconstruction of Boolean Models

We generate realisations of Poisson distributed cubes of different length and spheres of varying radii for a range of phase fractions ϕ. From any single image we determine the global measures S, H, χ for different values of particulate phase fraction ϕ and from (1) evaluate the equivalent local grain measures for that image, ρ_{im} and V_i^{im}. In Table 4 we give the error in the estimation of ρ_{im} and V_i^{im} for a system of cubes of size $\lambda = 8$ on a lattice of $L = 400^3$. In all cases the densities and the local measures, i.e., the mean values of the grain predicted via (1) vary $<< 1\%$ from the analytic results. Remarkably, even at high volume fractions where most grains overlap and resolving the local grain ensemble is difficult, the predictions of the local measures are still good.

Now the grains also have random shape and we generate a very heterogeneous isotropic mixture of five polyhedra (5-mix): a system made up of sticks (40x1x1), plates (20x20x1), cubes (8x8x8), and two other rectangular prisms (10x5x2) and (16x8x4) in the ratio $10\% : 10\% : 40\% : 20\% : 20\%$. The analytic local measures derived from (2) are given in Table 5 along with the error E_i in the estimates via a simulated sample (single snapshot) of the model at $L = 500$ for three different ϕ. The error in all the

Table 4. Errors in the local grain measures derived from a single snapshot of a complex systems of Poisson distributed cubes and spheres on a lattice of (400^3). Analytical values for the local grain measures V_i are given along with the error $E_i = |V_i - V_i^{im}|/V_i$ in the prediction of the local grain morphology. V_i^{im} are derived via the measurement of the global morphology S, H, χ and use of (1) at three different ϕ. ρ in the table is the analytic Poisson density from (1) and ρ_{im} gives the density derived from analysing the image.

	V_0^{im}	V_1^{im}	V_2^{im}		
$Cube, \lambda = 8$	V_0	V_1	V_2		
Analytic	512	64	8		
ϕ	$E_0[\%]$	$E_1[\%]$	$E_2[\%]$	$\rho[10^{-3}]$	$\rho_{im}[10^{-3}]$
0.2000	.106	.110	.117	.4340	.4359
0.5000	.084	.093	.093	1.376	1.354
0.8000	.456	.413	.273	3.195	3.143

local grain measures is again $< 1\%$ despite the complexity of the mixture. This proves again that the 'method of intensities' is very good and it indicates that one can define an equivalent local 'grain ensemble' from a single 3D snapshot of an arbitrary complex Boolean model.

Having accurately estimated the parameters ρ and the equivalent grain ensemble defined by V_i^{im} from a given data set, i.e., from a single image at any ϕ one can use this to reconstruct equivalent complex materials. A difficulty is here that a Boolean model which matches the measured V_i^{im} is however not unique. We choose to match the measures with the simplest matching mixture of Boolean particles. For example, one can match the V_i^{im} derived from the complex mixture at $\phi = 0.5$ (Table 2) by a Boolean model with a single equivalent grain defined by ($\lambda_1 = 17.047, \lambda_2 = 6.516, \lambda_3 = 3.341$). We wish to generate equivalent *voxelated* Boolean models; we therefore define grains with integral λ_j. The best match of a Boolean model with a single equivalent voxelated 'grain' to the 5-mix system ($\lambda_1 = 23, \lambda_2 = 4, \lambda_3 = 4$) gives a poor match to the V_i^{im} in Table 6. We therefore determine the best two-grain match to the observed V_i^{im}. As shown in Table 6 the fit to the 5-grain model is now very accurate.

Having defined an equivalent voxelated Boolean grain ensemble, the question remains; does this equivalent system exhibit similar *physical properties* to the original system for all densities? To investigate this we consider the percolation properties of the original and equivalent Boolean model. We vary the density of the original complex

Table 5. Local morphological measures of the 5-mix system of 10% sticks (40x1x1), 10% plates (20x20x1), 40% cubes (8x8x8), 20% each of rectangular prisms of size (10x5x2) and (16x8x4) on a 500^3 lattice. The theoretical values of the mixture are given in the first line, the others are the predictions from the simulations taken at different particle fractions ϕ.

	V_0	V_1	V_2	
Analytic	371.2	63.23	8.966	
ϕ	$E_0[\%]$	$E_1[\%]$	$E_2[\%]$	ρ_{im}
0.2000	.089	.075	.049	$5.731 \cdot 10^{-4}$
0.5000	.075	.121	.213	$1.754 \cdot 10^{-3}$
0.8000	.472	.450	.188	$4.405 \cdot 10^{-3}$

Table 6. (a) Match of an equivalent single- and two-grain Boolean model to the 5-mix system defined in Table 5. The errors in the local measures are given. A single equivalent-grain model does not match the measured V_i^{im} well. To accurately match the morphology one must consider a two-particle system. (b) Comparison of the percolation thresholds for the complex 5-mix model and the equivalent two-grain Boolean model.

Model	Mixture	E_ν
Single Boolean grain	$23 \times 4 \times 4$	$E_0 = 0.86\%$ $E_1 = 5.43\%$ $E_2 = 15.2\%$
Two Boolean grain	$84.2\% : 17 \times 6 \times 2$ $15.8\% : 18 \times 14 \times 5$	$E_0 = 0.026\%$ $E_1 = 0.040\%$ $E_2 = .0041\%$

Phase	$\phi_c^{original}$	$\phi_c^{stochastic}$
Particulate	.143	.146
Inverse	.112	.100

5-grain system defined in Table 5 and the equivalent Boolean two grain ensemble given in Table 6 and numerically determine the percolation thresholds ϕ_c for both phases of the two models. For the determination of the percolation threshold we use finite size scaling techniques [55]. Results are summarized in Table 6. The error in the predictions of ϕ_c for the equivalent stochastic model is $\simeq 0.2\%$ for the particulate phase and $\simeq 1\%$ for the inverse phase. This ability to closely predict ϕ_c of *either* phase from an image at *a single* phase fraction underlines the power of a morphological characterisation and reconstruction based on integral geometric measures.

3.2 Reconstruction of Sandstone Images

Having illustrated the ability of the method to reconstruct complicated Boolean models, we now attempt to accurately reconstruct the experimental 3D microtomographic images of sandstone cores. We note that the microstructure of a sandstone is a result of a complex physical process, which can include consolidation, compaction and cementation of an original grain packing. More complex and realistic models of sandstones have been derived [4, 40]. These methods require however, the simulation of the generating process including primary grain sedimentation followed by a diagenetic process such as compaction and cementation. This process is both computationally expensive and requires several fitting parameters. Reconstructing the microstructure of sandstones by the simple Boolean model may therefore not lead to the excellent match observed in the previous section. However, a Boolean model with spherical grains has been proposed as a model which gives a reasonable representation of consolidated sandstone and yields good qualitative information on structure/property relationships. Moreover, also other work [2, 27, 52, 58] has shown that the model of overlapping spheres gives a very good match to Fontainebleau sandstone data.

In the remainder of this section the reconstruction method based on the Boolean model is applied to the cross-bedded and Fontainebleau sandstone samples. Grain ensembles will be determined which match the global morphological Minkowski functionals of the rock images. We compare the goodnes-of-fit of three different Boolean models: overlapping spheres with matching 2-point information, the equivalent Boolean ensemble defined by local morphological measures V_i, and a recent model based on the probability density of covering spheres [58]. Differences between the sandstone samples and the reconstructions are discussed.

Fontainebleau sandstone. The Fontainebleau data sets were described in Sect. 1.2. Here we derive a number of appropriate Boolean ensembles based on different morphological measures.

In the first approach the diameters for an overlapping grain model with a deterministic spherical grain are estimated by means of the two-point correlation function introduced in Sect. 2.3 (IOS^C model). The sphere sizes are derived for each of the four images of the Fontainebleau sandstone. In the second approach, the diameters are estimated based on the 'method of intensities', i.e., the single sphere can also be determined by using the reconstruction technique (see (1)) based on the local grain measures V_ν ($\mathrm{IOS}^{(1)}$ model). Note: all model structures based on the reconstruction method outlined in this section carry a numeric index noting the number of different grains used for the reconstruction. The local grain measures of the Fontainebleau sandstone samples are summarised in Table 7, and the best matching IOS$^{(1)}$ models in Table 8. Interestingly, the size of the best matching IOS$^{(1)}$ sphere is significantly smaller than the IOS$^{(C)}$ sphere. As was observed previously for the heterogeneous 5-mix system with a single prismatic grain, a Boolean model with deterministic spheres does not give a good match to the data (Table 8).

To gain some degrees of freedom, we attempt to fit a single equivalent spheroidal grain (ROS$^{(1)}$ model) to the local V_ν, but observe only a slight improvement in the prediction. It is therefore necessary to use more complex models. Our first attempt is a Boolean model with random spherical grains to accurately match the local measures. This is done first for a two-point diameter distribution, i.e., for two different spheres (IOS$^{(2)}$). This leads to a better match, but the error in V_2 is still considerable. We then derive the best two-particle match for Boolean spheroids (ROS$^{(2)}$ model). The match to the local V_ν of the experimental image is now excellent (Table 8). It is interesting to note that the unique information obtained from the integral geometric measures leads to a quite complex equivalent stochastic model for the Fontainebleau sandstone. The ROS$^{(2)}$ model is composed of two very different particle sizes (Table 8).

Alternative methods to generate a Boolean ensemble exist for model microstructures based on spheres. Very recently Thovert et al. [58] introduced a reconstruction technique for a sandstone based on a model of overlapping spheres where the sphere size distribution is defined by the probability density of the covering radius for spheres (OSC model). This method is employed to generate a third appropriate Boolean model for the four sandstone images. The median of the radius distributions defined in [58] is given in Table 8. This median is comparable to the size of the larger particles in the two-particle reconstructions but is significantly smaller than either single-particle IOS fits. The original sandstone microstructure for the sample fb15 and various reconstructions via Boolean models are illustrated in Fig. 14. Visual inspection suggests that the ROS$^{(2)}$ and OSC model more closely resemble the original microtomographic image.

We now compare the intensities of the global Minkowski functionals for the various models. An analysis of the full 480^3 cubic subsets would only give a single value for the porosity of each of the four samples and would provide little data to compare to stochastic models. However, as shown in Sect. 1.2, the samples are reasonably heterogeneous in the pore volume fraction. Due to this natural heterogeneity and by appropriately choosing different window sizes on the image it is possible to measure morphological parameters for the sandstone images across a range of pore volume fractions ϕ. This gives us a

Table 7. Mean values of the local Minkowski functionals for the four Fontainebleau sandstone samples of size 480^3 given in Sect. 1.2 ($V_3 = 1$ for convex grains).

Sample	V_0 $[10^{-3}mm^3]$	V_1 $[10^{-3}mm^2]$	V_2 $[mm]$
fb7.5	.4019	3.994	.0473
fb13	.3619	3.752	.0466
fb15	.4506	4.520	.0523
fb22	.3361	3.965	.0509

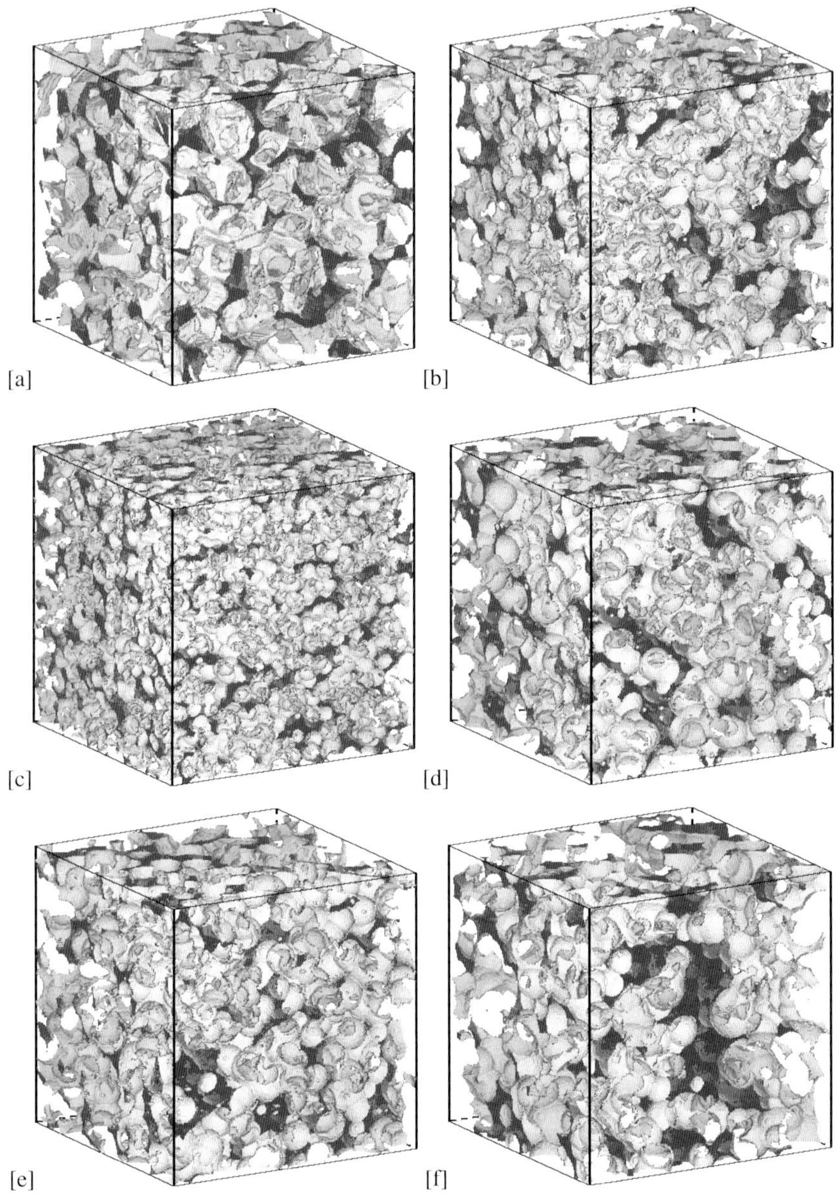

Fig. 14. Visual comparisons of reconstructions to a 240^3 subset of the original Fontainebleau sandstone sample fb15 ($\phi = 18.2\%$). [a] Fontainebleau sandstone, [b] IOS^C, [c] $\mathrm{ROS}^{(1)}$, [d] $\mathrm{IOS}^{(2)}$, [e] $\mathrm{ROS}^{(2)}$, and [f] OSC. Visual inspection suggests that the $\mathrm{ROS}^{(2)}$ and OSC model more closely resemble the original microtomographic image.

Table 8. Parameters for the Boolean models of the Fontainebleau sandstone and errors of the morphological matches. The first IOS model (IOS^C) was matched using the void-void correlation function. The other models are matched using the reconstruction based on estimating the local grain shape of a Boolean model. IOS is generalised to more than one sphere, all spheroids are randomly oriented (ROS). Note the large errors E_{V_i} in the local measures for the 1-particle IOS model. For all two-particle models one of the particles is very small and has a width of the order of a few voxels. a, b, c note the length of the half-axes with the exception of the OSC model, where a notes the median of the radius distribution $G(r_c) = 1/2$.

Model	Core	p	a [μm]	b [μm]	c [μm]	E_{V_0}	E_{V_1}	E_{V_2}
IOS^C	fb7.5	1	71.3			3.4	3.6	2.5
	fb13	1	65.8			2.4	2.6	1.8
	fb15	1	69.1			3.1	3.4	2.5
	fb22	1	59.0			1.6	1.8	1.3
IOS$^{(1)}$	fb7.5	1	45.8			$5.3 \cdot 10^{-3}$	.71	.022
	fb13	1	44.6			.016	.66	$6.8 \cdot 10^{-3}$
	fb15	1	47.4			$4.1 \cdot 10^{-3}$	.058	.84
	fb22	1	43.2			$2.5 \cdot 10^{-3}$	.44	.67
IOS$^{(2)}$	fb7.5	.865	16.6			$4.8 \cdot 10^{-3}$	$6.6 \cdot 10^{-4}$	.022
		.135	88.3					
	fb13	.864	16.6			$3.9 \cdot 10^{-5}$	$1.8 \cdot 10^{-4}$	$6.8 \cdot 10^{-3}$
		.136	84.8					
	fb15	.817	13.3			$7.5 \cdot 10^{-5}$	$1.1 \cdot 10^{-4}$	.018
		.183	83.3					
	fb22	.798	16.5			$2.2 \cdot 10^{-5}$	$3.0 \cdot 10^{-4}$	$8.7 \cdot 10^{-3}$
		.202	72.4					
ROS$^{(2)}$	fb7.5	.848	12.5	11.9	8.52	$1.5 \cdot 10^{-7}$	$7.5 \cdot 10^{-6}$	$1.5 \cdot 10^{-4}$
		.152	83.5	80.7	77.2			
	fb13	.863	15.9	11.4	9.09	$9.1 \cdot 10^{-8}$	$5.4 \cdot 10^{-7}$	$3.0 \cdot 10^{-5}$
		.137	84.6	76.7	76.1			
	fb15	.794	15.3	12.5	8.52	$1.8 \cdot 10^{-7}$	$2.7 \cdot 10^{-6}$	$8.9 \cdot 10^{-5}$
		.206	84.6	80.1	76.7			
	fb22	.751	16.5	10.8	8.52	$2.1 \cdot 10^{-7}$	$2.0 \cdot 10^{-6}$	$1.6 \cdot 10^{-5}$
		.249	74.4	67.0	63.6			
OSC	fb7.5		94.8			7.4	5.2	2.4
	fb13		86.0			12	8.6	4.1
	fb15		85.9			5.3	3.9	2.0
	fb22		67.7			3.2	2.4	1.4

more comprehensive data set with which to compare experimental images to equivalent Boolean models. For the Fontainebleau samples we consider cubic blocks of 480^3, 240^3 and 120^3. This provided in all cases a good spread of porosities across different sampling volumes. The measured morphological properties S, H, $\chi(\phi)$ are summarised in Fig. 15.

We note with interest that the fourth Fontainebleau data set fb22 exhibits very different measures to the first three sets. This could indicate the potential presence of heterogeneity, as seen in Fig. 5.(c) in Thovert et al. [58]. Another explanation may be that another diagenetic process occured during consolidation of this sample.

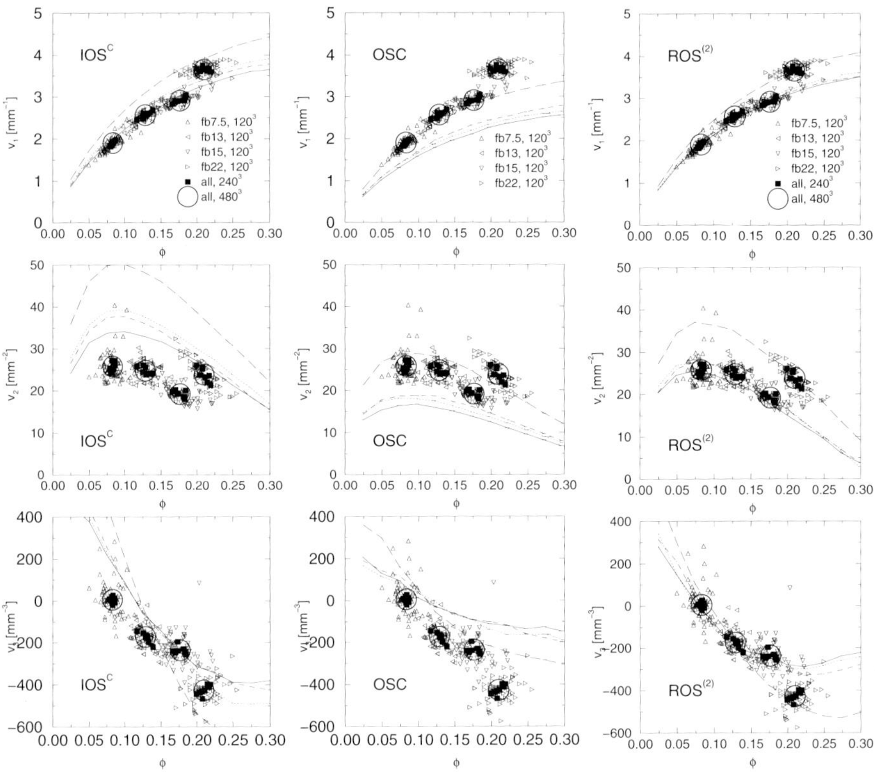

Fig. 15. Minkowski functionals over fraction for each of the four Fontainebleau sandstone samples compared to the matching Boolean models. [a] (left) IOS^C, [b] (middle) OSC, and [c] (right) $\mathrm{ROS}^{(2)}$. For H and χ the measures for the highly connected neighbourhood are given.

As indicated by the relative error in the local integral geometric measures (Table 8) the $\mathrm{ROS}^{(2)}$ model almost exactly matches the local measures of the sandstone samples. The IOS^C and OSC model perform poorly, particularly for H, where IOS^C overestimates, and OSC underestimates the image data (see Fig. 15). Another measure of the accuracy of the reconstruction is based on a comparison of the variation of the pore volume fraction per slice. This gives one a measure of the heterogeneity of the sample. The two best Boolean models based on the reconstruction by local Minkowski functionals are compared to the common IOS^C model and the OSC model in Fig. 16. It is difficult to distinguish which reconstruction gives the best match. The models $\mathrm{ROS}^{(2)}$ and IOS^C both do well, while the OSC model gives enhanced heterogeneity. An attempt to quantify the differences is given in Table 9 by measuring the variability of the pore volume per slice for the Fontainebleau sandstone and its reconstructions. No firm conclusions can be drawn, as the measures themselves vary too strongly (see the column for the x,y,z-variability of the Fontainebleau data in Table 9. The OSC model is too heterogeneous and gives the worst match, while $\mathrm{ROS}^{(2)}$ and IOS^C give reasonable agreement.

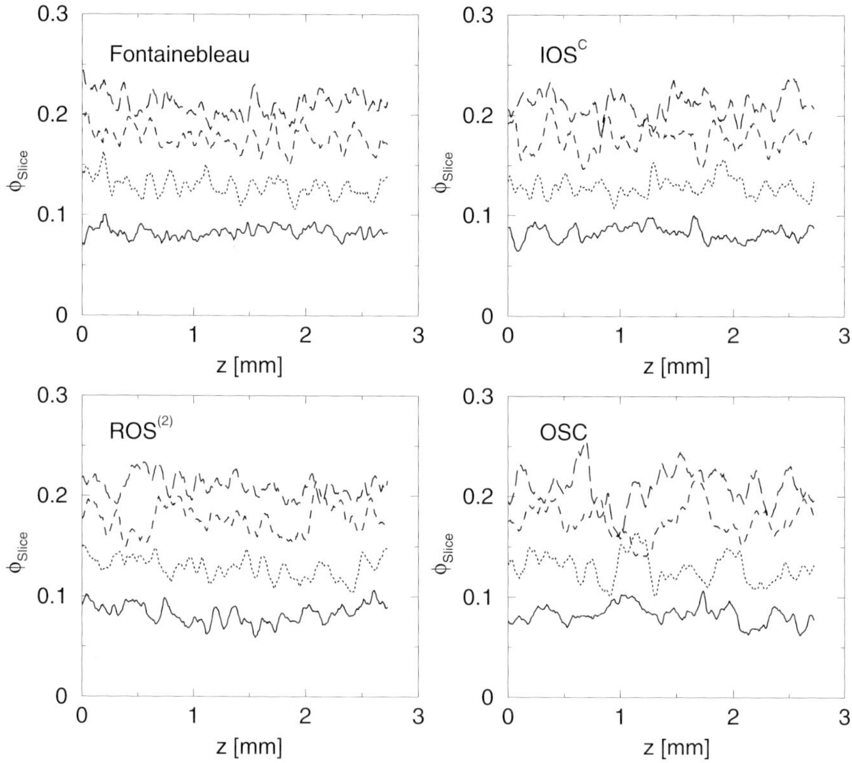

Fig. 16. Comparison in the variation of the porosity distribution along the z-axis of the Fontainebleau samples and reconstructions. Sample size is the same as in Fig. 6.

Cross-bedded sandstone. We adopt a similar procedure to analyse the cross bedded sandstone data set described in Sect. 1.2. Two cubic blocks of 300^3, 16 blocks at 150^3, 54 blocks at 100^3 and 250 blocks at 60^3 were obtained from the full sample volume. This provided a spread of porosities across different sampling volumes. With decreasing sampling volume the variability of the measures increases, but the values are consistent with the data for the larger volumes, suggesting that for the smaller blocks a meaningful

Table 9. Variability $\sigma_z(\phi_{\text{sl}})$ [%] of the pore volume per slice along the x,y,z-directions of the Fontainebleau samples and the z-directions of its reconstructions (see also Table 1, the values are from a single realisation at 480^3).

Sample	Fontainebleau	IOS^C	ROS$^{(2)}$	OSC
fb7.5	0.78/0.91/0.54	0.71	0.94	0.92
fb13	0.76/0.95/0.95	0.90	1.01	1.39
fb15	1.16/0.95/1.03	1.26	1.37	1.60
fb22	1.02/1.13/1.09	1.19	1.03	1.83

average is determined. The global morphological measures of the solid fraction of the sandstone are $\phi = 0.863$, $S = 1.82\ \mathrm{mm}^{-1}$, $H = 15.5\ \mathrm{mm}^{-2}$, and $\chi = 2.73\ \mathrm{mm}^{-3}$.

An IOS^C model microstructure is generated by matching two-point correlation information as described in the previous section. The equivalent local grain measures V_ν for the first 300^3 sample of the sandstone are $V_0 = 0.00464\ \mathrm{mm}^3$, $V_1 = 0.0340\ \mathrm{mm}^2$, and $H = 0.199\ \mathrm{mm}$. Attempts to fit a single spherical grain to these values are again not successful (Table 10). The best two-particle matches for Boolean spheroids are given in Table 10. The IOS$^{(2)}$ ensemble gives an excellent match to the average volume of the grains, but less so for average surface area and mean curvature. The ROS$^{(2)}$ system leads to an excellent match of all V_ν of the experimental image. The equivalent OSC model is also given. Results for the morphological measures are summarised in Table 10 and Fig. 17.

Again the IOS^C model and the OSC model do not give a good match to the local grain measures V_ν, leading to a poor match of the global measures. In particular, IOS^C overestimates S and H, while OSC underestimates S, H, and χ. ROS$^{(2)}$ closely represents the curved interface of the original structure. This is consistent with a visual comparison of the cross-bedded sandstone to its reconstructions by Boolean models given in Fig. 18.

We also compare the porosity variation of the original image and its reconstructions. In Fig. 19 the porosity variation along the x-axes of the two 300^3 cross-bedded sandstone samples are compared to the porosity variations of the reconstructions. The reconstructed images based on the two-grain models exhibit heterogeneity which is not observed in the IOS model (see Fig. 19 and Table 11). The OSC model gives a good match to the heterogeneity of the cross-bedded sandstone along this direction. This is different to the Fontainebleau sandstone match discussed previously, where OSC overpredicts heterogeneity; larger heterogeneity may be an inherent feature of the OSC model. None of the isotropic reconstructed images are as heterogeneous as the original rock. This is not surprising. Cross-bedding is a result of a complex geological deposition process in a fluvial environment, where i.e. flow channel geometries change over time, and long-range heterogeneity and correlations result. The Boolean models cannot mimic the complex physical processes which lead to the observed structure of a cross-bedded sandstone.

Table 10. Parameters for the Boolean models of the cross-bedded sandstone. The first IOS model (IOS^C) was matched using the void-void correlation function. The other models are matched using the reconstruction by local Minkowski functionals. IOS is generalised to more than one sphere, all spheroids are randomly oriented (ROS). a, b, c note the length of the half-axes with the exception of the OSC model, where a notes the median of the radius distribution $G(r_c) = 1/2$.

Model	p	$a[\mu m]$	$b[\mu m]$	$c[\mu m]$	E_{V_0}	E_{V_1}	E_{V_2}
IOS^C	1	94.2			0.37	0.29	0.15
IOS$^{(1)}$	1	103.4			$4.2 \cdot 10^{-3}$	$1.9 \cdot 10^{-3}$	$5.6 \cdot 10^{-2}$
IOS$^{(2)}$	.598	94.8			$6.4 \cdot 10^{-6}$	$2.0 \cdot 10^{-3}$	$3.6 \cdot 10^{-2}$
	.402	114					
ROS$^{(2)}$	.711	149	145	71	$4.6 \cdot 10^{-4}$	$9.3 \cdot 10^{-4}$	$4.7 \cdot 10^{-4}$
	.289	43	39	36			
OSC		106.7			0.28	$6.9 \cdot 10^{-3}$	$7.7 \cdot 10^{-2}$

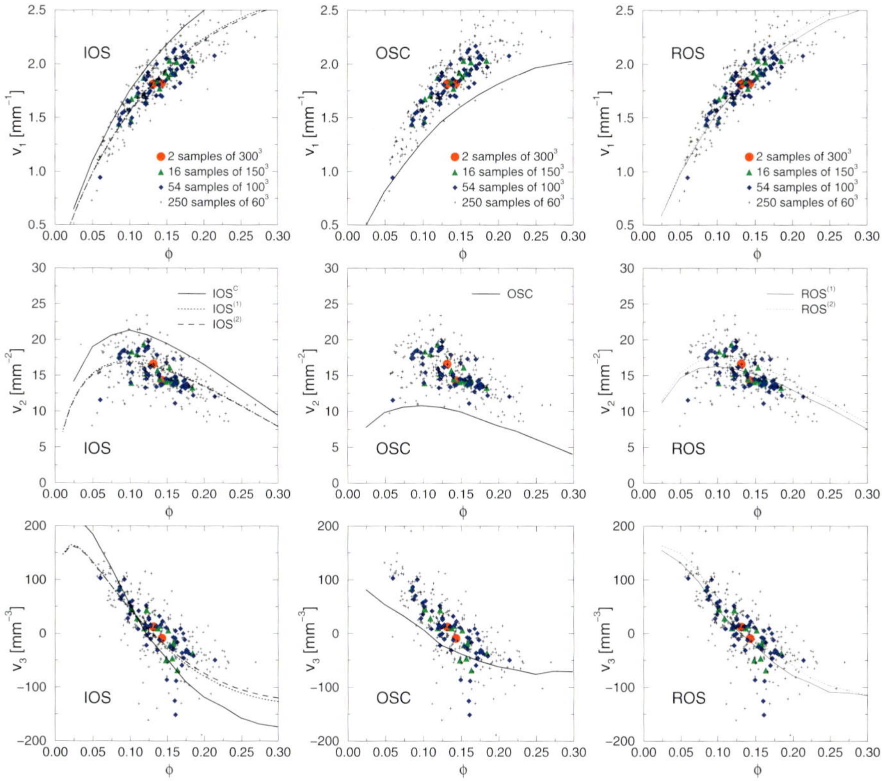

Fig. 17. Minkowski functionals over fraction for a cross-bedded sandstone compared to the matching Boolean models. [a] (left) IOS model matched using the void-void auto-correlation function as well as reconstructions by local Minkowski functionals with one or two spheres. [b] (middle) OSC model. [c] (right) Boolean models of randomly oriented spheroids.

Table 11. Variability $\sigma(\phi_{\rm sl})$ [%] of the pore volume per slice along different directions of the 300^3 cross-bedded sandstone samples and its reconstructions (see also Table 2). The reconstruction reflecting heterogeneity best is the OSC model.

Sandstone	IOS^C	$ROS^{(2)}$	OSC
x: 1.51/1.85			
y: 1.71/2.22	1.09 ± 0.08	1.48 ± 0.26	1.91 ± 0.08
z: 3.01/2.30			

4 Characterisation via Minkowski Functionals of Parallel Surfaces

In the Sect. 2 we introduced the Minkowski functionals as a complimentary method for characterizing of material morphology. While strong differences were observed between classes of models, in some cases Minkowski functionals were unable to discriminate

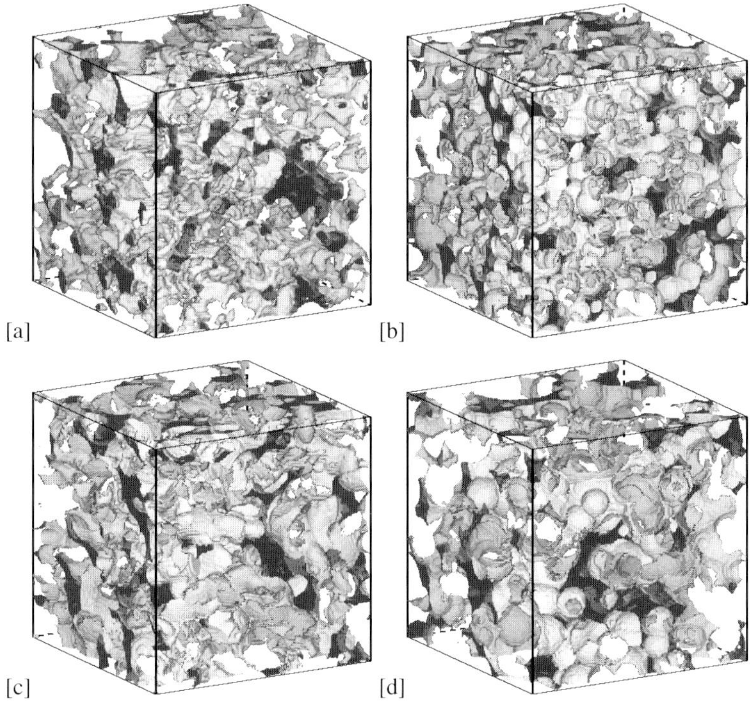

Fig. 18. Visual comparisons of reconstructions to a 150^3 subset of the original cross-bedded sandstone sample ($\phi = 18\%$). [a] Cross-bedded sandstone, [b] IOS^C, [c] $\mathrm{ROS}^{(2)}$, [d] OSC. The domain size is comparable to the 240^3 Fontainebleau datasets shown in Fig. 14.

measures for the same model class (e.g. Boolean models in Fig. 9). In this section we show that a more powerful discriminator of morphology is based on the evolution of the Minkowski functionals during erosion and dilation operations. This methology has been used previously to discriminate between different cosmological models of the early universe [30]. In these cases, the Minkowski functionals of parallel bodies of a point pattern (a convex set) was considered. The point pattern was dilated to obtain detailed morphological information where the parallel distance ϵ was used as the diagnostic parameter.

In the current section we use the evolution of the Minkowski functionals during erosion and dilation operations on complex *non-convex* morphologies. It is shown that these Minkowski functions (Mfs) gives a discriminating signature of structure. We also consider the morphology of the reconstructions of Fontainebleau sandstone discussed in Sect. 3.2 and use the method to discriminate the best match to the sandstone morphology.

4.1 Erosion-Dilation Operations

One may define the parallel body

$$A_\epsilon = \{\boldsymbol{x} \mid d(A, \boldsymbol{x}) \leq \epsilon\} \tag{5}$$

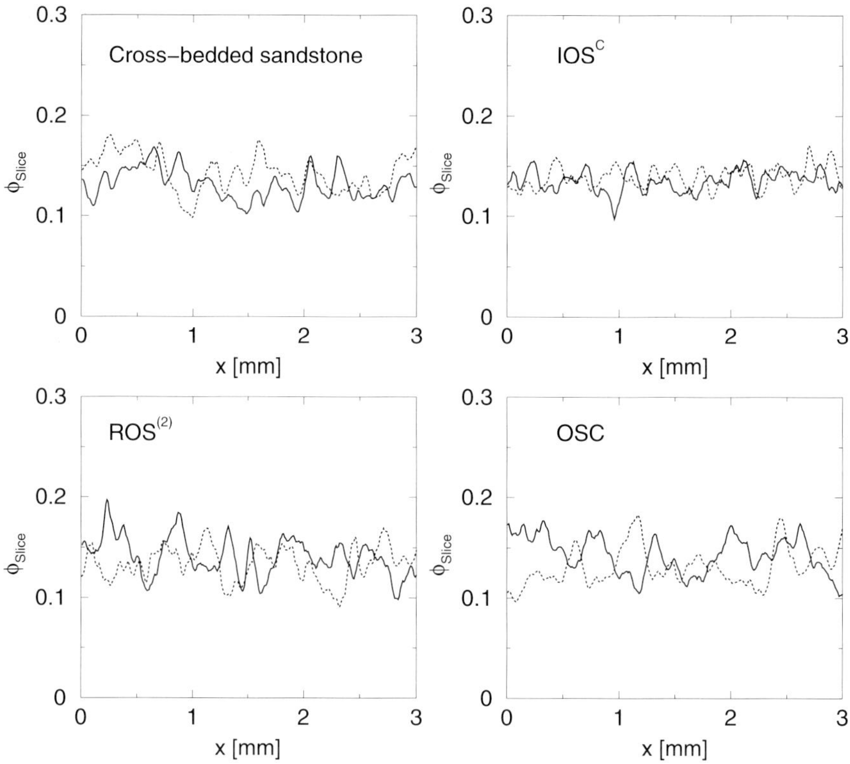

Fig. 19. Comparison in the variation of the porosity distribution along the x-axis of the cross-bedded sandstone and reconstructions. Sample size is the same as in Fig. 8.

of a structure A as the set of all points $\boldsymbol{x}$ with distances $d(A, \boldsymbol{x})$ less than ϵ to A. Here, $d(A, \boldsymbol{x}) = \text{Min}\,(|\boldsymbol{x} - \boldsymbol{y}|; \boldsymbol{y} \in A)$ denotes the minimal Euclidean distance where $|\boldsymbol{x}|$ is the standard norm in a d-dimensional Euclidean space. Changing ϵ corresponds to dilation $\epsilon > 0$ and erosion $\epsilon < 0$ of the spatial structure A. For negative values of ϵ the parallel body is defined formally by $A_{\epsilon<0} = \{\boldsymbol{x}; d(A^c, \boldsymbol{x}) \leq \epsilon\}^c$, i.e., as the complement set of all points with distances less than ϵ to the complement A^c of A. Alternatively, one may define the parallel body

$$A_\epsilon = \cup_{\boldsymbol{x} \in A} B_\epsilon(\boldsymbol{x}) \tag{6}$$

by the union of all spheres $B_\epsilon(\boldsymbol{x})$ of radius ϵ and centers $\boldsymbol{x} \in A$ inside of A. For $\epsilon < 0$ the parallel body is given by the set of all centers $\boldsymbol{x}$ so that the union $A = \cup_{\boldsymbol{x} \in A_\epsilon} B_\epsilon(\boldsymbol{x})$ equals A.

A direct implementation of (6) is computationally very inefficient as it requires one to place spheres of different size along all voxels at the interface at each erosion/dilation step. Our algorithm instead utilizes the definition given by (5) using an Euclidean Distance Map (EDM). From the EDM the erosion/dilation is trivially implemented by stepping through the different Euclidean distances from the interface.

In previous work [8] approximations to the EDM have been employed due to the large computational overheads associated with calculating the map. We have developed fast algorithms for calculating the EDM exactly which are similar to one published previously by Saito [49] with extensions to include periodic boundary conditions. The exact EDM for each voxel is calculated by growing (discretised) spheres at each voxel x_i until it reaches an interfacial voxel on A and returns the corresponding Euclidean distance d. We define the digital sphere of radius r by all voxels radiating from the central voxel that are separated by a Euclidean distance $d < r$ with the sphere center defined as the center of the voxel. The algorithm speed is dramatically accelerated by noting that the Euclidean distance to A any voxel x_j neighboring voxel x_i will be given by $\|d(A, x_i) - d(A, x_j)\| \leq 1$ (in units of the voxel size). One needs not step through all possible radii for each voxel, but instead one can use $d(A, x_j) = d(A, x_i) - 1$ as an initial guess. We increment $d(A, x_j)$ by a minimum of 0.1 at each step and generate the equivalent digitized sphere for that radius with an upper limit of $r = 50$. This gives up to 400 different radii, for which the integral geometric measures can be evaluated. We compute the measures on a minimum lattice size of 128^3 for a minimum of 50 realizations.

4.2 Discrimination of Model Composites

To illustrate the powerful discrimination of morphology which comes from measuring the Minkowski functions of parallel bodies we follow Roberts [47] and select two *model* images and generate a range of equivalent model morphologies via matching first and second order statistical properties. This method was used in [47] to illustrate the relatively strong signature provided by the chord-length distribution function(CDF). We illustrate the strong discrimination which is possible when using Minkowski functions for these systems.

In [47] Gaussian random field (GRF) systems are considered; the single cut model and the symmetric two-cut model (Berk's original model [6]). Other models considered can be formed from any number n of intersection I^n and union U^n sets of the GRF models. One then considers a specific GRF model with known statistical properties as the "*experimental*" data set. Reconstructions are generated by matching the volume fractions and by minimising the difference in the 2-point correlation functions via (4). The minimisation algorithm defines r_c, ξ and d for each model. All the models are able to provide an excellent fit to the data based on matching the two-point information. In [47] the CDF is used to attempt to better discriminate the best matching model morphology. We use the Minkowski functions $\phi, S, H, \chi(\epsilon)$ to measure the best morphological fit to the data.

In the first case the experimental data set was based on a symmetric two-cut model (N_1) at $\phi = 20\%$ obtained from the auto (field-field) correlation function [47]:

$$g(r) = e^{(-r/\ell_0)^2}; \quad \rho(k) = \frac{\ell_0^3}{(4\pi)^{2/3}} e^{(k\ell_0/2)^2} \tag{7}$$

with $\ell_0 = 2\mu\text{m}$. Attempts to match the experimental data set were made for models N_0, N_1, U_1^2, I_1^2. The resultant values of r_c, ξ and d for four different models are given in Table I of [47].

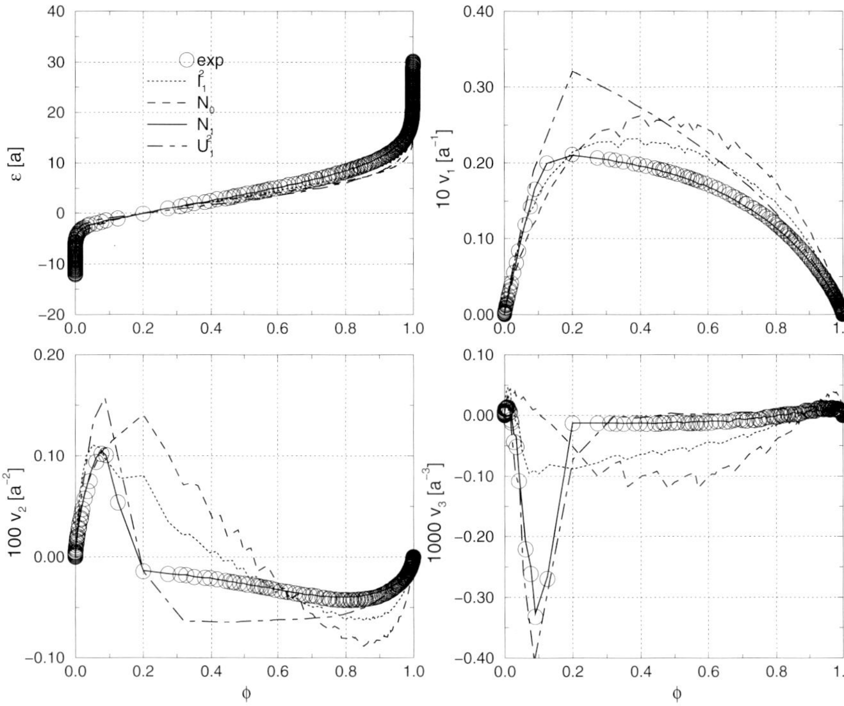

Fig. 20. Reconstruction of an experimental Gaussian model (N_1): Minkowski functionals of the erosion/dilation process for the experimental data set and the corresponding Minkowski functions for the matching model data sets defined in Table I of [47]. exp corresponds to the "experimental" image, and the lines give the best fit to the model morphology.

We show in Fig. 20 the Minkowski functions for the parallel sets of each model morphology and the original experimental data set. As in the examples discussed previously we consider the morphological measures $v_\nu(\phi)$ as a function of the porosity $\phi = \phi(\epsilon)$ with $0 < \phi < 1$.

In Table 12 we quantify the error in the measure of the Minkowski functionals for the different model morphologies by employing a relative least squares error:

$$E_i = \frac{\int_0^1 (\Delta v_i(\phi))^2 d\phi}{\int_0^1 (v_i(\phi))^2 d\phi}. \tag{8}$$

Since the porosity values are not identical for all ϵ, linear interpolation is used to do the resampling. We also show in Table 12 the errors $E\rho_1$ and $E\rho_2$ in the chord distribution function (CDF) measure from [47]. We observe that the Minkowski functions provide a far stronger signature of morphology than the CDF. While the CDF does discriminate between some models, it is impossible to choose between model N_1 and U_1 as the best reconstruction for the two-level cut model. For all four measures of the Minkowski functions the actual model morphology is more than *two orders of magnitude* more accurate than any other. It is important to note that we could discriminate morphology

Table 12. Integral error of the Minkowski functionals over fraction ((8)) for the models defined in Table I of [47]. The last two columns show the error in the CDF from [47].

M	E_1	E_2	E_3	$E\rho_1$	$E\rho_2$
N_0	.096	7.6	1.48	.25	.62
N_1	$2.5 \cdot 10^{-4}$	.0074	.0019	.04	.11
I_1^2	.023	.35	1.11	.20	.48
U_1^2	.121	1.60	.082	.02	.15

from one- or two-dimensional information (S and H) as readily as from χ. Therefore the method can be used to discriminate morphology from two-dimensional micrographs of the material or from one dimensional chord length measurements.

In the second case the *experimental* image was a 1-level cut model (N_0) and attempts to match this data set were made for models N_0, U_0^2, I_0^2. Again the Minkowski functions give a very strong signature of morphology and allow one to discriminate the correct morphology (see Fig. 21 and Table 13).

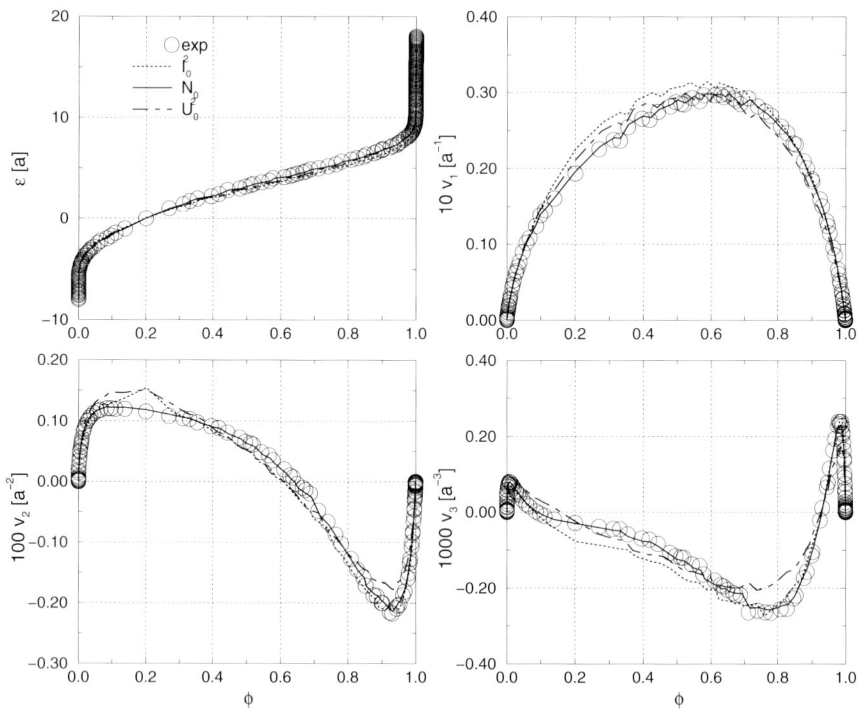

Fig. 21. Reconstruction of an experimental Gaussian model (N_0): Minkowski functionals of the erosion/dilation process for the experimental data set and the corresponding Minkowski functions for the matching model data sets defined in Table II of [47].

The Minkowski functions appear to provide a strong signature of composite morphology and hence may provide a method for selecting an accurate reconstruction of a disordered material. In the next section we attempt to do this for the experimental tomographic images of a termite nest and the Fontainebleau sandstone data sets.

Fontainebleau sandstone. In Sect. 3.2 three different reconstructions of a Fontainebleau sample were derived based on estimating the local Minkowski functionals of a Boolean model. One system gave an excellent match to the global Minkowski functionals of the sandstone samples. Here the three different Boolean ensembles (IOS^C, $\mathrm{ROS}^{(2)}$, and OSC) and a fourth stochastic model for Fontainebleau sandstone, based on a Gaussian reconstruction given by 5 intersecting 1-level-cut Gaussian models [44], are compared for the accuracy of reconstruction by evaluating the Minkowski functions. The Gaussian kernel used for the intersection reconstruction is given by the Fourier transform pair $g(x)$ and $\rho(k) = -F'(k)/(4\pi k)$ with

$$g(x) = e^{-x/\xi}(1 + x/\xi)\frac{\sin 2\pi x/d}{2\pi x/d} \tag{9}$$

$$F(k) = \frac{d}{2\pi}\left(\tan^{-1} c_- + \tan^{-1} c_+ + \frac{c_+}{1+c_+^2} + \frac{c_-}{1+c_-^2}\right), \tag{10}$$

$$c_\pm = \xi\left(\frac{2\pi}{d} \pm k\right), \tag{11}$$

where $\xi = 51.9$ μm and $d = 272$ μm [44]. To preserve the variability of the data and allow a direct comparison to the Fontainebleau sandstone samples, all reconstructions were generated at the same resolution and domain size; the 480^3 (20.3 mm^3) Fontainebleau samples are used.

The comparison of the Minkowski function data from the tomographic image of Fontainebleau sandstone to the reconstructions is given in Fig. 22. Large jumps are

Table 13. Integral error of the Minkowski functionals over fraction ((8)) for the models given in Table II of [47].

M	E_1	E_2	E_3	$E\rho_1$	$E\rho_2$
N_0	$6.74 \cdot 10^{-5}$	$5.30 \cdot 10^{-4}$	.00118	.001	.003
I_0^2	$6.73 \cdot 10^{-3}$	.0231	.056	.137	.036
U_0^2	$3.77 \cdot 10^{-3}$	.0376	.078	.008	.127

Table 14. Integral error of the Minkowski functions for the models of the tomographic image of the Fontainebleau sandstones calculated according to (8).

Model	E_1	E_2	E_3
Gaussian	.0031	.0710	.218
IOS^C	.0365	.180	.416
OSC	.075	.261	.430
$\mathrm{ROS}^{(2)}$	.0087	.0668	.207

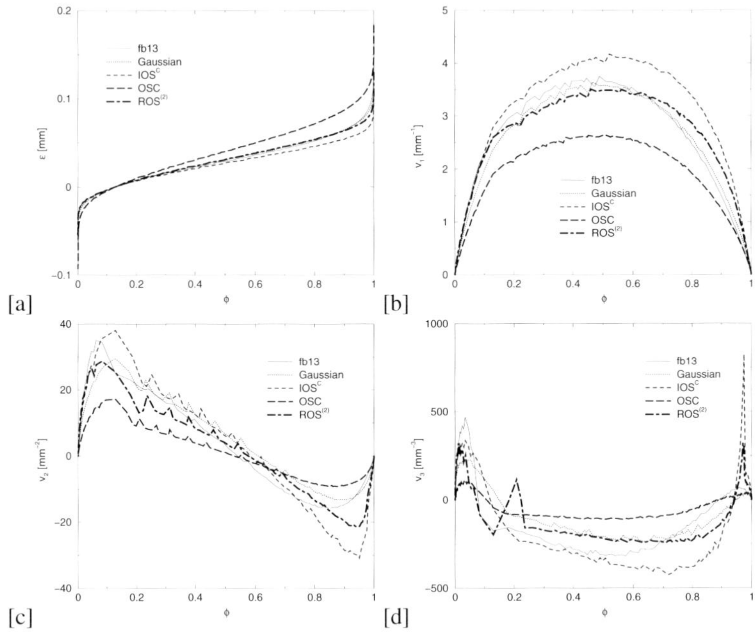

Fig. 22. Fontainebleau sandstone, sample fb13: Minkowski functionals over fraction ($\phi_0 = 13\%$).

observed in the Minkowski function data for some morphologies, particularly near the original image porosity ϕ_0. These jumps are due primarily to discretisation effects. While homogeneous, we have resolution limitations leading to discretisation errors in ϵ. All images contain features on the scale of a small number of voxels. When one erodes/dilates by a small ϵ, the step size is large and the effect on the morphological measures is large. Discretisation errors in ϵ therefore happen for all images; less so for features much larger than the voxel size.

This effect is particularly pronounced for Boolean sphere models. For more irregular particles, the 2-spheroid matches, the peaks become less pronounced. The same applies to the OSC model, where a continuous distribution of sphere sizes is present and this effect is smeared out.

The integral error of the Minkowski functions ((8)) for the four models is given in Table 14. We note that the Gaussian and ROS$^{(2)}$ models do best while IOS^C and OSC perform poorly. It is surprising that despite the large errors in the Minkowski functions of the Boolean reconstructions due to the peaks at ϕ_0 and $\phi \to 1$ the ROS$^{(2)}$ model is still good. Note, however, that the Boolean models were matched individually to the four Fontainebleau sandstone samples. The Gaussian model of reference [44] was matched to a single sample at $\phi = 15.4\%$ of Fontainebleau sandstone from another study. Thus the reconstruction using the Gaussian scheme does remarkably well. The OSC model, which gives the worst match to the Minkowski functions of the Fontainebleau sandstone sample, performs significantly better for the higher porosity samples. It is also remarkable, that the qualitative behaviour of most curves (e.g. the H curves in Fig. 22) are matched by the different models, even though none of the reconstruction schemes is actually based on parallel surface information.

5 Conclusions

In this paper we have used the Minkowski functionals to characterise, reconstruct and discriminate different complex material microstructures. We consider three distinct classes of digitised complex microstructure: particle based composites, structures generated by level-cuts through Gaussian fields, and models based on a Voronoi tesselation of space. The models are chosen to represent a wide range of composite structure. Particle models with different grain shapes and mixtures of grains are also employed. We also consider several experimental data sets generated from X-ray-CT imaging; samples include a suite of Fontainebleau sandstone, and a heterogeneous cross-bedded sandstone. We show first how the Minkowski functionals can discriminate different classes of structure.

We subsequently show that for the Boolean model of random composite media one can define a set of measures from a single image at any phase fraction ϕ which allows one to accurately reconstruct the medium for all other phase fractions. The accuracy of the reconstruction is illustrated by accurately predicting the percolation threshold p_c for the reconstructed data sets. We have illustrated the method for experimental sandstone images and showed that, in general, the reconstruction based on integral geometric measures of a Boolean model gives a better match to the morphology than the commonly used IOS^C and the recently developed OSC model. Moreover, unlike the IOS^C and OSC models, the current method is not limited to spherical inclusions, but can be used to generate more complex inclusion shapes.

We note that the morphological reconstruction of sandstones via Boolean models does not give as accurate a match to the experimental data as we observe when reconstructing a complex Boolean model. Clearly a Boolean process cannot mimic sandstone formation (sedimentation, diagenesis, compaction, etc.). The reconstruction method can be improved: one development that will further highlight the utility of the proposed technique is based on conditioning the equivalent ensemble to better mimic the local morphology of the medium. This could be done by conditioning to local curvature measures or chord-length distribution measurements which will limit the choice of grain "shape" to use in the equivalent Boolean ensemble. In oil recovery from petroleum reservoir rocks, an area of particular interest to the authors, recovery depends crucially on the mean curvature of the surfaces where immiscible phases meet at a contact angle. Conditioning an equivalent Boolean ensemble to exhibit the same distribution of local mean curvature should lead to a better prediction of multiphase flow properties on reconstructed images. The current work is based on deriving equivalent local measures from an image for Boolean models; (1). Extension of the methodology to more general systems; hard sphere mixtures, soft sphere models and models based on Gaussian random fields representing a wider range of complex materials including ceramics, composite materials and fibrous media is now being considered.

In the final section new measures were derived for the characterisation of the morphology of disordered systems based on the evolution of the Minkowski functions (Mfs) during erosion and dilation operations on non-convex morphologies. The method can be used to discriminate morphology from 1D (chord-length), 2D (micrographs), and 3D (tomographs) data sets. It is shown that the use of parallel sets and Minkowski functions leads to a very accurate discrimination of morphology. The application to experimental data of Fontainebleau sandstone shows that, while the reconstruction models considered work reasonable, none stands out as "brilliant" – and more realistic morphologies need to be considered to get a good match as in Sect. 4.2. This is left as future work, where we will consider process based models as developed in [39].

Acknowledgements

MAK thanks the Australian Research Council for financial support. KRM acknowledges support from Deutsche Forschungsgemeinschaft (DFG Grant No. Me1361/6 and Me1361/7). We thank the A.N.U. Supercomputing Facility for generous allocations of computer time. The CT image of the crossbedded sandstone was obtained by Richard Ketcham at the High Resolution CT facility at the University of Texas.

References

1. Adler, P., Jacquin, C., and Thovert, J.-F. (1992): The formation factor of reconstructed porous media. *Water Resources Research*, **28**, 1571–1576.
2. Arns, C. H., Knackstedt, M. A., and Pinczewski, W. V. (2002): $v_p : v_s$ relationships for model sandstones. *Geophys. Res. Lett.*, **29 (8)**, 10.1029/200/GEO13788.
3. Arns, C. H., Knackstedt, M. A., Pinczewski, W. V., and Mecke, K. R. (2001b): Euler-poincaré characteristics of classes of disordered media. *Phys. Rev. E*, **63**, 031112:1–13.
4. Bakke, S. and Oren, P. (1997): 3-d pore-scale modelling of sandstones and flow simulations in the pore networks. *SPE Journal*, **2**, 136–149.
5. Bentz, D. P., Quenard, D. A., Kentzel, H., Baruchen, F., Kunzel, Baruchel, J., Peyrin, F., Martys, N. S., and Garboczi, E. J. (2000): Microstructure and transport properties of porous building materials: Ii three dimensional x-ray tomographic studies. *Materials and Structures*, **33**, 147–153.
6. Berk, N. F. (1987): Scattering properties of a model bicontinuous structure with a well defined length scale. *Phys. Rev. Lett.*, **58**, 2718–2721.
7. Berk, N. F. (1991): Scattering properties of the leveled-wave model of random morphologies. *Phys. Rev. A*, **44**, 5069–5079.
8. Borgefors, G. (1986): Distance transformations in digital images. *Computer Vision, Graphics, and Image Processing*, **34**, 344–371.
9. Bryant, S., Mellor, D., and Cade, C. (1993): Physically representative network models of transport in porous media. *AIChE Journal*, **39**, 387–396.
10. Cahn, J. W. (1965): Phase separation by spinodal decomposition in isotropic systems. *J. Chem. Phys.*, **42**, 93–99.
11. D. Stoyan, W. S. K. and Mecke, J. (1995): *Stochastic Geometry and its Applications*. John Wiley & Sons, Chichester.
12. Dunsmuir, J. H., Ferguson, S. R., and D'Amico, K. L. (1991): Design and operation of an imaging x-ray detector for microtomography. *IOP Conference Series*, **121**, 257–261.
13. Flannery, B. P., Deckman, H. W., Roberge, W. G., and D'Amico, K. L. (1987): Three-dimensional x-ray microtomography. *Science*, **237**, 1439–1444.
14. Fredrich, J. T., Greaves, K. H., and Martin, J. W. (1993): Pore geometry and transport properties of Fontainebleau sandstone. *Intl. J. Rock Mech. Min. Sci.*, **30**, 691–697.

15. Gibson, L. and Ashby, M. (1988): *Cellular Solids: Structure and Properties*. Pergamon Press, Oxford.
16. Hilfer, R. (1958): *Transport and Relaxation Phenomena in Porous Media*, volume XCII of *Advances in Chemical Physics*, pages 299–424. John Wiley & Sons, Inc., New York.
17. Hyde, S. T., Barnes, I. S., and Ninham, B. W. (1990): Curvature energy of surfactant interfaces confined to the plaquettes of a cubic lattice. *Langmuir*, **6**, 1055–1062.
18. Jacobs, K., Herminghaus, S., and Mecke, K. R. (1998): Thin polymer films rupture via defects. *Langmuir*, **14**, 965–969.
19. Jernot, J. P. and Jouannot, P. (1993): Euler-poincaré characteristic of a randomly filled three-dimensional network. *J. Microsc.*, **171**, 233–237.
20. Joshi, M. (1974): *A class of stochastic models for porous materials*. Ph.D. thesis, University of Kansas, Lawrence.
21. Knackstedt, M. A. and Roberts, A. P. (1996): Morphology and macroscopic properties of conducting polymer blends. *Macromolecules*, **29**, 1369–1371.
22. Lindquist, W. B., Lee, S.-M., Coker, D. A., and Jones, K. (1996): Medial axis analysis of three dimensional tomographic images of drill core samples. *J. Geophys. Res.*, **101B**, 8297–8310.
23. Lindquist, W. B. and Venkatarangan, A. (1999): Investigating 3d geometry of porous media from high resolution images. *Phys. Chem. Earth (A)*, **25**, 593–599.
24. Lindquist, W. B., Venkatarangan, A., Dunsmuir, J., and Wong, T. F. (2000): Pore and throat size distributions measured from synchrotron x-ray tomographic images of Fontainebleau sandstones. *J. Geophys. Res.*, **105B**, 21508.
25. Lohmann, G. (1998): *Volumetric Image Analysis*. Wiley-Teubner, Chichester, New York.
26. Manwart, C., Torquato, S., and Hilfer, R. (2000): Stochastic reconstruction of sandstones. *Phys. Rev. E*, **62**, 893–899.
27. Martys, N. S., Torquato, S., and Bentz, D. P. (1994): Universal scaling of fluid permeability for sphere packings. *Phys. Rev. E*, **50**, 403.
28. Mecke, J. and Stoyan, D. (2001): The specific connectivity number of random networks. *Adv. Appl. Prob.*, **33**, 576–583.
29. Mecke, K. (1997): Morphology of spatial patterns: Porous media, spinodal decomposition and dissipative structures. *Acta Physica Polonica B*, **28**, 1747–1781.
30. Mecke, K. R. (1994): *Integralgeometrie in der Statistischen Physik: Perkolation, komplexe Flüssigkeiten und die Struktur des Universums*, pages 61–68. Harry Deutsch, Frankfurt.
31. Mecke, K. R. (1996): Morphological characterization of patterns in reaction diffusion systems. *Phys. Rev. E*, **53**, 4794–4800.
32. Mecke, K. R. (1998): Integral geometry and statistical physics. *International Journal of Modern Physics B*, **12**, 861–899.
33. Mecke, K. R. and Seyfried, A. (2002): Strong dependence of percolation thresholds on polydispersity. *Europhys. Lett.*, **58**, 28–34.
34. Mecke, K. R. and Stoyan, D., eds. (2000): *Statistical Physics - The Art of Analyzing and Modeling Spatial Structures*, volume 554 of *Lecture Notes in Physics*. Springer, Berlin.
35. Monnereau, C. and Vignes-Adler, M. (1998): Optical tomography of real three-dimensional foams. *Journal of Colloid and Interface Science*, **202**, 45–53.
36. Nelder, J. A. and Mead, R. (1965): A simplex method for function minimization. *Computer Journal*, **7**, 308–313.
37. Oh, W. and Lindquist, W. B. (1999): Image thresholding by indicator kriging. *IEEE Transactions on Pattern Analysis and Machine Intelligence*, **21**, 590–602.
38. Ohser, J. and Mücklich, F. (2000): *Statistical Analysis of Materials Structures*. J. Wiley & Sons, Chichester, New York.
39. Øren, P. E. and Bakke, S. (2001): Process-based reconstruction of sandstones and prediction of transport properties. *Transport in Porous Media*, **-**, submitted.

40. Øren, P. E., Bakke, S., and Arntzen, O. J. (1998): Extending predictive capabilities to network models. *SPE Journal*, **3**, 324–336.
41. Po-zen Wong, J. K. and Tomanic, J. P. (1984): Conductivity and permeability of rocks. *Phys. Rev. B*, **30**, 6606–6614.
42. Quiblier, J. (1984): A new three-dimensional modeling technique for studying porous media. *J. Colloid Interface Sci.*, **98**, 84–102.
43. Roberts, A. P. (1997a): Morphology and thermal conductivity of model organic aerogels. *Phys. Rev. E*, **55**, R1286–R1289.
44. Roberts, A. P. (1997b): Statistical reconstruction of three-dimensional porous media from two-dimensional images. *Phys. Rev. E*, **56**, 3203–3212.
45. Roberts, A. P., Bentz, D. P., and Knackstedt, M. (1996): Correlating microstructure to the petrophysical properties of porous rocks. In *Proceedings of the SPE Asia Pacific Oil and Gas Conference*, pages 551–559. Soc. Petrol. Eng., Adelaide, Australia.
46. Roberts, A. P. and Knackstedt, M. (1995): Mechanical and transport properties of model foamed solids. *J. Materials Sci. Letters*, **14**, 1357–1359.
47. Roberts, N., Reed, M., and Nesbitt, G. (1997): Estimation of the connectivity of a synthetic porous medium. *J. Microsc.*, **187**, 110–118.
48. Rosenfeld, A. (1976): *Digital Picture Processing*. Academic Press, New York.
49. Saito, T. and Toriwaki, J.-I. (1994): New algorithms for euclidean distance transformation of an n-dimensional digitized picture with applications. *Pattern Recognition*, **27**, 1551–1565.
50. Santaló, L. A. (1976): *Integral Geometry and Geometric Probability*. Addison-Wesley, Reading.
51. Scheidegger, A. (1974): *The Physics of Flow through Porous Media*. University of Toronto Press, Toronto.
52. Schwartz, L. M., Martys, N., Bentz, D. P., Garboczi, E. J., and Torquato, S. (1993): Cross-property relations and permeability estimation in model porous media. *Phys. Rev. E*, **48**, 4584–4591.
53. Serra, J., ed. (1992): *Image Analysis and Mathematical Morphology*, volume 1,: Theoretical Advances. Academic Press, New York.
54. Spanne, P., Thovert, J., Jacquin, J., Lindquist, W. B., Jones, K., and Coker, D. (1994): Synchotron computed microtomography of porous media: topology and transports. *Phys. Rev. Lett.*, **73**, 2001–2004.
55. Stauffer, D. and Aharony, A. (1994): *Introduction to Percolation Theory*. Taylor & Francis, London, 2nd edition.
56. Teubner, M. (1991): Level surfaces of gaussian random fields and microemulsions. *Europhys. Lett.*, **14**, 403–408.
57. Teubner, M. and Strey, R. (1987): Origin of the scattering peak in microemulsions. *J. Chem. Phys.*, **87**, 3195–3200.
58. Thovert, J.-F., Yousefian, F., Spanne, P., Jacquin, C. G., and Adler, P. M. (2001): Grain reconstruction of porous media: Application to a low-porosity Fontainebleau sandstone. *Phys. Rev. E*, **63**, 061307.
59. Torquato, S. (2002): *Random Heterogeneous Materials: Microstucture and Macroscopic Properties*. Springer, New York.
60. Yeong, C. L. Y. and Torquato, S. (1998): Reconstructing random media. *Phys. Rev. E*, **57**, 495–506.

Topological Characterization of Porous Media

Hans-Jörg Vogel

University of Heidelberg, Institute of Environmental Physics,
INF 229, 69120 Heidelberg, Germany

Abstract. It is an attractive approach to predict flow and in based on direct investigations of their structure. The most crucial property is the of the structure because it is difficult to measure. This is true both at the pore scale, which may be represented as a binary structure, and at a larger scale defined by continuous macroscopic state variables as phase density or . At the pore scale a function is introduced which is defined by the as a function of the pore diameter. This function is used to generate of the porous structure that allow to predict bulk hydraulic properties of the material. At the continuum scale the structure is represented on a grey scale representing the porosity of the material with a given resolution. Here, topology is quantified by a connectivity function defined by the Euler characteristic as a function of a porosity threshold. Results are presented for the structure of natural soils measured by . The significance of topology at the continuum scale is demonstrated through numerical simulations. It is found that the effective permeabilities of two heterogeneous having the same auto-covariance but different topology differ considerably.

1 Introduction

Flow and transport within porous media are governed by their porous structure. This is true at a small scale of single pores but also at a larger scale where the structure of continuous fields of hydraulic properties such as the hydraulic permeability becomes relevant. It is an old scientific challenge to predict the behavior of porous media from some structural properties which are directly observable so that no expensive experiments are required.

Nowadays, more and more techniques are available to directly investigate the structure of porous media. Especially X-ray tomography has become an attractive technique capable to detect pore structure with high resolution of some 0.01mm [21] which is also presented by Arns et al. [2] in this volume. At a larger scale the structure of density or porosity at resolutions of some 1.0 mm can be measured by X-ray tomography [39]. Nanotomography as a 3D-image tool by scanning microscopy is presented in this volume by Magerle [22]. In this paper, soil as one of the most complex porous media is considered.

To bridge the gap between structure and function the quantification of relevant structural properties is required. At the pore scale, the size distribution of pores is important while at the larger continuum scale the pore-scale effects may be smeared out and the autocorrelation of permeabilities may play a crucial role. Besides these metric properties, also the topology of the structure is highly relevant for any kind of transport, meaning the way the structural units are interconnected in space. This was demonstrated at the pore-scale [20, 43] and at the continuum scale for the permeability field of an aquifer [45].

Moreover, the shape of structural units may be quantified using Minkowski functionals, see [2, 3] in this volume. An overview on quantitative methods for the characterization of porous media can be found in the precurser to this volume [26].

The focus of this contribution is on topology where quantitative information is still hard to get. This is because a topological description of a three dimensional structure requires information at a resolution corresponding to the size of the structural units, e.g. pores, in all three dimensions. There are no stereological methods to infer 3D topological characteristics from two dimensional sections. Hence, measurements of connectivity typically requires a complete three dimensional analysis of the structure or at least a representative elementary volume (REV). This can be achieved for simple isotropic media as sandstones where a few cubic millimeters may be sufficient to fully characterize the porous structure. Here, local percolation probabilities [15] can successfully applied to characterize topology. In more heterogeneous porous media where the structural units are extremely anisotropic, e.g. root channels, cracks and earthworm borrows in , a representative three dimensional sample is extremely expensive and most topological measures are afflicted with unacceptable uncertainties at the edges.

In this paper connectivity functions for the pore scale and for continuous fields are introduced which are based on the Euler characteristic. An important advantage of this approach is the fact that an unbiased estimation of the Euler characteristic is possible from local measurements, meaning there are no edge effects and the shape of the required three dimensional sample is irrelevant.

The paper is organized as follows. In the first section the connectivity function for binary pore structures is defined as the Euler characteristic as a function of the pore size. In the second part it is demonstrated how to adapt a pore network model to a given and connectivity function. Numerical simulations of effective hydraulic properties based on network models are compared to direct experimental findings to demonstrate the effect of topology. In the third section the same concept is applied to continuum fields, where the connectivity function is defined as the Euler characteristic in dependency of the threshold applied to the continuous field. This approach is demonstrated for the structure of porosity in a Loess soil measured by X-ray tomography. Finally, the significance of topology is demonstrated for water flow in random permeability fields having the same but different topology.

2 Pore Scale Topology

Besides the porosity and the size distribution of pores, the topology of the pore space is a crucial property regarding flow and transport in porous media. The way how pores are interconnected may be even more important than their number and size.

The basic topological properties are the number N of isolated objects, i.e. pores, and the number of redundant connections C within the pores space, and, in three dimensions, the number of completely enclosed cavities H [14]. For a detailed discussion of these measures see the contribution of Robins [34] in this volume. The Euler characteristic χ combines the basic topological measures to

$$\chi = N - C + H \; . \tag{1}$$

Hence, the Euler characteristic is a measure of connectivity which gives positive values for poorly connected structures where $N > C$ and negative values for more connected structures where $N < C$. For the topology of a network of pores in natural porous media, the quantity H is of minor importance since solid particles completely surrounded by pores are rare. An essential advantage for the measurement of χ is the fact that an unbiased estimation of χ is possible for a 3D cutout of the porous media. Thereby, the basic quantities N, C and H are not counted explicitely but the Euler characteristic may be deduced from the local geometry of the structure [38, 8]. This is demonstrated for the application to digitized images in the contribution of Ohser et al. [27] in this volume.

The corresponding algorithms for binary digitized images are straightforward [28, 40]. Consequently, we may obtain χ from two parallel serial sections as a minimum three dimensional sample also referred to as disector [38, 8].

This is of practical importance in such media where a full three dimensional representation is hard to get and this is rather the rule than the exception. The complete structure or a representative elementary volume in terms of topology is typically not available. In most cases we are restricted to some small subsamples for which the possible determination of N, C and H would be unacceptable due to the uncertainty at the edges of the investigated cutout. To estimate the Euler characteristic, however, a small stack of serial sections may be sufficient. This was demonstrated for the pore structure of a soil [41]. Thereby, the number of serial sections can be reduced to the benefit of the size of the individual sections in order to sample the structure more representatively and thus, to arrive at a more reliable estimation of χ [44]. This aspect is illustrated in Fig. 1.

Another approach towards the quantification of topology in porous media is the measure of percolation probabilities denoting the probability to find a continuous path through a sample of a given size. Percolation probabilities can be evaluated as a function of sample size and porosity. This leads to connectivity functions [16, 17] which may be related to effective properties of the material [46]. However, a prerequisite of this approach is the knowledge of the complete structure or at least a representative subsample. Hence, the same problems as for the determination of N, C and H are encountered. For isotropic structures as e.g. sandstone, however, percolation probabilities provide the topological information required [18].

In the following, a connectivity function is introduced which is based on the Euler characteristic χ and can thus be determined from local measurements using a small stack of serial sections. Figure 1 shows an example of the heterogeneous structure of pores in a silty clay soil. For a detailed description of this material and the sampling technique see [41].

Topological characteristics of porous media depend on the spatial resolution of the available data. This is true if pores exist which are in the size range of the resolution or smaller. Then, the connectivity of the porous structure is expected to change with resolution because more connections or more isolated features appear.

The idea of the connectivity function proposed here is to measure the Euler characteristic as a function of pore size. Starting with the largest pores, which may be isolated and not connected, the smaller pores are successively added which change the topology in a characteristic way, e.g. in soil we expect an increase of connectivity. This kind of connectivity function yields an integral description of the overall topology and should

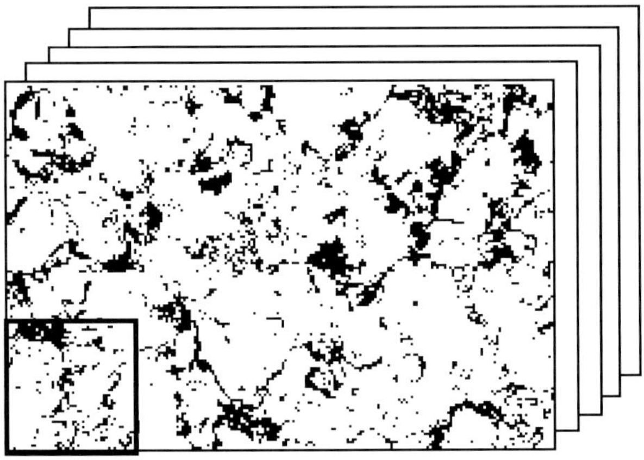

Fig. 1. One of 20 serial sections through undisturbed silty clay soil used to determine the Euler characteristic. Natural width is 20 mm, resolution is 0.013 mm/pixel, separation of serial sections is 0.4 mm, pores are black. The square in the lower left corner represents the size of a cube having the same volume as a pair of serial sections.

reflect the hydraulic behavior of the material because it contains information on the way how pores of different size are interconnected.

As a prerequisite, a criteria of pore size is required. Morphological granulometry [37] provides an excellent tool to filter pores which are smaller than a given threshold. It is based on morphological erosions and dilations using a spherical structuring element. Based on a binary three dimensional sample $\mathbb{S}$ including the subset of pores $\mathbb{P} \subset \mathbb{S}$, the eroded set $\mathbb{P}_{\mathrm{erod}}$ is obtained as

$$\mathbb{P}_{\mathrm{erod}} = \{x : \mathbb{B}_x^d \subset \mathbb{P}\} = \mathbb{P} \ominus \mathbb{B}^d \tag{2}$$

where $\mathbb{B}_x^d$ is the sphere with diameter d centered at point x. Consequently $\mathbb{P}_{\mathrm{erod}}$ describes the set of all locations within $\mathbb{P}$ where the structuring element fits completely into the pore space. The morphological dilation is defined as

$$\mathbb{P}_{\mathrm{dil}} = \{x : \mathbb{B}_x^d \cap \mathbb{P} \neq \emptyset\} = \mathbb{P} \oplus \mathbb{B}^d \tag{3}$$

The morphological opening combines and :

$$\mathbb{P}_{\mathrm{op}} = (\mathbb{P} \ominus \mathbb{B}^d) \oplus \mathbb{B}^d \tag{4}$$

which removes all pores smaller than the structuring element $\mathbb{B}^d$. Starting with a three dimensional image recorded at a given resolution λ, the application of n structuring elements $\mathbb{B}^d$ with increasing diameter $d = n\lambda$ produces realizations $\mathbb{P}_{\mathrm{op}}(d)$ of the porous structure where only pores $> d$ are considered. Hence, a cumulative pore size distribution can be defined as

$$F(d) = \Phi(\mathbb{P}_{\mathrm{op}}(d)) \tag{5}$$

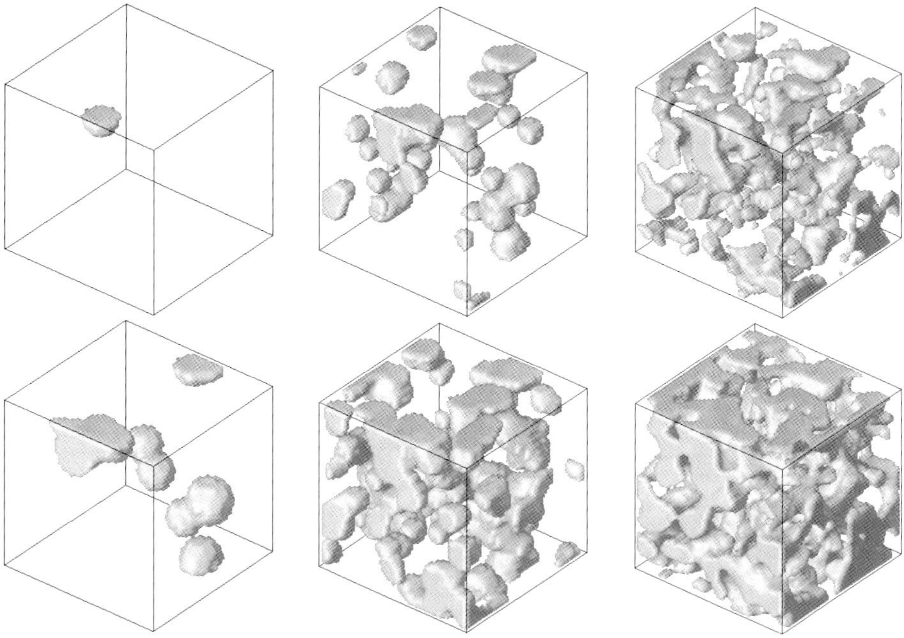

Fig. 2. Simulated pore structures from excursion sets according to different thresholds $t = 100$ upper row and $t = 120$ lower row. From left to right: isosurfaces for pores $\mathbb{P}_{\rm op}(d)$ with decreasing minimum diameter d. The related connectivity functions are shown in Fig. 3.

where $\Phi()$ means 'porosity of'. The size criteria used here reflects almost perfectly our idea of the 'hydraulic diameter' of a pore. Now, the Euler characteristic χ can be determined as a function of d by calculating χ for the different $\mathbb{P}_{\rm op}(d)$ to obtain the connectivity function $\chi(d)$. It should be noted that for the calculation of $\chi(d)$ a three dimensional sample is required where all dimension exceed the maximum pore diameter so that $\mathbb{P}_{\rm op}(d)$ can be determined. Hence, a number of serial sections is required and a disector is not sufficient. Moreover it should be noted that the opened set $\mathbb{P}_{\rm op}(d)$ may include short necks that are smaller than the diameter of the structuring element. This could be avoided by using $\mathbb{P}_{\rm erod}(d)$.

To demonstrate the behavior of $\chi(d)$ for a simple isotropic structure, a three dimensional porous media was modeled as an excursion set of a Gaussian random field [30, 32]. The random field was generated according to a predefined auto-covariance function. It was then segmented into pores and solid according to a given threshold t meaning all values $< t$ are considered to be pores. In Fig. 2 (right) the obtained porous structure is illustrated as an isosurface according to two different thresholds t. Clearly, the structure is more intensely connected for higher values of t.

The corresponding connectivity functions are shown in Fig. 3. Generally, the connectivity increases with decreasing pore size d and with increasing porosity. The latter is controlled by the threshold t of the Gaussian random field. For high porosities ($t = 150$) connectivity decreases for small d because there are completely enclosed cavities H

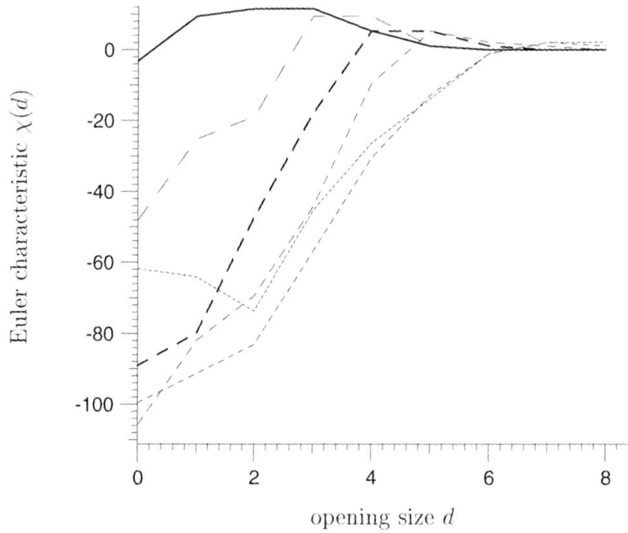

Fig. 3. Connectivity functions of excursion sets related to different threshold values $t \in 100, 110, 120, 130, 140, 150$ (decreasing dash length) applied to the Gaussian random field . Thick lines are the results for t = 100 and t = 120 which are illustrated in Fig. 2.

within the pore space, see (1). In natural porous media, however, this behavior is not expected, since solid clusters completely surrounded by pores are rare.

3 Pore Network Modeling

Network models are idealized representations of the complex pore geometry which may be used to calculate effective hydraulic properties of porous media. This approach was introduced by [9, 10, 11] and subsequently developed in the field of petroleum engineering [7, 6] but also in the field of soil physics and hydrology [4, 5].

The calculation of effective permeabilities or the pressure-saturation relation of network models is straightforward. Hence network modeling is an efficient tool to investigate the effect of geometrical aspects on the effective behavior of porous media [20, 12, 13]. If network models are intended to provide predictions of hydraulic properties [47, 31], the geometry of the network has to be adopted to some relevant observables of the porous structure. Thereby, topology is typically not considered explicitely, because direct measurements are notorious difficult.

However, the connectivity function $\chi(d)$ defined in the previous section is a quantitative handle on topology. Using $\chi(d)$ together with the pore size distribution $F(d)$ as obtained from morphological openings, a network model may be generated which mimics the measured structure in terms of pore size and topology. The first question is, how to generate a network model that fulfills a predefined $\chi(d)$ and $F(d)$? And the second crucial question is, to what extend does such a network model reproduce the hydraulic behavior of the material which may be answered by comparing network simulations with direct experiments. This work was presented in detail by [42, 44] and is reviewed here only briefly.

The basic geometry of the network is a face-centered cubic grid with grid constant λ and a coordination number $Z = 12$. The bonds represent cylindrical pores with radius r while the nodes have no extra volume. To reach a given Euler characteristic and porosity only a part of all available bonds $Z_{\mathrm{eff}} < Z$ is used while the residual bonds $Z - Z_{\mathrm{eff}}$ are interpreted to represent 'background porosity', i.e. small pores with diameter below the resolution of the available data. As demonstrated in [42] the unknown parameters λ and Z_{eff} can be directly calculated from the input functions $\chi(d)$ and $F(d)$. Then, the probability distribution of different classes of pore radii within the network can be directly obtained from $F(d)$.

The realization of $\chi(d)$, however, is more demanding. The Euler characteristic of a network model is simply obtained by

$$\chi = N_n - N_b \tag{6}$$

where N_n is the number of nodes to which at least one bond is connected, and N_b the total number of bonds; see also [23]. In order to realize a predefined connectivity function $\chi(d)$, we start with the generation of the largest pores. During the generation process the actual Euler characteristic χ_{act} of the network is continously updated using (6). Prior to attributing a radius d to a bond the difference between $\chi(d)$ and $\chi_{\mathrm{act}}(d)$ is evaluated. If it is negative only an isolated bond can be chosen, otherwise only bonds connected to already existing pores are considered. In this way, a predefined connectivity function is realized. In Fig. 4, an example of such a network model including 8^3 nodes is shown. To simulate hydraulic properties 64^3 nodes were used.

Once the network geometry is generated effective hydraulic properties can be simulated. First, the pressure-saturation relation $\theta(\psi)$ is computed describing the water saturation θ in dependency of the capillary pressure ψ and second, the relative hydraulic conductivity function $K_r(\psi)$ describing the hydraulic conductivity relative to the conductivity at water saturation as a function of the capillary pressure ψ. To determine the fluid saturation for a given capillary pressure ψ each bond above a critical radius r_c is drained, provided that it is in contact with the non-wetting phase:

$$r_c > 2\sigma\psi^{-1} \tag{7}$$

This is in accordance with the Young-Laplace equation where σ is the surface tension and complete wetting is assumed, i.e. the contact angle between water and solid is zero. In Fig. 4 the distribution of the wetting and non-wetting phase at a certain capillary pressure is illustrated.

To calculate the hydraulic conductivity at different water saturations we assume Poiseuille flow within the cylindrical bonds of the network. Given a pressure gradient across the ends of the network, this leads to a system of linear equations which was solved by a conjugate gradient method [29].

In Fig. 5 simulation results are compared to experimental findings in a silty clay soil, for details see [42]. Considering the fact that the simulated results are only based on directly measured morphological data, i.e. $\chi(d)$ and $F(d)$, the agreement of the shape of $\theta(\psi)$ and $K_r(\psi)$ was found to be reasonably good. This means that the hydraulic

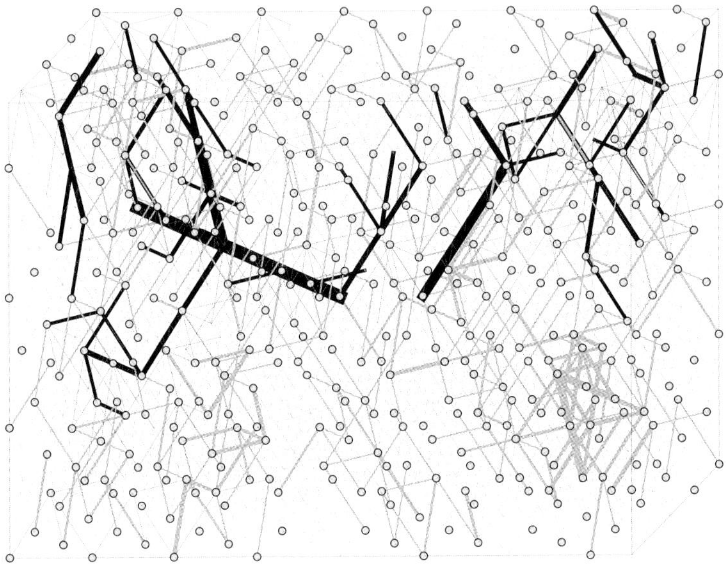

Fig. 4. Example of a networkmodel of 8^3 nodes at a certain capillary pressure where some pores are filled with water (grey) and others with air (black).

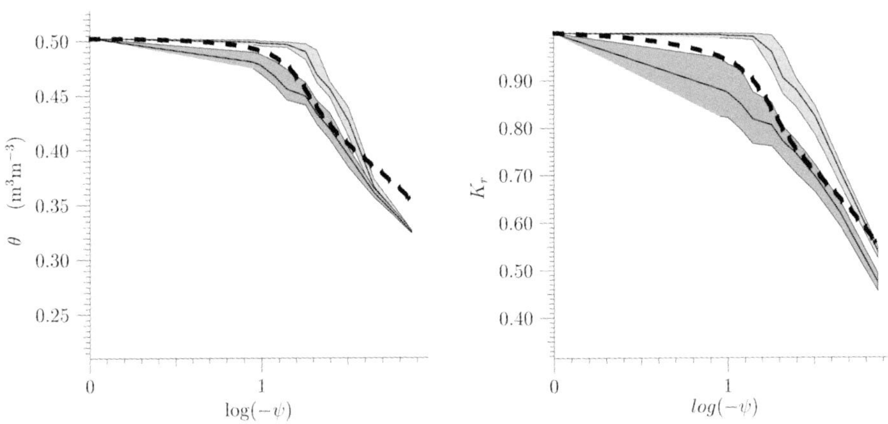

Fig. 5. Water characteristic $\theta(\psi)$ (left) and hydraulic conductivity function $K(\psi)$ (right) obtained from network models with random topology (light grey) and measured topology of a Loess soil (dark grey). The shaded areas enclose the maximum and minimum values of 20 realizations. The dashed line represents the experimental results obtained for the Loess soil. The capillary pressure ψ is in hPa.

properties are mainly governed by the pore size distribution and topology and that the quantitative approach to describe these properties is promising - at least for the material investigated here.

The significance of topology becomes evident in case the topological characteristics are ignored for the network simulations (Fig. 5). Then, the pressure-saturation relation exhibits a marked air entry point at a critical capillary pressure which corresponds to the percolation threshold of a random network. Accordingly, the shape of the relative hydraulic conductivity function is changed.

4 Topology of Continuous Fields

At the pore scale it is evident that the topological characteristics of the pore space is essential in terms of flow and transport. But also at a larger scale, where individual pores are no more visible, topology may play a crucial role regarding transport in porous media. At this large scale the binary image of the pore structure disappears and is replaced by fields of continuous variables as porosity or permeability. Obviously, the continuity and connectivity of various conductive regions is significant in terms of flow and transport.

The effect of spatial variability is commonly studied through numerical models that solve the governing equations for fluid flow and transport. Such models require an explicit description of the spatial patterns of hydraulic properties which must be deduced from some kind of observations or measurements.

One important aspect is the distribution function of hydraulic properties, as e.g. the hydraulic permeability, which is often found to be lognormal. Another important measure is the correlation scale of the heterogeneous field which may be quantified by the auto-covariance function. These quantities may be inferred from a set of point measurements even in plane sections and can be used for the generation of continuous correlated random fields [33, 36].

Thereby, however, the topological characteristic of the generated spatial pattern is merely an implicit consequence of the generation process. For an explicit consideration of topology, some quantitative criteria are required that address the connectivity of continuous fields. However, the same problems as encountered at the pore scale are to be solved: the requirement of continuous information on the spatial pattern at a spatial resolution corresponding to the size of the structural units which are relevant. As an example, the connectivity of a narrow tortuous band of high permeability cannot be detected by a set of point measurements but requires a continuous map at a resolution below the width of that band. This kind of structure is crucial for flow and transport and is rather the rule than the exception in natural porous media as fractured rocks, sediments and soil. In isotropic media as sandstone however, such features typically do not exist.

Nowadays, there are techniques available which allow a continuous measurement of the structure at high spatial resolution. For the analysis of porous media at a small scale X-ray tomography has become a very efficient tool [19, 35, 39, 21]. On a larger scale different geophysical methods are available. In the following, a connectivity function for continuous fields is introduced which is based on the same concepts as for binary structures. It is demonstrated for the structure of a soil sample obtained from X-ray tomography.

Figure 6 (left) shows a horizontal section through undisturbed soil obtained by a medical X-ray tomograph. Basically, X-ray tomography measures the local X-ray absorption coefficient μ_x which is related to the density of electrons. Hence, assuming homogeneity of the solid phase, i.e. absorption coefficient $\mu_s = const$, and a homogeneous fluid within the pore space characterized by the constant μ_w , the measured

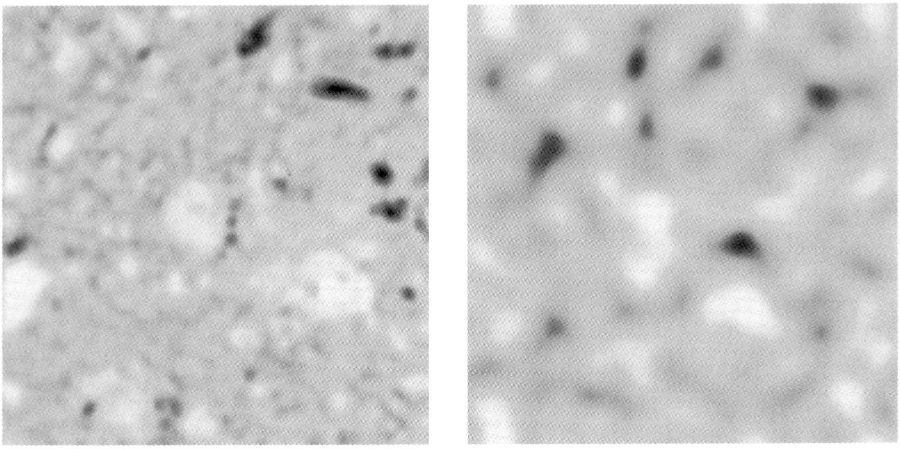

Fig. 6. 2D density image of a Loess soil (left, natural height 110 mm, resolution 0.44 mm) and a Gaussian random field (right) with equivalent grey level distribution and correlation length.

absorption at each location x is given by

$$\mu_x = [1 - \Phi_x]\mu_s + \Phi_x \mu_w \ . \tag{8}$$

Consequently, the porosity Φ_x at each location x can be obtained directly from measurements of μ_x and the image in Fig. 6 may be interpreted as a map of porosity $\Phi(x)$. Moreover, for a given material, the porosity is expected to be closely related to permeability.

The classical approach of modeling this kind of structure is based on the distribution function of Φ expressed by the cumulative density function (cdf) $G(\Phi)$ and the spatial correlation. The latter is quantified by the auto-covariance function $C_\Phi(h)$ which is calculated as

$$C_g(h) = \langle[\Phi(x) - \langle\Phi(x)\rangle][\Phi(x+h) - \langle\Phi(x)\rangle]\rangle \tag{9}$$

where $\langle\rangle$ denotes expectation. The measured structural properties $G(\Phi)$ and $C_\Phi(h)$ are shown in Fig. 7. From (9) the correlation length λ was obtained as

$$\lambda = \frac{1}{\sigma^2}\int_0^\infty C_\Phi(h)\mathrm{d}h \tag{10}$$

where σ^2 is the variance of Φ.

Following the classical approach, a Gaussian random field $f_0(x)$ was generated [1, 33, 36]. Thereby , the same correlation length as measured in the X-ray image was realized. Then, the resulting field $f_0(x)$ was transformed to $f_1(x)$ so that the cumulative density function F of the random field matches the measured distribution $G(\Phi)$:

$$f_1(x) = G^{-1}[F(f_0(x))] \tag{11}$$

As a consequence of such a histogram matching, also referred to as 'nonlinear filtering' [30] the spatial correlation of $f_1(x)$ is changed compared to $f_0(x)$. Thus, the correlation

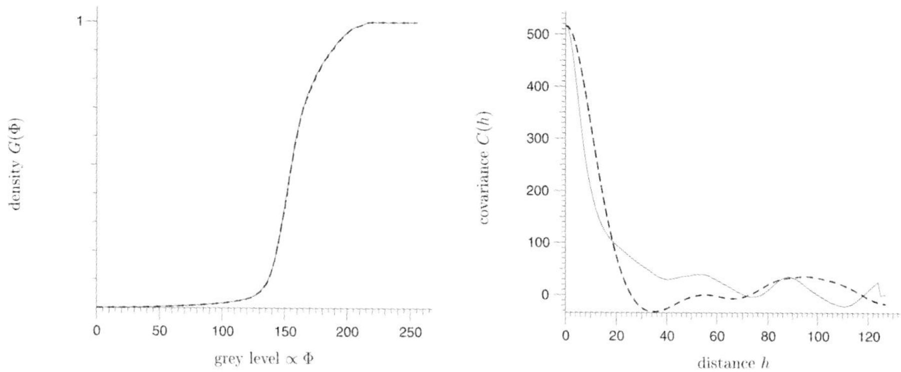

Fig. 7. Density function of porosity $G(\Phi)$ (left) and auto-covariance $C_\Phi(h)$ (right) for the natural structure of soil (thin line) and the correlated random field (dashed line) shown in Fig. 6.

length λ which was used to generate f_x was altered iteratively so that λ of $f_1(x)$ matches that of $g(x)$.

In this way the correlated random field shown in Fig. 6 (right) was generated. As shown in Fig. 7 the distribution function of the random field is identical compared to the original image while the correlation matches only in terms of the correlation length λ. From a qualitative point of view, the modeling approach seems to be appropriate also with respect to connectivity for this two dimensional example.

In order to quantitatively describe the topological characteristics of the continuous fields, the Euler characteristic χ was calculated as a function of a threshold value t. For each t we obtain a binary image (Fig. 9) and a corresponding Euler characteristic which yields the connectivity function $\chi(t)$ for continuous fields. The results for the two-dimensional examples of Fig. 6 are shown in Fig. 8 where the Euler numbers are plotted relative to the maximum value. Clearly, the absolute values of $\chi(t)$ differ considerably because of the smoothness of the random field. However the shapes of the different connectivity functions are rather similar. Starting from low values of t, $\chi(t)$ increases which indicates a number of isolated dark islands, i.e. islands of high porosity (see also Fig. 6). While increasing t further, the structure becomes more and more connected and reaches high negative values which means a considerable number of loops or, from the perspective of the background, a considerable number of isolated islands of low porosity. This behavior is illustrated in Fig. 9 where the extreme values of $\chi(t)$ are represented together with the case for $\chi(150) \approx 0$ which may be related to a percolation threshold [24].

As already mentioned, the random field resembles the original image quite well and one would not expect considerable differences for effective hydraulic conductivity in case the grey values of the two images in Fig. 6 are assumed to represent permeabilities. However, we are looking at a two dimensional cut out from a three dimensional sample and one could easily imagine that the 2D islands of high porosity are intensely connected in three dimensions as could be expected from root holes or earthworm burrows in soil.

In Fig. 10 a three dimensional isosurface ($t = 30$) for the original X-ray image and a 3D Gaussian random field is shown. Again, the correlation lengths and the distribution

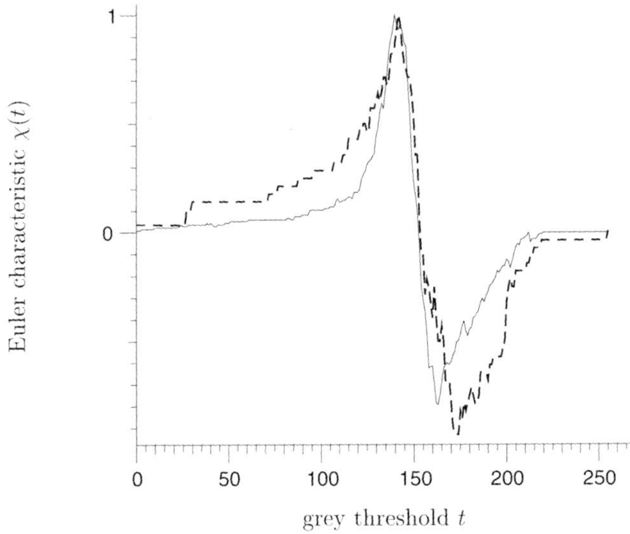

Fig. 8. 2D Connectivity functions of the porosity image (thin line) and the correlated random field (dashed line) shown in Fig. 6. The Euler numbers are plotted relative to the maximum values.

functions are the same for both structures. However the topology in 3D is considerably different. In Gaussian random fields, the extreme values are always located on islands while the highest porosities in the undisturbed soil are related to large pores which typically originate from roots or earthworms and hence, are intensely connected.

In analogy to the two dimensional case discussed above, the 3D connectivity functions $\chi(t)$ was calculated for the two structures which came out to be considerably different as well (Fig. 11). The results clearly confirm the qualitative findings that the random field generates a large number of isolated objects of high porosity $(t < 120)$ while the natural structure is more connected. The same is true for the other extremity of low porosity $(t > 190)$ where the connectivity of low porosity is lower for the random field. In contrast to the two dimensional case, the values for $\chi(t)$ are positive for

Fig. 9. Gaussian random field of Fig. 6 binarized according to different thresholds $t = 140, 150, 160$ (from left to right)

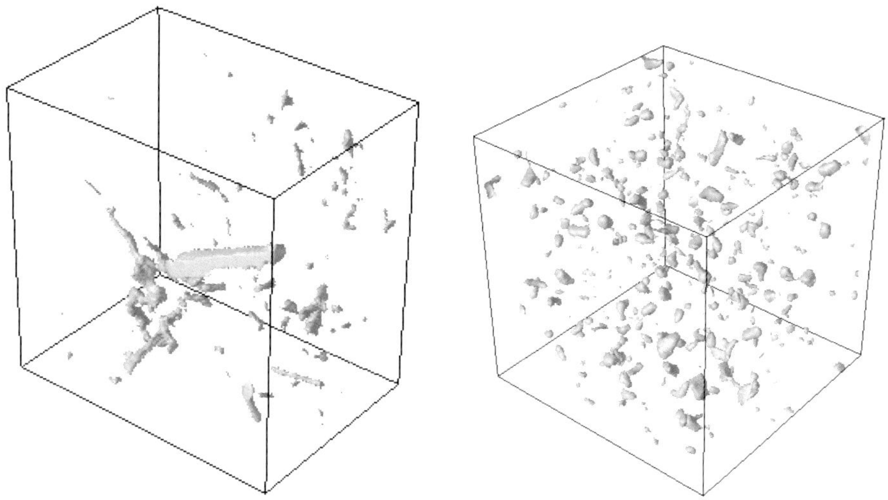

Fig. 10. 3D porosity image of a Loess soil (left, natural height 110 mm, resolution 0.44 mm) and Gaussian random field (right) with equivalent grey level distribution and correlation length. Grey levels below 40 indicating high porosities are marked by isosurfaces.

high values of t because the number of completely enclosed cavities are counted with a positive sign according to (1).

5 Numerical Simulations

To demonstrate the significance of the topological characteristics water flow was simulated in 2-dimensional random permeability fields. The aim was to generate different structures having the same auto-covariance function but different topology.

We start with a Gaussian random field f_0 with correlation length λ_0 and variance σ_0. As already discussed in the previous section, the extreme values of f_0 are located on islands which are not connected. In contrast, the values around the expectation of f_0 are located on connected bands. Thus, the values of f_0 were resorted in a way that the extreme values are located on these connected bands. This was achieved for each location x by

$$f_1(x) = |2f_0(x) - f_{0,\mathrm{max}}| \tag{12}$$

where $f_{0,\mathrm{max}}$ is the maximum value of f_0. Obviously, the histogram of the resulting random field f_1 is affected by this transformation except the values of f_0 are equally distributed in $[f_{0,\mathrm{mim}}, f_{0,\mathrm{max}}]$. For this reason the distribution function of f_0 was tranformed to an equal distribution using (11) prior to the application of (12).

Moreover, also the spatial correlation is affected by (12). In the original random field f_0 the spatial distance between minimum and maximum values is considerably larger as compared to f_1. Consequently, to arrive at two random fields with comparable auto-covariance but different topology, we generated another Gaussian random field f_2 having the same covariance as f_1.

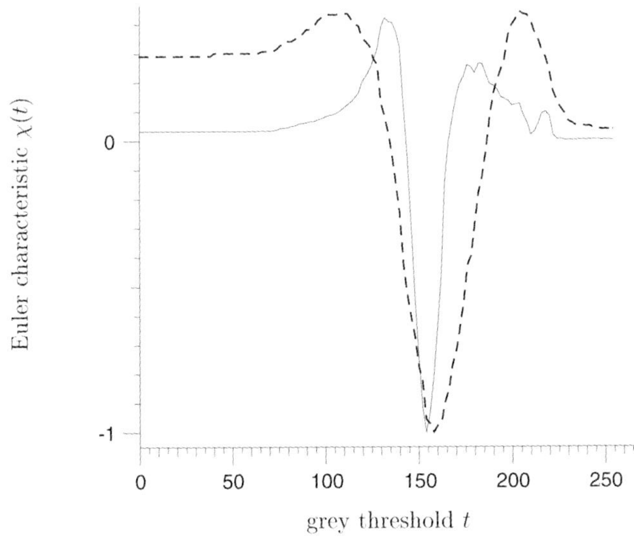

Fig. 11. 3D connectivity function $\chi(t)$ for the 3D X-ray image of soil (line) and the Gaussian random field (dashed line). The Euler characteristics are plotted relative to the maximum value to allow a better comparison of the two curves. The images are shown in Fig. 10

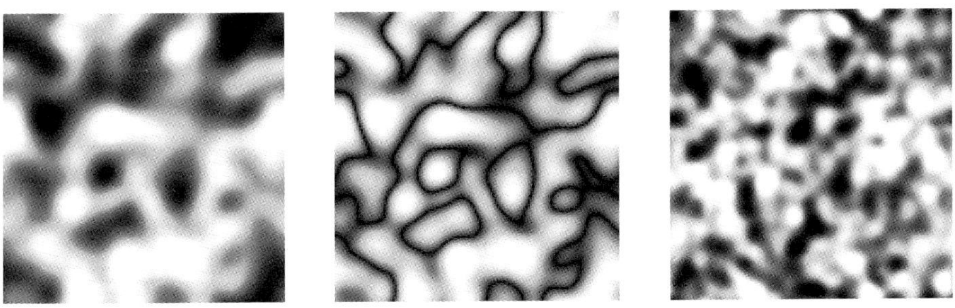

Fig. 12. Gaussian random field f_0 (left) after redistribution according to (12), f_1 (middle) and the Gaussian random field f_2 with the same auto-covariance function as f_1 but different topology.

In Fig. 12 the original random field f_0, the result after redistribution, f_1, and the random field f_2 which was adapted to the auto-covariance of f_1 are shown. The corresponding covariance and connectivity functions are shown in Fig. 13. The topological difference between f_1 and f_2 is clearly reflected by the connectivity function. Increasing the threshold t for the random field f_1 leads to an early drop of the Euler characteristic to slightly negative values indicating the number of loops within the connected dark band through the structure. In contrast, $\chi(t)$ for the Gaussian random fields f_0 and f_2 reflects the isolated clusters of the extreme values.

The next step was to investigate the effect of the topological characteristics on effective properties of the heterogeneous fields. Therefore, the grey values of the random fields were considered to represent the logarithms of the permeability, $\log(K)$. Based

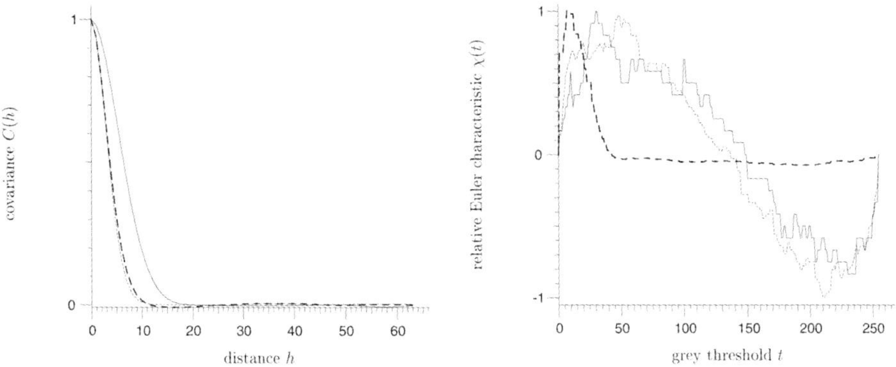

Fig. 13. Covariance and connectivity function for the random fields f_0 (line), f_1 (dashed line), and f_2 (dotted line). The connectivity functions are normalized by the maximum value of $\chi(t)$ for better comparison.

on the resulting structure of permeabilities the velocity field was calculated according to Darcy's law

$$\boldsymbol{u} = K \cdot \nabla p \tag{13}$$

together with the mass balance

$$\nabla \cdot \boldsymbol{u} = 0 \tag{14}$$

which was numerically solved using a finite element scheme. Thereby a pressure gradient from the left to the right edge was simulated with no-flux boundaries at top and bottom.

The resulting velocity fields are presented in Fig. 14. As expected, the velocity field of f_1 is dominated by the continuous band of high permeability. Also for f_2 water flow is concentrated to one single path along which the distances between regions of high permeability are minimal. The effective permeability K_{eff} of f_1 was by a factor of 71.5 higher compared to K_{eff} of f_2. This clearly shows that topology may be critical for effective properties of heterogeneous fields.

6 Conclusions

Topological characteristics of porous media proved to be significant for hydraulic properties and transport both at the pore scale and at a larger scale described by a continuous field of effective material properties. At the pore scale, the proposed connectivity function based on the Euler characteristic together with the pore size distribution allowed a reasonable good prediction of the effective hydraulic properties of a Loess soil.

At the continuum scale, numerical simulations of effective permeabilities for heterogeneous permeability fields evidence the effect of topology also at this scale. Hence, to predict effective properties of porous media in terms of flow and transport, the autocovariance function which is typically used to characterize the spatial heterogeneity is not sufficient. Modeling and quantifying spatial heterogeneity should be supplemented by a topological description. The Euler characteristic as a function of a threshold value

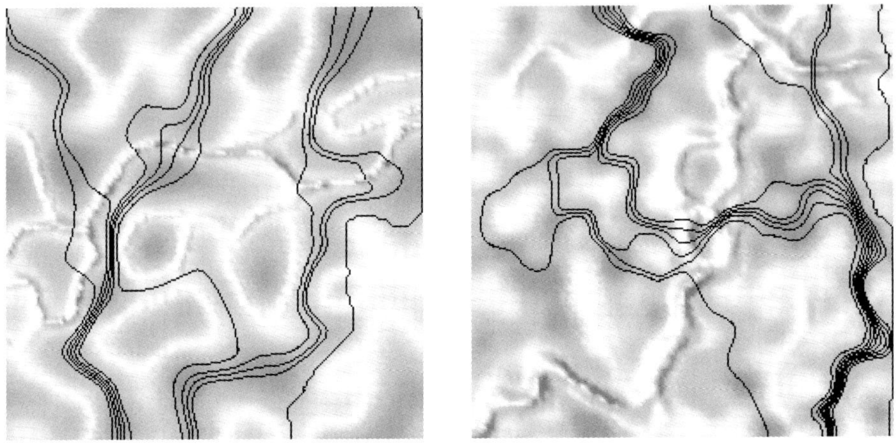

Fig. 14. Field of permeabilities (grey level) and velocity (elevation) together with iso pressure lines for stationary convective flow through the random fields f_1 (left) and f_2 (right). The effective permeability K_{eff} of f_1 is by a factor of 71.5 larger compared to K_{eff} of f_2.

provides a useful connectivity function which quantifies the topological characteristics of continuous fields.

It is a challenge for future research to generate random fields of predefined autocovariance and predefined topology and to systematically investigate the relation of topology and effective properties.

References

1. P. M. Adler, C. G. Jacquin, J. A. Quiblier, Flow in simulated porous media, Int. J. Multiphase Flow 16 (1990) 691–712
2. C. Arns, M. Knackstedt, K. Mecke, Characterising the morphology of disordered materials, Lecture Notes in Physics (2002), this volume
3. C. Beisbart, R. Dahlke, K. Mecke, H. Wagner, Vector- and tensor-valued descriptors of spatial patterns, Lecture Notes in Physics (2002), this volume
4. B. Berkowitz, I. Balberg, Percolation theory and its application to groundwater hydrology, Water Resources Res. 29 (1993) 775–794
5. M. A. Celia, P. C. Reeves, L. A. Ferrand, Recent advances in pore scale models, Rev.Geophys. 33 (1995) 1049–1057
6. R. Chandler, J. Koplik, K. Lerman, J. F. Willemsen, Capillary displacement and percolation in porous media, J. Fluid Mech. 119 (1982) 249–267
7. I. Chatzis, F. A. L. Dullien, Modeling pore structure by 2-D and 3-D networks with application to sandstone, J. Can. Pet. Technol. 16 (1977) 97–108
8. R. T. DeHoff, Use of the disector to estimate the Euler characteristic of three dimensional microstructures, Acta Stereol. 6 (1987) 133–140
9. I. Fatt, The network model of porous media. I. Capillary pressure characteristics, Pet. Trans. AIME 207 (1956) 144–159
10. I. Fatt, The network model of porous media. II. Dynamic properties of a single size tube network, Pet. Trans. AIME 207 (1956) 160–163

11. I. Fatt, The network model of porous media. III. Dynamic properties of networks with tube radius distribution, Pet. Trans. AIME 207 (1956) 164–181
12. L. A. Ferrand, M. A. Celia, The effect of heterogeneity on the drainage capillary pressure-saturation relation, Water Resour. Res. 28 (3) (1992) 859–870
13. S. Friedman, N. Seaton, On the transport properties of anisotropic networks of capillaries, Water Resour. Res. 32 (1996) 339–347
14. H. Hadwiger, Vorlesung über Inhalt, Oberfläche und Isoperimetrie, Springer-Verlag, Berlin, 1957
15. R. Hilfer, Geometric and dielectric characterization of porous media, Phys. Rev. B 44 (1991) 60
16. R. Hilfer, Local-porosity theory for flow in porous media, Physical Review B 45 (1992) 7115–7121
17. R. Hilfer, Transport and relaxation phenomena in porous media, Adv. Chem. Phys. XCII (1996) 299–424
18. R. Hilfer, T. Rage, B. Virgin, Local percolation probabilities for a natural sandstone, Physica A 241 (1997) 105–110
19. J. W. Hopmans, M. Cislerova, T. Vogel, X-ray tomography of soil properties, In: Tomography of Soil-Water-Root Processes. Eds.: Ande S.H. and J.W. Hopmans. ASA, SSSA Madison, Wisconsin,USA (1994) 17–28
20. G. R. Jerauld, S. J. Salter, The effect of pore-structure on hysteresis in relative permeability and capillary pressure: pore-level modeling, Transort in Porous Media 5 (2) (1990) 103–151
21. M. A. Knackstedt, A. P. Sheppard, M. Sahimi, Pore network modelling of two-phase flow in porous rock: the effect of correlated heterogeneity, Adv. Water Res. 24 (2001) 257–277
22. R. Magerle, Nanotomography, Lecture Notes in Physics (2002), this volume
23. J. Mecke, D. Stoyan, The specific connectivity number of random networks, Adv. Appl. Prob. (SGSA) 33 (2001) 576–583
24. K. Mecke, H. Wagner, Euler characteristic and related measures for random geometric sets, J. Stat. Phys. 64 (1991) 843
25. K. Mecke, Additivity, convexity, and beyond: applications of Minkowski Functionals in statistical physics, Lecture Notes in Physics 554 (2000) 111–184
26. K. Mecke, D. Stoyan, Statistical physics and spatial statistics - the art of analyzing and modeling spatial structures and pattern formation, Lecture Notes in Physics 554.
27. J. Ohser, W. Nagel, K. Schladitz, The Euler number of discretized sets, Lecture Notes in Physics (2002), this volume
28. J. Ohser, W. Nagel, The estimation of the Euler-poincaré characteristic from observations on parallel sections, J. Microsc. 184 (1996) 117–126
29. W. H. Press, S. A. Teukolsky, W. T. Vetterling, B. P. Flannery, Numerical Recipes in C. The Art of Scientific Computing, second edition Edition, Cambridge University Press, Cambridge, 1992
30. J. Quiblier, A new three dimensional modeling technique for studying porous media, J. Colloid Interface Sci. 98 (1984) 84–102
31. H. Rajaram, L. A. Ferrand, M. A. Celia, Prediction of relative permeabilities for unconsolidated soils using pore-scale network models, Water Resour. Res. 33 (1997) 43–52
32. A. P. Roberts, S. Torquato, Chord-distribution functions of three-dimensional random media: approximate first-passage times of gaussian processes, Phys. Rev. E 59 (1999) 4953–4963
33. M. J. L. Robin, A. L. Gutjahr, E. A. Sudicky, J. L. Wilson, Cross-correlated random field generation with the direct fourier transform method, Water Resour. Res. 29 (1993) 2385–2397
34. V. Robins, Computational topology for point data: Betti numbers and α-shapes, Lecture Notes in Physics (2002), this volume

35. H. Rogasik, J. W. Crawford, O. Wendroth, I. M. Young, M. Joschko, K. Ritz, Discrimination of soil phases by dual energy X-ray tomography, Soil Sci. Soc. Am. J. 63 (1999) 741–751
36. K. Roth, Steady state flow in an unsaturated, two-dimensional, macroscopically homogeneous, miller-similar medium, Water Resources Res. 31 (1995) 2127–2140
37. J. Serra, Image Analysis and Mathematical Morphology, Academic Press, London, 1982
38. D. C. Sterio, The unbiased estimation of number and sizes of arbitrary particles using the disector, J. of Microsc. 134 (1984) 127–136
39. V. C. Tidwell, L. C. Meigs, T. Christian-Frear, C. M. Boney, Effects of spatially heterogeneous porosity on matrix diffusion as investigated by X-ray absorption imaging, J. Contam. Hydrol. 42 (2000) 285–302
40. H. J. Vogel, Digital unbiased estimation of the Euler-Poincaré characteristic in different dimensions, Acta Stereol. 16/2 (1997) 97–104
41. H. J. Vogel, Morphological determination of pore connectivity as a function of pore size using serial sections, Europ. J. Soil Sci. 48 (1997) 365–377
42. H. J. Vogel, K. Roth, A new approach for determining effective soil hydraulic functions, Europ. J. Soil Sci. 49 (1998) 547–556
43. H. J. Vogel, A numerical experiment on pore size, pore connectivity, water retention, permeability, and solute transport using network models, Europ. J. Soil Sci. 51 (2000) 99–105
44. H.-J. Vogel, K. Roth, Quantitative morphology and network representation of soil pore structure, Adv. Water Res. 24 (2001) 233–242
45. A. W. Western, G. Blöschl, R. B. Grayson, Toward capturing hydrologcally significant connectivity in spatial patterns, Water Resour. Res. 37 (2001) 83–97
46. J. Widjajakusuma, C. Manwart, B. Biswal, R. Hilfer, Exact and approximate calculations for the conductivity of sandstones, Physica A 270 (1999) 325–331
47. W. R. Wise, A new insight on pore structure and permeability, Water Resour. Res. 28 (1992) 189–198

Nanotomography: Real-Space Volume Imaging with Scanning Probe Microscopy

Robert Magerle

Physikalische Chemie II, Universität Bayreuth, 95440 Bayreuth, Germany

Abstract. Nanotomography is a general procedure for high-resolution volume imaging based on scanning probe microscopy (SPM). The approach is similar to sputter depth profiling. The specimen under study is eroded step by step and the remaining material is imaged with scanning probe microscopy at each freshly exposed surface. From the resulting series of images the specimen's three-dimensional microstructure can be reconstructed with methods adopted from computed tomography. This approach is expected to be a simple and versatile means for real-space volume imaging of various materials and physical properties with micron, nanometer, and even atomic resolution.

1 Introduction

The relationship between microstructure and macroscopic properties is a central issue in materials science. In this context, the role of computer simulations of processing, structure formation, and physical properties of materials is constantly increasing. Most modern materials have complex three-dimensional structures on the micron and nanometer scale. Unfortunately, most microscopy techniques available today are limited to two-dimensional (2D) imaging and the choices for obtaining high-resolution three-dimensional (3D) volume images are rather limited, at least in materials science. This considerably hinders a detailed study of the structure-property relationship and renders a direct comparison between simulation results, the materials microstructure, and its physical properties difficult, if not impossible. In particular, quantitative experimental data is needed as input for reliable computer simulations and for benchmarking of models.

2 Status of the Field

Computed tomography (CT) uses X-rays and the spatial resolution with table-top setups is limited to about 20 μm [1]. With synchrotrons this limit has been pushed to 0.15 μm in selected cases [2] and an example of state of the art CT microscopy is given by C. Arns *et al.* in this volume. Magnetic resonance imaging (MRI) is also limited to about 10–20 μm resolution [3]. In most materials typical heterogeneties, such as the grain structure of metallic and ceramic materials, have features much smaller then 10 μm, which is the smallest voxel diameter in common CT and MRI.

With laser scanning confocal microscopy a much better resolution is achievable. Using 488-nm light the smallest voxel size is about $150 \times 150 \times 450$ nm^3. (For two recent examples from polymer science see [4, 5].) This technique, however, is limited to optically transparent and rather homogenous materials that do not scatter too much. These limitations rule out materials such as filled polymers as well as most metallic and ceramic materials.

In many fields of materials science, scanning electron microscopy (SEM) and transmission electron microscopy (TEM) are the most common microscopy techniques next to optical microscopy. Common SEM and TEM are both limited to 2D imaging of surfaces and (ultra) thin sections. Volume imaging with 20–50 nm resolution can be achieved with TEM by reconstruction from serial sections [6, 7] This technique requires the specimen to be machinable into a series of thin, usually 20–50 nm thick, slices. This rules out most materials like metals, ceramics, or semiconductors and has not been proven to be successful in polymer science. Another option is electron tomography [8, 9] which is commonly used for 3D imaging of viruses and proteins. It requires usually a large number of identical objects, such as viruses, to obtain projection images from many different viewpoints which are needed for the stereoscopic reconstruction procedures. In case of a single specimen a tilt series, i.e., images from many different viewpoints, are needed which are often difficult to obtain from a single specimen because of electron beam damage.

At the utmost high-resolution edge of volume imaging techniques is the '3D atom probe' [10, 11]. With this technique the 3D structure of metallic alloys can be imaged with atomic and nanometer resolution. The technique is, however, limited to this small length scale and to metallic materials which are machinable into sharp field emission tips. (Although, other materials have also been studied in few cases.) As a result, no widely applicable volume imaging technique covering the micron and nanometer range exist for materials science today.

3 Principle of Nanotomography

Since 1982 the scanning tunneling microscope [12] and its offspring forming the family of scanning probe microscopy (SPM) made their mark, next to optical microscopy and electron microscopy, in many different fields of physics, chemistry, and material science as well as biology. The high spatial resolution achievable, the technique's simplicity, and the ability to map physical properties not detectable by other means on such fine scales opened up new avenues for research and enabled numerous discoveries.

The basic idea of SPM is to bring a fine tip (the probe) close to the surface and to keep the distance between tip and surface constant while scanning the tip along the surface line by line (Fig. 1). In this way the shape $z(x, y)$ of the surface S is measured. $z(x, y)$ is also called height image. At the same time one or more physical properties $P(x, y)$ are measured at the tip position (x, y) with high spatial resolution. Both data sets can be combined to a map $P(x, y, z(x, y))$ which is defined on the curved surface $S = \{(x, y, z(x, y))\}$. The coordinates (x, y) form a regular grid with typically 256×256 or 512×512 grid points. Depending on the type of microscope and its spatial resolution, the scan range can be as small as 10×10 nm^2 or can range up to (typically) 100×100 µm^2.

Various means have been invented to keep the tip-sample distance constant and to measure physical properties at the tip position. The probably most common techniques

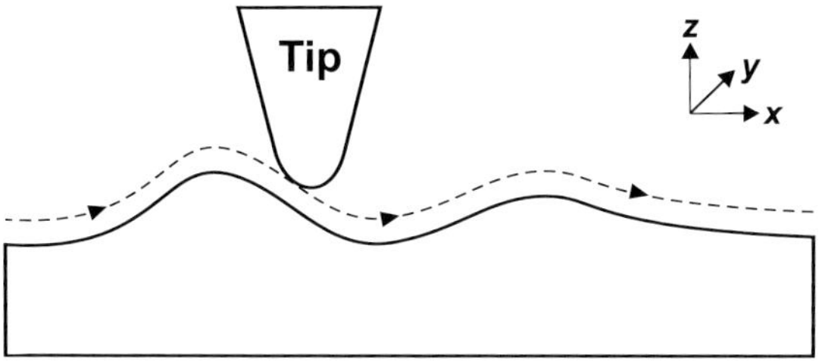

Fig. 1. Basic principle of scanning probe microscopy. A fine tip is scanned along a surface line-by-line while keeping the tip-surface distance constant

today are those introduced first by Gerd Binnig, Heinrich Rohrer, and their coworkers in 1982 and 1986. With scanning tunneling microscopy [12] (STM) a tunneling current between the tip and the conductive specimen is used to detect the vicinity of the surface. With scanning force microscopy (SFM), which is also called atomic force microscopy [13] (AFM), the force between the tip and the surface is measured. Since then, numerous variants and combinations of these two have been developed [14, 15, 16].

Traditionally, however, SPM has been limited to surface studies and shed little light on a material's inner structure. A common approach to overcome the limitation of 2D imaging techniques is reconstruction from serial sections, introduced first by Gustav Born in 1876 [17, 18]. The specimen under study is either physically sectioned into a series of thin slices or images of thin layers are obtained non-destructively by suitable measuring techniques like in computed tomography, magnetic resonance imaging, and laser scanning confocal microscopy. Then the series of 2D images is combined by a computer to a 3D picture of the studied object.

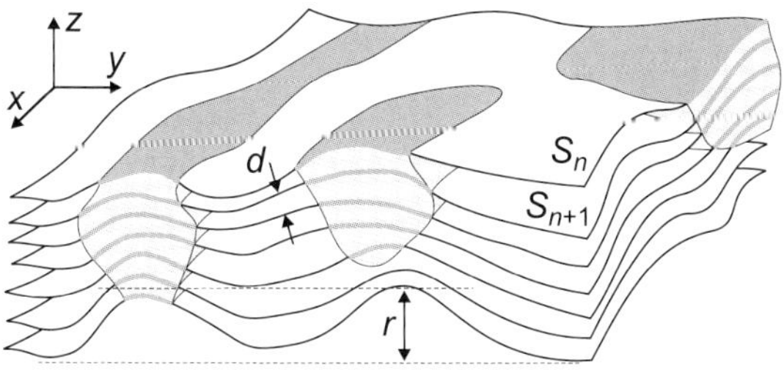

Fig. 2. Schematic illustration of the principle of volume reconstruction from a series of scanning probe microscopy images. (Reprinted from: R. Magerle, 'Nanotomography', *Physical Review Letters* **85**, 2749 (2000); ©2000 by the American Physical Society)

Nanotomography [19, 20] employs the same basic idea to expand SPM to volume imaging. However, rather than sectioning the specimen into slices, thin layers are stripped off step-by-step and the remaining material is imaged with SPM at each stage (Fig. 2). This approach is similar to sputter depth profiling, for instance with secondary ion mass spectroscopy [21] (SIMS), which is a common technique in the thin-film and semiconductor area. The lateral resolution of imaging SIMS is limited by the sputter beam diameter which can be about 100 nm in best cases. Also depth resolution of sputter depth profiling can be severely deteriorated by surface roughness, which causes signals from different depths to be detected simultaneously [22]. Therefore, sputter depth profiling has been traditionally limited to rather homogenous materials with homogenous erosion rates, such as model semiconductor specimens.

A key feature of Nanotomography is that both the surface properties under study as well as the actual (and changing) shape of the surface is measured with SPM after each erosion step. As a result the 2D maps used with Nanotomography are curved in general (Fig. 2), which solves the problem of (sputtering induced) surface roughness and renders the depth resolution of the method to be rather independent of surface roughness. This effect is one of the major advantages of Nanotomography and means that neither homogenous etching nor flat sample surfaces are required for the method, ensuring its application to a large class of heterogeneous and thereby interesting materials, such as composites, alloys, and sub-micron electronic devices.

4 Proof of Concept: Block Copolymer Microdomain Structure

As first example and proof of concept the core of a dislocation within the three-dimensional spatial microdomain structure of the block copolymer polystyrene-*block*-polybutadiene-*block*-polystyrene (SBS) has been imaged with about 10 nm resolution [20]. Another study has revealed an unusual arrangement of microdomains in thin films of block copolymers [23].

SBS is a prototype of a modern synthetic rubber and is widely used as thermoplastic elastomer. The material consist of a mixture of polystyrene (PS) and polybutadiene (PB) molecules covalently linked together to larger block copolymer molecules. At room temperature the PS and PB blocks are not miscible and phase separate, like oil and water, into microdomains which form a regular superstructure with length scales given by molecular size. In the particular material studied the PS forms hard (glassy) cylinders about 24 nm in diameter embedded in a soft (liquid) PB matrix forming a hexagonally ordered array of PS cylinders (Fig. 3). The hard PS domains act as physical crosslinks of SBS molecules which causes a rubber-like behavior of the material.

The lateral distribution of PS and PB domains can be mapped with TappingMode scanning force microscopy [24, 25, 26] (TM-SFM) which has been established in recent years as a standard tool to investigate surfaces of soft materials, such as polymers. Its excellent lateral resolution together with its potential to distinguish different materials without further staining has made TM-SFM an attractive alternative to established techniques such as transmission electron microscopy. In TM-SFM a vibrating tip is used and the vicinity of the surface is detected via the damping of the tip oscillation. The tip is scanned along the surface while keeping the amplitude of the tip oscillation constant. Simultaneously with the height image $z(x, y)$ the phase φ between the exciting and the resulting tip-oscillation is measured. The resulting phase image $\varphi(x, y)$ reflects lateral differences in the mechanics of the tip-sample interaction which differs considerable between the hard PS and soft PB domains [26].

To simplify the experiment a 140 nm thick SBS film was prepared on a polished silicon wafer about 1 cm^2 in size. For equilibration and long-range ordering of the

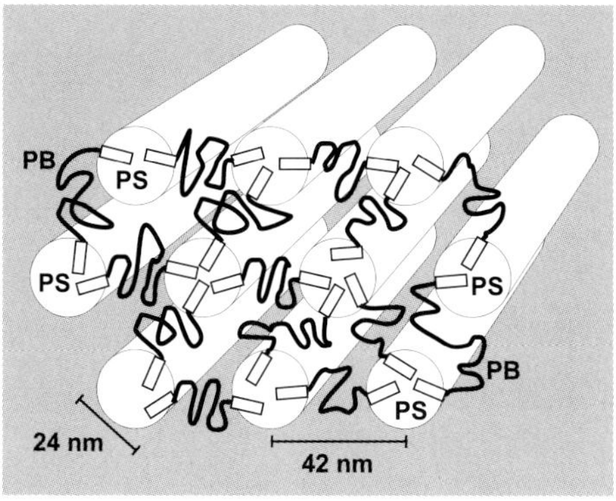

Fig. 3. Schematic illustration of the microdomain structure of polystyrene-*block*-polybutadiene-*block*-polystyrene (SBS)

SBS microdomain structure the thin film specimen was exposed overnight to toluene vapor. After this procedure PS cylinders tend to orient parallel to the substrate [27] and the film forms terraces with thicknesses smaller and larger than the original thickness, corresponding to a integer multiple of layers of PS cylinders (Fig. 4a). After annealing in toluene vapor the TM-SFM phase image shows a stripe-like pattern, indicating that the PS cylinders are oriented parallel to the film surface (Fig. 4b).

For stepwise erosion ex situ plasma etching with air has been used which removed $\langle d \rangle = 7.5 \pm 0.2$ nm (on average) thick layer during each (20 s lasting) etching step. After each etching step the same spot of the specimen was imaged with TM-SFM. A spot was chosen with characteristic landmarks within and nearby the imaged area which help in finding the same spot again after ex situ plasma etching. The resulting series of height and phase images is shown in Fig. 5. The phase images $\varphi_n(x, y)$ [Figs. 5(h)–(n)] can be interpreted as maps of the PS-to-PB ratio at the exposed sample surface after the n-th etching step. Bright (dark) areas indicate a high (low) phase which corresponds to a high PS (PB) volume fraction. Together with the corresponding height images $z_n(x, y)$ [Figs. 5(a)–(g)] describing the curved surfaces, they can be combined to phase maps $\varphi_n(x, y, z_n(x, y) - n\langle d \rangle)$ which are defined on curved surfaces S_n, as indicated in Fig. 2.

Since the TM-SFM images can not be taken exactly at the same spot, the images of the series have to be aligned (registered) prior to volume reconstruction. Image registration is widely used in CT, MRI, and other areas of science [28]. In this work the correlation coefficient between two subareas of neighboring height images was maximized. After image registration a volume map of the phase φ can be reconstructed since the phase is sampled at the points $(x, y, z_n(x, y) - n\langle d \rangle)$ which span a regular mesh within the original specimen.

A visualization of the 3D phase distribution is shown in Fig. 6 where the isosurface enclosing the volume with normalized TappingMode phase > 0.11 is displayed. The enclosed volume can be interpreted as PS cylinders within a $200 \times 160 \times 45$ nm^3 large portion of the SBS film. With a computer and suitable data visualization software (AVS/Express) the 3D data set and thereby the materials inner structure can be explored

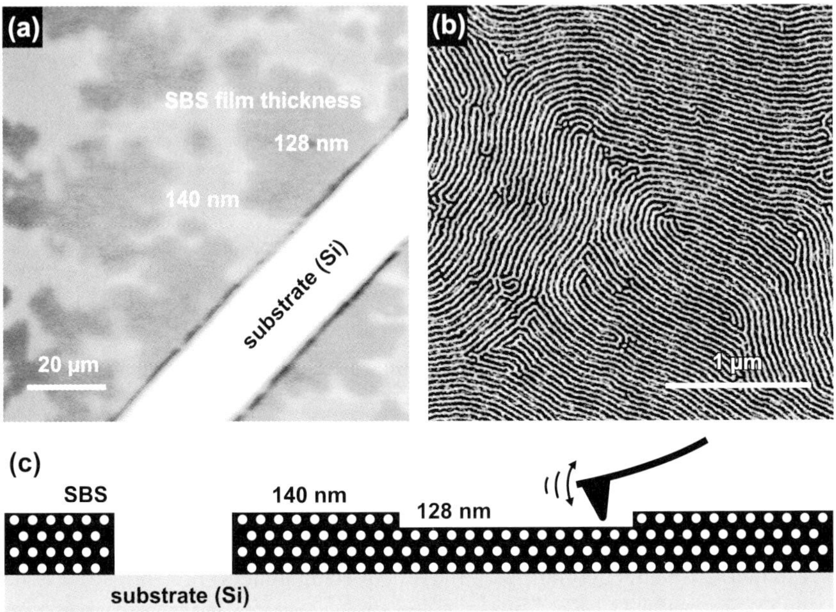

Fig. 4. (**a**) Bright field optical micrograph of the SBS thin film specimen. After annealing in toluene vapor terraces of homogenous thickness are formed, which correspond to a integer multiple of layers of PS cylinders. The terraces appear as bright and dark areas due to a interference effect of illuminating light. (**b**) Schematic cross section of the SBS film. (**c**) TappingMode SFM phase image of the surface of the SBS film prior to plasma etching and Nanotomography imaging

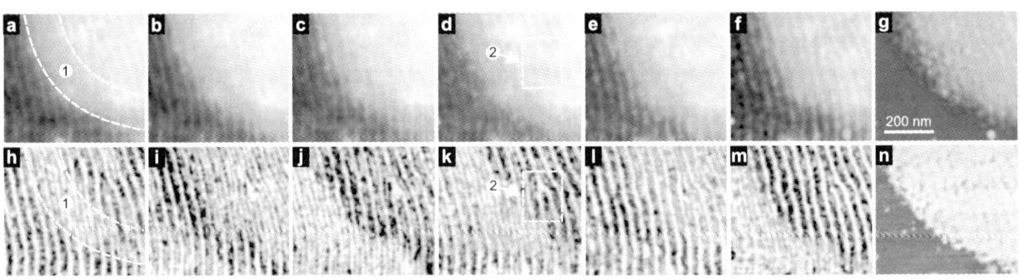

Fig. 5. Series of TappingMode SFM images from a thin SBS film. Height [(**a**)-(**g**)] and the corresponding phase images [(**h**)-(**n**)] are shown. Between each pair of images a 7.5 ± 0.2 nm (on average) thick layer has been removed from the sample by plasma etching. Region (2) is displayed as 3D image in Fig. 6 (Reprinted from: R. Magerle, 'Nanotomography', *Physical Review Letters* **85**, 2749 (2000); ©2000 by the American Physical Society)

in detail by selecting different threshold values for isosurface calculation, different color codes and viewpoints, etc. In this way it was discovered that in the dislocation core the branching PS cylinder 1 is connected with all other four PS cylinders 2 and 3 as well as 4 and 5, the latter being not expected form the original set of 2D images.

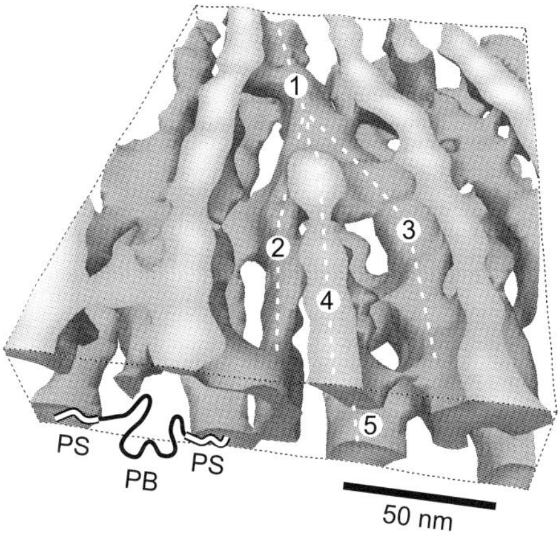

Fig. 6. 3D image of the isosurface enclosing the volume with normalized TappingMode phase > 0.11 reconstructed from a series of TappingMode phase and height images. The enclosed volume can be interpreted as PS cylinders within a $200 \times 160 \times 45$ nm^3 large portion of the SBS film. For size comparison and illustration of the molecular structure of the material, a SBS molecule bridging two PS domains is sketched (Reprinted from: R. Magerle, 'Nanotomography', *Physical Review Letters* **85**, 2749 (2000); ©2000 by the American Physical Society)

These results demonstrate the feasibility and possibilities of the Nanotomography approach. In the present case the resolution achieved is about 10 nm in all three dimensions and is only limited by the tip radius which was 10 nm in this particular case. With sharper tips (2 nm radius are commercially available today) a higher resolution should be achievable.

5 Potential Applications

Nanotomography offers novel possibilities for high-resolution volume imaging of many types of materials and physical properties that have not been accessible to volume imaging before. In the following some of these potential applications will be briefly sketched. Additional examples and details are given in Refs. [19, 20].

5.1 Polymer Films and Coatings

A common type of coating is a polymeric film on a solid substrate. Often the polymer is much softer than the substrate which makes it quite difficult to prepare thin cross sections for TEM imaging and preparing serial sections is practically impossible. Also the substrate is often too thick and not transparent for electrons so that electron tomography is not applicable.

Nanotomography might be a solution for this microscopy problem. In the example given above the specimen was an about 140 nm thin polymer film on a solid substrate, which can be considered as a prototype of a coating. In Fig. 6 a small portion of its entire volume is displayed. The total data set covers a $1.5 \times 1.5\,\mu m^2$ large area of the film which can be visualized and 'explored' with a suitable computer. In another example [23] we have imaged and displayed the entire volume of an about 50 nm thick block copolymer film—from the original film surface down to the silicon substrate. Our results revealed the influence of the free surface on the microdomain morphology of block copolymers, which we found to deviate at the surface from the bulk morphology—a behavior similar to the phenomenon of surface reconstruction of classical crystals [29].

Commercial coatings are complex formulations with many different components. For thin films of homopolymer blends as well as block copolymers it is well known (from experiments and theory) that the presence of surfaces and interfaces can alter the local composition and structure of the film. A common behavior is the enrichment of the component with the lowest interfacial energy at the film-substrate interface or the film surface. (For a recent review see [30].) Nanotomography offers the possibility to investigate in detail the influence of surfaces and interfaces on the 3D spatial structure and chemical composition of modern coatings which is difficult, if not impossible, to access with other techniques.

Coatings and plastic films are two prominent applications of polymer based materials. Both have in common that the typical thickness of the final product is in the range of $\sim$ 10 μm which is within the scan range of scanning probe microscopy. As a result, the 3D microstructure of the complete, macroscopic, product as it is used in real applications could be imaged with Nanotomography. The data on the 3D structure can be compared with modeling results of the structure formation process, such as extrusion, foaming, crystallization, and texture formation. They might also be used as input for modeling the mechanical behavior and transport properties of the material as well as the final product. Goals for future research might be a better understanding of surface, thin film, and interface structure, the structure formation processes, and the resulting macroscopic film properties.

5.2 Metals

Metallic materials are probably the largest class of materials studied and numerous types of microscopy techniques exist for imaging their microstructure. Many specimen preparation techniques employed in metallography use polishing or etching as last processing step. Implementing the Nanotomography approach appears therefore as straightforward. In particular in-situ SPM imaging during electrochemical etching [32] would have the advantage that the specimen does not need to be removed from the SPM for etching. High-speed imaging in liquids has been demonstrated with scanning tunneling microscopy [31] as well as TappingMode scanning force microscopy [33, 34, 35]. This is would speed-up the volume imaging process considerably resulting in image acquisition rates of seconds (and less) per layer. Local variations of compositions could be mapped by various scanning probe techniques, such as scanning tunneling microscopy, force modulation microscopy, or tunneling atomic force microscopy, just to mention a few.

A variant of the Nanotomography approach might be the combination of optical microscopy, interference microscopy for topography measurements, and in-situ electrochemical etching. It might expand the well established field of optical metallography to high-resolution volume imaging of non-transparent metallic materials.

5.3 Ceramics

Due to their insulating nature ceramics are often difficult to study with electron microscopy and other techniques employing charged particles, in particular when a high spatial resolution is required. Here, Nanotomography might provide an alternative approach. Semiconductor technology offers a vast number of recipes for etching various types of chemically rather inert compounds such as oxides and nitrides [36]. For erosion a plasma etching variant might be used. Specimen preparation techniques for electron microscopy also often involve (selective) etching for providing a chemical contrast in electron microscopy. A further possibility is locally well defined etching with a focused ion beam as it is used for the repair of modern photolithography masks.

5.4 Composites and Filled Polymers

Composites, such as fibreglass, and polymeric materials with inorganic filler particles are often a challenge for microscopy. In many cases it is extremely difficult or even impossible to prepare thin sections for TEM imaging or smooth surfaces. Nanotomography offers the advantage that neither flat surfaces nor homogenous erosion rates are needed for high-resolution imaging. This opens up new possibilities. For instance two different kinds of etchants (or etching techniques) might be used successively to erode different components of a composite, such as an inorganic filler embedded in a polymeric matrix. In this case the changing topography and the locally varying erosion rate could be used to identify and map different components in 3D.

5.5 Semiconductor Devices

Modern submicron devices are build-up from various materials, such as semiconductors, metals, isolators, densely packed with intriguing complexity. Despite all advancements of semiconductor technology no 3D volume imaging technique exists today for imaging these tiny objects with nanometer resolution for failure analysis and quality control. In future this might be achieved with Nanotomography which is not hindered by uneven surfaces and different (and varying) erosion rates. Moreover, numerous etching and polishing procedures are known in semiconductor technology [36]. In particular, reactive ion beam etching and chemo-mechanical polishing appear as promising candidates. Also the development into a fully automated system appears to be straightforward since both fully automated SPM as well as etching are state-of-the-art semiconductor technology today.

With scanning capacitance microscopy [37] (SCaM) and scanning spreading resistance microscopy [38] (SSRM) high-resolution 2D maps of doping densities can be measured on cross sections of submicron devices. For these measurements the last specimen preparation step is chemo-mechanical polishing. Therefore combining SCaM or SSRM with the Nanotomography approach should be straightforward. This could enable '3D dopant profiling' with nanometer resolution—a task for which no solution is known today although its urgent need is defined in the 'Roadmap for Semiconductors' since 1997 [39, 40].

5.6 Possible Impact to Other Fields

It may appear trivial, solid objects are made from materials and solid objects are studied in many fields of science. Thus, Nanotomography, as a new technique for imaging materials, might enable discoveries in other areas of science too. The following examples illustrate this.

- **Mineralogy and geology.** Information about geological processes could be gained via the microstructure of minerals.
- **Archeology.** The materials that have been used in previous times are of such tremendous importance that entire periods have been named after them, such as the Stone Age, the Bronze Age, and the Iron Age. The microstructure of artifacts gives insight into ancient material processing techniques. It might also reveal the artifacts age.
- **Paleontology.** 3D volume images of microfossils and 'nannofossils' might reveal new information about biogenesis.
- **Biology.** Starting from freeze fractured surfaces the 3D ultrastructure of cells could be imaged with advanced scanning probe techniques. For instance, infrared scattering from a specific chemical compound and mapping with 10 nm resolution has been recently demonstrated [41]. This might lead to 3D maps of the distribution of different chemical compounds within a cell.

6 Perspectives: Seeing, Feeling, and Understanding Materials

With the success of SPM in mind, volume imaging with SPM promises new insights into the physics of materials on a nanometer scale. Vision and reception of spatial structures is one of man's most important senses. Therefore, visualization of complex 3D structures, enhanced by virtual reality and rapid prototyping techniques, is an extremely valuable tool for literally seeing and feeling the 3D microstructure of a material. This means that volume imaging and visualization alone are already expected to provide a better understanding of microstructure, structure formation processes, and structure-property relationships.

The second step of understanding is to link observations to physical models. This is illustrated in Fig. 7. The (simplified) production scheme of a material starts with synthesis of raw materials. Then, by processing a mixture of them a material with a certain microstructure and macroscopic physical properties is formed. A main theme of materials science is the design of materials with tailored properties by control of components and processing conditions on one hand side and the understanding of this processing-microstructure-property relationship on the other. In this context, the role of computer simulations is constantly increasing. One type of modeling aims to simulate the processing of a material in order to predict its microstructure. Another type of modeling aims to predict the macroscopic properties of a material from its microstructure. Here often a simplified model structure, for instance a Boolean model, is used as input. With Nanotomography a much closer link could be formed between the materials microstructure and the related modeling steps.

One link is to modeling of structure formation processes. Today, the primary data produced by many models and computer simulations are irregular spatial structures and

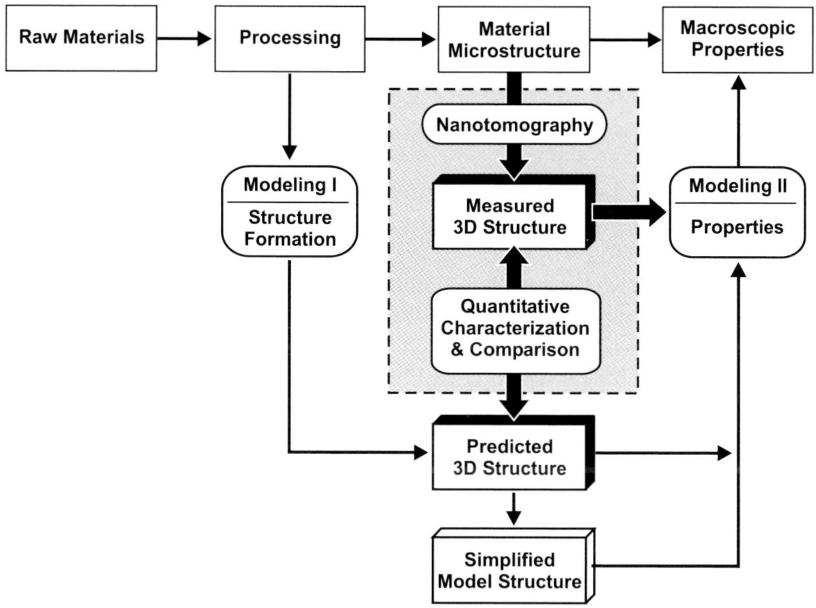

Fig. 7. Illustration of the possibilities offered by Nanotomography in the context of materials science and modeling of materials

volume data sets. A direct comparison in real space with the materials 3D microstructure would result in a much closer link between between experimental observations and models. Nanotomography could help to provide this often missing direct link. It could be used for a better and more detailed testing of existing (and future) models. This, in turn, is expected to lead to a better understanding of underlying microstructure formation processes and improved models thereof.

Essential for achieving this goal are advanced methods for quantitative characterization of images and spatial structures, see, e.g., [42, 43, 44]. Also this volume gives an impression of the status and developments in this field. Nanotomography is expected to provide volume images of new types of materials and physical properties. Therefore, the methods for image and structure characterization have to be evaluated and (if needed) modified in order to meet the specific requirements of the studied microstructure as well as the employed imaging technique. One subtask is 2D image processing and analysis of SPM images. Another one is image registration of series of 2D images in lateral and vertical direction. Here methods developed for CT and MRI imaging [28] are expected to be applicable with little (if any) modifications. A novel aspect in this context might be the fact that with Nanotomography the 2D 'slices' are curved. Furthermore, the influence of the specific imaging artifacts introduced by the employed etching and SPM technique on the employed methods for 2D and 3D image characterization should be evaluated.

The other link is to models which aim to predict from the microstructure of a material its macroscopic properties, such as transport and mechanical properties. Here, volume images measured with Nanotomography could be used as input and replace modeling results and simplified model structures, which are often used as substitute when data on the real 3D structure are not available. Furthermore, by comparing results obtained with measured and model structures different models and model structures could be tested and benchmarked.

7 Summary

In general, Nanotomography is a novel combination of established techniques and procedures. It offers a large number of new and often unique opportunities for high-resolution real-space volume imaging of heterogeneous materials and physical properties. Grinding, polishing, and etching—which are all basically erosion techniques—are common specimen preparation techniques for most types of microscopy. For imaging the chemical composition of a specimen as well as its topography a number of microscopy techniques exist, ranging from optical and interference microscopy, electron microscopy to the family of scanning probe microscopies. Finally, for image registration (alignment), reconstruction of 3D volume data from a series of 2D maps, display of 3D volume data, as well a characterization of 3D structures numerous procedures exist, in particular in the medical imaging area. The next step to turn this novel approach into a routine tool is to build machines which integrate all three steps: automatic erosion, serial imaging, and 3D reconstruction.

Acknowledgments

I thank G. Krausch, M Konrad, A. Knoll, N. Rehse, and M. Hund for help and discussions. The presented work was partly supported by the Deutsche Forschungsgemeinschaft (SFB 481).

References

1. for instance from SCANCO Medical AG (Switzerland), http://www.scanco.ch.
2. McNulty, I., W.S. Haddad, J.E. Trebes, E.H. Anderson (1995): 'Soft x-ray scanning microtomography with submicrometer resolution'. *Rev. Sci. Instr.* **66**, pp. 1431–1433.
3. Blümler, P., B. Blümich, R.E. Botto, E. Fukushima (eds.) (1998): *Spatially Resolved Magnetic Resonance: Methods, Materials, Medicine, Biology, Rheology, Geology, Ecology, Hardware*. (Wiley)
4. White, W.R., P. Wiltzius (1995): 'Real Space Measurement of Structure in Phase Separating Binary Fluid Mixtures'. *Phys. Rev. Lett.* **75**, pp. 3012–3015.
5. Jinnai, H., T. Koga, Y. Nishikawa, T. Hashimoto, S.T. Hyde (1997): 'Curvature Determination of Spinodal Interface in a Condensed Matter System'. *Phys. Rev. Lett.* **78**, pp. 2248–2251.
6. Sjöstrand, F.S. (1958): 'Ultrastructure of retinal rod synapses of the guinea pig eye as revealed by 3-dimensional reconstructions from serial sections'. *J. Ultrastructure Research* **2**, pp. 122–170.
7. Gaunt, W.A., P.N. Gaunt (1978): *Three Dimensional Reconstruction in Biology*. (Pitman Medical, Kent)

8. Spontak, R.J., M.C. Williams, and D.A. Agard (1988): 'Three-dimensional study of cylindrical morphology in a styrene-butadiene-styrene block copolymer'. *Polymer* **29**, pp. 387–395.
9. Frank, J. (1996): *Three-dimensional electron microscopy of macromolecular assemblies.* (Academic Press, San Diego)
10. Miller, M.K., G.D.W. Smith (1989): *Atom-Probe Microanalysis: Principles and Applications to Materials.* (Materials Research Society, Pittsburgh)
11. Miller, M.K. (2000): *Atom Probe Tomography.* (Plenum, New York)
12. Binnig, G., H. Rohrer, C. Gerber, E. Weibel (1982): 'Surface Studies by Scanning Tunneling Microscopy'. *Phys. Rev. Lett.* **49**, pp. 57–61.
13. Binnig, G., C.F. Quate, C. Gerber (1986): 'Atomic Force Microscope'. *Phys. Rev. Lett.* **56**, pp. 930–933.
14. Wiesendanger, R. (ed.) (1995): *Scanning Probe Microscopy and Spectroscopy: Methods and Applications.* (Cambridge University Press)
15. Wiesendanger, R. (ed.) (1998): *Scanning Probe Microscopy: Analytical Methods.* (Springer)
16. Bonnell, D.A. (ed.) (2000): *Scanning Probe Microscopy and Spectroscopy: Theory, Techniques, and Applications.* (John Wiley & Sons)
17. Born, G. (1876): 'Ueber die Nasenhölen und den Thränennasengang der Amphibien'. *Morphologisches Jahrbuch* **2**, pp. 578–580.
18. Born, G. (1883): 'Plattenmodellirmethode'. *Archiv für mikroskopische Anatomie* **22**, pp. 585–599.
19. Magerle, R. (1999): 'Nanotomography'. Patent publications DE 198 59 877 A 1 and WO 00/39569.
20. Magerle, R. (2000): 'Nanotomography'. *Phys. Rev. Lett.* **85**, pp. 2749–2752.
21. Wittmaack, K. (1991): In: *Sputtering by Particle Bombardment III*, ed. by R. Behrisch and K. Wittmaack (Springer, Berlin), pp. 161–256.
22. Carter, G., B. Navinšek, J.L. Whitton (1983): In: *Sputtering by Particle Bombardment II*, ed. by R. Behrisch (Springer, Berlin), pp. 231–269.
23. Konrad, M., A. Knoll, G. Krausch, R. Magerle (2000): 'Volume Imaging of an Ultrathin SBS Triblock Copolymer Film'. *Macromolecules* **33**, pp. 5518–5523.
24. Zhong, Q., D. Inniss, K. Kjoller, V.B. Elings (1993): 'Fractured polymer/silica fiber surface studied by tapping mode atomic force microscopy'. *Surf. Sci.* **290**, pp. L688–L692.
25. Quist, A.P., J. Ahlbom, C.T. Reimann, B.U.R. Sundqvist (1994): 'Scanning force microscopy studies of surface defects induced by incident energetic macromolecular ion''. *Nucl. Instr. Meth.* **B 88**, pp. 164–169.
26. Magonov, S.N., J. Cleveland, V. Elings, D. Denley, M.-H. Whangbo (1997): 'Tapping-mode atomic force microscopy study of the near-surface composition of a styrene-butadiene-styrene triblock copolymer'. *Surf. Sci.* **389**, pp. 201–211.
27. Kim, G., M. Libera (1998): 'Morphological Development in Solvent-Cast Polystyrene-Polybutadiene-Polystyrene (SBS) Triblock Copolymer Thin Films'. *Macromolecules* **31**, pp. 2569–2577.
28. Brown, L.G. (1992): 'A Survey of image registration techniques'. *ACM Computing Surveys* **24**, pp. 325–376.
29. Rehse, N., M. Konrad, A. Knoll, R. Magerle, G. Krausch (2000): 'Surface Reconstruction of an Ordered Fluid: An Analogy with Crystal Surfaces'. *Phys. Rev. Lett.* **87**, p. 035505.
30. Krausch, G. (1995): 'Surface induced self-assembly in thin polymer films'. *Mat. Sci. Eng. Rep.* **R14**, pp. 1–94.
31. Magnussen, O.M, M.R. Vogt (2000): 'Dynamics of Individual Atomic Kinks during Crystal Dissolution'. *Phys. Rev. Lett.* **85**, pp. 357–360.
32. Itaya, K. (1998): 'In situ scanning tunneling microscopy in electrolyte solutions'. *Progress in Surface Science* **58**, pp. 121–148.

33. Paloczi, G.T., B.L. Smith, P.K. Hansma, D.A. Walters, M.A. Wendmann (1998): 'Rapid imaging of calcite crystal growth using atomic force microscopy with small cantilevers'. *Appl. Phys. Lett.* **73**, pp. 1658–1660;
34. Viani, M.B., T.E. Schäffer, G.T. Paloczi, L.I. Pietrasanta, B.L. Smith, J.B. Thompson, M. Richter, M. Rief, H.E. Gaub, K.W. Plaxco, A.N. Cleland, H.G. Hansma, P.K. Hansma (1999): 'Fast imaging and fast force spectroscopy of single biopolymers with a new atomic force microscope designed for small cantilevers'. *Rev. Sci. Instr.* **70**, pp. 4300–4303.
35. Sulchek, T., R. Hsieh, J.D. Adams, S.C. Minne, C.F. Quate, D.M. Adderton (2000): 'High-speed atomic force microscopy in liquid'. *Rev. Sci. Instr.* **71**, pp. 2097–2099.
36. Köhler, M. (1998): *Ätzverfahren für die Mikrotechnik.* (Wiley-VCH, Weinheim)
37. Zavyalov, V.V., J.S. McMurray, C.C. Williams (1999): 'Scanning capacitance microscope methodology for quantitative analysis of *p-n* junctions'. *J. Appl. Phys.* **85**, pp. 7774–7783.
38. De Wolf, P., M. Geva, T. Hantschel, W. Vandervorst, and R.B. Bylsma (1998): 'Two-dimensional carrier profiling of InP structures using scanning spreading resistance microscopy'. *Appl. Phys. Lett.* **73**, pp. 2155–2157.
39. Semiconductor Industry Association (1997): *The National Technology Roadmap for Semiconductors*, p. 181.
40. Semiconductor Industry Association (1998): *International Technology Roadmap for Semiconductors*, p. 297.
41. Knoll, B., F. Keilmann (1999): 'Near-field probing of vibrational absorbtion for chemical microscopy'. *Nature* **399**, pp. 134–137.
42. Lohmann, G. (1997): *Volumetric Image Analysis.* (Wiley-Teubner, Stuttgart)
43. Ohser, J., F. Mücklich (2000): *Statistical Analysis of Microstructures in Materials Science.* (John Wiley & Sons, Chichester)
44. Mecke, K.R., D. Stoyan (eds.) (2000): *Statistical Physics and Spatial Statistics: The Art of Analyzing Spatial Structures and Pattern.* (Springer, Heidelberg)

Bicontinuous Surfaces in Self-assembling Amphiphilic Systems

Ulrich Schwarz[1] and Gerhard Gompper[2]

[1] Max-Planck-Institut für Kolloid- und Grenzflächenforschung, D-14424 Potsdam
[2] Institut für Festkörperforschung, Forschungszentrum Jülich, D-52425 Jülich

Abstract. Amphiphiles are molecules which have both hydrophilic and hydrophobic parts. In water- and/or oil-like solvent, they self-assemble into extended sheet-like structures due to the hydrophobic effect. The free energy of an amphiphilic system can be written as a functional of its interfacial geometry, and phase diagrams can be calculated by comparing the free energies following from different geometries. Here we focus on bicontinuous structures, where one highly convoluted interface spans the whole sample and thereby divides it into two separate labyrinths. The main models for surfaces of this class are triply periodic minimal surfaces, their constant mean curvature and parallel surface companions, and random surfaces. We discuss the geometrical properties of each of these types of surfaces and how they translate into the experimentally observed phase behavior of amphiphilic systems.

1 Surfaces in Self-assembling Amphiphilic Systems

The subject of this article is the theoretical description of amphiphilic systems, which are one example of *soft condensed matter*. Soft matter is material that has a typical energy scale of $k_B T \approx 4 \times 10^{-21} J$, where k_B is the Boltzmann constant and $T \approx 300K$ is room temperature. For such material, perturbations arising from the thermally activated movement of its molecular components are sufficient to induce configurational changes, and entropy is at least equally important as energy in determining its material properties. The understanding of the underlying mechanisms is essential for the application of many technologies in everyday life, including colloidal dispersions (paints, inks, food, creams, lotions), foams (beverages), liquid crystals (displays), polyelectrolyte gels (diapers) and soaps (washing and cleaning). Moreover, the term *soft matter* also includes most biomaterials, for example blood or cartilage. One particularly important biomaterial is the *biomembrane*, that is the protein carrying lipid bilayer which surrounds each cell and its organelles. A good model system for the structural properties of biomembranes are *lipid bilayers*, which form spontaneously in mixtures of water and lipids and which are one of the main subjects of this paper.

Soft matter systems are very often characterized by competing interactions (including entropic ones), which lead to structure formation on the length scale between tens and hundreds of nanometers (1 nm = 10^{-9} m). Since this length scale is not accessible by optical lithography, self-assembly in soft matter systems is one of the main concepts of nanoscience. With structure formation being so prominent in soft matter systems, their theoretical description often centers around their spatial structure in three dimensions. For a rough classification of the different approaches used, it is convenient to use an analogy to the field of random geometries, and to distinguish between point, fiber and surface processes. Examples of systems which are suited for approaches along these lines are colloidal dispersions, polymer networks and sheet-like structures in self-assembling amphiphilic systems, respectively. Here we will treat the latter case, but we will also show that surface dominated system can be related to Gibbs distributions for scalar fields, namely through the use of isosurface constructions.

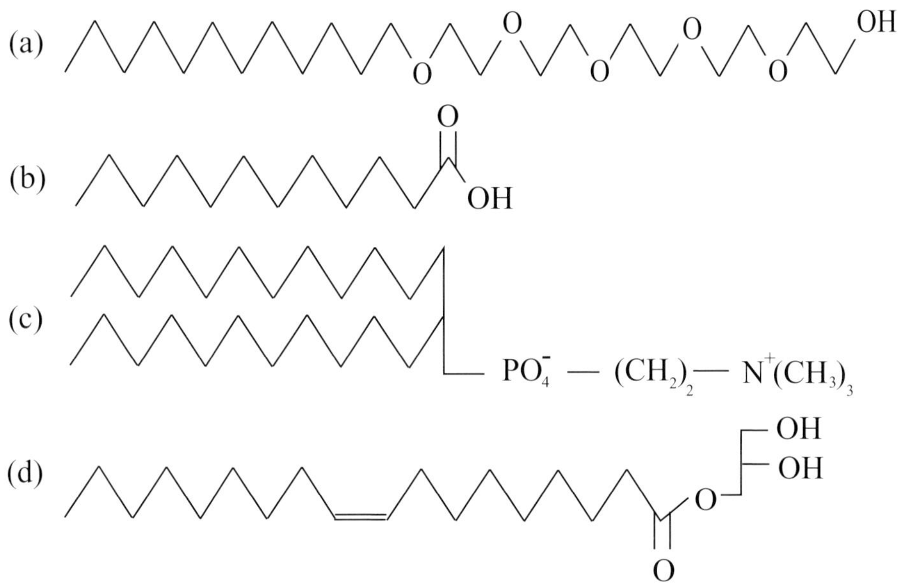

Fig. 1. Schematic representation of different amphiphilic molecules. Hydrophobic tails are to the left, hydrophilic heads to the right. (a) Pentaethylene glycol dodecyl ether, $C_{12}H_{25}(OCH_2CH_2)_5OH$ or in short $C_{12}E_5$, is a small surfactant. (b) Lauric acid, $CH_3(CH_2)_{10}COOH$ or in short LA. Fatty acids are the simplest amphiphiles and can be desolved to high amounts in phospholipid bilayers. (c) Dilauroyl phosphatidylcholine, or in short DLPC. Phosphatidylcholines have two hydrocarbon tails and a zwitterionic headgroup. They are abundant in animal cells and the best studied model system for biological lipids. (d) Monoolein, a monoacyl glycerol with one (unsaturated) hydrocarbon tail.

Amphiphilic systems are solutions of amphiphiles in suitable solvent. *Amphiphiles* are molecules which consist of two parts, one being hydrophilic (water-like, called the head) and one being hydrophobic (oil-like, called the tail). Well known classes of amphiphiles are *tensides* (used for washing and cleaning purposes) and *lipids* (the basic components of biomembranes). Figure fig:amphiphiles shows several examples for which phase diagrams are discussed below. Tensides are often called *surfactants*, due to their surface activity at interfaces between water-like and oil-like phases. Hydrophilic and hydrophobic molecules demix due to the hydrophobic effect: oil-like molecules are expelled from a region of water-like molecules since they disturb their network of hydrogen bonds. Hydrophilic and hydrophobic parts of an amphiphilic molecule cannot demix due to the covalent linkage between them, but the amphiphiles can self-assemble in such a way as to shield its hydrophilic parts from its hydrophobic ones and vice versa. Amphiphilic systems have many similarities with diblock copolymer systems, for which the term *microphase separation* is used to denote the molecular tendency for structure formation (see the contribution by Robert Magerle). However, in contrast to diblock copolymers, amphiphiles are small molecules, thus the entropy of their molecular configurations plays only a minor role in determining the overall structure. Moreover, whereas in diblock copolymer systems one component is sufficient to obtain stable mesophases, in amphiphilic systems usually the presence of an aequeous solvent is necessary to obtain well-pronounced structure formation. Both diblock copolymer and amphiphilic systems show structure formation on the nanometer scale, but the length scale is set by different control parameters: for diblock copolymers and amphiphiles, these are polymer size and solvent concentrations, respectively. Since the solvent in an amphiphilic system usually has no special physical properties by itself, its properties are mostly determined by its interfaces. This stands in marked contrast to the case of diblock copolymer systems, where the entropy of chain configurations in the regions away from the interfaces cannot be neglected in a physical description (for a review see [71] and references therein). As we will discuss in more detail below, interface descriptions have been very successful in describing the properties of amphiphilic systems (for reviews see [63, 40, 64, 104, 92]).

Amphiphilic systems can be classified according to the solvent used. A mixture of water and amphiphile is called a *binary* system, and a mixture of water, oil and amphiphile is called a *ternary* system. In binary systems, amphiphiles self-assemble into spherical, cylindrical or bilayer structure in such a way that the hydrophobic tails are shielded from the hydrophilic solvent; in ternary systems, amphiphiles self-assemble into monolayer structures in such a way that the hydrophobic tails and the hydrophilic heads face the hydrophobic and hydrophilic solvents, respectively (see Fig. 2). Irrespective of these differences, amphiphiles in binary and ternary systems assemble into similar geometries, because in both cases one deals with extended sheet-like structures (the same geometries also occur in diblock copolymer systems). At room temperature, amphiphiles in mono- and bilayers usually form a two-dimensional fluid, that is the molecular constituents show only short-ranged and no long-ranged order in the plane of the interface. Therefore the sheet-like structures in amphiphilic systems are called *fluid membranes* (in fact membrane fluidity is essential for the functioning of the protein machinery carried by biomembranes).

Since the physics of amphiphilic systems is mostly determined by their interfaces, different geometries have comparable free energies and small changes in external variables can induce phase transitions. In particularly, for amphiphilic systems different geometries are stable for different values of concentrations and temperature. In order to achieve systematic control of amphiphilic systems, a large effort has been invested into experimentally determining the phase diagram for many important amphiphilic systems. It has been found that despite their molecular diversity, the phase behavior of amphiphilic systems follows some general rules which result from their intrinsic tendency for self-assembly into sheet-like structures. For example, for increasing amphiphile concentration one usually finds the generic phase sequence micellar disordered - micellar

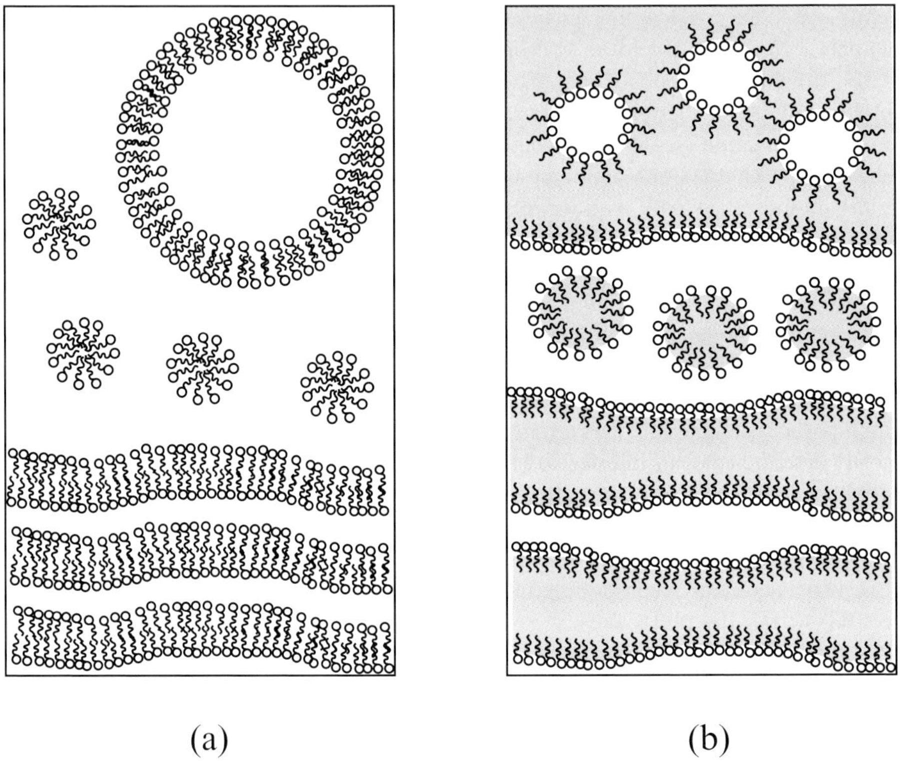

Fig. 2. Self-assembly of amphiphiles: (a) in binary systems, amphiphiles self-assemble into bilayers, micelles and vesicles in order to shield the hydrophobic tails from the aequeous solvent. (b) In ternary systems, the oil-like solvent swells the hydrophobic regions. If the oil-like solvent is the majority component, inverse structures occur. In ternary systems, all interfaces are amphiphilic monolayers.

ordered - hexagonal - cubic bicontinuous - lamellar. *Micellar* means spherical geometry; *micellar ordered* corresponds to an ordered (usually cubic) array of spherical micelles. *Hexagonal* denotes a two-dimensional packing of cylindrical aggregates. *Bicontinuous* means that one surface partitions space into two separate labyrinths, each of which can be used to traverse space. *Cubic bicontinuous* corresponds to a space-filling arrangement of one interface, which folds onto itself in a cubic arrangement. Sometimes non-cubic spacegroups are found, but most ordered bicontinuous structures in equilibrium are cubic. In some systems, bicontinuous yet disordered phases occur, which are called *sponge phases* and *microemulsions* for binary and ternary systems, respectively. *Lamellar* corresponds to a one-dimensional stack of interfaces. Ordered arrangements like hexagonal or cubic symmetries can be probed by scattering techniques, and bicontinuity by diffusion experiments (e.g. using nuclear magnetic resonance). Figure fig:geometries tries to visualize the different interface geometries. It is the case of cubic bicontinuous phases which is the main focus of this paper.

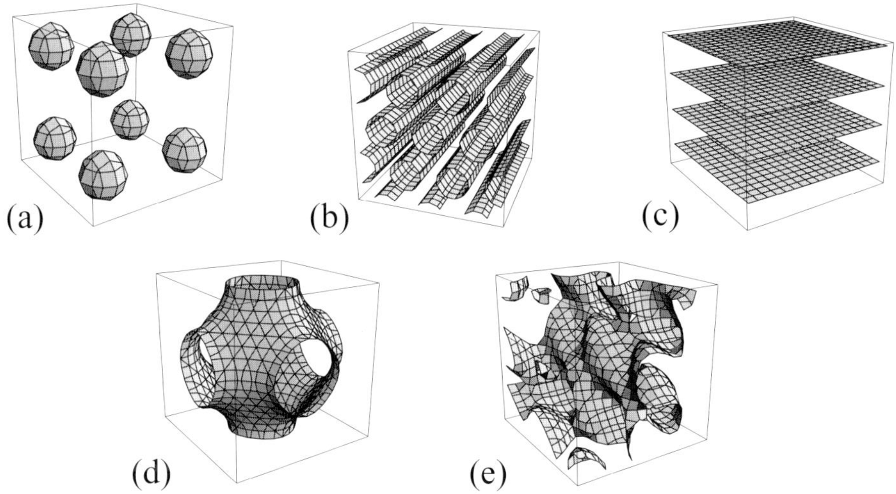

Fig. 3. Geometries of self-assembled interfaces: (a) micellar (spheres), (b) hexagonal (cylinders), (c) lamellar (planes), (d) cubic bicontinuous (cubic minimal surfaces) and (e) disordered bicontinuous (random surfaces). In amphiphilic systems, mono- or bilayers are draped onto the mathematical surfaces. *Bicontinuous* means that there is one surface which spans the whole sample, thereby separating it into two disjunct yet intertwined labyrinths (like the corresponding surface, each labyrinth is connected and spans the whole sample). Note that except for (e), all surfaces are ordered and have constant mean curvature.

There exists a large body of literature on the occurence of bicontinuous phases in amphiphilic systems (for reviews see [27, 68, 103]). In Fig. 4 we show experimental phase diagrams for binary [72, 108] and ternary [59, 76] systems with the small surfactant $C_{12}E_5$ (this is the molecule shown in Fig. 1a). For the binary system, Fig. 4a, at low temperature one sees the sequence hexagonal H_1 - cubic bicontinuous V_1 - lamellar L_α with increasing amphiphile concentration. For high temperature, the hexagonal and cubic bicontinuous phases disappear, the lamellar phase L_α expands (its interfaces unbind), and for very small amphiphile concentration a sponge phase L_3 occurs. For the ternary system, Fig. 4b, we recognize the same situation again close to the binary side water-amphiphile. For equal amounts of water and oil and not too large concentration of amphiphile, that is in the lower middle of the Gibbs triangle, one sees the following phase behavior: at low temperature an emulsification failure occurs, that is the micellar disordered phase L_1 coexists with an oily excess phase. At high temperature, the emulsification failure disappears and is replaced by a microemulsion (the central, triangular one phase region). We can conclude that at low temperature, strongly curved structures prevail (H_1, L_1), while at high temperature, those structures become stable which are locally flat (L_α, L_3, microemulsion). As we will show below, this behavior can be understood using the concept of a temperature dependent spontaneous curvature: spontaneous curvature is finite at low temperature, but vanishes at high temperature. The concept of finite spontaneous curvature also offers a natural explanation for the emulsification failure at low temperature. As we will argue below, the cubic bicontinuous phase V_1 is an intermediate structure between strongly curved and locally flat. It also follows from the phase behavior described that at low temperature, stable phases tend to be ordered (H_1, V_1), while disordered phases profit from increased temperature (L_3, microemulsion). We will discuss below that the disordered phases indeed have a larger configurational entropy. In this respect, it is the lamellar phase L_α which acts as an intermediate structure.

In Fig. 5 we show experimental phase diagrams for two different lipid-water mixtures. Figure fig:phasebehavior2a is for water and 2:1 lauric acid / dilauroyl phosphatidylcholine [110] and Fig. 5b for water and monoolein [88] (these are the molecules shown in Fig. 1b-d). Despite the molecular differences, the macroscopic phase behavior is surprisingly similar. At low temperature the membranes loose their fluidity. At intermediate temperatures, the lamellar phase is stable, and at high temperatures, it is replaced by the hexagonal phase. Whereas in the case of the surfactant $C_{12}E_5$ spontaneous curvature is finite at ambient temperatures and vanishes at high temperature, for the lipids spontaneous curvature vanishes at ambient temperatures and increases with temperature. Several cubic bicontinuous phases are stable for intermediate temperatures, in the sequence lamellar - G - D - P (here G, D and P stand for cubic bicontinuous structures which are discussed in more detail below) with increasing water content. The last phase in this sequence undergos an emulsification failure, that is it coexists with an excess water phase. We will show below that this phase behavior can be explained nicely if the lipid monolayers of the cubic bicontinuous phases are modeled as parallel surfaces to a cubic minimal midsurface.

In this article, we focus on the interface description of amphiphilic systems, that is the free energy of the system is written as a functional of its interface configuration. The interface free energy is introduced in the next section (Sect. 2), and in the rest of this article we will specify this free energy expression for different instances of cubic bicontinuous phases. In each case, we will discuss the relevant geometric properties and show how they relate to the resulting free energy expressions and phase diagrams. Our starting point are *triply periodic minimal surfaces* (TPMS) in Sect. 3, a subject well-known from differential geometry. These structures are expected to occur if the amphiphilic interface is symmetric in regard to its two sides and if temperature is not too high as to destroy the ordered state. TPMS are also the reference state for their parallel surfaces and CMC-companions, which are the adequate structural models for a detailed analysis of amphiphilic monolayers in cubic bicontinuous phases. In the case that the physical interface is not symmetric in regard to its two sides (that is if spontaneous curvature exists, like usually for an amphiphilic monolayer at a water-oil interface), the relevant mathematical representations are the constant mean curvature companions of the TPMS, that is *triply periodic surfaces of constant mean curvature* (CMC-surfaces) treated in Sect. 5. The case of lipid bilayers might be considered to be the composition of two such CMC-surfaces, but we will argue in Sect. 4 that in this case the relevant geometry is in fact the one of *parallel surfaces* to a TPMS. In all of these cases, we are interested in ordered structures, and therefore the main method will be minimization of the corresponding free energy functionals. This is different in Sect. 6, where we will discuss disordered bicontinuous phases, that is the sponge phases and microemulsions, which often occur at higher temperatures due to entropic effects. These structures are

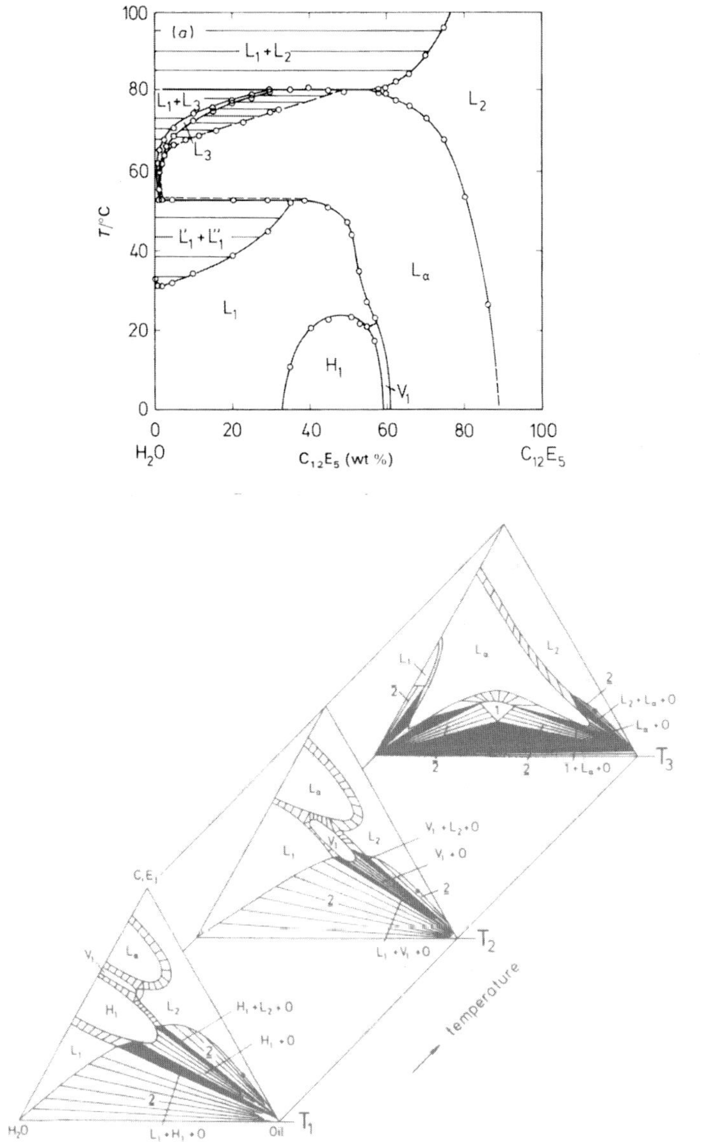

Fig. 4. Experimentally determined phase behavior for the surfactant $C_{12}E_5$: binary system ($H_20/C_{12}E_5$) [108] (top) and ternary system ($H_20/C_{12}E_5/C_{14}$) [76] (bottom). In the binary system, the hexagonal phase H_1, the cubic bicontinuous phase V_1, the lamellar phase L_α, the sponge phase L_3 and the micellar phases L_1 and L_2 are stable. In the ternary system, the same phases occur close to the binary side $H_20/C_{12}E_5$. At $T_1 = 5°T$ and $T_2 = 25°T$, L_1 coexists with excess oil (emulsification failure). At $T_3 = 48°T$ (balanced temperature), a bicontinuous microemulsion is stable in the middle of the Gibbs triangle.

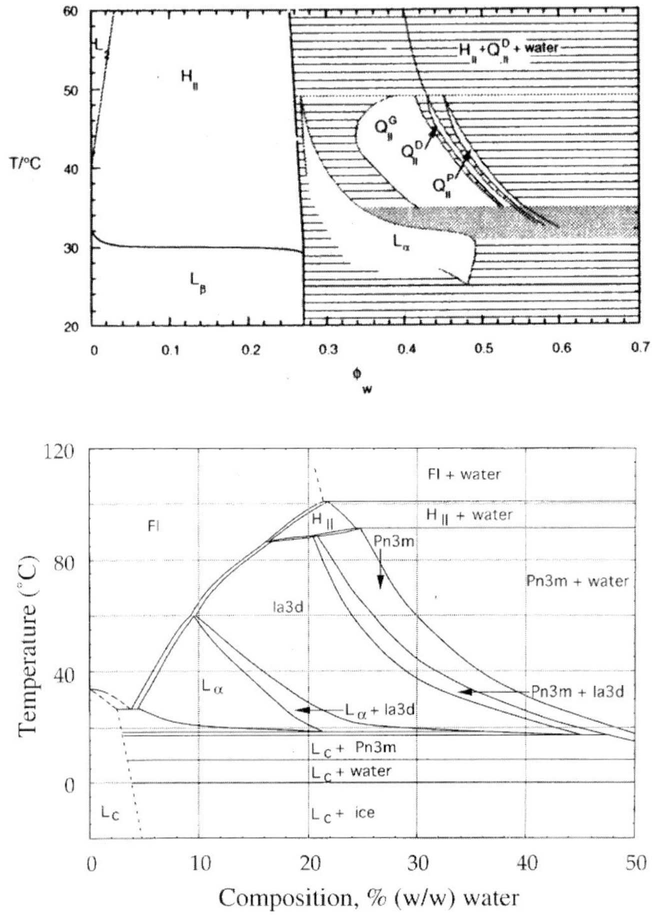

Fig. 5. Experimentally determined phase behavior for lipid-water mixtures: 2:1 lauric acid / dilauroyl phosphatidylcholine [110] (top) and monoolein [88] (bottom). In both phase diagrams, one sees the sequence lamellar - cubic bicontinuous - excess water (emulsification failure) at intermediate temperatures (around $40°C$) and with increasing water concentration. In (a), G, D and P are stable. In (b), G (Ia3d) and D (Pn3m) are stable.

modeled as *random surfaces*, thus the main model here will be Gibbs distributions, in particular the theory of Gaussian random fields and Monte Carlo simulations.

2 Free Energy Functionals

2.1 Interface Models

From the viewpoint of elasticity theory, amphiphilic interfaces with small curvatures can be considered to be thin elastic shells, which are known to have few fundamental modes of deformation: out-of-plane bending, in-plane compression and in-plane shearing [60]. However, since amphiphilic interfaces are fluid and nearly incompressible, in-plane strain is irrelevant and the most relevant deformation mode is bending. For small curvatures the free energy of an amphiphilic interface is a function only of its geometry [8, 47]:

$$F = \int dA \left\{ \sigma + 2\kappa (H - c_0)^2 + \bar{\kappa} K \right\} . \tag{1}$$

Here dA denotes the differential area element and H and K mean and Gaussian curvature, respectively. The latter two follow from the two principal curvatures k_1 and k_2 as $H = (k_1 + k_2)/2$ and $K = k_1 k_2$. The three material parameters introduced in (1) define the energy scales of the corresponding changes: σ is *surface tension* and corresponds to changes in surface area, κ is *bending rigidity* and corresponds to cylindrical bending, and $\bar{\kappa}$ is *saddle-splay modulus* (or *Gaussian bending rigidity*) and corresponds to changes in topology due to the Gauss-Bonnet-theorem for closed surfaces, $\int dAK = 2\pi\chi$, where χ denotes Euler characteristic. The *spontaneous curvature* defines the reference point for bending deformations. It has to vanish for symmetric amphiphilic sheets (like lipid bilayers), but in general is finite for non-symmetric ones (like monolayers). For monolayers, it is usually a linear function of temperature, $c_0 \sim (T - T_b)$. Therefore spontaneous curvature vanishes at the balanced temperature T_b, at which solvent properties make the monolayer symmetric in regard to bending.

For amphiphilic interfaces, surface area is proportional to the number of amphiphilic molecules, thus σ can also be interpreted as a chemical potential for amphiphiles. The bending rigidity κ has to be positive, otherwise the system would become instable to spontaneous convolutions. For surfactant and lipid systems, its values are of the order of 1 and 20 $k_B T$, respectively. Surface tension σ can assume negative values (favorable chemical potential for the influx of amphiphiles), as long as bending rigidity κ exists and has a positive value, in order to prevent an instability towards proliferation of interfacial area. The value of the saddle-splay modulus $\bar{\kappa}$ is often debated, but usually it is assumed to have a small negative value. This assumption is validated by the following argument [48]: for $\sigma = 0$ and $c_0 = 0$, we can rewrite (1) as

$$F = \int dA \left\{ \frac{1}{2} \kappa_+ (k_1 + k_2)^2 + \frac{1}{2} \kappa_- (k_1 - k_2)^2 \right\} \tag{2}$$

with

$$\kappa_+ = \kappa + \frac{\bar{\kappa}}{2}, \quad \kappa_- = -\frac{\bar{\kappa}}{2} . \tag{3}$$

From this we conclude that the Gaussian bending rigidity has to satisfy $\kappa_+ > 0$ and $\kappa_- > 0$, that is $-2\kappa < \bar{\kappa} < 0$, otherwise the system would become unstable. For $\bar{\kappa} < -2\kappa$ ($\kappa_+ < 0$), we would get $k_1 = k_2 \to \infty$, that is many small droplets, and for $\bar{\kappa} > 0$ ($\kappa_- < 0$), we would get $k_1 = -k_2 \to \infty$, that is a minimal surface with very small lattice constant.

From the mathematical point of view, it is interesting that for $\sigma = 0$ and $c_0 = 0$, the bending energy from (1) is not only invariant under rescaling with length, but also invariant under conformal transformations in general. This has intriguing consequences for vesicles [104] (vesicles are depicted in Fig. 2, but are not subject of this article) and phase transitions in systems without spontaneous curvature (as will be discussed in Sect. 6).

Since the energy scales involved are of the order of $k_B T$, thermal noise is sufficient to induce shape changes. Due to thermal fluctuations on smaller length scales, the effective values for the material parameters are changed (renormalized) at larger length scales [49, 79]. It has been shown in the framework of renormalization group theory that logarithmic corrections arise due the two-dimensional nature of the amphiphilic interfaces, so that [79, 17, 7]

$$\kappa_R(l) = \kappa - \frac{3k_B T}{4\pi} \ln \frac{l}{\delta} , \tag{4}$$

$$\bar{\kappa}_R(l) = \bar{\kappa} + \frac{5k_B T}{6\pi} \ln \frac{l}{\delta} \tag{5}$$

where l is the length scale on which the system is analysed and δ is a microscopic cutoff on the scale of the membrane thickness. Thus bending rigidity decreases, while Gaussian bending rigidity increases with increasing length scale l.

For the following, it is useful to introduce two dimensionless quantities which correspond to the two curvatures of a two-dimensional surface. For a structure with surface area A and volume V, the length scale V/A can be used to rescale its curvatures. In order to be able to convert easily from local to global quantities, we consider a triply periodic CMC-surface with Euler characteristic χ and integrated mean curvature $H_i = \int dAH$ per conventional (that is simple cubic) unit cell:

$$K \left(\frac{V}{A}\right)^2 = \frac{2\pi\chi V^2}{A^3} , \quad H \left(\frac{V}{A}\right) = \frac{H_i V}{A^2} . \tag{6}$$

These expressions motivate the definition of the *topology index* Γ and the *curvature index* Λ:

$$\Gamma = \left(\frac{A^{*3}}{2\pi|\chi|}\right)^{\frac{1}{2}} , \quad \Lambda = \frac{H_i^*}{A^{*2}} \tag{7}$$

where $A^* = A/V^{2/3}$ and $H_i^* = H_i/V^{1/3}$ are the scaled surface area and the scaled integral mean curvature per conventional unit cell, respectively. Note that these definitions can be applied to any triply periodic surface; here we used CMC-surfaces only for an heuristic motivation of the quantities defined. Both the topology index Γ and the curvature index Λ do not depend on scaling and choice of unit cell. They are universal geometrical quantities which characterize a surface in three-dimensional space. The topology index describes its porosity (the larger its value, the less holes the structure has) and its specific area content (the larger its value, the more inner surface the structure contains), and the curvature index describes how strongly the structure is curved (irrespective of the actual lattice constant). It is interesting to note that the two indices defined here correspond to the two isoperimetric ratios known from integral geometry. For a TPMS, the curvature index Λ vanishes and the topology index Γ is its most important geometric characteristic.

2.2 Ginzburg-Landau Models

A different but equally powerful approach to amphiphilic interfaces is the isosurface construction (also known as phase field method), which derives a two-dimensional surface from a three-dimensional scalar field $\Phi(\mathbf{r})$. For ternary amphiphilic systems, $\Phi(\mathbf{r})$ can be interpreted as the local concentration difference between water ($\Phi = 1$) and oil ($\Phi = -1$). The position of the amphiphilic monolayer can be identified with the isosurface $\Phi(\mathbf{r}) = 0$. For binary amphiphilic systems, $\Phi(\mathbf{r})$ can be interpreted as the local concentration difference between water on different sides of a bilayer, and the isosurface $\Phi(\mathbf{r}) = 0$ marks the position of the bilayer mid-surface. In both cases, an energy functional can be defined in the spirit of a Ginzburg-Landau theory for the scalar field $\Phi(\mathbf{r})$. A reasonable choice is the Φ^6-model introduced by Gompper and Schick [39, 40]:

$$\mathcal{F}[\Phi] = \int d\mathbf{r} \; \{(\Delta\Phi)^2 + g(\Phi)(\nabla\Phi)^2 + f(\Phi)\} \tag{8}$$

with the following choice for f and g:

$$f(\Phi) = (\Phi + 1)^2(\Phi - 1)^2(\Phi^2 + f_0) \; , \; g(\Phi) = g_0 + g_2\Phi^2 \; . \tag{9}$$

It has been shown that this model is similar to an interface description as given in (1), and prescriptions have been given how to calculate the parameters of the interface Hamiltonian from the given Ginzburg-Landau theory [41]. Due to the invariance under $\Phi \to -\Phi$, spontaneous curvature c_0 vanishes in this model. For given model parameters (g_0, g_2, f_0), this energy functional can be minimized for its spatial degrees of freedom, and a phase diagram can be calculated as a function of model parameters by identifying the absolute minimum at every point. A reasonable choice for amphiphilic systems is $g_0 < 0$, f_0 close to 0 and g_2 not too large. Then the formation of interfaces is favored and a lamellar phase becomes stable. Calculation of the corresponding parameters of the interface model then yields $\sigma < 0$, $\kappa > 0$ and $\bar{\kappa} < 0$. Note that $\sigma < 0$ favors the formation of interfaces, and that $\bar{\kappa} < 0$ favors the lamellar phase.

3 Triply Periodic Minimal Surfaces (TPMS)

Minimal surfaces and surfaces of constant mean curvature (see Sect. 5) in general have attracted a lot of attention both in mathematics and in physics, partially due to their intrinsic beauty, which might be considered to follow from the fact that they are solutions to variational problems [51]. It is important to note that the different fields are interested in different aspects of the same object: while for mathematicians the most central question is the existence proof for a minimal surface of interest, physicists are more interested in its representation, which can be used to derive physical properties of corresponding material systems. As Karcher and Polthier remark, outside mathematics only pictured minimal surfaces have been accepted as existent [56]. In fact this statement can also be reversed: not every structure which one can picture is necessarily a minimal surface. In this article we consider only established minimal surfaces. As we are motivated by physical considerations, we are interested only in embedded surfaces which do not self-intersect. In regard to mathematics, our main emphasis here will be on representations, since the physical properties of amphiphilic systems follow from their spatial structure.

Minimal surfaces are surfaces with $H = 0$ everywhere. If one varies a surface with a normal displacement $\delta\phi(u, v)$ (where u and v are the internal coordinates of the surface), it can be shown that the change in surface area is $\Delta A = 2\delta \int dA\phi(u, v)H(u, v) + O(\delta^2)$ [54], that is a minimal surface is a stationary surface for variations of surface area. Therefore surfaces under surface tension (like soap films), which try to minimize surface area, form minimal surfaces. Here we consider surfaces which are dominated by bending rigidity rather than by surface tension. However, for vanishing spontaneous curvature the resulting structures also correspond to minimal surfaces. (1) explains why: these surfaces have to minimize $\int dAH^2$ (they are so-called *Willmore surfaces*), and minimal surfaces are a special case of Willmore surfaces, since $H = 0$ is a trivial minimization of this functional.

If mean curvature $H = (k_1 + k_2)/2 = 0$, then Gaussian curvature $K = k_1 k_2 = -k_1^2 \leq 0$. Thus minimal surfaces are everywhere either saddle-like or flat, and nowhere convex. In fact it can be shown that the flat points with $k_1 = k_2 = 0$ are isolated. Moreover it follows from $K \leq 0$ that minimal surfaces without boundaries cannot be compact [54]. Until 1983, only two embedded minimal surfaces were known which are non-periodic: the plane and the catenoid. Then a new surface of this type was found, the Costa-surfaces, which from the distance looks like the union of a plane and a catenoid [56]. Today some more surfaces of this type are known, but the majority of all embedded minimal surfaces without boundaries are in fact periodic. There is only one simply-periodic minimal surface, the helicoid, and there are some doubly-periodic surfaces, with the Scherk-surface being most prominent. The majority of all known periodic minimal surfaces is triply-periodic. Here we will focus on cubic minimal surfaces and their physical realizations, which are cubic bicontinuous structures.

The best known triply periodic minimal surface (TPMS) is the P-surface found by Schwarz in 1867 and depicted in Fig. 3d. Schwarz and his students found 5 TPMS, including the cubic cases P, C(P) and D. Until 1970 no more examples were found, then Schoen described 13 more [96], including the cubic cases G, F-RD, I-WP, O,C-TO and C(D). He proved the existence of G by providing an explicit (Weierstrass) representation, and the existence of the others was proven in 1989 by Karcher [55]. More TPMS have been discovered by Fischer and Koch [23, 57] and others, and today many more can be generated by making controlled modifications to computer models of known TPMS, a method which was pioneered by Polthier and Karcher [56]. However, these surfaces tend to be rather complicated, and we will show below that only the simple (as quantified by the topology index) TPMS are relevant for amphiphilic systems. In Fig. 6 we show visualizations of 10 different TPMS with cubic symmetry. There representations have been obtained from the Φ^6-Ginzburg-Landau theory as will be explained below. For some of these structures, we also show line-like representations of the two labyrinths defined by the surface (*skeletal graphs*). For example, in the case of the Schwarz P-surface, the two skeletal graphs are the edges of the surrounding cube and the three lines through the origin.

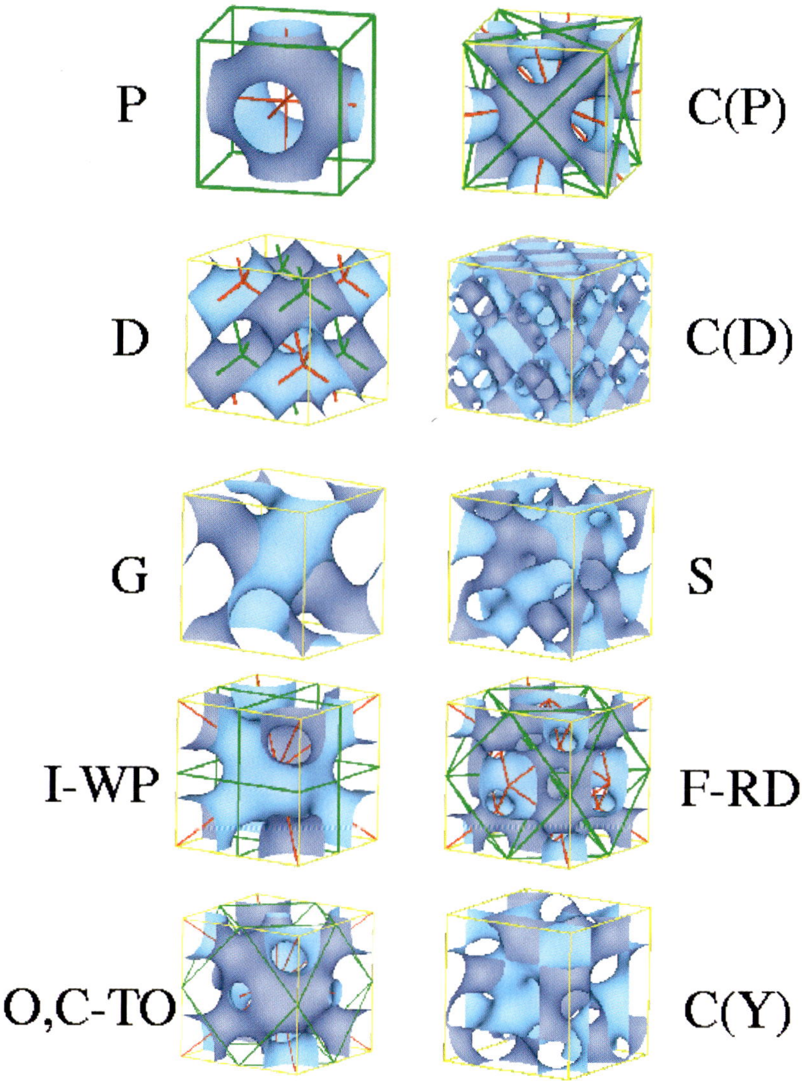

Fig. 6. Visualizations of 10 different cubic minimal surfaces. Every triply periodic surface divides space into two labyrinths, which in some cases are represented here by skeletal graphs. D and G are the basis for the double-diamond structure and the gyroid, respectively, which often occur in physical systems.

Table 1. Euler characteristic χ, scaled surface area A^* in the conventional unit cell and topology index $\Gamma = ({A^*}^3/2\pi|\chi|)^{1/2}$ for those TPMS, for which exact results are known from Weierstrass representations. Here $k_1 = K(1/2)/K(\sqrt{3}/2)$ where $K(k)$ is the complete elliptic integral of the first kind, and $k_2 = K(1/\sqrt{3})/K(\sqrt{2/3})$. Note that often a unit cell is chosen for D which is the eighth part of the one chosen here; then one has $\chi = -2$ and $A^* = 1.918893$.

	χ	A^*	Γ
G	-8	$3(1+k_1^2)/2k_1 = 3.091444$	0.766668
D	-16	$3/k_1 = 3.837785$	0.749844
I-WP	-12	$2\sqrt{3} = 3.464102$	0.742515
P	-4	$3k_1 = 2.345103$	0.716346
C(P)	-16	$3/k_2 = 3.510478$	0.655993

Different methods can be used to obtain representations of TPMS. Until recently, the main method were the Weierstrass representation formulae. For the cubic TPMS, they are known for P, D and G [25, 26] as well as for I-WP [62, 12]. For each of these TPMS, a fundamental domain can be identified, so that the rest of the surface follows by replicating it with the appropriate space group symmetries ($Im\bar{3}m$, $Pn\bar{3}m$, $Ia\bar{3}d$ and $Im\bar{3}m$, respectively). The Weierstrass representation is a conformal mapping of certain complicated regions within the complex plane onto the fundamental domain:

$$(x_1, x_2, x_3) = Re \int_0^{u+iv} dz\; R(z)\; (1 - z^2, i(1 + z^2), 2z) \tag{10}$$

where (u, v) are the internal (and conformal) coordinates of the minimal surface. The Weierstrass mapping can be understood as the inversion of the following composition: first the surface is mapped onto the unit sphere via its normal, and then the unit sphere is mapped onto the complex plane by stereographic projection. The geometrical properties of a surface follow from the Weierstrass representation as

$$dA(z) = |R(z)|^2\; (1 + |z|^2)^2\; dudv,\;\; H(z) = 0,\;\; K(z) = \frac{-4}{|R(z)|^2(1 + |z|^2)^4} \tag{11}$$

with $z = u + iv$. Obviously the (isolated) poles of $R(z)$ correspond to the flat points ($K = 0$) of the minimal surface. Only few choices of $R(z)$ yield embedded minimal surfaces. The ones for D and P have been known since the 19th century from the work of Schwarz: for D it is $R(z) = (z^8 - 14z^4 + 1)^{-\frac{1}{2}}$. P follows simply by the Bonnet transformation $R(z) \to e^{i\theta} R(z)$ with $\theta = 90°$. Equation (11) implies that P and D have the same metric and the same distribution of Gaussian curvature. However, since they map differently into embedding space, they have different space groups and lattice constants. The gyroid G was discovered in 1970 by Schoen [96] as another Bonnet transformation of D, with $\theta = 38.015°$. The Weierstrass representation for I-WP was found only recently [62, 12]. If one of its poles is chosen to be at infinity, one has $R(z) = (z(z^4+1))^{-\frac{2}{3}}$. In Table 1 we give exact results for geometrical properties which have been derived from Weierstrass representations (although for C(P) no Weierstrass representation is known, these values can be derived from its complementary relationship to P). Note that the topology index Γ establishes the hierarchy G - D - P within this

Bonnet family. This sequence corresponds to the connectivity of the labyrinths defined by the TPMS: G has 3-fold, D 4-fold and P 6-fold coordination, since higher coordinated structures have more holes.

It is interesting to note that the identification of these structures in material systems is a difficult enterprise, which depends on the availability of suitable representations. Physical representations of TPMS have been built since the 19th century, using soap films draped onto wire skeletons (without the wire skeletons surface tension would shrink these structures into collapse). In 1967 Luzzati and Spegt noted that certain phases in lipid systems are cubic bicontinuous [67], but it was only in 1976 that Sciven suggested the relevance of TPMS as structural models for cubic bicontinuous phases [102]. The Luzzati-Spegt structure was later identified with the gyroid structure G, and the diamond structure D was identified both in lipid [65] and diblock copolymer systems [114]. For diblock copolymer systems, the gyroid structure G was identified in 1994 [45], and little later it was noted that earlier identifications of cubic bicontinuous phases in diblock copolymer systems often mistook the gyroid structure G for the diamond structure D [46] (in fact, for diblock copolymer systems one should rather use the terms *double diamond* and *double gyroid*, since in this case two interfaces are arranged around the corresponding TPMS). Today, the structures G, D, P and I-WP seem to be non-ambiguously identified in amphiphilic system, with experimental evidence based mainly on small angle scattering experiments, electron transmission microscopy, and swelling and diffusion experiments [27, 68, 103].

Apart from the cases G, D, P and I-WP given above, no more Weierstrass representations are known for cubic TPMS, so for all other cases numerical methods have to be used. For example, the *Surface Evolver* is a software package written by Brakke which allows to minimize triangulated surfaces for different energy functionals (compare Sect. 8). Here we discuss the method of constructing the isosurfaces of a scalar field $\Phi(\mathbf{r})$, as introduced in Sect. 2.2. For our purpose, the usefulness of this model lies in its rugged energy landscape, which means that many more local minima exist than the absolute minima corresponding to the lamellar phase. In physical terms, these additional minima correspond to modulated phases which are metastable. If started with suitable initial conditions, the minimization procedure therefore yields representations which can be used to characterize the structural properties of these phases. Therefore this model has been used repeatedly in order to investigate bicontinuous cubic phases [41, 42, 43, 97]. In particularly, it has been found that the resulting representations are very close to TPMS [42, 43]. This finding can be explained as follows [97]: for a triply-periodic cubic structure, the free energy per unit volume $f = F/V$ follows from the interface description of (1) as

$$f = \frac{1}{a}(\sigma A^*) + \frac{1}{a^3}\left(2\kappa \int dA\, H^2 + 2\bar{\kappa}\pi\chi\right) \tag{12}$$

where a is the lattice constant of the conventional unit cell. Both terms in brackets are scale invariant, that is they do not depend on the lattice constant a. Since for the Ginzburg-Landau model at hand the first and second term in brackets is negative and positive, respectively (compare Sect. 2.2), a balance exists between the negative surface tension term, which favor small values of a, and the positive curvature contributions, which favors large a. If we make the assumption that the minimization problem can be decomposed in two independent minimizations, one for lattice constant a and one for the shape, we can first minimize f for a while considering A^* and $\int dA\, H^2$ to be constant:

$$a_{min} = \left(\frac{6\pi\bar{\kappa}\chi + 6\kappa\int dAH^2}{|\sigma|A^*}\right)^{\frac{1}{2}}, \quad f_{min} = -\left(\frac{4}{27}\right)^{\frac{1}{2}}\left(\frac{(|\sigma|A^*)^3}{2\bar{\kappa}\pi\chi + 2\kappa\int dAH^2}\right)^{\frac{1}{2}} \tag{13}$$

Similar mechanisms are always at work in amphiphilic systems and explain why in contrast to surface tension, bending rigidity does not necessitate the presence of a scaffold (like a wire skeleton for soap films) to prevent collapse. In a second step, we now minimize f_{min} for shape. If we assume A^* to be constant, we have reduced our problem to the Willmore problem, and this is why minimal surfaces with $H = 0$ appear in the framework of the Φ^6-theory. The final free-energy density can be written as

$$f_{min} = -\left(\frac{4}{27}\right)^{\frac{1}{2}}\left(\frac{|\sigma|^3}{|\bar{\kappa}|}\right)^{\frac{1}{2}} \Gamma \tag{14}$$

where Γ is the topology index defined in (7). We conclude that the stability of the different TPMS is governed by their topology indices and that the most favorable bicontinuous cubic phase is the gyroid G since it has the largest value for Γ (compare Table 1). Although this conclusion is based on some assumptions, it is corroborated by a detailed numerical analysis [97].

In order to obtain representations for a large list of TPMS, we used the Fourier approach and the theories of black and white space groups to obtain TPMS as local minima of the Φ^6-theory [97]. Black and white space groups are also known as magnetic or Shubnikov space groups and lead to a crystallographic classification of TPMS [23, 24]. Implementation of black and white symmetries leads to a considerable reduction in the degrees of freedoms of the Fourier series. Therefore this method is computationally cheap and its results are easy to document and to reuse. A triply periodic surface partitions space into two labyrinths, which can be considered to be colored black and white. In the framework of the Ginzburg-Landau theory, black and white correspond to $\Phi > 0$ and $\Phi < 0$, respectively. The surface is called *balanced* if there exists an Euclidean transformation α which maps the white labyrinth onto the black one and vice versa; otherwise it is called *non-balanced*. Examples for non-balanced surfaces are I-WP and F-RD, and the Fourier approach for the corresponding structure follows the usual rules for the respective space group [106]. However, if the surface is balanced, the structure is characterized by *two* space groups: if the colored structure has space group $\mathcal{H}$, the uncolored structure has space group $\mathcal{G} = \mathcal{H}\otimes\{1, \alpha\}$. $\mathcal{H}$ contains all symmetry operations of $\mathcal{G}$ that do not interchange the two labyrinths. It is a subgroup of $\mathcal{G}$ of index 2, that is the quotient group is isomorphic to the cyclic group of order 2, $\mathcal{G}/\mathcal{H} \cong \mathbb{Z}_2 \cong \{1, \alpha\}$. For example, for the balanced cubic surface defined by $\cos x + \cos y + \cos z = 0$, the operation interchanging the two labyrinth is the translation by half a body diagonal, and we have $\mathcal{H} = Pm\bar{3}m$ and $\mathcal{G} = Im\bar{3}m$.

Since $\mathcal{H}$ has index 2 in $\mathcal{G}$, it follows from the theorem of Hermann [105] that $\mathcal{H}$ has either the same point group or the same Bravais lattice as $\mathcal{G}$. If $\mathcal{H}$ and $\mathcal{G}$ have the same point group, their Bravais lattices have to be different. Thus the Euclidean operation α has to be a translation which extends one cubic Bravais-lattice into another. For the cubic system, there are only three different Bravais lattices (simple cubic P, body-centered cubic I and face-centered cubic F), and only two possibilites to extend one into the other by a translation $\mathbf{t}_\alpha$: for $\mathbf{t}_\alpha = a(\mathbf{x}+\mathbf{y}+\mathbf{z})/2$ a P-lattice becomes a I-lattice, and for $\mathbf{t}_\alpha = a\mathbf{x}/2$ a F-lattice becomes a P-lattice. The condition $\Phi(\mathbf{r}+\mathbf{t}_\alpha) = -\Phi(\mathbf{r})$ leads to reflection conditions for the reciprocal vectors. For Miller indices (h,k,l), one finds $h+k+l = 2n+1$ for $P \to I$, and $h,k,l = 2n+1$ for $F \to P$. Note that these reflection conditions are similar to the well-known ones $h+k+l = 2n$ for I and $h+k, h+l, k+l = 2n$ for F. The P-surface has space group $Pm\bar{3}m$, but a black and white symmetry with supergroup $Im\bar{3}m$, therefore it is an example for the case $P \to I$. Thus we have $h+k+l = 2n+1$ and the first Fourier mode is $(1,0,0)$. Note that the double P structure, which one would expect for a diblock copolymer system, has no black and white symmetry, but space group $Im\bar{3}m$, thus one has $h+k+l = 2n$ and the first mode is $(1,1,0)$.

If $\mathcal{H}$ and $\mathcal{G}$ have the same Bravais lattice, their point groups have to be different. Thus the Euclidean operation α has to be a point group operation which extends one cubic point group into another one. There are five cubic point groups and six ways to extend one of them into another by some P_α. The condition $\Phi(P_\alpha \mathbf{r}) = -\Phi(\mathbf{r})$ leads to more complicated rules than in the case of identical points groups [97]. However, there are few relevant examples from this class, the most interesting one being the gyroid G, where P_α is the inversion and the resulting rules are rather simple again (the even part of the Fourier series vanishes).

Fischer and Koch have completely enumerated all 34 cubic group-subgroup pairs $\mathcal{G} - \mathcal{H}$ with index 2 which are compatible with cubic balanced TPMS [23]. Although there is no way to completely enumerate all TPMS belonging to a given pair $\mathcal{G} - \mathcal{H}$ in general, this is possible for a certain subset, that is for all cubic balanced TPMS which contain straight lines which form a three-dimensional network [23]. This list reads P, C(P), D, C(D), S and C(Y). In our work, we implemented Fourier series for these structures as well as for the gyroid G (which is balanced, but does not contain any straight lines) and the non-balanced structures I-WP and F-RD. In some cases like G, D and P, the first mode of the Fourier ansatz implementing the correct black and white symmetry is already sufficient to obtain a representation which is topologically correct (*nodal approximation*). In all other cases investigated, the addition of one more mode (with the relative weight fixed by visual inspection) is sufficient. Nodal approximations for bicontinuous structures have been discussed first by von Schnering and Nesper [115]. In Table 2 and Table 3 we give nodal approximations and the corresponding space group information for some of the balanced and non-balanced structures investigated.

Implementing the complete Fourier series and numerically minimizing the Φ^6-functional from (8) with nodal approximations as initial conditions, we arrived at *improved nodal approximations* which were tabulated in [97] with up to six Fourier modes. The isosurface construction is easily implemented using the marching cube algorithm. The resulting triangulations can then be used to investigate geometrical properties of these surfaces. In particular, widely used mathematics programs (like Mathematica, Maple or Matlab) can be easily used to obtain useful representations of TPMS from (improved) nodal approximations. Our numerical work shows that the

Table 2. Group - subgroup pairs $\mathcal{G} - \mathcal{H}$, the relation $\mathcal{H} \to \mathcal{G}$ between them and nodal approximations for some balanced cubic minimal surfaces. $\mathcal{G}$ and $\mathcal{H}$ differ either in Bravais lattice or in point group, but not in both. Nodal approximations only consider the space group information given by $\mathcal{G} - \mathcal{H}$. For the simple cases, the first mode of the corresponding Fourier series is sufficient to obtain the correct topology.

	$\mathcal{H}$	$\mathcal{G}$	$\mathcal{H} \to \mathcal{G}$	nodal approximations
G	$I4_132$	$Ia\bar{3}d$	$432 \to m\bar{3}m$	$\sin(x)\cos(y) + \sin(y)\cos(z) + \sin(z)\cos(x)$
D	$Fd\bar{3}m$	$Pn\bar{3}m$	$F \to P$	$\cos(x-y)\cos(z) + \sin(x+y)\sin(z)$
P	$Pm\bar{3}m$	$Im\bar{3}m$	$P \to I$	$\cos(x) + \cos(y) + \cos(z)$
C(P)	$Pm\bar{3}m$	$Im\bar{3}m$	$P \to I$	$\cos(x) + \cos(y) + \cos(z) + 3\cos(x)\cos(y)\cos(z)$

Table 3. Nodal approximations for some non-balanced cubic minimal surfaces, for which $\mathcal{H} \equiv \mathcal{G}$. In these cases, more than one mode is needed.

	$\mathcal{G}$	nodal approximations
I-WP	$Im\bar{3}m$	$2[\cos(x)\cos(y) + \cos(y)\cos(z) + \cos(z)\cos(x)]$ $-[\cos(2x) + \cos(2y) + \cos(2z)]$
F-RD	$Fm\bar{3}m$	$4\cos(x)\cos(y)\cos(z)$ $-[\cos(2x)\cos(2y) + \cos(2y)\cos(2z) + \cos(2z)\cos(2x)]$

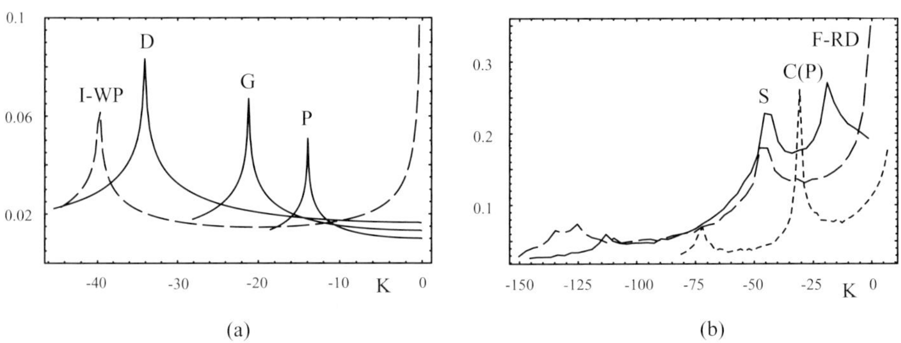

Fig. 7. Distribution $f(K)$ of Gaussian curvature K over the cubic minimal surfaces investigated in the framework of the Φ^6-model. (a) For the structure G, D, P and I-WP, the data is obtained from their exact Weierstrass representations; the agreement with the numerical results from the Φ^6-model is good (not shown). (b) For the structures F-RD, S and C(P), no Weierstrass representation is known, and the data is obtained from the Φ^6-model.

six modes of the improved nodal approximations are sufficient to decrease the deviation of the curvature properties from the real TPMS by one order of magnitude compared to the nodal approximations. Also we measured for the first time the distribution $f(K) = \int dA(u,v)\,\delta(K - K(u,v))$ of Gaussian curvature K over the surfaces. For this purpose, we used up to 100 Fourier modes. The distributions $f(K)$ for different structures are plotted in Fig. 7. In the cases for which Weierstrass representations are known, the same data has been derived from the exact representations, and the resulting agreement was very good. In general, we found that the surfaces C(D), C(P), F-RD, S and C(Y), for which no Weierstrass representations are known, are much more complicated than the cases G, D, P and I-WP. This is evidenced by larger values of the topology index Γ, multi-modal distributions $f(K)$ of Gaussian curvature K, and the larger widths of these distributions. The latter property can be quantified by defining a dimensionless variance

$$\Delta = \frac{\langle (K - \langle K \rangle)^2 \rangle}{\langle K \rangle^2} \tag{15}$$

where $\langle \ldots \rangle$ means area average. In Table 4 we give the variance Δ of the different distributions. Its value is the same for G, D and P due to the existence of a Bonnet-transformation between them. This means that G, D and P have the narrowest distribution of K (they are most uniformely curved), and all other structures have a considerably wider one, with I-WP being the next best structure.

Table 4. Variance Δ of the distributions of Gaussian curvature $f(K)$. The values for Δ are the same for G, D and P due to the existence of a Bonnet-transformation between them.

	G, D, P	I-WP	S	F-RD	C(P)
Δ	0.218702	0.482666	0.586079	0.649801	0.842022

4 Parallel Surfaces

We now consider the case of cubic bicontinuous phases in lipid-water mixtures. The two examples for the experimental phase behavior of such systems given in Fig. 5 show obvious similarities, and we will show now that a theoretical description can nicely explain these [99, 100]. It has been shown by a thorough analysis of electron density maps derived from X-ray data that the mid-surfaces of the cubic bicontinuous structures in these systems are very close to minimal surfaces [68]. Indeed if one considers the lipid bilayer as one entity, it has no spontaneous curvature by symmetry and one expects a minimal surface shape for the midsurface. Each of the two monolayers of the bilayer has unequal sides and therefore finite spontaneous curvature. For lipids monolayers, spontaneous curvature increases towards the water side as a linear function of temperature. From the interface point of view, one expects each monolayer to form a CMC-surface. However, then the distance to the minimal mid-surface would vary with position and the amphiphilic tails would have to stretch in order to fill the internal space of the membrane [3, 10]. It can be shown in the framework of a simple microscopic model that the relative importance of stretching to bending contributions to the free energy of the bilayer scales as $(a/\delta)^2$, where a is lattice constant and δ is tail length [100]. Therefore stretching is probibitively expensive and the two monolayers do not form CMC-surfaces as expected from the curvature energy of (1), but rather parallel surfaces to the minimal mid-surface. The free energy of the cubic bicontinuous phases then follows by specifying (1) with spontaneous curvature for two monolayers which are parallel surfaces to a given TPMS. Since the parallel surface geometry does not allow to completely relax the bending energy, the overall structure is called *frustrated* [3, 10, 20].

In principle, the analysis in the framework of the parallel surface model is rather simple, since there exist exact formulae which express the geometrical properties of the parallel surface as a function of the geometrical properties of the minimal surface:

$$dA^{\delta} = dA\,(1 + K\delta^2)\,,\, H^{\delta} = \frac{-K\delta}{1 + K\delta^2}\,,\, K^{\delta} = \frac{K}{1 + K\delta^2} \qquad (16)$$

where δ is the distance between the two surfaces (that is amphiphilic chain length). Note that these formulae are an extension of Steiner's theorem from integral geometry to non-convex bodies, which is valid as long as the distance δ is smaller than the smallest radius of curvature of the surface (otherwise the surface will self-intersect). Since mean curvature H vanishes on the reference surface, the bending energy now becomes a function only of its Gaussian curvature K. However, as we have seen above, K is distributed over the TPMS in a non-trivial way, which makes the detailed analysis rather complicated. Nevertheless, one central conclusion can be already made at this point: using (16) in (1) leads to an effective curvature energy for the lipid bilayer (to second order in δ) [85]

$$F_{bi} = \int dA\,\{4c_0^2\delta^2\kappa + (2\bar{\kappa} + 8c_0\delta\kappa + 4c_0^2\delta^2\kappa)K + 4\kappa\delta^2K^2\}\ . \qquad (17)$$

Thus we see that although effective spontaneous curvature vanishes due to the bilayer symmetry, the effective saddle-splay modulus, $\bar{\kappa}_{bi} = 2\bar{\kappa} + 8c_0\delta\kappa + O(\delta^2)$, is corrected to higher positive values due to the presence of the monolayer spontaneous curvature c_0. We conclude that as long as $c_0\delta \gtrsim -\bar{\kappa}/4\kappa$, the bicontinuous cubic phases are favored over the lamellar phase, since the preferred curvature of the monolayers translates into a topological advantage of saddle-type bilayer structures. We also see that the correction term in $\bar{\kappa}_{bi}$ is linear in c_0 and therefore in temperature T. Therefore bicontinuous phases become more favorable with increasing temperature in general (compare also Sect. 6 on disordered bicontinuous structures).

For a detailed analysis of cubic bicontinuous phases made from bilayers, we first note that the volume fraction of the lipid tails (the hydrocarbon volume fraction, which for simplicity we identify with the lipid volume fraction, since the lipid heads are rather small) can be calculated as

$$v = \frac{1}{a^3}\int_{-\delta}^{\delta} d\delta' \int dA^{\delta'} = 2A^*\left(\frac{\delta}{a}\right) + \frac{4\pi}{3}\chi\left(\frac{\delta}{a}\right)^3 \qquad (18)$$

where again we have used the Gauss-Bonnet theorem. (18) can be inverted to give a/δ, the lattice constant a in units of the chain length δ, as a function of hydrocarbon volume v. For small v we find

$$\frac{a}{\delta} = \frac{2A^*}{v}\left(1 + \frac{1}{12\,\Gamma^2}v^2 + \mathcal{O}(v^4)\right)\ . \qquad (19)$$

and one can check numerically that $a/\delta = 2A^*/v$ is an excellent approximation for $v \lesssim 0.8$. For larger values of v, the surfaces begin to self-intersect and our model becomes unphysical. Combining (1) and (16) yields for the bending energy per unit volume of the two monolayers (we use a factor $\delta/4\kappa c_0^2$ to write this expression dimensionless and a factor δ to write c_0 dimensionless)

$$f = v \left\{ \int \frac{dA^*}{A^*} \left(1 - \Xi(K^*) \left(\frac{v}{\Gamma}\right)^2\right)^{-1} \right. \\ \left. \times \left(1 - \frac{1+c_0}{c_0} \Xi(K^*) \left(\frac{v}{\Gamma}\right)^2\right)^2 + \frac{r}{4c_0^2} \left(\frac{v}{\Gamma}\right)^2 \right\} \tag{20}$$

where $K^* = Ka^2$ is scaled Gaussian curvature, δ/a has been replaced by $v/2A^*$, $\Xi(K^*)$ has been defined as $K^* A^*/8\pi\chi$ (this can be considered to be a local analogue of Γ^2), and r as $-\kappa/2\bar{\kappa}$ ($0 \leq r \leq 1$ due to the restrictions on $\bar{\kappa}$). The free-energy density f now is a function of lipid volume fraction v, spontaneous curvature c_0, the ratio r of the two bending constants, and the distribution $f(K)$ of Gaussian curvature K. The (numerical) evaluation of this expression is only possible with the knowledge of the $f(K)$ for all TPMS of interest, which have been numerically obtained in [97] as described above.

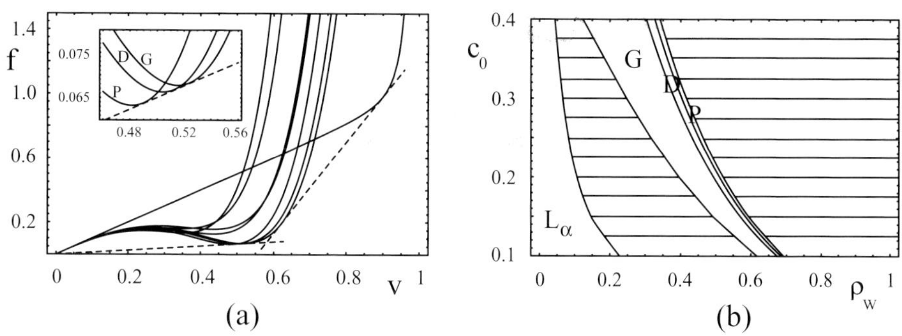

Fig. 8. For cubic bicontinuous phases in lipid-water mixtures, the lipid monolayers can be modeled as parallel surfaces to a minimal mid-surface. (a) Free energy densities f as a function of lipid volume fraction v for several cubic bicontinuous phases and the lamellar phase. (b) Theoretical phase diagram as a function of water volume fraction and (dimensionless) spontaneous curvature.

In the limit of a planar mid-surface (lamellar phase), (20) simplifies to $f = v$ - the larger the lipid volume fraction, the more frustrated bending energy per volume accumulates. For a full analysis, one also has to include the effect of thermal fluctuations. For the lamellar phase, they lead to a steric repulsion between the interfaces, which lead to an additional term $\sim v^3/(1-v)^2$ for the lamellar phase. For the cubic bicontinuous phases, steric repulsion is irrelevant since the lateral restriction on the scale of a lattice constant leads only to small perpendicular excursions. However, here the renormalization of the saddle splay modulus $\bar{\kappa}$ becomes relevant, and the formula given in (5) has to be incorporated. The renormalization of bending rigidity κ and spontaneous curvature c_0 is irrelevant here, since thermal fluctuations occur mainly on the level of the lipid bilayer, for which mean curvature H vanishes. For the lamellar phase, Gaussian curvature K vanishes as well, and the renormalization of all material parameters is irrelevant. Putting everything together, we can numerically calculate phase diagrams from the free energy densities of the different phases by using the Maxwell construction (construction of convex hull). In Fig. 8 we show both the free energy densities as a function of lipid volume fraction v (for fixed spontaneous curvature) and the theoretical phase diagram as a function of water volume fraction and spontaneous curvature. The main results are in excellent agreement with the experimental phase diagram shown in Fig. 5: from the many TPMS considered, only G, D and P are stable, their regions of stability have the shape of shifted parabolae, and they occur in the sequence L - G - D - P - emulsification failure. These results can be understood as follows: G, D and P can achieve the least frustration since they have the narrowest distribution of Gaussian curvature as measured by the variance Δ given in Table 4. This prediction has been stated before by Helfrich and Rennschuh [50], but at that time hardly any data was known to support it. The lipid volume fraction at which they achieve this can be estimated by setting the dimensionless mean curvature averaged over the parallel surface

$$\langle H^\delta \rangle_\delta \delta = \frac{\int dA^\delta \, H^\delta \delta}{\int dA^\delta} = \frac{(v/\Gamma)^2}{4 - (v/\Gamma)^2} \tag{21}$$

equal to the spontaneous curvature c_0. Therefore c_0 as a function of v essentially scales as $\sim (1 - \rho_W)^2/\Gamma^2$, where $\rho_W = 1 - v$ is the water volume fraction. This explains the characteristic shape of the bicontinuous stability regions in both the theoretical and experimental phase diagrams. Note that at high temperature, that is large spontaneous curvature, eventually the hexagonal phase will become stable, which cannot be treated in the framework presented here. The optimal lipid volume fraction v follows from (21) as

$$v = \left(\frac{4c_0}{1 + c_0} \right)^{\frac{1}{2}} \Gamma \,. \tag{22}$$

Therefore the structures G, D and P become stable in the sequence of their geometry index Γ, that is as G - D - P. If the water content corresponds to a average curvature of P which is larger than the optimal (spontaneous) curvature, some of the water is simply expelled from the structure, in order to keep the optimal curvature (emulsification failure). Finally it should be noted that the Bonnet-transformation connecting G, D and P causes the three structures to be stable along a triple line (see inset of Fig. 8). This means that the stability of D and P is very delicate and can easily be destroyed by additional physical effects. Therefore we conclude that in the framework of our interfacial approach, from all TPMS considered the gyroid G is the only phase which has a robust stability in lipid-water mixtures, since among the phases with favorable distribution of Gaussian curvature, its geometrical properties are closest to the ones of the lamellar phase.

5 Surfaces of Constant Mean Curvaturve (CMC)

Surfaces of constant mean curvature (CMC-surfaces) have $H = const$ everywhere, so minimal surfaces are a special case of CMC-surfaces. In contrast to minimal surfaces, CMC-surfaces with finite H can be compact, but the only CMC-surface which is compact and embedded is the sphere (for a long time, the sphere was believed to be the only compact CMC-surface, but in 1986 Wente found the first CMC-torus). Non-compact embedded CMC-surfaces are cylinder and unduloid, and all other known CMC-surfaces are doubly or triply periodic.

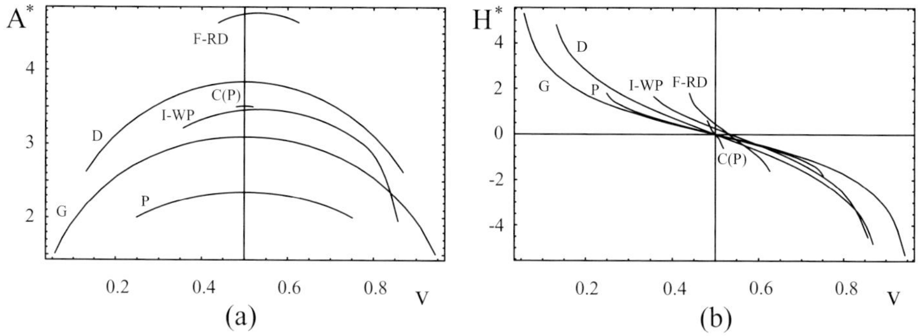

Fig. 9. Geometrical data for triply periodic surfaces of constant mean curvature: (a) scaled surface area A^* and (b) scaled mean curvature H^* as a function of volume fraction v. The two branches are symmetrical for G, D, P and C(P) since their minimal surface members are balanced.

CMC-surfaces are solutions to the variational problem of minimal surface area under a volume constraint. This can be shown as follows: we first introduce a Lagrange parameter for volume, that is pressure p. The corresponding energy is $-pV$. For normal variations $\delta\phi(u,v)$, we have $\Delta F = 2\delta\sigma \int dA\phi(u,v)H(u,v) - p\delta \int dA\phi(u,v) + O(\delta^2)$. If the surface is required to be stationary in regard to variations in δ, we obtain the Laplace equation $H = p/2\sigma$ and H is constant over the whole surface. This explains why soap bubbles and liquid droplets are spheres (respectively spherical caps when bound by a surface). In amphiphilic systems, CMC-surface arise in the presence of spontaneous curvature: like minimal surfaces minimize the Willmore functional $\int dAH^2$, CMC-surfaces with $H = c_0$ minimize the functional $\int dA(H - c_0)^2$. In any case, since H is constant over the surface, surface area and volume now vary in the same way, and we have $dA/dV = 2H$.

Table 5. Values for $c = dv(H^*)/dH^*|_{H^*=0}$, where $v(H^*)$ is the volume fraction of one of the two labyrinths for the corresponding family of surfaces of constant mean curvature. The measured values follow from our spline interpolation of the numerical data of [2]. In the second row we give $-{A_0}^2/2\pi\chi$, a new estimate for c derived in the text.

	G	D	I-WP	P	F-RD	C(P)
measured	0.2191	0.1411	0.1385	0.2117	0.0665	0.0466
approximation	0.1901	0.1465	0.1592	0.2188	0.0906	0.1226

All simple TPMS which are of interest for physical reasons are members of a family of CMC-surfaces [55]. Each of these families consists of two branches, corresponding to positive and negative mean curvatures, which are separated by the minimal surface member. Like the minimal surface member, each of the the triply periodic CMC-surfaces of the family partitions space into two intertwined, yet separate labyrinths, with volume fractions v and $1-v$. Each family of CMC-surfaces, and therefore the volume fractions of the two labyrinths and the surface area A, is parametrized by H. If the TPMS-member is balanced, then the operation α maps one branch of the family onto the other, in particular $v(H) = 1 - v(-H)$, $v(H = 0) = v_0 = 0.5$ and $A(H) = A(-H)$. For an amphiphilic monolayer of thickness δ in a ternary system, v can be identified with the hydrocarbon volume fraction (that is oil and amphiphilic tails), $1 - v$ with the water volume fraction (where we neglect the contributions of the amphiphilic heads), and $2A\delta$ with the amphiphile volume fraction. In the seminal work by Anderson, Davis, Nitsche and Sciven from 1990, the CMC-families were numerically constructed for P, D, I-WP, F-RD and C(P) using finite element methods [2]. In 1997, Grosse-Brauckmann numerically constructed the G-family in a similar way (using the software package *Surface Evolver*) [44]. For these families, the following data has been tabulated: volume fraction v as a function of scaled mean curvature H^* and scaled surface area A^* as a function of v. Rearrangement and interpolation with cubic splines provides smooth functions $H^*(v)$ and $A^*(v)$. As v is varied away from the value v_0 for the TPMS ($= 0.5$ for balanced structures), one has

$$
\begin{aligned}
H^*(v) &= -\frac{(v - v_0)}{c} + \mathcal{O}\left((v - v_0)^2\right) \\
A^*(v) &= A_0 - \frac{(v - v_0)^2}{c} + \mathcal{O}\left((v - v_0)^3\right)
\end{aligned}
\tag{23}
$$

where the numerical value of c can be extracted from the cubic splines (compare Table 5). Note that the two relationships are not independent due to the relation $dA/dV = 2H$. They are related through c, whose values are well approximated by the analogous values for the parallel surface companions to the TPMS, which can be derived as follows [98]: if δ denotes the perpendicular distance from the minimal to its parallel surface, to lowest order in δ the volume fraction v and the mean curvature H^*, averaged over the surface in the unit cell, are given by $v = v_0 + A^*\delta$ and $H^* = 2\pi\chi\delta/A^*$, respectively. Thus $H^* = 2\pi\chi(v - v_0)/A^{*2}$ and $c = -A^{*2}/2\pi\chi$ for the parallel surface case. The corresponding numbers are given in Table 5; except for C(P), the overall agreement with the numerical data for c for the CMC-surfaces is remarkably good.

Equation (23) is a useful approximation for CMC-surfaces close to the TPMS, where they behave similar to parallel surfaces. However, as mean curvature grows, the numerical data starts to deviate from these approximations, changes in v and A^* become slower, and a turning point is reached, where v reaches an extremal value and starts to decrease again as a function of H. Beyond the turning point, the surfaces correspond to nearly spherical regions connected by small necks which resemble pieces of unduloids. Finally these necks disappear and each branch terminates in an assembly of sphere, which might be close-packed or self-intersecting. Figure fig:cmc shows the geometrical data for the families G, D, P, C(P), I-WP, F-RD as tabulated in the literature [2, 44]. We do not show the parts of the data beyond the turning point, as these correspond to surfaces whose structure is unphysical.

We now consider a ternary mixture of water, oil and amphiphile, where amphiphiles self-assemble into monolayers with spontaneous curvature c_0. In such a system, all relevant phases (micellar, hexagonal, lamellar, cubic bicontinuous) can be modeled as CMC-surfaces (spheres, cylinders, planes, triply periodic CMC-surfaces). This approach has first been used by Safran and coworkers [94, 95, 116], who in particular discussed the surfaces from the D-family. In our work [98], we extended this analysis to all families of interest, including the G-family, for which the relevant data has become available only recently and which features very prominently in experimental systems. A ternary mixture has two independent degrees of freedom for concentration, which we choose to be the hydrocarbon volume fraction v and the ratio w of amphiphile to hydrocarbon volume fraction. For the formulae given later it is in fact useful to modify the definition of w and to scale it with the dimensionless spontaneous curvature c_0: $w = \rho_A / v c_0 \delta$. The analysis of (1) for CMC-surfaces is simplified considerably by the fact that as mean curvature H is constant, there is no difference between local and global curvature properties and the integral over the surface becomes trivial. We use a factor $2\kappa c_0^3$ to rewrite the free-energy density in the dimensionless form

$$f = \frac{A}{c_0 V} \left(\frac{H}{c_0} - 1 \right)^2 - \frac{2\pi \chi r}{c_0{}^3 V} \,. \tag{24}$$

For phases with lamellar, cylindrical and spherical aggregates, this free-energy density can easily be expressed as a function of the concentration degrees of freedom:

$$f_L(w,v) = wv , \tag{25}$$

$$f_C(w,v) = wv \left(\frac{w}{4} - 1 \right)^2 , \tag{26}$$

$$f_S(w,v,r) = wv \left[\left(\frac{w}{3} - 1 \right)^2 - \frac{r w^2}{9} \right] . \tag{27}$$

Note that only f_S depends on r, since the other two structures have no Gaussian curvature. Since the aggregates are disconnected, the dependence on v is trivial. The phase boundaries $S - C$, $S - L$, $C - L$ and the emulsification failure are obtained from (25), (26) and (27) to be $w = 24/(7 - 16r)$, $w = 6/(1 - r)$, $w = 8$ and $w = 3/(r - 1)$, respectively. For example, for $r = 0$ and increasing w, we find the phase sequence L - C - S - emulsification failure, which is typical for amphiphilic systems (for simplicity, we identify phase transitions with crossing points of the free-energy-density curves, and use the Maxwell construction only for the emulsification failure). In Fig. 4 one sees that experimentally the emulsification failure indeed occurs at constant w (straight line through water apex). For increasing r, the spherical phase becomes more favorable and finally supresses the cylindrical phase.

The structures based on CMC-surfaces discussed here are very different from the lipid bilayer structures discussed in the preceding section: now only one monolayer is present, and one of the two labyrinths is filled with hydrocarbon. In the following, they are called *single structures*. It should be noted that for non-balanced structures, it makes a difference which of the two labyrinth is filled with hydrocarbon; for example, the I-WP-family generates two single structures, which we call I and WP. In ternary amphiphilic systems there also exists an analogue to the bilayer structures discussed before, which we call *double structures* and mark with an index I. Double structures have the same geometries like cubic bicontinuous phases in diblock copolymer systems. They can be considered to be TPMS-based bilayer structures where the inner part of the bilayer has been swollen with suitable solvent. Since each of the two monolayers has the same spontaneous curvature c_0, a double structure can be modeled as the combination of the two surfaces of a CMC-family with $H = c_0$ and $H = -c_0$. Since for not too large

H these two CMC-surfaces essentially correspond to the shrinkage of one of the two separate labyrinths defined by the minimal surface member, the two interfaces of a double structure do not intersect. Note that in principle there exists another class of structures, that is double structures where oil and water (and therefore amphiphile orientation and mean curvature) have been reversed. However, since finite spontaneous curvature selects only phases with curvature towards one specific side, they are suppressed for energetic reasons by the lamellar phase.

In our work [98], we considered 8 different single structures, which exist for the volume intervals $[0.056, 0.944]$ for G, $[0.131, 0.869]$ for D, $[0.249, 0.751]$ for P, $[0.481, 0.519]$ for C(P), $[0.357, 0.857]$ for I, $[0.143, 0.643]$ for WP, $[0.439, 0.625]$ for F and $[0.375, 0.561]$ for RD. We also considered 6 double structures, which exist for the volume intervals $[0.112, 1.0]$ for G_I, $[0.262, 1.0]$ for D_I, $[0.498, 1.0]$ for P_I, $[0.962, 1.0]$ for $C(P)_I$, $[0.624, 1.0]$ for I-WP_I and $[0.818, 1.0]$ for F-RD_I. Note that the gyroid structures cover the largest intervals in v for their respective class: there is no other structure which can incorporate so extreme volume fractions like the gyroid. Together with the 3 non-cubic phases treated above, we considered 17 different phases. Since the cubic phases consist of one connected aggregate, the scaling of the free energy density f with hydrocarbon volume fraction v will be non-trivial. For a given value of v, the mean curvature $H(v,a) = H^*(v)/a$ and the surface area $A(v,a) = A^*(v)a^2$ within a unit cell are determined by the curves plotted in Fig. 9. The amphiphile concentration $\rho_A = A(v,a)/a^3 = A^*(v)/a$ fixes a, so that $a = A^*(v)/\rho_A = A^*(v)/(wvc_0)$. From (24) we can then derive

$$f_{BC}(w,v,r) = wv\left[\Lambda(v)\, wv - 1\right]^2 + r\,\frac{(wv)^3}{\Gamma(v)^2} \tag{28}$$

where we have used the definition for the curvature index Λ and the topology index Γ from (7). Note that now the indices are v-dependent. For the single structures, f follows by combining using the data shown in Fig. 9 with the definitions in (7). For the double structures, the procedure is somehow more complicated. However, for balanced double structures, it becomes simple again, one then can use

$$\Lambda_I(v) = \Lambda(v/2)/2, \Gamma_I(v) = 2\Gamma(v/2), f_{BC,I}(w,v,r) = 2f_{BC}(w, \frac{v}{2}, r) \tag{29}$$

in (28).

For $r = 0$ (vanishing saddle splay modulus $\bar{\kappa}$), the analysis of cubic bicontinuous phase behavior becomes rather simple: in contrast to the case of the parallel surface model discussed in the last section, now the bending energy given in (28) can be completely relaxed, namely by satisfying $w(v) = 1/v\Lambda(v)$. This leads to lines of vanishing frustration in the Gibbs triangle. For the free-energy density of the micellar and hexagonal phases ((27) and (26), respectively), the lines of vanishing frustration follow as $w = 3$ and $w = 4$, respectively. All lines of vanishing frustration are plotted in Fig. 10a. Obviously for this case phase behavior is very complex and degenerated, since every structure considered has some region of stability around its line of vanishing frustration. This type of degeneracy caused by the bending energy has been discussed before [6] and leads to the conclusion that additional physical effects have to be operative, as it is not observed in experimental systems.

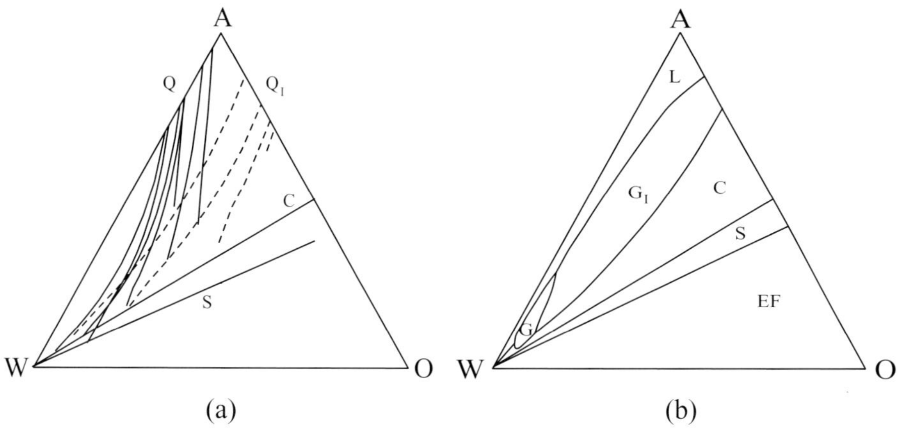

Fig. 10. (a) Lines of vanishing frustration for $r = 0$. With increasing amphiphile concentration, the structures S - C- double cubic - single cubic are stable. (b) Phase behavior for $r = 1/15$. Now all cubic phases have been supressed except the two gyroid phases. In (a) and (b), $c_0 = 1/6$, which corresponds to $H_2O/C_{14}/C_{12}E_5$ at $T = 20°C$.

Using (23), a simple approximation can be derived for the lines of vanishing frustration of the cubic phases:

$$w = -cA_0/v_0(v - v_0), w_I = -4cA_0/(v - 1)\,. \tag{30}$$

Thus the minimal surface case corresponds to the stable solutions for $w \gg 1$ at $v = v_0$ and $v = 1$, respectively. For the balanced single structures and the double structures the hierarchy of the different phases within the band-like region occupied by a certain structural type is thus determined by the values of cA_0. Using the approximation $c \approx -A^{*2}/2\pi\chi$ derived above, we find $cA^* \approx \Gamma^2$, where A^* and Γ correspond to the minimal surface members. Therefore the phase sequence is approximately determined by the topology index of the minimal-surface member of each family. In particular, for a given structural type we expect to find the sequence G - D - P as a function of either v or w.

For $r > 0$ (negative saddle splay modulus $\bar{\kappa}$), spheres become more favorable, cubic phases less favorable and cylinders and lamellae experience no change in free energy. Therefore the cubic phases will finally disappear, but our numerical analysis shows that cubic phases can persist up to $r = 0.2$, in contrast to earlier work, which predicted $r = 0.1$ [116]. The reason for this becomes clear in Fig. 10b where we show the full phase diagram for $r = 1/15 = 0.07$: the only cubic phases stable here are the two gyroid structures, which have not been considered before. There are two main reasons for their outstanding performance: (28) shows that their large values for the topology index Γ reduces the energetic penalty caused by r, and since they can accomodate extreme volume fractions, they can compete with other structures at all relevant concentrations. By comparing the theoretical phase diagram from Fig. 10 with the experimental one from Fig. 5, we conclude that the experimental system should correspond to a rather large value of r. Earlier work moreover suggests that incorporation of thermal fluctuations would favor micellar phases near the water-amphiphilic side and lamellar phases near the water apex [101]. Such a modification is expected to considerably improve the agreement between the two phase diagrams.

6 Random Surfaces

6.1 Microemulsion and Sponge Phases

When κ/k_BT, the bending rigidity in thermal units, becomes sufficiently low (that is temperature T sufficiently high), bicontinuous structures can now longer maintain their long-range crystalline order and melt into a disordered phase, which is characterized by an exponential decay of correlations in the interfacial positions. Such phases have been observed experimentally for a long time in many binary and ternary amphiphilic systems. *Microemulsions* are macroscopically homogeneous and optically isotropic mixtures of oil, water and amphiphiles. On a mesoscopic scale, they consist of two multiply connected and intertwined networks of oil- and water-channels, which are separated by an amphiphilic monolayers. Free-fracture microscopy, where the sample is quickly frozen, cut, and then studied with an electron microscope, reveals the intriguing structure of this phase [52], see Fig. 11. A similar phase, the *sponge phase*, appears in binary systems of water and amphiphile, where now the two labyrinths are occupied by water, which are separated by an amphiphilic bilayer. The pictures obtained by free-fracture microscopy [107] are even more suggestive in this case, because the sample has a preference to break along the bilayer mid-surface, so that the three-dimensional structure of the membrane becomes visible. An example is shown in Fig. 12, which clearly shows the saddle-like geometry of the amphiphile film. Therefore, the intuitive picture of microemulsion and sponge phases as fluid versions of bicontinuous cubic phases is strongly supported by these experiments.

These phases have been investigated experimentally in considerable detail over many years. In particular, their phase behavior and scattering intensities have been studied carefully. A theoretical understanding of the statistical mechanics of membranes, however, is only beginning to emerge in recent years [75, 40, 78, 18, 35]. This is no surprise, since the statistical mechanics of a surface, which can not only change its shape, but also its topology in all possible ways, is extremely complicated. In principle, a partition function of the form

$$Z = \sum_{topologies} \int' \mathcal{D}\mathbf{R}(\tau) \exp\{\mathcal{H}[\mathbf{R}(\tau)]/k_BT\} \tag{31}$$

has to be calculated, where $\mathcal{D}\mathbf{R}(\tau)$ denotes an integration over all possible shapes with parametrization $\mathbf{R}(\tau)$ of the surface at fixed topology, where τ is a two-dimensional coordinate system on the surface. However, this integral cannot be just over all possible parametrization $\mathbf{R}(\tau)$ of a surface of fixed topology, but has to be restricted to those parametrizations, which lead to physically different shapes in the embedding space; this is indicated by the prime. Finally, the contributions off all different topologies have to be summed over. It is clear that this problem is sufficiently complex that no exact solution will be found anytime soon. Therefore, approximations have to be made in order to get some insight into the behavior of these phases.

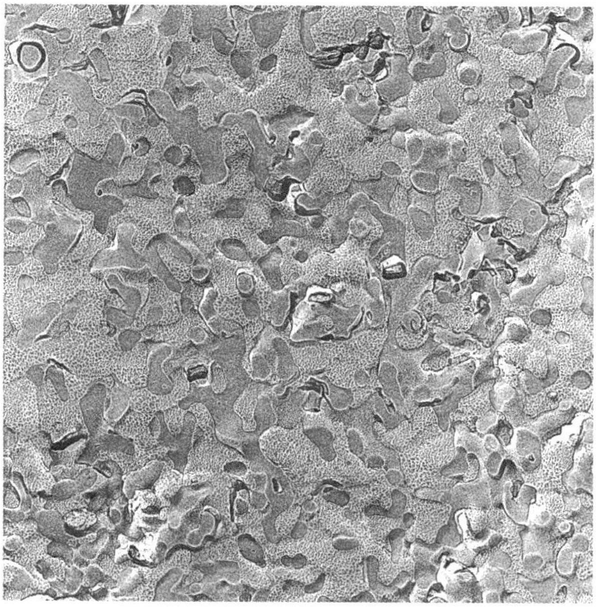

Fig. 11. Freeze-fracture microscopy picture of a balanced microemulsion phase. From [52].

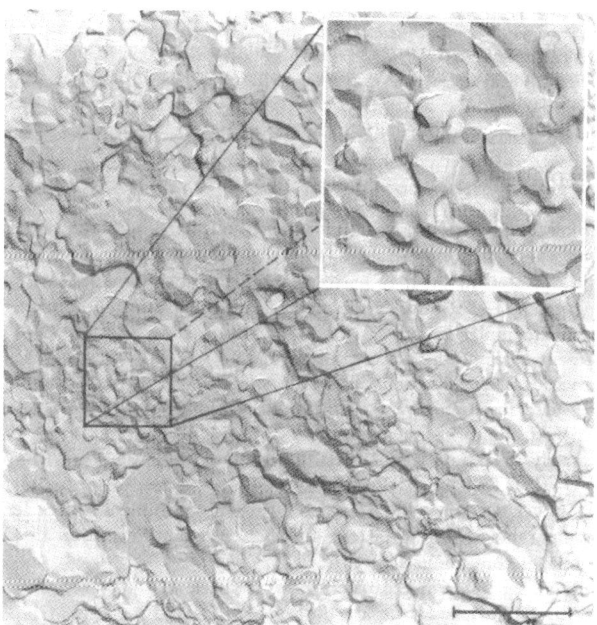

Fig. 12. Freeze-fracture microscopy picture of a sponge phase. From [107].

6.2 Gaussian Random Fields

A very useful approach is to describe the interfaces as isosurfaces of *Gaussian random fields* (GRF). This corresponds to a Ginzburg-Landau model as discussed in Sect. 2.2, in which the free-energy functional is taken to be quadratic in the scalar field $\Phi(\mathbf{r})$. This Gibbs distribution is homogeneous and isotropic, and all the functional integrals are Gaussian and can be performed exactly. In fact, there is a considerable mathematical literature on the isosurfaces of GRF, see e.g. the book by Adler [1]. In order to comply with the conventions of this field, we now make two changes to our notation. In the following, area content is denoted by S (rather than by A) and amphiphilic (hydrocarbon) volume fraction by Ψ (rather than by v).

The amphiphilic monolayers in microemulsions have been modelled as level surfaces of GRF by Berk [4, 5], Teubner [112], Pieruschka and Marcelja [81], and Pieruschka and Safran [82, 83]. Here the starting point is a Gaussian free-energy functional of the general form

$$\mathcal{H}_0[\phi] = \frac{1}{2}\int d^3q\, \nu(\mathbf{q})^{-1}\Phi(\mathbf{q})\Phi(-\mathbf{q}) \,. \tag{32}$$

The average geometry of the $\Phi(\mathbf{q}) = \alpha$ level surfaces can be calculated for arbitrary spectral density $\nu(\mathbf{q})$. For the surface density, S/V, the mean curvature H, the Gaussian curvature K, and the mean curvature squared H^2, the following averages are obtained [112]:

$$\frac{S}{V} = \frac{2}{\pi}\exp\left[-\frac{\alpha^2}{2}\right]\sqrt{\frac{1}{3}\langle q^2\rangle}\,, \tag{33}$$

$$\langle K\rangle = -\frac{1}{6}\langle q^2\rangle(1-\alpha^2)\,, \tag{34}$$

$$\langle H\rangle = \frac{1}{2}\alpha\sqrt{\frac{\pi}{6}\langle q^2\rangle}\,, \tag{35}$$

$$\langle H^2\rangle = \langle K\rangle + \frac{1}{5}\frac{\langle q^4\rangle}{\langle q^2\rangle} \tag{36}$$

where

$$\langle q^n\rangle = \int \frac{d^3q}{(2\pi)^3} q^n \nu(\mathbf{q}) \,. \tag{37}$$

The value α of the level cut can be used to describe the preferred curvature of the membrane, as well as the volume fractions of oil and water. Since $\langle H\rangle$ is a linear function of α, compare (35), this parameter is proportional to the spontaneous curvature c_0. In particular, for $\alpha = 0$ the mean curvature of the surface vanishes; this applies to a balanced system, where $c_0 = 0$.

For balanced systems, it follows from (33,34) that the topology index $\Gamma = \sqrt{8}/\pi = 0.9003$ (compare (7)). This value is only slightly larger than the ones for cubic bicontinuous phases (compare Table 1), since the balanced random sponge features only few disconnected parts (note that Γ doubles when the structure is duplicated). For small curvatures (that is small α), one can use (33,34,35) to derive a relationship between topology index Γ and curvature index Λ which is independent of spectral density $\nu(\mathbf{q})$ and α:

$$\Gamma = \frac{\sqrt{8}}{\pi}\left(1 + \frac{2^8}{\pi^6}\Lambda^4 + \dots\right) . \tag{38}$$

When the curvature index Λ increases since the sponge's interfaces gain curvature, the topology index increases, too, since disconnected parts proliferate.

The GFF-approach is most predictive when the Gaussian model of random interfaces is related to the statistical mechanics of membranes by a variational approximation [82, 83]. In this case, the spectral density $\nu(\mathbf{q})$ in the functional (32) is determined by the requirement that the $\phi(\mathbf{r}) = 0$ level surfaces mimic the behavior of interfaces controlled by the curvature Hamiltonian (1) as close as possible. The usual variational approach employs the Feynman-Bogoljubov inequality,

$$F \leq F_0 + \langle \mathcal{H} - \mathcal{H}_0 \rangle_0 \tag{39}$$

where $\mathcal{H}$ and F are the Hamiltonian and the free energy of the system of interest, respectively, and $\mathcal{H}_0$ and F_0 the same quantities of the reference system. In this way, an upper bound for the true free energy is obtained. This is more complicated in the case of random surfaces, because the GRF-Hamiltonian is defined everywhere in space, while the curvature Hamiltonian is only defined on the level surface. Therefore, the curvature energy does not restrict fluctuations of the field $\Phi(\mathbf{r})$ away from the level surface. In order to suppress such fluctuations, one usually makes the *mean-spherical approximation*, that is the constraint $\langle \Phi(\mathbf{r})^2 \rangle = 1$ is introduced.

With this variational approach, Pieruschka and Safran [82] have been able to derive the following form for the spectral density

$$\nu(\mathbf{q}) = \frac{a}{q^4 + bq^2 + c} \tag{40}$$

and to relate the parameters a, b and c to the curvature elastic moduli κ and $\bar{\kappa}$ and the surface density S/V. The first interesting result is that the spectral density is found to be *independent* of $\bar{\kappa}$. Exact expressions for the parameters can be found in [22]. To leading order in $k_B T/\kappa$, the parameters simplify to

$$a = \frac{15\pi^2}{16}\frac{k_B T}{\kappa}\frac{S}{V} , \tag{41}$$

$$b = \frac{3}{2}\pi^2 \left(\frac{S}{V}\right)^2 , \tag{42}$$

$$c = \left(\frac{3\pi^2}{4}\right)^2 \left(\frac{S}{V}\right)^4 . \tag{43}$$

The spectral density (40) is equivalent to the scattering intensity in bulk contrast. Its Fourier transform yields the correlation function

$$\langle \Phi(\mathbf{r})\Phi(\mathbf{r}') \rangle = \int d^3q \, e^{i\mathbf{q}\cdot(\mathbf{r}-\mathbf{r}')} \nu(\mathbf{q}) = \frac{A}{r} \exp[-r/\xi] \sin(kr) \tag{44}$$

where the second equality holds for $|b| < 2\sqrt{c}$. Thus, the correlation function is characterized by two length scales, the correlation length ξ and the typical domain size $d = 2\pi/k$ of the oil- of water channels, which are obtained from (40) and (44) to be [113]

$$\xi^{-1} = \frac{1}{2}\sqrt{2\sqrt{c}+b}\,, \tag{45}$$

$$k = \frac{1}{2}\sqrt{2\sqrt{c}-b}\,. \tag{46}$$

With the results (41), (42) and (43), the asymptotic behavior for small k_BT/κ of the dimensionless product $k\xi$ is found to be

$$k\xi = \frac{64}{5\sqrt{3}}\frac{\kappa}{k_BT}\,. \tag{47}$$

The free energy of the sponge phase can also be calculated from the GRF approach. To leading order in an expansion in k_BT/κ, the free-energy density $f = F/V$ is found to be [83]

$$f = \frac{\pi^2}{40}[2\kappa - 5\bar{\kappa}]\left(\frac{S}{V}\right)^3 - \frac{k_BT}{12}\ln\left(\delta\frac{S}{V}\right)\,. \tag{48}$$

This implies that for small membrane volume fractions $\Psi = \delta S/V$ (where again δ is the thickness of the amphiphilic interface), the entropic term dominates over the energy term, and that the sponge phase becomes unstable in this regime.

We want to end this section with a short discussion of the reliability of the predictions of the GRF model as a variational approximation for membrane ensembles. The weak point of this approach is that it is not clear whether the calculated entropy is actually equivalent to the physical conformational entropy of the membranes [74]. The main problem is that the curvature energy only controls the shape of the $\Phi(\mathbf{r}) = 0$ level surface, while the values of the scalar field Φ at all other points in space are not affected by it. The fluctuations of Φ in these oil- and water-regions are mainly determined by the mean-spherical constraint $\langle\Phi^2(\mathbf{r})\rangle = 1$. Obviously, an appreciable contribution to the total entropy arises from the fluctuations of Φ these 'bulk' regions. This would not affect the predictions of the model as long as the 'bulk' contributions were independent of the interface positions. Unfortunately, there is no argument so far that this is indeed the case.

6.3 Phase Behavior of Random Surfaces

From here on we consider only the case of vanishing spontaneous curvature, $c_0 = 0$, that is balanced microemulsions and sponge phases. Then the phase behavior is controlled by the bending rigidity κ, the saddle-splay modulus $\bar{\kappa}$, the density of membrane area per volume, S/V, and a microscopic cutoff δ, which can be identified with the thickness of the amphiphilic interface. It is very important to realize that for vanishing spontaneous curvature, phase transition as a function of the amphiphile volume fraction cannot be understood on the basis of the curvature energy alone — i.e. without considering the effect of thermal fluctuations [85]. The reason is that the curvature Hamiltonian is conformally invariant in three spatial dimensions, which implies in particular that it is invariant under a simultaneous rescaling of all length scales. Since the curvature energy is scale invariant, the energy *density* scales as the third power of an inverse length, i.e. as $(S/V)^3$. Therefore, the curvature energy of any given structure — spherical, cylindrical, lamellar, cubic, or random — scales in exactly the same way with decreasing amphiphile volume fraction, and their relative order is maintained. Therefore, thermal fluctuations are crucial for these phase transitions.

It has been suggested by several authors [93, 9, 73, 28] that the free energy of the sponge phase can be obtained by integrating out the membrane fluctuations on scales less than the typical domain size. This integration over small-scale fluctuations leads to renormalized, scale-dependent curvature moduli $\kappa_R(l)$ and $\bar{\kappa}_R(l)$ as given by (4) and (5), respectively, with $l/\delta = \Psi^{-1}$, where Ψ is the membrane volume fraction. This implies that the curvature Hamiltonian (2) has to be replaced by

$$F = \int dA \left\{ \frac{1}{2}\kappa_{+,R}(\Psi^{-1})(k_1 + k_2)^2 + \frac{1}{2}\kappa_{-,R}(\Psi^{-1})(k_1 - k_2)^2 \right\} \tag{49}$$

with

$$\kappa_{+,R}(\Psi^{-1}) = \kappa_+ + \frac{k_B T}{3\pi} \ln \Psi \,, \tag{50}$$

$$\kappa_{-,R}(\Psi^{-1}) = \kappa_- + \frac{5 k_B T}{12\pi} \ln \Psi \,. \tag{51}$$

The stability arguments used in Sect. 2.1 imply that $\kappa_{+,R}$ and $\kappa_{-,R}$ have to be positive for the free energy (49) to be stable against collapse of the structure to molecular scales. Therefore, there are instabilities at $\kappa_{+,R}(\Psi^{-1}) = 0$ and $\kappa_{-,R}(\Psi^{-1}) = 0$. The latter instability can be identified with the emulsification failure of the sponge phase, so that the phase boundary is predicted to occur at [73]

$$\ln \Psi = \frac{6\pi}{5} \frac{\bar{\kappa}}{k_B T} \,. \tag{52}$$

This result can be understood intuitively as follows. For sufficiently large membrane volume fraction, both $\kappa_{+,R}$ and $\kappa_{-,R}$ are positive. Therefore, the system tries to minimize $(k_1 + k_2)^2$ and $(k_1 - k_2)^2$. This can be achieved by decreasing both k_1 and k_2, i.e. by swelling a given structure as much as possible — the lamellar or sponge phase is stable at this value of Ψ. On the other hand, as soon as $\kappa_{+,R}$ or $\kappa_{-,R}$ become negative at some small value of Ψ, the free energy can be reduced by collapsing the structure. With decreasing length, however, $\kappa_{+,R}$ and $\kappa_{-,R}$ increase and finally become positive. Therefore, the collapse stops at a length scale which is exactly the length scale determined by (52).

It is worth mentioning that a similar result follows from the calculation of the shapes and free energies of passages in lamellar phases [31]. Passages are catenoid-like connections between adjacent lamellae which proliferate close to phase transitions to disordered bicontinuous phases. From a detailed calculation, which takes into account the Gaussian membrane fluctuations, the density, ρ, of passages per unit base area of the stack can be obtained to be

$$\rho = \frac{A_0(\kappa/k_BT)}{\delta^2}\left(\frac{d}{\delta}\right)^{4/3} \exp[4\pi\bar{\kappa}/k_BT] \tag{53}$$

Here, $A_0(\kappa/k_BT)$ is a function which has to be calculated numerically; it is predicted to increase monotonically with decreasing κ/k_BT in [31]. Therefore, the density of passages increases with increasing lamellar spacing $d/\delta = \Psi^{-1}$. When the average distance between passages shrinks to the average membrane separation, a transition to the sponge phase can be expected to occur. This happens at

$$\frac{10}{3}\ln \Psi/\Psi^* = 4\pi\frac{\bar{\kappa}}{k_BT} \tag{54}$$

where Ψ^* is a function of κ/k_BT. Equations (52) and (54) agree perfectly.

The consideration of the renormalization of κ and $\bar{\kappa}$ implies that the free-energy density, f, of the sponge phase should behave as [89, 87, 90]

$$f = (A + B\ln\Psi)\Psi^3 \ . \tag{55}$$

As explained above, the overall scaling with Ψ^3 derives from the conformal invariance of the curvature energy. A represents the bending energy without thermal fluctuations, and is a linear function of both κ and $\bar{\kappa}$. B represents the logarithmic corrections from the renormalization and is a linear function of temperature. Both A and B depend on the detailed geometrical structure of the sponge phase, and thus cannot be obtained from simple scaling arguments. It is important to note that the functional dependence of the free energy (55), which is based on the renormalization of the curvature elastic moduli due to *small-scale* membrane fluctuations, does not agree with the free energy (48) of the Gaussian random field model, which includes the topological entropy of a disordered bicontinuous phase.

6.4 Monte Carlo Simulations of Triangulated Surfaces

In order to go beyond the approximations discussed in Sects. 6.2 and 6.3, discretized surface models can be investigated by Monte Carlo simulations [35, 34, 36, 37]. Surface triangulations provide the best way of discretizing a surface as uniformly as possible. The model consists of vertices, which are connected by bonds in such a way that the bonds form a triangular network. Several Hamiltonians have been suggested for triangulated surfaces, such that their shapes and fluctuations are governed in the continuum limit by the curvature energy from (1) [35]. Two examples are Gaussian-spring models, in which neighboring vertices — the vertices connected by bonds — interact with harmonic spring potentials, and tether-and-bead models, in which the interaction $V(r)$ between neighboring vertices is defined the potential

$$V(r) = \begin{cases} \infty & 0 \le r < \sigma_0 \\ 0 \quad \text{for} & \sigma_0 < r < \ell_0 \\ \infty & \ell_0 < r \end{cases} \tag{56}$$

where σ_0 is the hard-core diameter of the beads, and ℓ_0 is the tether length. For tether lengths $\ell_0 < \sqrt{3}\sigma_0$, the network is self-avoiding, because a bead of some distant part of the membrane does not fit through the hole between the three beads of a triangle, even for maximally stretched tethers.

A Monte Carlo simulation of sponge phases requires three types of Monte Carlo moves [36]. The first step is to displace selected beads by a random vector chosen uniformly in the cube $[-s, s]^3$. Here, s is determined by the criterion that roughly 50% of trial moves is accepted according to the Boltzmann weight. The second step is to vary the connectivity of the bond network, in order to allow for diffusion and fluidity of the membrane. This dynamic triangulation is performed for any two adjacent triangles. The bond which forms the common edge of the two triangles is cut, and a new bond is inserted, which connects the two previously unconnected beads. The bond can only be cut if every bead retains bonds to at least three other beads. This procedure guarantees that the network remains two-dimensionally connected, no holes open up in the network, and the topology of the network does not change. Finally, the third step is to change the topology of the surface. This is done by removing two nearby triangles from the surface, and by connecting the corresponding vertices by a prism of six new triangles. Of course, the inverse step is also possible, with a passage six triangles being removed, and two new triangles inserted to close the surfaces. The acceptance of both the bond-flip and topology-change moves is determined by the Boltzmann weight and is therefore controlled by the curvature energy. We want to mention parenthetically that some care has to be taken to find a good discretization of the bending energy [33].

A typical configuration of a triangulated surface in a cubic box with parameters κ, $\bar{\kappa}$ and Ψ chosen in the stability region of the microemulsion or sponge phase is shown in Fig. 13. This configuration nicely demonstrates the bicontinuous structure of balanced microemulsions and sponge phases. The saddle-like geometry of the membrane can also be easily seen. Finally, the figure shows that *locally* the structure of the sponge phase strongly resembles the cubic phases discussed above. Therefore, a sponge phase should indeed be considered as the molten state of the crystalline cubic phase.

A more quantitative comparison with the theoretical approaches of Sects. 6.2 and 6.3 above can be made by determining the phase diagram of the randomly-triangulated surface model, and by calculating the osmotic pressure, p, in the simulations as a function of the membrane volume fraction Ψ. From (55), we obtain

$$\begin{aligned} p\delta^3/k_BT &\equiv \frac{1}{k_BT}[\Psi\partial f/\partial\Psi - f] \\ &= [(2A+B) + 2B\ln\Psi]\,\Psi^3 \end{aligned} \tag{57}$$

i.e. the same functional dependence as the free-energy density itself. This dependence of the osmotic pressure is indeed nicely confirmed by the simulation data [36]. The simulations therefore provide strong evidence for the renormalization of the elastic moduli of the curvature model and for the dependence (55) of the free energy on the membrane volume fraction.

The phase diagram for fixed bending rigidity κ is shown as a function of $\bar{\kappa}$ and Ψ in Fig. 14. The simulation data are compared with the prediction (52) for the phase boundary. Since the slopes of the phase boundaries in this logarithmic plot agree very well, not only the exponential dependence of the membrane volume fraction at the transition on the saddle splay modulus is confirmed, but also the value of the universal prefactor in (52) is strongly supported.

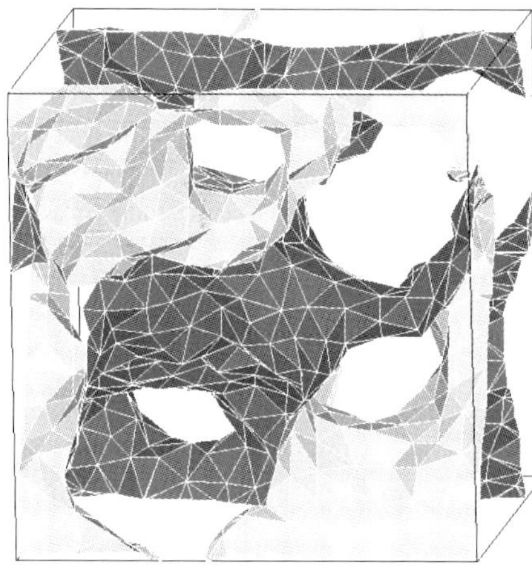

Fig. 13. A typical membrane configuration in a sponge phase for bending rigidity $\kappa/k_BT \simeq 1.6$. The two sides of the membrane are shaded differently in order to emphasize the bicontinuous structure of this phase. From [36].

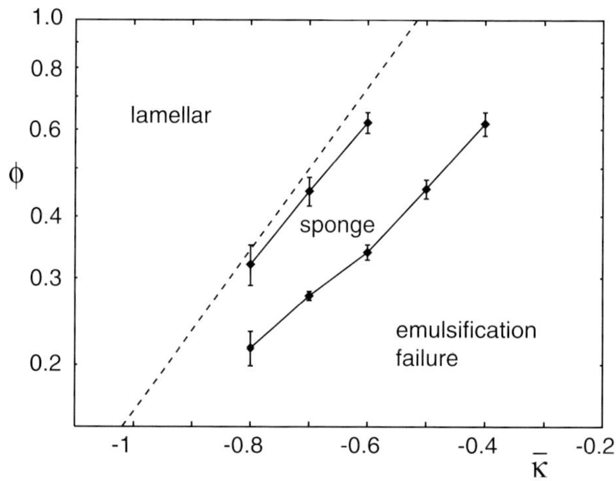

Fig. 14. The phase diagram as a function of membrane volume fraction Φ and saddle-splay modulus $\bar{\kappa}$, for bending rigidity $\kappa/k_BT \simeq 1.6$. Note the logarithmic scale of the abscissa. The dashed line shows the theoretical prediction (52). From [36].

6.5 Comparison with Experiments

Experimentally, phase diagrams and scattering intensities have been studied systematically for many different surfactant molecules. Qualitatively, the agreement with the theoretical approaches is very reasonable. For example, the scattering curve in bulk contrast shows a peak at non-zero wave vector in the microemulsion phase, which moves out and decreaeses in height with increasing surfactant concentration. A quantitative comparison, however, is much more difficult. We want to discuss here three different classes of experiments, where a such a quantitative comparison has been made.

The first type of experiments are scattering studies in bulk and film contrast. In bulk contrast, the scattering intensity $S_{ww}(\mathbf{q})$ is proportional to the spectral density $\nu(\mathbf{q})$ of the Gaussian random field model, compare (40), for wave vectors q which are not much larger than the characteristic wave vector k of the domain structure, compare (46). The functional dependence of (40) describes the scattering data in this regime very well, as was first noted by Teubner and Strey [113], who derived this result on the basis of a Ginzburg-Landau model, very similar to that introduced in Sect. 2.2. We want to mention parenthetically, that for wave vectors $k \ll q \ll 1/\delta$, the intensity is dominated by the scattering from sharp, planar interfaces; in this limit, the famous Podod law [84, 112] predicts $I(q) \sim (S/V)q^{-4}$.

In the limit of wave vector $q \to 0$, the scattering intensity in film contrast is given by [87]

$$I(q \to 0) \sim \Psi \left(\frac{\partial p}{\partial \Psi} \right)^{-1} \tag{58}$$

where p is the osmotic pressure of (57). For the free energy (55), this implies

$$\left[\Psi I(q \to 0) \right]^{-1} \sim const + \ln \Psi \tag{59}$$

Such a behavior has indeed be observed experimentally in [87]. However, this result has been questioned by Daicic et al. [14, 13]. This has lead to a intensive debate, with arguments against [15, 16] and in favor [91, 86] of the existence of a logarithmic renormalization of elastic moduli in the sponge phase.

In a second type of experiment, information about the average geometry of the surfactant film can be extracted from the scattering intensity in the regime $k < q \ll 1/\delta$. This information is contained in the corrections to the asymptotic $1/q^4$ law for smaller values of the wave vector [111]. Experimentally, the average Gaussian curvature is found to be [11]

$$(V/S)^2 \langle K \rangle = 1.25 \pm 0.10 \tag{60}$$

which is in excellent agreement with the Gaussian random field result $(V/S)^2 \langle K \rangle = 1/\Gamma^2 = -\pi^2/8 = 1.23$ and simulations of the Φ^6-Ginzburg-Landau model [32, 30].

The third type of experiment concerns the phase behavior of mixtures of water and non-ionic surfactant as a function of temperature and surfactant concentration. As mentioned in Sect. 4, the saddle-splay modulus $\bar{\kappa}$ is a linear function of temperature in this case. Therefore, the concentrations at the phase boundaries of the lamellar and the sponge phase are expected from (52) to depend exponentially on temperature. A logarithmic plot of the phase diagram of $C_{12}E_5$ in water [109], compare Fig. 4, is indeed consistent with this expectation.

The most detailed information about both the scattering intensities and phase behavior in these systems has been obtained very recently in ternary amphiphilic systems of water, oil and non-ionic surfactant C_iE_j, to which small traces of an amphiphilic block copolymer has been added [53, 21, 22, 29, 38]. The results obtained in this system provide further, strong evidence for the existence of a logarithmic renormalization of κ and $\bar{\kappa}$ in microemulsions and sponge phases.

7 Summary and Outlook

In this contribution we discussed the geometrical properties of surfaces which can be used as structural models for cubic bicontinuous phases in amphiphilic systems: triply periodic minimal surfaces, their parallel surfaces and constant mean curvature companions, and bicontinuous random surfaces. For each class of surfaces, we showed how the geometrical properties translate into physical properties of bicontinuous phases in amphiphilic systems in the framework of an interfacial description. The surprising success of this approach relates to the fact that in amphiphilic systems, the solvent has little physical properties by itself and the free energy is essentially determined by the interfaces. Although there are several physical effects which have been neglected in our treatment of amphiphilic systems, including van der Waals and electostatic interactions, it can be concluded that the most essential aspects of phase behavior are now well understood in terms of the properties of the underlying geometries.

During the last years, the interest in bicontinuous phases has increased due to some promising applications in the nano- and biosciences. For example, amphiphilic self-assembly has been used to synthesize *mesoporous systems* [58], that is porous material with amorphous walls (usually silica-based) and pore sizes on the nanometer scale. As a matter of fact, the combination of amphiphilic self-assembly and crystallization also seems to be a basic aspect of biomineralization [70], and there are many algae whose mineral skeletons look similar to cubic bicontinuous phases [51]. There is also a large effort underway to synthesize bicontinuous structures from graphitic material (*Schwarzites*), which offers the advantage of atomically smooth walls [69]. Porous material on the nanometer scale is not only interesting for its structural properties (for example use as sieves or catalysts), but also for its electronic properties (for example use as photonic bandgap material).

In the biosciences, two recent developments involved cubic bicontinuous phases. It was found that the gyroid phase in the monoolein-water system provides a functional environment for the crystallization of integral membrane proteins like bacteriorhodopsin [77, 66]. Membrane proteins are notoriously difficult to crystallize in a three-dimensional array, which however is necessary for structure determination by X-ray scattering. And an extensive analysis of transmission electron micrographs of many biological specimen has shown that cubic bicontinuous structures are locally formed in several regions of the cells [61, 19], including endoplasmatic reticulum, Golgi apparatus and mitochondria. These are extended lipid bilayer systems in the cell which carry a protein machinery responsible for processing and packaging material inside the cell. In fact it is easy to imagine that biological cells regulate their lipid composition and therefore their spontaneous curvature as to achieve geometries favorable for the task at hand. In general, cubic bicontinuous phases seem to be ideal space partitioners if a large amount of active surface area is needed while simultaneously providing good access for incoming and outgoing material through the space away from the interface.

What problems are left regarding the subjects of this article ? From the viewpoint of mathematics, there are two interesting aspects. On the one hand, the physical motivation leads to certain variational problems for which little is known about the corresponding solutions. One example is the energy functional $\int dA(H-c_0)^2$ with a volume constraint; in principle this problem was subject of Sect. 5, but in lack of adequate mathematical knowledge (this problem has been treated for vesicles [104], but not for bicontinuous structures), we used CMC-surfaces as structural models. On the other hand, some classes of surfaces are well investigated, but still representations are missing which could be used in physical modelling. Therefore additional progress in regard to Weierstrass representations for TPMS, Weierstrass-like representations for their CMC-companions, and other (numerical) representations would be most welcome. With a larger repertoire of representations, it should also become easier to treat non-local effects like van der Waals and electrostatic interactions; here integral geometry might be helpful in reducing the dimensions of the corresponding integrals, and sophisticated summation techniques are needed to account for the infinite extension of the bulk phases.

From the viewpoint of material sciences, it seems worth to study the role of boundaries and inclusions. The term *boundaries* includes grain boundaries between different, yet adjacent phases, for example between a hexagonal and a gyroid phase; until now, this problem has been investigated only for simple geometries. It also includes the boundary to an external surface (which might even be curved or chemically structured) and free boundaries. For example, it has been recently observed experimentally that the bicontinuous cubic phase in the binary system water - $C_{12}E_6$ is the only known material system which shows crystal faces with high Miller indices (*devil's staircase*) [80]. The term *inclusions* includes several object of biological and technological relevance, in particular integral membrane proteins, polymers anchored in the membrane and (functionalized) nanoparticles. In all of these cases, the framework described here should provide a good starting point for future research, which has to bridge the gap between the mathematical knowledge on the geometrical properties and the physics knowledge of the material basis of these fascinating structures.

8 WWW Resources

Here we list some internet addresses which relate to the subject of this contribution:

- `http://www-sfb288.math.tu-berlin.de/~konrad/`:
  Homepage of Konrad Polthier, one of the pioneers for numerical representations and visualizations of minimal surfaces.
- `http://people.math.uni-bonn.de/kgb/`:
  Homepage of Karsten Grosse-Brauckmann, a mathematician working on CMC-surfaces, in particular on the G-family.
- `http://www.susqu.edu/facstaff/b/brakke/evolver/`:
  The *Surface Evolver* is a public domain software package written by Ken Brakke

for the propagation of triangulated surfaces. In particular, Ken Brakke shows many examples of TPMS obtained with the *Surface Evolver*.

- `http://www.msri.org/publications/sgp/SGP/indexc.html`:
  The Scientific Graphics Project offers stunning visualizations of minimal and CMC-surfaces, obtained mostly in the framework of Weierstrass representations.
- `http://www.gang.umass.edu/`:
  The Center for Geometry, Analysis, Numerics and Graphics (GANG), where a lot of work on CMC-surfaces takes place.
- `http://www.mpikg-golm.mpg.de/th/people/schwarz/`:
  More information on the work described here, including color pictures of all structures investigated and (improved) nodal approximations.

References

1. Adler, R. J., ed. (1981): *The Geometry of Random Fields* (Wiley, Chichester)
2. Anderson, D. M., H. T. Davis, L. E. Scriven, J. C. C. Nitsche (1990): 'Periodic surfaces of prescribed mean curvature', *Adv. Chem. Phys.* **77**, pp. 337–397
3. Anderson, D. M., S. M. Gruner, S. Leibler (1988): 'Geometrical aspects of the frustration in the cubic phases of lyotropic liquid crystals', *Proc. Natl. Acad. Sci. USA* **85**, pp. 5364–5368
4. Berk, N. F. (1987): 'Scattering properties of a model bicontinuous structure with a well defined length scale', *Phys. Rev. Lett.* **58**, p. 2718
5. Berk, N. F. (1991): 'Scattering properties of the leveled-wave model of random morphologies', *Phys. Rev. A* **44**, p. 5069
6. Bruinsma, R. (1992): 'Elasticity and excitations of minimal crystals', *J. Phys. II France* **2**, pp. 425–451
7. Cai, W., T. C. Lubensky, P. Nelson, T. Powers (1994): 'Measure factors, tension and correlations of fluid membranes', *J. Phys. II France* **4**, pp. 931–949
8. Canham, P. B. (1970): 'The minimum energy of bending as a possible explanation of the biconcave shape of the human red blood cell', *J. Theor. Biol.* **26**, p. 61
9. Cates, M. E., D. Roux, D. Andelman, S. T. Milner, S. A. Safran (1988): 'Random surface model for the L_3-phase of dilute surfactant solutions', *Europhys. Lett.* **5**, p. 733
10. Charvolin, J., J. F. Sadoc (1990): 'Structures built by amphiphiles and frustrated fluid films', *J. Phys. (Paris) Colloq.* **51**, pp. 83–96
11. Chen, S.-H., D. D. Lee, K. Kimishima, H. Jinnai, T. Hashimoto (1996): 'Measurement of the Gaussian curvature of the surfactant film in an isometric bicontinuous one-phase microemulsion', *Phys. Rev. E* **54**, pp. 6526–6531
12. Cvijovic, D., J. Klinowski (1994): 'The computation of the triply periodic I-WP minimal surface', *Chem. Phys. Lett.* **226**, pp. 93–99
13. Daicic, J., U. Olsson, H. Wennerström, G. Jerke, P. Schurtenberger (1995): 'Light scattering from the L_3 (sponge) phase: Evidence against logarithmic corrections to ideal scaling', *Phys. Rev. E* **52**, pp. 3266–3269
14. Daicic, J., U. Olsson, H. Wennerström, G. Jerke, P. Schurtenberger (1995): 'Thermodynamics of the L_3 (sponge) phase in the flexible surface model', *J. Phys. II France* **5**, pp. 199–215
15. Daicic, J., U. Olsson, H. Wennerström, G. Jerke, P. Schurtenberger (1996): 'Reply to "comment on 'thermodynamics of the L_3 (sponge) phase in the flexible surface model' "', *J. Phys. II France* **6**, pp. 95–96

16. Daicic, J., U. Olsson, H. Wennerström, G. Jerke, P. Schurtenberger (1997): 'Reply to "comment on 'light scattering from the L_3 (sponge) phase: Evidence against logarithmic corrections to ideal scaling' "', *Phys. Rev. E* **56**, pp. 1278–1279
17. David, F. (1989): 'Geometry and field theory of random surfaces and membranes', in Nelson, D., T. Piran, S. Weinberg, eds., 'Statistical Mechanics of Membranes and Surfaces', (World Scientific, Singapore), pp. 157–223
18. David, F., P. Ginsparg, J. Zinn-Justin, eds. (1996): *Fluctuating geometries in statistical mechanics and field theory* (Elsevier, Amsterdam)
19. Deng, Y., M. Marko, K. F. Buttle, A. Leith, M. Mieczkowski, C. A. Mannella (1999): 'Cubic membrane structure in amoeba (chaos carolinensis) mitochondria determined by electron microscopic tomography', *J. Struct. Biol.* **127**, pp. 231–239
20. Düsing, P. M., R. H. Templer, J. M. Seddon (1997): 'Quantifying packing frustration energy in inverse lyotropic mesophases', *Langmuir* **13**, pp. 351–359
21. Endo, H., J. Allgaier, G. Gompper, B. Jakobs, M. Monkenbusch, D. Richter, T. Sottmann, R. Strey (2000): 'Membrane decoration by amphiphilic block copolymers in bicontinuous microemulsions', *Phys. Rev. Lett.* **85**, pp. 102–105
22. Endo, H., M. Mihailescu, M. Monkenbusch, J. Allgaier, G. Gompper, D. Richter, B. Jakobs, T. Sottmann, R. Strey, I. Grillo (2001): 'Effect of amphiphilic block copolymers on the structure and phase behavior of oil-water-surfactant mixtures', *J. Chem. Phys.* **115**, pp. 580–600
23. Fischer, W., E. Koch (1987): 'On 3-periodic minimal surfaces', *Z. Kristallogr.* **179**, pp. 31–52
24. Fischer, W., E. Koch (1996): 'Spanning minimal surfaces', *Phil. Trans. R. Soc. Lond. A* **354**, pp. 2105–2142
25. Fodgen, A., S. T. Hyde (1992): 'Parametrization of triply periodic minimal surfaces. I. Mathematical basis of the construction algorithm for the regular class', *Acta Cryst. A* **48**, pp. 442–451
26. Fodgen, A., S. T. Hyde (1999): 'Continuous transformations of cubic minimal surfaces', *Eur. Phys. J. B* **7**, pp. 91–104
27. Fontell, K. (1990): 'Cubic phases in surfactant and surfactant-like lipid systems', *Colloid Polym. Sci.* **268**, pp. 264–85
28. Golubović, L. (1994): 'Passages and droplets in lamellar fluid membrane phases', *Phys. Rev. E* **50**, pp. R2419–R2422
29. Gompper, G., H. Endo, M. Mihailescu, J. Allgaier, M. Monkenbusch, D. Richter, B. Jakobs, T. Sottmann, R. Strey (2001): 'Measuring bending rigidity and spatial renormalization in bicontinuous microemulsions', *Europhys. Lett.* **56**, pp. 683–689
30. Gompper, G., J. Goos (1994): 'Fluctuating interfaces in microemulsion and sponge phases', *Phys. Rev. E* **50**, pp. 1325–1335
31. Gompper, G., J. Goos (1995): 'Fluctuations and phase behavior of passages in a stack of fluid membranes', *J. Phys. II France* **5**, pp. 621–634
32. Gompper, G., M. Kraus (1993): 'Ginzburg-Landau theory of ternary amphiphilic systems II: Monte Carlo Simulations', *Phys. Rev. E* **47**, p. 4301
33. Gompper, G., D. M. Kroll (1996): 'Random surface discretizations and the renormalization of the bending rigidity', *J. Phys. I France* **6**, pp. 1305–1320
34. Gompper, G., D. M. Kroll (1997): 'Fluctuations of polymerized, fluid and hexatic membranes: continuum models and simulations', *Curr. Opin. Colloid Interface Sci.* **2**, pp. 373–381
35. Gompper, G., D. M. Kroll (1997): 'Network models of fluid, hexatic and polymerized membranes', *J. Phys. Condens. Matter* **9**, pp. 8795–8834
36. Gompper, G., D. M. Kroll (1998): 'Membranes with fluctuating topology: Monte carlo simulations', *Phys. Rev. Lett.* **81**, pp. 2284–2287

37. Gompper, G., D. M. Kroll (2000): 'Statistical mechanics of membranes: Freezing, undulations, and topology fluctuations', *J. Phys. Condens. Matter* **12**, pp. 29–37
38. Gompper, G., D. Richter, R. Strey (2001): 'Amphiphilic block copolymers in oil-water-surfactant mixtures: Efficiency boosting, structure, phase behavior, and mechanism', *J. Phys. Condens. Matter* **13**, pp. 9055–9074
39. Gompper, G., M. Schick (1990): 'Correlation between structural and interfacial properties of amphiphilic systems', *Phys. Rev. Lett.* **65**, pp. 1116–1119
40. Gompper, G., M. Schick (1994): 'Self-assembling amphiphilic systems', in Domb, C., J. L. Lebowitz, eds., 'Phase transitions and critical phenomena', volume 16 (Academic Press, London), pp. 1–176
41. Gompper, G., S. Zschocke (1992): 'Ginzburg-Landau theory of oil-water-surfactant mixtures', *Phys. Rev. A* **46**, pp. 4836–4851
42. Góźdź, W., R. Hołyst (1996): 'High genus periodic gyroid surfaces of nonpositive Gaussian curvature', *Phys. Rev. Lett.* **76**, pp. 2726–2729
43. Góźdź, W., R. Hołyst (1996): 'Triply periodic surfaces and multiply continuous structures from the Landau model of microemulsions', *Phys. Rev. E* **54**, pp. 5012–5027
44. Große-Brauckmann, K. (1997): 'On gyroid interfaces', *J. Colloid Interface Sci.* **187**, pp. 418–428
45. Hajduk, D. A., P. E. Harper, S. M. Gruner, C. C. Honeker, G. Kim, E. L. Thomas, L. J. Fetters (1994): 'The gyroid: a new equilibrium morphology in weakly segregated diblock copolymers', *Macromolecules* **27**, pp. 4063–4075
46. Hajduk, D. A., P. E. Harper, S. M. Gruner, C. C. Honeker, E. L. Thomas, L. J. Fetters (1995): 'A reevaluation of bicontinuous cubic phases in starblock copolymers', *Macromolecules* **28**, pp. 2570–2573
47. Helfrich, W. (1973): 'Elastic properties of lipid bilayers: theory and possible experiments', *Z. Naturforsch. C* **28**, pp. 693–703
48. Helfrich, W. (1981): 'Amphiphilic mesophases made of defects', in Balian, R., M. Kleman, J.-P. Poirier, eds., 'Physics of Defects', Les Houches Summer School 1980 (North Holland, Amsterdam)
49. Helfrich, W. (1985): 'Effect of thermal undulations on the rigidity of fluid membranes and interfaces', *J. Phys. France* **46**, pp. 1263–1268
50. Helfrich, W., H. Rennschuh (1990): 'Landau theory of the lamellar-to-cubic phase transition', *J. Phys. (Paris) Colloq.* **51**, pp. 189–195
51. Hildebrandt, S., A. Tromba (1996): *Kugel, Kreis und Seifenblasen. Optimale Formen in Geometrie und Natur* (Birkhäuser Verlag, Basel)
52. Jahn, W., R. Strey (1988): 'Microstructure of microemulsions by freeze fracture microscopy', *J. Phys. Chem.* **92**, p. 2294
53. Jakobs, B., T. Scottmann, R. Strey, J. Allgaier, L. Willner, D. Richter (1999): 'Amphiphilic block coloymers as efficiency boosters for microemulsions', *Langmuir* **15**, pp. 6707–6711
54. Jost, J. (1994): *Differentialgeometrie und Minimalflächen* (Springer-Verlag, Berlin)
55. Karcher, H. (1989): 'The triply periodic minimal surfaces of Alan Schoen and their constant mean curvature companions', *Manuscripta Math.* **64**, pp. 291–357
56. Karcher, H., K. Polthier (1996): 'Construction of triply periodic minimal surfaces', *Phil. Trans. R. Soc. Lond. A* **354**, pp. 2077–2104
57. Koch, E., W. Fischer (1988): 'On 3-periodic minimal surfaces with non-cubic symmetry', *Z. Kristallogr.* **183**, pp. 129–152
58. Kresge, C. T., M. E. Leonowicz, W. J. Roth, J. C. Vartulli, J. S. Beck (1992): 'Ordered mesoporous molecular sieves synthesized by a liquid-crystal template mechanism', *Nature* **359**, pp. 710–712

59. Kunieda, H., K. J. Shinoda (1982): 'Phase behavior in systems of nonionic surfactant/water/oil around the hydrophile-lipophile-balance-temperature (HLB-temperature)', *J. Dispersion Sci. Technol.* **3**, p. 233
60. Landau, L. D., E. M. Lifshitz (1970): *Theory of elasticity*, volume 7 of *Course of Theoretical Physics* (Pergamon Press, Oxford), 2nd edition
61. Landh, T. (1995): 'From entangled membranes to eclectic morphologies: cubic membranes as subcellular space organizers', *FEBS Letters* **369**, pp. 13–17
62. Lidin, S., S. T. Hyde, B. W. Ninham (1990): 'Exact construction of periodic minimal surfaces: the I-WP surface and its isometries', *J. Phys. France* **51**, pp. 801–813
63. Lipowsky, R. (1991): 'The conformation of membranes', *Nature* **349**, pp. 475–481
64. Lipowsky, R., E. Sackmann (1995): *Structure and Dynamics of Membranes*, volume 1A and 1B of *Handbook of biological physics* (Elsevier, Amsterdam)
65. Longley, W., T. J. McIntosh (1983): 'A bicontinuous tetrahedral structure in a liquid-crystalline lipid', *Nature* **303**, pp. 612–614
66. Luecke, H., H.-T. Richter, J. K. Lanyi (1998): 'Proton transfer pathways in bacteriorhodopsin at 2.3 Ångstrom resolution', *J. Struct. Biol.* **121**, pp. 82–91
67. Luzzati, V., P. A. Spegt (1967): 'Polymorphism of lipids', *Nature* **215**, pp. 701–704
68. Luzzati, V., R. Vargas, P. Mariani, A. Gulik, H. Delacroix (1993): 'Cubic phases of lipid-containing systems', *J. Mol. Biol.* **229**, pp. 540–551
69. Mackay, A. L., H. Terrones (1991): 'Diamond from graphite', *Nature* **352**, p. 762
70. Mann, S., G. A. Ozin (1996): 'Synthesis of inorganic materials with complex form', *Nature* **382**, pp. 313–318
71. Matsen, M. W., F. S. Bates (1996): 'Unifying weak- and strong-segregation block copolymer theories', *Macromolecules* **29**, pp. 1091–1098
72. Mitchell, D. J., G. J. T. Tiddy, L. Waring, T. Bostock, M. P. McDonald (1983): 'Phase behaviour of polyoxyethylene surfactants with water', *J. Chem. Soc. Faraday Trans.* **79**, pp. 975–1000
73. Morse, D. C. (1994): 'Topological instabilities and phase behavior of fluid membranes', *Phys. Rev. E* **50**, pp. R2423–R2426
74. Morse, D. C. (1997): 'Entropy and fluctuations of monolayers, membranes, and microemulsions', *Curr. Opin. Coll. Interface Sci.* **2**, pp. 365–372
75. Nelson, D., T. Piran, S. Weinberg, eds. (1989): *Statistical Mechanics of Membranes and Surfaces* (World Scientific, Singapore)
76. Olsson, U., U. Würz, R. Strey (1993): 'Cylinders and bilayers in a ternary nonionic surfactant system', *J. Phys. Chem.* **97**, pp. 4535–4539
77. Pebay-Peyroula, E., G. Rummel, J. P. Rosenbusch, E. M. Landau (1997): 'X-ray structure of bacteriorhodopsin at 2.5 Ångstroms from the microcrystals grown in lipidic cubic phases', *Science* **277**, pp. 1676–1681
78. Peliti, L. (1996): 'Amphiphilic membranes', in David, F., P. Ginsparg, J. Zinn-Justin, eds., 'Fluctuating Geometries in Statistical Mechanics and Field Theory', (North-Holland, Amsterdam), pp. 195–285
79. Peliti, L., S. Leibler (1985): 'Effects of thermal fluctuations on systems with small surface tension', *Phys. Rev. Lett.* **54**, pp. 1690–1693
80. Pieranski, P., P. Sotta, D. Rohe, M. Imperor-Clerc (2000): 'Devil's staircase-type faceting of a cubic lyotropic liquid crystal', *Phys. Rev. Lett.* **84**, pp. 2409–2412
81. Pieruschka, P., S. Marčelja (1992): 'Statistical mechanics of random bicontinuous phases', *J. Phys. II France* **2**, p. 235
82. Pieruschka, P., S. A. Safran (1993): 'Random interfaces and the physics of microemulsions', *Europhys. Lett.* **22**, p. 625
83. Pieruschka, P., S. A. Safran (1995): 'Random interface model for sponge phases', *Europhys. Lett.* **31**, p. 207

84. Porod, G. (1951): 'Die Röntgenkleinwinkelstreuung von dichtgepackten kolloiden Systemen 1', *Kolloid Z. Z. Polym.* **124**, p. 83–114
85. Porte, G. (1992): 'Lamellar phases and disordered phases of fluid bilayer membranes', *J. Phys.: Condens. Matter* **4**, pp. 8649–8670
86. Porte, G., J. Appell, J. Marignan (1997): 'Comment on "light scattering from the L_3 (sponge) phase: Evidence against logarithmic corrections to ideal scaling"', *Phys. Rev. E* **56**, pp. 1276–1277
87. Porte, G., M. Delsanti, I. Billard, M. Skouri, J. Appell, J. Marignan, F. Debeauvais (1991): 'Scaling laws for some physical properties of the L_3 (sponge) phase', *J. Phys. II France* **1**, p. 1101
88. Qiu, H., M. Caffrey (2000): 'The phase diagram of the monoolein/water system: metastability and equilibrium aspects', *Biomaterials* **21**, pp. 223–234
89. Roux, D., M. E. Cates, U. Olsson, R. C. Ball, F. Nallet, A. M. Bellocq (1990): 'Light scattering from a surfactant 'sponge' phase: Evidence for a hidden symmetry', *Europhys. Lett.* **11**, p. 229
90. Roux, D., C. Coulon, M. E. Cates (1992): 'Sponge phases in surfactant solutions', *J. Phys. Chem.* **96**, p. 4174
91. Roux, D., F. Nallet, C. Coulon, M. E. Cates (1996): 'Comment on "thermodynamics of the L_3 (sponge) phase in the flexible surface model"', *J. Phys. II France* **6**, pp. 91–93
92. Safran, S. A. (1999): 'Curvature elasticity of thin films', *Adv. Phys.* **48**, pp. 395–448
93. Safran, S. A., D. Roux, M. E. Cates, D. Andelman (1986): 'Origin of middle-phase microemulsions', *Phys. Rev. Lett.* **57**, p. 491
94. Safran, S. A., L. A. Turkevich (1983): 'Phase diagrams for microemulsions', *Phys. Rev. Lett.* **50**, pp. 1930–1933
95. Safran, S. A., L. A. Turkevich, P. Pincus (1984): 'Cylindrical microemulsion: a polymer-like phase?', *J. Physique - LETTRES* **45**, pp. L69–L74
96. Schoen, A. H. (1970): 'Minimal surfaces', NASA Technical Note D-5541, Washington, D.C.
97. Schwarz, U. S., G. Gompper (1999): 'A systematic approach to bicontinuous cubic phases in ternary amphiphilic systems', *Phys. Rev. E* **59**, pp. 5528 – 5541
98. Schwarz, U. S., G. Gompper (2000): 'Stability of bicontinuous cubic phases in ternary amphiphilic systems', *J. Chem. Phys.* **112**, pp. 3792 – 3802
99. Schwarz, U. S., G. Gompper (2000): 'Stability of inverse bicontinuous cubic phases in lipid-water mixtures', *Phys. Rev. Lett.* **85**, pp. 1472–1475
100. Schwarz, U. S., G. Gompper (2001): 'Bending frustration of lipid-water mesophases based on cubic minimal surfaces', *Langmuir* **17**, pp. 2084–2096
101. Schwarz, U. S., K. Swamy, G. Gompper (1996): 'The lamellar-to-isotropic transition in ternary amphiphilic systems', *Europhys. Lett.* **36**, pp. 117–122
102. Scriven, L. E. (1976): 'Equilibrium bicontinuous structure', *Nature* **263**, pp. 123–125
103. Seddon, J. M., R. H. Templer (1995): 'Polymorphism of lipid-water systems', in Lipowsky, R., E. Sackmann, eds., 'Structure and dynamics of membranes - from cells to vesicles', volume 1A of *Handbook of biological physics* (Elsevier, Amsterdam), pp. 97–160
104. Seifert, U. (1997): 'Configurations of fluid membranes and vesicles', *Adv. Phys.* **46**, pp. 13–137
105. Senechal, M. (1990): *Crystalline symmetries - an informal mathematical introduction* (Adam Hilger, Bristol)
106. Shmueli, U., ed. (1996): *International tables for crystallography. Volume B: Reciprocal space* (Kluwer Academic Publishers, Dordrecht)
107. Strey, R., W. Jahn, M. Skouri, G. Porte, J. Marignan, U. Olsson (1992): 'Fluid membranes in the water/NaCl-AOT system: A study combining small-angle neutron scattering, electron microscopy and NMR self-diffusion', in Chen, S.-H., J. S. Huang, P. Tartaglia, eds.,

'Structure and Dynamics of Strongly Interacting Colloids and Supramolecular Aggregates in Solution', (Kluwer, Dordrecht), pp. 351–363
108. Strey, R., R. Schomäcker, D. Roux, F. Nallet, U. Olsson (1990): 'Dilute lamellar and L_3 phases in the binary water - $C_{12}E_5$ system', *J. Chem. Soc. Faraday Trans.* **86**, pp. 2253–2261
109. Strey, R., R. Schomäcker, D. Roux, F. Nallet, U. Olsson (1990): 'On the dilute lamellar and L_3 phases in the binary water-$C_{12}E_5$ system', *J. Chem. Soc. Faraday Trans.* **86**, p. 2253
110. Templer, R. H., J. M. Seddon, N. A. Warrender, A. Syrykh, Z. Huang, R. Winter, J. Erbes (1998): 'Inverse bicontinuous cubic phases in 2:1 fatty acid/phosphatidylcholine mixtures. the effect of chain length, hydration, and temperature', *J. Phys. Chem. B* **102**, pp. 7251–7261
111. Teubner, M. (1990): 'Scattering from two-phase random media', *J. Chem. Phys.* **92**, p. 4501
112. Teubner, M. (1991): 'Level surfaces of gaussian random fields and microemulsions', *Europhys. Lett.* **14**, p. 403
113. Teubner, M., R. Strey (1987): 'Origin of the scattering peak in microemulsions', *J. Chem. Phys.* **87**, p. 3195
114. Thomas, E. L., D. B. Alward, D. J. Kinning, D. C. Martin, D. L. Handlin, L. J. Fetters (1986): 'Ordered bicontinuous double-diamond structure of star block copolymers - a new equilibrium microdomain morphology', *Macromolecules* **19**, pp. 2197–2202
115. von Schnering, H. G., R. Nesper (1991): 'Nodal surfaces of Fourier series: fundamental invariants of structured matter', *Z. Phys. B* **83**, pp. 407–412
116. Wang, Z.-G., S. A. Safran (1990): 'Curvature elasticity of ordered bicontinuous structures', *Europhys. Lett.* **11**, pp. 425–430

Morphology of Langmuir Monolayer Phases

Mathias Lösche and Peter Krüger

Leipzig University, Institute of Experimental Physics I, Linnéstr. 5, D-04103 Leipzig, Germany

Abstract. Micrometer-scale domains formed during the first-order 'main' phase transition within Langmuir monolayers show characteristic shapes that are a consequence of the molecular interactions within the surface film. The relevant Hamiltonian includes the (one-dimensional) line tension, long-range dipole-dipole repulsion and a spontaneous curvature deriving from the chirality of the constituent molecules. Experimental observations have been compared with computer models in which domains were approximated by finite segments and their energy minimized with respect to the layout of the segments within the plane. Several levels of sophistication have been analyzed: (a) Elongated domains that are isolated within the film area, such that domain-domain interactions may be neglected; (b) isolated domains with non-trivial shapes; (c) elongated domains in a crowded situation where domain-domain interactions have been taken into account. Since nucleation events lead to a random distribution of individual domains within an ensemble in the latter case, the experimental results and the segmented models are at best *similar* in their pattern formation. These similarities are quantitatively compared using *Minkowski* functionals.

1 Introduction

Biological membranes – or 'biomembranes' – are complex, quasi-twodimensional lipid-protein structures that form in aqueous media by self-assembly from molecular building blocks and constitute arguably the most important elements that help distinguish biological cells from their non-biological environment. Moreover, biomembranes form compartments within cells and are key ingredients in a broad range of processes that keep life going – from biosynthesis to signal processing. Although biomembranes contain a host of different molecular species, the main types of molecules that define their physical properties are (a) lipids which provide the structural basis of their organization and (b) proteins that supply them with a large variety of functions [40]. For a quantitative understanding of the physics that underlies the functioning of cells an assessment of structural and dynamical aspects is indispensable, as is the quantification of the molecular interactions within the membrane. The full complexity of biomembranes – in which dozens of similar yet distinct lipid species and, depending on their function, a similar number of different proteins may interact – is by far too large for a direct physical characterization. Hence, membrane models [22] have been developed that contain typically only a few of the key ingredients, in chemically pure form, yet are hoped to help reveal the basic physical principles that underlie the properties of complex biomembranes. A particularly simple membrane model is a lipid surface layer on an aqueous subphase [1, 39], a so-called Langmuir film [19]. Due to the amphiphilic character of lipid molecules – one end of which is hydrophilic and the other hydrophobic – they form insoluble surface films that are strictly monomolecular. Lipids are co-aligned within the film with their hydrophobic chains pointing toward the gas phase and their hydrophilic headgroups pointing toward the aqueous subphase. Langmuir films are easy to prepare and to control in their physical properties, typically in a flat container with hydrophobized (Teflon, PTFE) walls – a Langmuir trough. For example, the area per molecule, A, within the surface film may be readily controlled by means of a simple sliding barrier that sits on the trough's edges and confines the total surface area. As the film forms, it reduces the surface tension of the free water surface, γ_0, to a value γ and thus exerts a 'lateral' (one-dimensional, 1D) pressure, $\pi = \gamma_0 - \gamma$. Langmuir films of pure compounds, such as phosphatidylcholines, phosphatidylethanolamines or phosphatidic acids with fatty acid chains esterified to their glycerol backbones, e.g., L-α-dipalmitoylphosphatidylcholine (DPPC), undergo phase transitions that are clearly revealed from $\pi - A$ isotherms [1]. In Langmuir films of simple, one-chain amphiphiles, such as fatty acids or fatty esters, there exists a host of phases – closely related to liquid crystalline phases [6] – that are distinguished by the degree of order and the symmetry of the molecular arrangements of their hydrophobic chains [14]. In comparison, the phase diagram of phospholipids is simpler (c.f. Fig. 1), with a gaseous (G) phase at large A, followed by a disordered ('liquid expanded', LE) phase, a phase of hexatic order ('tilted condensed', TC) and an ordered phase ('solid condensed', SC) as the monolayer is compressed and/or cooled [7].

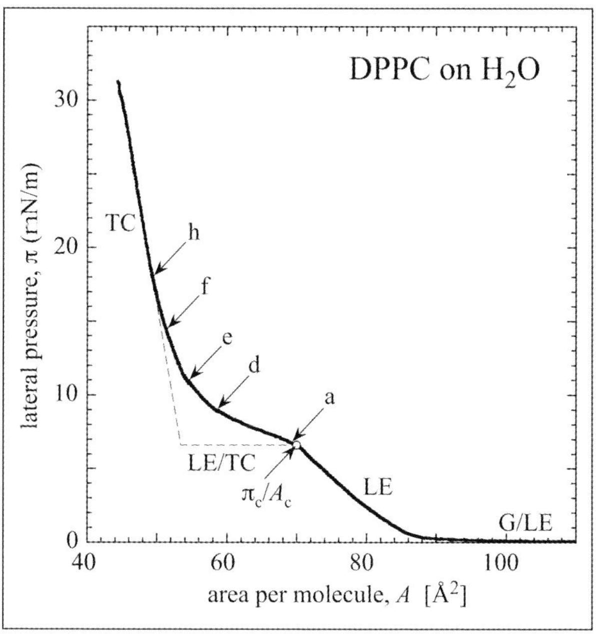

Fig. 1. π-A isotherm of L-α-DPPC near room temperature. Indicated are some of the locations where the fluorescence micrographs presented in Fig. 2 have been obtained. Also shown is the kink point π_c/A_c that marks the onset of the LE/TC phase transition.

As shown by fluorescence microscopy of Langmuir monolayers [23] both the G/LE and LE/TC phase transitions are of first order [24]. In the case of the LE/TC transition, this is revealed from the observation that mesoscopic, i.e. μm scale, domains of the phase of higher intrinsic order coexist with areas of phase of lower intrinsic order in the course of these transitions [16, 42]. Contrast in such images is achieved due to the fact that the fluorescently labeled lipid marker, while water-insoluble by itself and thus confined to the surface monolayer together with the majority component of the lipid, is dissolved in much higher concentration into a disordered phase than into a phase of higher intrinsic order. Thus, images such as those presented in Fig. 2 have to be interpreted such that the dark areas correspond to the more ordered phase (TC) while the interconnecting brightly fluorescent matrix consists of phase areas with higher intrinsic disorder, i.e., they are of the LE phase. The experimentally observed shapes of the phase coexistence domains have been the subject of numerous studies [29] in which the line tension between the coexisting 2D phases and the electrostatic repulsion between collinear dipoles were identified as the main interactions within the monolayers [9, 30]. Descriptive assessments of the domain forms in monolayers of DPPC [27] and related lipids [28] close to the equilibrium have been performed in recent work. These studies have determined that the most probable, 'fundamental' [27] form observed upon slow compression and/or long-term annealing is a characteristic bean shape with a subtle asymmetry for which the molecular chirality was held responsible. In fact, molecular chirality has been earlier identified as a further determinant of domain shapes [10, 42, 43], but only recently has its impact been quantitatively described [17]. The present paper summarizes recent work on the evaluation of domain shapes in phospholipid Langmuir monolayers and focuses particularly on their quantification by unleashing the power of integral geometry [3, 36, 41].

2 Experimental

Chloroform was from Merck (Darmstadt, Germany) and methanol from Sigma-Aldrich (Deisenhofen, Germany). Both were of p. a. grade. The synthetic phospholipids dipalmitoylphosphatidylcholine and dimyristoylphosphatidic acid (L-α-DPPC and L-α-DMPA), as well as cholesterol were from Sigma-Aldrich. They were specified 99%+ pure and used as received. NaCl, NaOH and ethylenediamine tetraacetic acid (EDTA) were from Sigma-Aldrich (p. a. grade). Pentylcyanobiphenyl (5CB) was from Merck. The chain-labeled fluorescent lipid NBD-PC from Molecular Probes (Leiden, The Netherlands) was used at a concentration of 0.5 mol% with respect to total lipid. Lipids were spread from 3:1 (chloroform:methanol) solutions in concentrations of $\approx$ 0.5 mg/mL on pure water subphases. The water was filtered using a Milli-Q device (Millipore, Eschborn, Germany) and had a resistivity of $>$ 18 MΩcm. Both the fluorescence and Brewster angle microscopy setup are of local design [18, 20]. Fluorescence microscopy data are directly stored into computer memory (928 $\times$ 640 pixels) and processed using the public domain software NIH Image, v. 1.62, and routines written in the yorick 1.4 programming language [17].

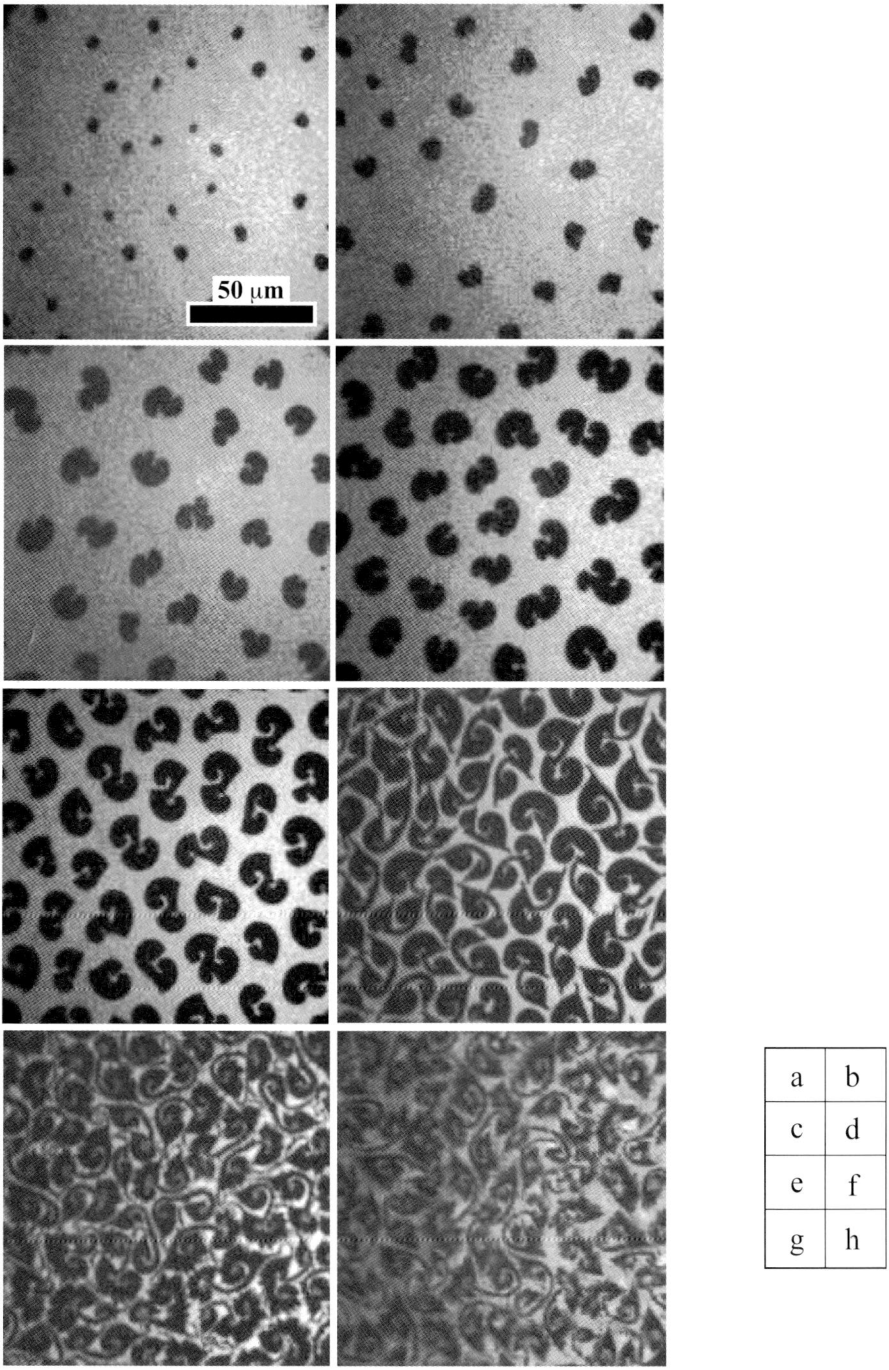

Fig. 2. Development of monolayer textures of L-α-DPPC on pure water as revealed from fluorescence micrographs (0.5 mol% of NBD-PC) obtained along the isotherms as indicated in Fig. 1.

3 Generic Results on Mesoscopic Monolayer Textures

The room temperature π-A isotherm of L-α-DPPC on H_2O is shown in Fig. 1. It displays four distinct regions, denoted as [7, 14] G/LE, LE, LE/TC, and TC. While the pure gas phase, G, is located at molecular areas that are much larger as the ones shown, a higher order phase, SC, completes the sequence of phases along the isotherm at surface pressures π in excess of 30 mN/m. At areas around $A \approx 100$ Å^2/molecule, a first-order phase transition from the G to the LE phase is almost completed upon compression. It was shown earlier [24] that the mesoscopic texture of the monolayer in that regime, as revealed by fluorescence microscopy, is characteristic of a 2D foam structure [4, 8]. The next section of the isotherm is characterized by a homogeneous phase, LE, which turns into yet another phase transition regime, LE/TC, upon further compression of the surface film. Within this regime, dashed lines in Fig. 1 indicate the course of the first-order phase transition as it is expected to appear in an ideal 3D system. In contrast, the actual course of the isotherm is much more sloped – due to non-vanishing interactions within the anisotropic monolayer – and approaches the regime of low compressibility of the TC phase asymptotically at the termination of the LE/TC phase transition. During this phase transition, fluorescence microscopy shows the development of *mesoscopic*, i.e., μm scale, phase coexistence textures that are far from being trivial in their 2D morphology and morphogenesis (c.f. Fig. 2). Arrows along the isotherm in Fig. 1 indicate the locations where representative fluorescence micrographs in the respective panels of Fig. 2 have been taken.

As soon as the surface pressure increases beyond the kink point in the isotherm, π_c/A_c, that marks the onset of the LE/TC phase transition, TC phase nuclei appear in the fluorescence images, as shown in Fig. 2a. They grow in size as the film is further compressed (Fig. 2b - e). While the nucleation events occur rather uncorrelated in space (Fig. 2a), the growing domains repel each other strongly upon further compression, such that they are located on a (disordered) pseudo-hexagonal lattice as the phase transition is halfway completed (Fig. 2d). In the mature texture (Fig. 2e), long-range electrostatic repulsion has become quite obvious, since the neighboring domains keep a minimum distance of ≈ 10 μm, and the domain boundaries have adjusted to form non-trivial shapes such that this minimum distance is retained everywhere within the monolayer. Upon approaching the terminus of the phase transition (Fig. 2f - g), the domain shapes become progressively more elongated and pointed, thus filling efficiently voids between adjacent domains while keeping as much of a mutual distance as possible under the condition of a large proportion of TC phase areas. Close to the terminus (Fig. 2g), no thermodynamic difference between the dark and bright areas does any longer exist and redistribution of the dye occurs: The domains start to dissolve and do so progressively more (Fig. 2h) until the appearance of the monolayer is again homogeneous. This, however, occurs on a rather long time scale since the interdiffusion within the TC phase is slow.

The sequence of images shown in Fig. 2 is typical for L-α-DPPC monolayer films. However, the detailed shape depends critically on the purity of the compound. Since surface active impurities often accumulate at the domain perimeter, thus reducing the line tension between the coexisting phases, most additives will change the appearance of the LE/TC phase transition. A well-known example is cholesterol which reduces the line tension greatly, resulting in the formation of striped phases [43]. Why do we believe the DPPC samples forming monolayer textures such as those presented in Fig. 2 are particularly pure (as compared to DPPC samples from other commercial suppliers which we have found to produce significantly different phase transition textures)? – First, inspection of micrographs of the fully developed domains such as those shown in Fig. 2e shows that the texture is dominated by a limited set of prototypical, non-trivial shapes (c.f. Fig. 3a) that account for almost all of the domains observed in a particular compression state. Secondly, the monolayer shows the same appearance homogeneously over its entire area, as exemplified in Fig. 3b - e in a few images taken at the same pressure as Fig. 2e (which is a section of the image shown in Fig. 3d). This argument is based on the following reasoning: As we will show below, the fully developed texture is a consequence of a small number of well-defined energy terms – due to line tension, long-range dipolar repulsion and molecular chirality – that determine the energy of the system. It may thus be expected that near thermodynamic equilibrium *particular* domain shapes are formed which minimize the total free energy of the monolayer with respect to all other possible shapes. If chemical impurities interfere with this process, it is frequently observed that the texture depends largely on the area of observation within the monolayer presumably since these impurities are inhomogeneously distributed. Typically, one observes then neighboring regions in which the domain size varies widely within a small distance of each other. As Fig. 3 demonstrates, this is not the case in the shown experiment. This implies that important details of the observed domain shapes near equilibrium bear their origin in molecular properties of the system – which is the motivation and starting point for the more quantitative work undertaken in this study.

Another feature of the surface textures is contained in Figs. 3 d,e, which show snapshots of the monolayer, taken at different times, that overlap partially. The grayed disks indicate the same regions within the surface film, and a closer inspection shows that the individual domains oscillate rotationally under thermal excitation around their centers of gravity within the potential exerted by their neighbors. From Figs. 2 and 3 it is obvious that the DPPC domains formed in the course of the LE/TC phase transition are affected by an inherent, spontaneous and *unidirectional* curvature. The arms of all domains bend to the left-hand side if viewed from the respective centers of nucleation. This mesoscopic property has been linked to the *molecular* chirality of the phospholipid [17, 42]: By replacing L-α-DPPC with its oppositely handed isomer, D-α-DPPC, in the experiment the directionality of the bent is reversed. Thus, not only the line tension - which relates to the difference in free energy of lipid molecules inside an ordered domain and of those at its border – and the electrostatic energy – which derives from the different in dipole moment density of the two phases – but also the in-plane curvature of the domains is related to a *molecular* property of the system.

The finding that monolayer domains of highly pure, synthetic lipids show a limited set of prototypical domain shapes – as shown in Figs. 2 and 3 – and that the microscopic, molecular interactions determine the mesoscopic appearance of the monolayer texture has further reaching implications. Let us revisit the monolayer texture of the matured phase separation texture shown in Figs. 2e and 3. Beyond their undisputed esthetical value, these micrographs are fascinating as viewed from a Physics standpoint because they visualize directly the *microscopic* interactions operative between the constituent molecules within the surface film that is only on the order of 2 nm thick. The dipole

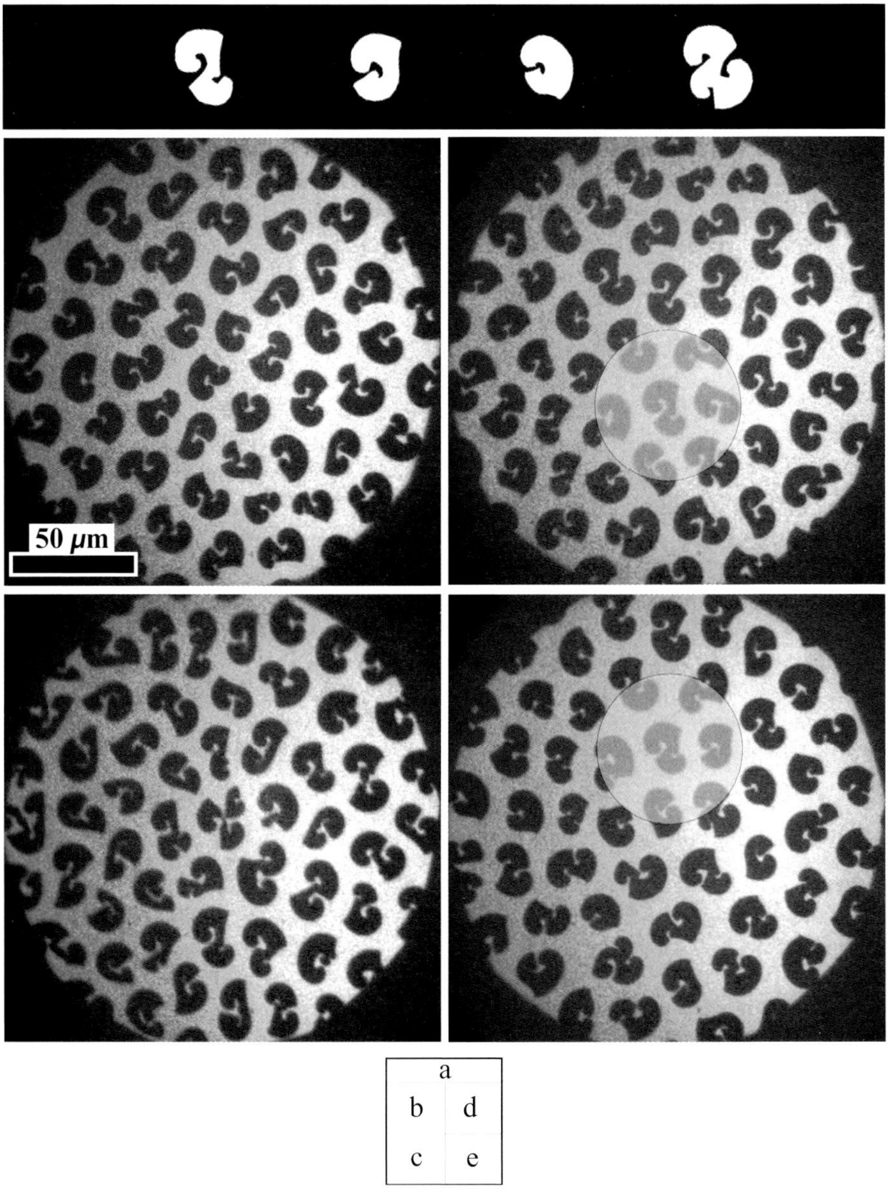

Fig. 3. Characteristic domain shapes (a) observed in representative fluorescence micrographs of L-α-DPPC, shown in (b-e), taken at various locations within the monolayer in the vicinity of the location marked 'e' in the isotherm of Fig. 1. The monolayer is the same as the one shown in Fig. 2: Fig. 2e is a clip from panel (d). The shaded circular disks in (d) and (e) are centered on the same domain within the monolayer and are intended to mark the same regions of the monolayer in a partially overlapping view. These two snapshots have been obtained in short sequence (on the order of 20 sec.).

moment density in the TC phase – due to the higher order and thus higher molecular density of that phase as compared to LE – results in (a) the elongation of a domain of given area against the action of the line tension and (b) a strong repulsion between adjacent domains [15, 17, 30]. These may be conceived as rather rigid island floating in a continuous, liquid film of the more disordered phase, LE. But not only that: The *growth* of the domains from small sizes, such as those shown in Fig. 2a, to the mature texture reflects also these molecular interactions. Thus adjacent, mature domains fit into each other like pieces of a complex jigsaw puzzle – although they are separated by channels of LE phase which keep them at a consistent minimal distance of $\approx 7\ \mu$m. The fascinating conclusion is: By just looking at, and quantitatively analyzing, the monolayer textures, one should get a quantitative handle on the molecular interactions!

4 Free Energy of Monolayer Textures

As worked out by McConnell and others, the main contributions to the free energy of the system stem from dipolar forces and line tension [15, 29]. Briefly, the lipid molecules bear a (relatively small) dipole moment, $\underline{\mathbf{1}}$, on their hydrophobic alkane chains, which incorporates contributions of the methyl at one chain end and of the carbonyl at the other. This small dipole moment dominates the much larger dipole moment, $\underline{\mathbf{2}}$, located on the hydrophilic headgroup, for the following reasons. First, the headgroups are embedded in a medium of higher dielectric constant, ϵ, than the acyl chains. Secondly, the orientations of the image dipoles for both classes of dipole moments, brought about by the jump in ϵ at the hydrophilic/hydrophobic interface, differ fundamentally with respect to the 'real' molecular dipoles they are deriving from. ($\underline{\mathbf{1}}$) The normal components of the acyl chain's dipole moment and its image dipole are collinear while their in-plane components oppose each other. The net dipole moment at the interface that derives from the hydrophobic part of the molecules is thus normal to the surface, such that neighboring dipole moments within the film lead to a long-range electrostatic potential. ($\underline{\mathbf{2}}$) Conversely, the in-plane components of the headgroup dipole moments and their image dipoles align while their normal components cancel each other, leading to a net in-plane dipole moment due to each individual molecule. Since neighboring molecules may be misaligned with respect to their azimuthal orientation, this yields only a short-range contribution to the electrostatic potential.

We have recently introduced a third term to the Hamiltonian that describes the structure formation in these systems and takes account of the propensity of neighboring molecules to misalign with respect to their tilt azimuth due to their molecular chirality [17]. Technically, this contribution to the free energy of the system takes the form of a spontaneous curvature within the film and is motivated through the asymmetric nature of the effective pair potential [38] between adjacent molecules [17]. This curvature term, while being indirectly revealed in images such as those shown in Figs. 2 and 3, is more immediately observed in Brewster angle [12, 13] micrographs, as demonstrated in recent work [17] (c.f. Fig. 1 there). The complete Hamiltonian that we propose to determine the shape-dependent contribution to the system energy is [17]:

$$
\begin{aligned}
E &= E_{line} + E_{dipolar} + E_{chiral} \\
&= \lambda \oint_{\delta D} ds + \frac{\mu^2}{2} \int\int_{D} \frac{dAdA'}{|\boldsymbol{r}-\boldsymbol{r}'|^3} + \int_{D} \Delta\mathbf{k}^T \mathbf{H} \Delta\mathbf{k} \, dA
\end{aligned}
\tag{1}
$$

where D is the domain area, δD its perimeter, λ the line tension, μ the effective dipole moment difference between the unit areas of the coexisting phases, $\mathbf{r} = (r_1, r_2)^T$ denotes an in-plane location within the area, A, $\mathbf{\Delta k} = (k_{1,0} - k_1, k_{2,0} - k_2)^T$ are local deviations from the spontaneous curvatures $\mathbf{k} = (k_{1,0}, k_{2,0})^T$ in two dimensions, and $\mathbf{H}$ is the Hesse matrix that contains the second term of the Taylor expansion of the potential associated with the twisting of a molecule around an axis normal to the interface within the steric potential originating from its neighbors in the surface film.

5 Quantification of Experimentally Determined Domain Shapes

If our hypothesis is correct that the domain shapes are *entirely* determined by the molecular parameters characterized in (1), it should be possible to find their values for a particular system by a quantitative analysis of the experimental results by solving the minimization problem, $\mathrm{grad}_{\lambda,\mu,\mathbf{k}_0,\mathbf{H}} E = 0$. Unfortunately, (1) is a rather demanding integral-differential form which we have been unable to solve either analytically or numerically using standard procedures. In contrast, however, it is comparably easier to calculate a set of shapes by solving the variation problem, $\delta E = 0$ with fixed parameters λ, μ, $\mathbf{k}_0$, and $\mathbf{H}$ and compare the solutions with the experimentally observed shapes. One advantage of this approach is that one escapes easily from difficulties arising due to the divergence of the dipolar energy as $\mathbf{r} \to \mathbf{r}'$. In continuum theories of the problem, this divergence is usually dissolved by an *ad hoc* assumption of a lower limit of the validity of the model at distances that approach the intermolecular spacings. In our approach, we need not be concerned with this difficulty. A major disadvantage of the chosen approach, however, is that the absolute length scale of the system is lost, such that molecular parameters may only be determined with respect to each other. Within the variational approach to energy minimization, various levels of sophistication may be envisaged [17]. (a) If domains are elongated, such as those observed for binary phospholipid-cholesterol mixtures [10, 11, 43], it may be permissible to treat them effectively as onedimensional objects. In consequence, all three contributions in (1) are reduced by one spatial dimension. Thus, the line tension becomes shape-*independent*, i.e., for a single domain it reduces to $E_{line} = 2\lambda l$, where l is the domain length. (b) If domains are well separated within the monolayer area, such that domain-domain interaction may be neglected, the task reduces to finding the equilibrium shape of a single domain (elongated or with non-trivial, twodimensionally extended shape). We have so far sequentially developed approaches for minimizing the shape of elongated, isolated domains; treating isolated complex domain shapes; and finally solving the case of elongated, crowded domains that interact with each other strongly. Thus far, we have not been able to approach the most complex case – i.e., the one exemplified in Figs. 2 and 3 – in which an ensemble of domains with non-trivial shapes interact strongly with one another.

(A) Spatially Isolated Domains

As described earlier [17], we found a monolayer system – the binary mixture of DPPC and pentylcyanobiphenyl (5CB) – that shows the characteristic phospholipid phase behavior at low 5CB concentrations. However, within the phase transition regime domains are formed that are elongated with a very small aspect ratio and show a clear signature of chirality in that they form delicate spirals. In addition, the domains keep a large distance to each other during the most part of the phase transition, such that this system is indeed a good choice for quantitative studies with the simplest case imaginable. Figure 4 shows a fluorescence micrograph of L-α-DPPC with 19 mol% 5CB (and 0.5 mol% of NBD-PC). On the right, simulations overlay magnified clippings from such images which show a one-lobed (top) and a two-lobed (bottom) domain. The model contours are formed by a succession of 100 straight segments. For a particular parameter set, (ζ, k) - in which $\zeta = \mu^2/2H_{11}$ is a reduced dipole moment density, H_{11} is the term of the Hesse matrix that survives the reduction of dimensionality and k is a onedimensional spontaneous curvature – , the energy minimization starts with three straight segments. Subsequently, their number is consecutively increased and the energy minimized – using a steepest decent algorithm – until the resulting contour is sufficiently smooth for a quantitative comparison with experimentally observed shapes. This simple procedure works well for isolated domains with any number of lobes. As recognized from Fig. 4, the correspondence between model and experimental result is almost perfect. Not surprisingly, the model parameters for the description of the one-lobed and the two-lobed domains work out to be identical. This is another indication that the mesoscopic domain form is entirely determined by the molecular properties.

The modeling of domains that are not reduced to elongated shapes encounters problems in that fully developed, mature domains are always in a distance of each other where interdomain interactions may not be neglected. Furthermore, the algorithm to minimize the domain energies we have developed [17] allows only simple cases to be treated. We thus confine ourselves to treating DPPC domains in the early stage of the phase transition. As already pointed out earlier [27], the domains are *not* of a trivial, i.e. circular, shape once their size is sufficiently large to be observed with optical resolution. Figure 5 shows an experimental result (left side) with DPPC domains just above $\pi \approx 6$ mN/m at room temperature. Three domains are depicted which are on their way to developing into 'bean' shapes. In agreement with McConlogue's observations [27], they are not entirely symmetrical, which is again a signature of the molecular chirality. As shown on the right, the molecularly motivated modelling approach captures these non-trivial disk shapes faithfully. For the domain in the upper left of the micrograph, this is directly demonstrated by overlaying the experimental result with the model contour.

(B) Crowded Domain Ensembles: Minkowski Function Assessment of Similarity of Textures [3, 34]

The most complex form of structure formation in lipid monolayers occurs when domains get so close to one another that electrostatic interdomain interactions become substantial. In a number of experimental studies, situations have been reported in which this occurs at relatively small values of λ, such that the resulting domains are elongated, forming

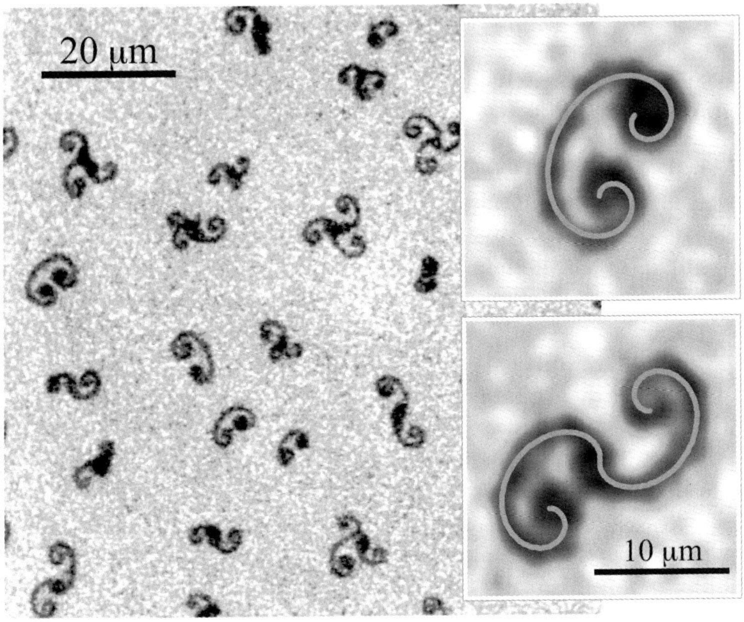

Fig. 4. Background image: Fluorescence micrograph (0.5 mol% NBD-PC) of an L-α-DPPC monolayer with 19 mol% 5CB on water (T $\approx$ 20°C, $\pi \approx$ 14 mN/m). Overlay panels: Comparison of 1D models (see text) with experimental data similar to those shown in the background image. Top: one-lobed domain; bottom: two-lobed domain. Adapted from [17].

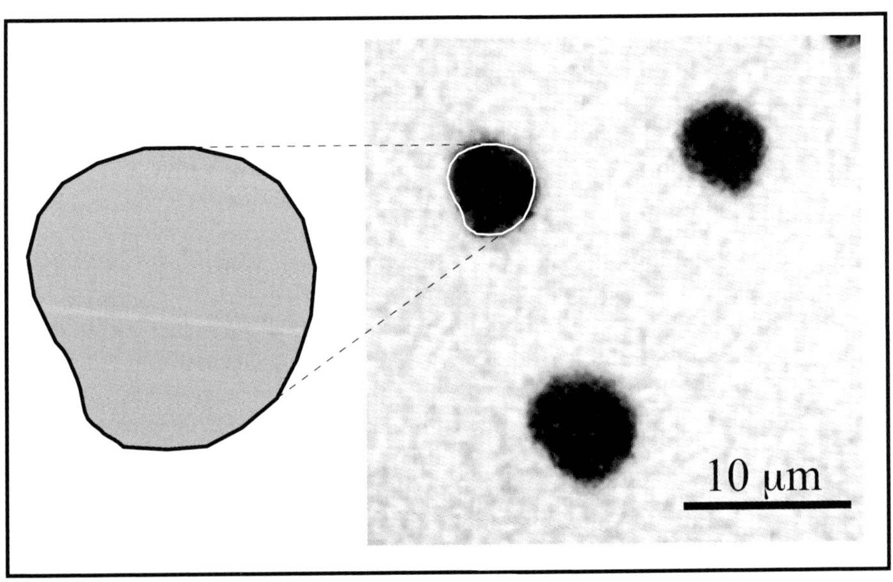

Fig. 5. Comparison of a 2D model (left; see text) with experimental data (right: L-α-DPPC on water shortly after the onset of the phase transition). Adapted from [17].

effectively a 1D lamellar phase ("stripe phase") [15, 43]. In a few cases, the continuous development of such stripe phases upon film compression or temperature variations has been reported [10, 11]. Textures that occur in such situations, e.g., in monolayers of DPPC or DMPA mixtures with cholesterol, are distinguished from the results observed in the DPPC/5CB system, Fig. 4, in the following respects:

- The area fraction of the ordered phase is much larger in the mature DPPC / cholesterol system than in DPPC/5CB;
- growth in the crowded situation leads to interdigitation of arms from different domains;
- depending on the system and experimental conditions, growth in the crowded situation may lead to branching of the domains. Such branching has been extensively studied by various groups [21, 25, 26, 37].

In order to assess this situation in model calculations, the energy minimization needs to be only marginally extended: (1) is modified such that it includes both intradomain *and* interdomain electrostatic interactions,

$$E_{dipolar} = \sum_{i,j} \frac{\mu^2}{2} \iint_{D_{i,j}} \frac{dAdA'}{|\boldsymbol{r} - \boldsymbol{r}'|^3} \tag{2a}$$

which reduces to

$$E_{dipolar} = \sum_{i,j} \frac{\mu^2}{2} \iint_{\gamma_{i,j}} \frac{dsds'}{|\boldsymbol{r} - \boldsymbol{r}'|^3} \tag{2b}$$

in the case of a striped phase with γ denoting the 1D shape of a domain. We have placed a domain ensemble into an external potential with soft walls to enable simulations of a finite sample size. For the implementation, a number of domain nuclei is distributed at random within the potential well. As in the simulation of isolated domain shapes, each nucleus has initially three segments attached with random orientations and fixed kink angles between adjoining segments within the plane. The number of segments is incremented and the local kink angle α at the joints of the segments adjusted to minimize the total energy of the arrangement of domains until a preset length of the domains, corresponding to a particular area fraction of the ordered monolayer phase, has been reached. As a deficiency of the implementation, the algorithm is susceptible to domain cross-over, particularly at the well boundaries. In addition, highly curved structures may occur close to these boundaries. Consequently, a comparison of the models with experimental results is limited to the central sections of the well. Another significant complication in the comparison of modeling results with experimental observations arises in that both types of results depend largely on the starting condition, i.e., the locations of the randomly occurring or randomly chosen domain nucleation sites in the experimental situation or the model, respectively. Yet both classes of starting conditions determine sensitively the detailed arrangements of the domain ensembles. Since these starting conditions never coincide, a direct comparison – similar to the one afforded in the previous sections – is not feasible. Instead, one has to compare *similarity classes* of domain ensembles in order to quantify the correspondence between a computed morphology and a given experimental result. We have applied integral geometric methods [2, 32, 34, 35] to afford a quantitative comparison of the morphology patterns formed by domain ensembles.

A coordinate which characterizes the content of an image containing an arbitrary irregular pattern is defined by a tripel of intensities, s, ν and χ. To access these intensities for an image that contains essentially two gray scale levels – such as the ones analyzed here –, the image is sliced while the threshold is placed at the minimum of the bimodal gray level distribution. Since we are concerned with elongated domains, the width of these domains is of minor importance for the assessment of the underlying pattern. Consequently, we skeletonized the images by standard techniques of mathematical morphology (i.e. erosion and dilation, c.f. Fig. 6). The resulting pattern of curves – one pixel wide – is well suited for a quantitative comparison with the simulated patterns. For such a quantification, these curves were stepwise dilated with disks of increasing radii, ρ. For the resulting 2D regions, three intensity functions of ρ were determined. These functions are (i) the normalized portion of black pixels, $s(\rho)$; (ii) the normalized length of the border between black and white regions, $\nu(\rho)$; and (iii) the normalized Euler function, $\chi(\rho)$, i.e., the difference of the number of the connected black and white regions. In concert, these functions characterize the pattern quite sensitively and depend only weakly on the chosen part of the image. Scaling the image size by a factor n leads to the following rescaling laws for the *Minkowski* functionals: $s_1 = s_2$; $\nu_1 = \nu_2/n$; $\chi_1 = \chi_2/n^2$. To quantify the similarity of two images, their morphological distance has to be determined. This distance is defined as [33, 34]

$$a = \int_0^\infty (s_1(\rho) - s_2(\rho))^2 d\rho + \int_0^\infty (\nu_1(\rho) - \nu_2(\rho))^2 d\rho + \int_0^\infty (\chi_1(\rho) - \chi_2(\rho))^2 d\rho \qquad (3).$$

We observed that all three Minkowski intensities converge uniformly to zero as the shapes of patterns converge toward each other and have thus confined the quantification to analysis of only $\chi(\rho)$. Since the scaling between simulated and experimental images is *a priori* unknown, we have calculated a as a function of the scaling factor n of the simulated images. Two images were then accepted to be similar if their morphological distance falls below the square of the modulus a_0 of the unscaled experimental image at least for one n:

$$a_\chi(n) = \int_0^\infty (\chi_{exp}(\rho) - \chi_{mod}(\rho, n))^2 d\rho \leq 0.1 \int_0^\infty \chi_{exp}^2(\rho) d\rho \qquad (4).$$

As a side effect, one also obtains the proper value for the best-match scaling.

Figure 7 shows two examples of experimental data for DMPA with 4 mol% cholesterol on an aqueous solution of 5 mM NaCl and 10 μM EDTA, adjusted to pH = 11 with NaOH. The experimental data shown in Fig. 7c and d have been taken at T = 10^oC/$\pi \sim$ 14 mN/m and T = 4^oC/$\pi \sim$ 8 mN/m, respectively. They are compared to energy minimized ensemble arrangements that have been computed using input parameters that were found to result in a reasonable agreement with the data. The thickness of the bodies of the computed domains is chosen to obtain a close visual correspondence to the observed experimental shapes. It has to be emphasized that the shown synthetic ensemble textures

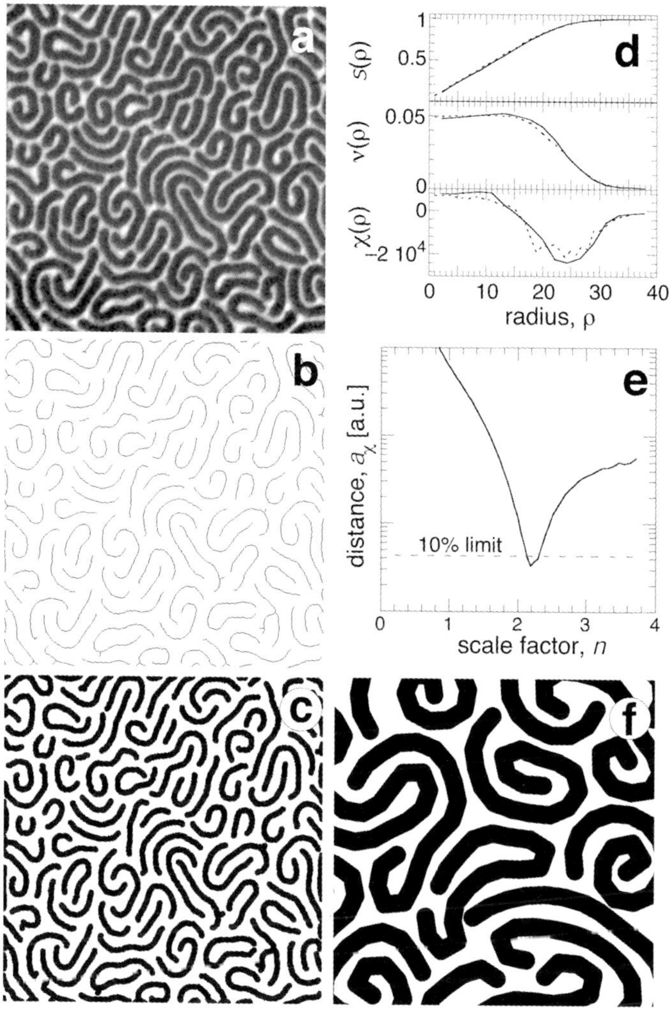

Fig. 6. Synopsis of the manipulations applied for the quantification of the similarity of growth patterns with simulated data through *Minkowski* functionals. We start with a gray-scale image (a) which consists, however, essentially of two gray levels. The raw data have been corrected for a slightly inhomogeneous illumination within the plane of the image. The gray-scale image is sliced and then skeletonized to yield connected trains of the domain's center lines that are just one pixel wide (b). Finally, disks with radii ρ are placed on each black pixel along the trace of the domain's skeletons to obtain the pattern shown in (c). The result is compared with a synthetic model. Panel (f) shows a synthetic pattern determined to be 'similar' to the experimental example. To reveal this simularity visually, the domain bodies have also been widened to match the experimental stripe width. For the displayed examples, the three *Minkowski* functionals, $s(\rho)$, $\nu(\rho)$ and $\chi(\rho)$, are given in (d) as solid lines for the synthetic pattern and dotted lines for the experimental pattern. To determine the *Minkowski* distance between the two images, the quadratic difference, a_χ, of the Euler functions $\chi(\rho)$ are calculated according to (4) as a function of a scale factor, n (e). Two patterns are assumed to be similar if the minimum of a_χ with respect to n falls below a threshold (chosen to be 10% of the modulus of the experimental image). Adapted from [17].

are not necessarily the best possible matches: Due to limited computing power, we have not systematically screened parameter space on a close-meshed grid; moreover, even under identical parameters different explicit textures will be found depending on the starting conditions. However, already a visual inspection shows that the correspondence between the two pairs of patterns is striking: In the data set at T = 10°C, the relatively short domain bodies are organized on pseudo grid lines that form a striped pattern which extends from the top to the bottom of the micrograph. This feature is faithfully reproduced in the synthetic pattern, Fig. 7a. Moreover, as indicated by the two dashed lines overlaid on both the experimental and simulated data, the sequence of neighboring domain shapes may develop very similar in both cases. Clearly, the close correspondence in the depicted example is fortuitous – under slightly different starting conditions, it could have occurred in a significantly different manner, or it could have been found in a mirror-symmetric arrangement – , however, its occurrence suggests strongly that the images compared in Figs. 7 are certainly *similar* to each other. On a more quantitative basis we note that the distance of the *Euler* functions χ of the two images is much smaller than the square of the modulus of the experimental image: $a_\chi \sim 0.088 a_{\chi,0}$. At a lower temperature, T = 4°C (Fig. 7d), the textures in the monolayer have generally elongated, transforming the phase morphology into a striped pattern. The pseudo rows observed in Fig. 7c have disappeared and neighboring domain bodies interdigitate deeply alongside each other. However, the correlation length of the striped pattern is extremely short as the texture incorporates a large number of defect points. Again, these features are faithfully resembled in the simulated image, Fig. 7b, which shows hairpin shapes formed by individual domains as well as elongated shapes that are squeezed in between more compact domains. Quantitatively, the *Euler* functions of the two images are close to each other, $a_\chi \sim 0.09 a_{\chi,0}$ (for comparison, the Minkowski distance between the images Fig. 7a and 7b is $a_\chi \sim 50 a_{\chi,0}$).

(C) Determination of Molecular Quantities from Simulated Domain Shapes

What can be learned from such simulations in terms of the molecular properties of the system? Clearly, the observed mesoscopic structures reflect details of the molecular interactions which might be quantified upon systematically studying the molecular parameters used to model the experimental results. Unfortunately, since the computation of equilibrium shapes derives from a continuum theory, the absolute length scale has been lost and the molecular parameters may only be determined in relation to one another. While the absolute magnitudes may be determined using thermodynamic relations [31] or micromanipulation [5, 44], it is also interesting to observe the trends that the system follows upon continuous variation of thermodynamic parameters. As shown in earlier work [11], the DMPA/cholesterol system at high pH undergoes a continuous transition from a pseudo hexagonal phase to a striped phase upon lowering the temperature T at a fixed area A. We have repeated such measurements for a quantitative evaluation. Consistent with the earlier work we find an elongation of the domains – indicative of a reduction of the line tension λ – upon decreasing T. Two representative images from this measurement are shown in Fig. 7 b,d. Figure 8 visualizes a correlation of the experimental results with simulated data. In both cases, the width-to-length ratio ψ has

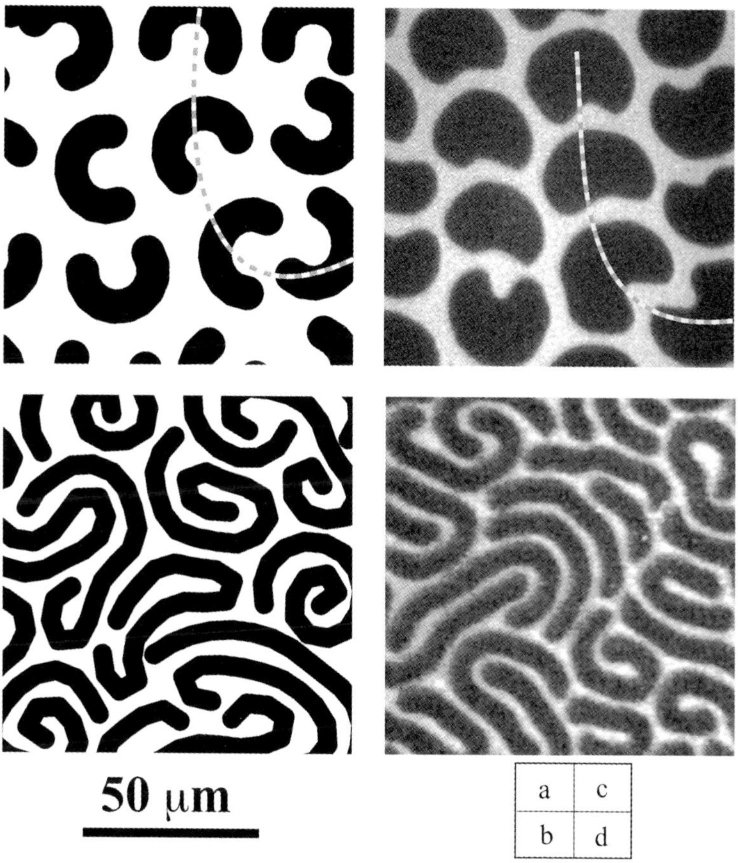

Fig. 7. Comparison of 1D models (a,b) with experimental data (c,d) obtained with L-α-DMPA + 4 mol% cholesterol and 0.5 mol% NBD-PC on an aqueous subphase at pH = 11, 5 mM Na^+, 10 μM EDTA, $\pi \approx 14$ mN/m at T $\approx 10^o$C (c) and T $\approx 4^o$C (d). The Minkowski distances between the simulated and the experimentally observed data are $a_\chi \approx 0.088 a_{\chi,0}$ (a,c) and $a_\chi \approx 0.090 a_{\chi,0}$ (b,d). The dashed lines in (a) and (c) indicate a succession of domains that happened to develop quite similarly in the experiment and the simulation – beyond the general similarity of the underlying patterns. Adapted from [17].

been determined as a quantitative measure of domain elongation, and ψ has been plotted as a function of λ for the simulation results (in arbitrary units, since the modeling of the domain shapes yields only the ratio $\lambda^2/2\mu$). The relation $\psi = \psi(\lambda)$ is clearly sigmoidal, showing a large drop in ψ upon lowering λ within a narrow regime while ψ stays essentially constant above and below this transition regime. In a second step, ψ has been determined as a function of T for the experimental data. Finally, by using the implicit interdependence, λ has been mapped to T. From this mapping we derive an exponential relation, $\lambda = \lambda_0 e^{aT}$ (see inset in Fig. 8). In the main body of the figure, this linear mapping has been utilized to compare the simulation results with the rescaled experimental observations, and it is revealed that the correspondence between the two is almost perfect. At low λ, the models suggest ψ ratios that are systematically too large. This is attributed to inherent difficulties of the simulation algorithm. For reference, some of the simulated domain shapes are included in Fig. 8 and related to the corresponding data points. As indicated on the left hand side of the main panel, the domain shape emerging from the 2D modeling with vanishing λ is identical to that emerging from the 1D models if one fixes the latter with the appropriate domain width.

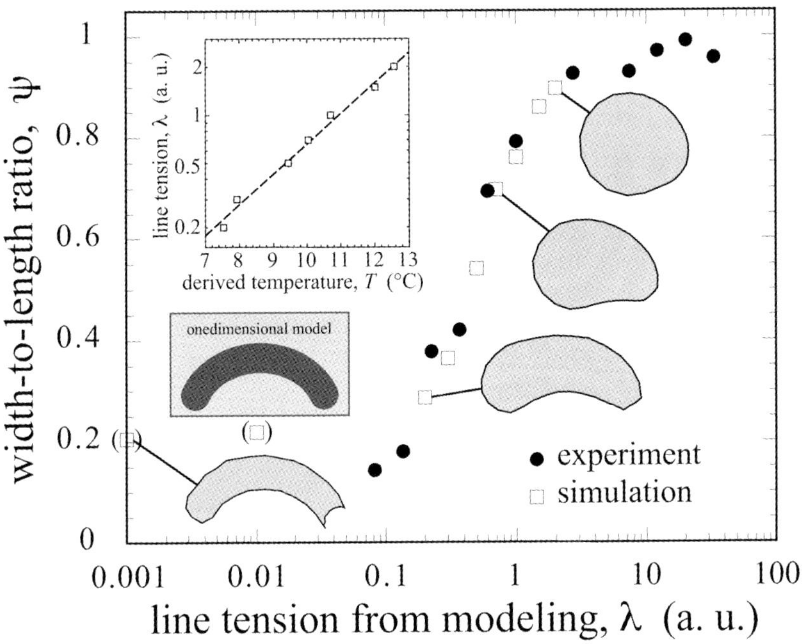

Fig. 8. Correlation, via the width/length ratio ψ of the domains, of the temperature-dependent shape transformations observed for L-α-DMPA/4 mol% cholesterol with 2D simulations in which only the line tension λ has been adjusted and all other parameters were kept constant. Some of the experimental data have been shown in Fig. 7. The experimental data have been collected between $T = 17^o$C (highest λ) and $T = 3^o$C (lowest λ). The inset graph shows the correlation between T, from the experiment, and λ, from the simulations. The sketches at the left hand side demonstrate that the 1D and 2D simulations at low ψ yield the same results. The simulation results shown in parenthesis at low λ are believed to lie systematically too high in ψ due to problems with the algorithm. Adapted from [17].

6 Conclusions

We have shown that the molecular chirality of phospholipids determines the morphology of liquid-condensed domains in aqueous surface monolayers sensitively and have developed a concept for the calculation of such morphologies which is directly related to the molecular asymmetry. The contribution of chirality to the Hamiltonian of the system is sizable for species such as DPPC for which the chirality is paired with a large size of the headgroup in relation to the cross-section of the aliphatic chains; this leads to a spontaneous curvature of the domains within the plane of the monolayer. With this concept, we have been able to simulate realistically the domain shapes in a number of experimental situations with progressively higher complexity under equilibrium conditions. In two-dimensional simulations of domain shapes we have observed that a reduction of the line tension λ - normalized to the electrostatic repulsion between collinear dipole moments on the molecules – is a necessary, but not a sufficient condition for the formation of a striped phase: Elongated domains under equilibrium conditions appear only if the film-forming molecules are chiral. This is consistent with experimental observation: To our knowledge, stripes of solid and liquid-condensed phases formed by achiral amphiphiles have thus far not been reported in the literature.

Acknowledgements

This work has been supported by the DFG through the Sonderforschungsbereich (SFB) 294, TP F3, and the priority program 'Wetting and structure formation at interfaces' under contract no. Lo352/7-2. We also thank the Fonds der Chemischen Industrie, Frankfurt, for support.

References

1. Albrecht, O., H. Gruler, E. Sackmann (1978): 'Polymorphism of phospholipid monolayers', *J. Physique* (France) **39**, pp. 301–313
2. Arns, C., M. Knackstedt, K.R. Mecke (2002): 'Characterising the morphology of disordered materials'. In: *Lecture Notes in Physics*, ed. by K.R. Mecke, D. Stoyan (Springer, Berlin), this volume
3. Beisbart, C., R. Dahlke, K.R. Mecke, H. Wagner (2002): 'Vector- and tensor-valued descriptors for spatial patterns'. In: *Lecture Notes in Physics*, ed. by K.R. Mecke, D. Stoyan (Springer, Berlin), this volume
4. Berge, B., A.J. Simon, A. Libchaber (1990): 'Dynamics of gas-bubbles in monolayers', *Phys. Rev. A* **41**, pp. 6893–6900
5. Bevegnu, D.J., H.M. McConnell (1992): 'Line tension between liquid domains in lipid monolayers', *J. Phys. Chem.* **96**, pp. 6820–6824
6. Bibo, A.M., C.M. Knobler, I.R. Peterson (1991): 'A monolayer phase miscibility comparison of long-chain fatty acids and their ethyl esters', *J. Phys. Chem.* **95**, pp. 5591–5599
7. Cadenhead, D.A., F. Müller-Landau, B.M.J. Kellner (1980): 'Phase transitions in insoluble one and two-component films at the air/water interface'. In: *Ordering in Two Dimensions*, ed. by S.K. Sinha (Elsevier North Holland, Amsterdam), pp. 73–81
8. Dennin, M., C.M. Knobler (1997): 'Experimental studies of bubble dynamics in a slowly driven monolayer foam', *Phys. Rev. Letters* **78**, pp. 2485–2488
9. Fischer, A., M. Lösche, H. Möhwald, E. Sackmann (1984): 'On the nature of the lipid monolayer phase transition', *J. Physique Lett.* **45**, pp. 785–791
10. Heckl, W.M., M. Lösche, D.A. Cadenhead, H. Möhwald (1986): 'Electrostatically induced growth of spiral domains in the presence of cholesterol', *Eur. Biophys. J.* **14**, pp. 11–17
11. Heckl, W.M., H. Möhwald (1986): 'A narrow window for observation of spiral lipid crystals', *Ber. Bunsenges. Phys. Chem.* **90**, pp. 1159–1163

12. Hénon, S., J. Meunier (1991): 'Microscope at the Brewster angle: Direct observation of first-order phase transitions in monolayers', *Rev. Sci. Instrum.* **62**, pp. 936–939
13. Hönig, D., D. Möbius (1991): 'Direct visualization of monolayers at the air-water interface by Brewster angle microscopy', *J. Phys. Chem.* **95**, pp. 4590–4592
14. Kaganer, V.M., H. Möhwald, P. Dutta (1999): 'Structure and phase transitions in Langmuir monolayers', *Rev. Mod. Phys.* **71**, pp. 779–819
15. Keller, D.J., H.M. McConnell, V.T. Moy (1986): 'Theory of superstructures in lipid monolayer phase transitions', *J. Phys. Chem.* **90**, pp. 2311–2315
16. Knobler, C.M. (1990): 'Seeing phenomena in flatland: Studies of monolayers by fluorescence microscopy', *Science* **249**, pp. 870–874
17. Krüger, P., M. Lösche (2000): 'Molecular chirality and the domain shapes in lipid monolayers on aqueous surfaces', *Phys. Rev. E* **62**, pp. 7031–7043
18. Krüger, P., M. Schalke, Z. Wang, R.H. Notter, R.A. Dluhy, M. Lösche (1999): 'Effect of hydrophobic surfactant proteins SP-B and SP-C on binary phospholipid monolayers. I. Fluorescence and dark-field microscopy', *Biophys. J.* **77**, pp. 903–914
19. Langmuir, I. (1917): 'The constitution and fundamental properties of solids and liquids. II. Liquids', *J. Am. Chem. Soc.* **39**, pp. 1848–1906
20. Lautz, C., T.M. Fischer, M. Weygand, M. Lösche, P.B. Howes, K. Kjaer (1998): 'Determination of alkyl chain tilt angles in Langmuir monolayers: A comparison of Brewster angle autocorrelation spectroscopy and x-ray diffraction', *J. Chem. Phys.* **108**, pp. 4640–4646
21. Lee, K.-Y.C., H.M. McConnell (1993): 'Quantized shape transitions in lipid monolayer domains', *J. Phys. Chem.* **97**, pp. 9532–9539
22. Lipowsky, R., E. Sackmann, eds. (1995): 'Structure and Dynamics of Membranes'. In: *Handbook of Biological Physics* ed. by A.J. Hoff (Elsevier, Amsterdam), Vol. 1
23. Lösche, M., H. Möhwald (1984): 'Fluorescence microscope to observe dynamical processes in monomolecular layers at the air/water interface', *Rev. Sci. Instrum.* **55**, pp. 1968–1972
24. Lösche, M., E. Sackmann, H. Möhwald (1983): 'A fluorescence microscopic study concerning the phase diagram of phospholipids', *Ber. Bunsenges. Phys. Chem.* **87**, pp. 848–852
25. Mayer, M.A., T.K. Vanderlick (1994): 'Calculation of shapes of dipolar domains in insoluble monolayers: Analysis of shape stability and transitions', *J. Chem. Phys.* **100**, pp. 8399–8407
26. Mayer, M.A., T.K. Vanderlick (1995): 'Calculation of shapes of dipolar domains in two-dimensional films: Effect of dipole tilt', *J. Chem. Phys.* **103**, pp. 9788–9794
27. McConlogue, C.W., T.K. Vanderlick (1997): 'A close look at domain formation in DPPC monolayers', *Langmuir* **13**, pp. 7158–7164
28. McConlogue, C.W., T.K. Vanderlick (1999): 'Molecular determinants of lipid domain shape', *Langmuir* **15**, pp. 234–237
29. McConnell, H.M. (1991): 'Structures and transitions in lipid monolayers at the air-water interface', *Annu. Rev. Phys. Chem.* **42**, pp. 171–195
30. McConnell, H.M., V.T. Moy (1988): 'Shapes of finite two-dimensional domains', *J. Phys. Chem.* **92**, pp. 4520–4525
31. McConnell, H.M., P.A. Rice, D.J. Bevegnu (1990): 'Brownian motion of lipid domains in electrostatic traps in monolayers', *J. Phys. Chem.* **94**, pp. 8965–8968
32. Mecke, K.R. (1996): 'Morphological characterization of patterns in reaction-diffusion systems', *Phys. Rev. E* **53**, pp. 4794–4800
33. Mecke, K.R. (1998): 'Integral geometry in statistical physics', *Int. J. Mod. Phys.* **12**, pp. 861–899
34. Mecke, K.R. (2000): 'Additivity, convexity, and beyond: Applications of Minkowski functionals in Statistical Physics'. In: *Lecture Notes in Physics*, ed. by K.R. Mecke, D. Stoyan (Springer, Berlin), pp. 72–184
35. Mecke, K.R., T. Buchert, H. Wagner (1994): 'Robust morphological measures for large-scale structure in the universe', *Astron. Astrophys.* **288**, pp. 697–704

36. Mecke, K.R., D. Stoyan, eds. (2000): 'Statistical Physics and Spatial Statistics', Lecture Notes in Physics (Springer, Berlin), Vol. 554
37. Miranda, J.A. (1999): 'Closed form results for shape transitions in lipid monolayer domains', *J. Phys. Chem. B* **103**, pp. 1303–1307
38. Nandi, N., B. Bagchi (1997): 'Prediction of the senses of helical amphiphilic assemblies from effective intermolecular pair potentials: Studies on chiral monolayers and bilayers', *J. Phys. Chem. A* **101**, pp. 1343–1351
39. Philips, M.C., D. Chapman (1968): 'Monolayer characteristics of saturated 1,2-diacyl phosphatidylcholines (lecithins) and phosphatidylethanolamines at air-water interface', *Biochim. Biophys. Acta* **163**, pp. 301–313
40. Singer, S.J., G.L. Nicolson (1972): 'The fluid mosaic model of the structure of cell membranes', *Science* **173**, pp. 720–731
41. Stoyan, D., A. Davtyan, D. Turetayev (2002): 'Statistics for randem domains and particles'. In: *Lecture Notes in Physics*, ed. by K.R. Mecke, D. Stoyan (Springer, Berlin), this volume
42. Weis, R.M., H.M. McConnell (1984): 'Two-dimensional chiral crystals of phospholipid', *Nature* **310**, pp. 47–49
43. Weis, R.M., H.M. McConnell (1985): 'Cholesterol stabilizes the crystal-liquid interface in phospholipid monolayers', *J. Phys. Chem.* **89**, pp. 4453–4459
44. Wurlitzer, S., P. Steffen, T.M. Fischer (2000): 'Line tension of Langmuir monolayer phase boundaries determined with optical tweezers', *J. Chem. Phys.* **112**, pp. 5915–5918

Spatial Order in Liquid Crystals: Computer Simulations of Systems of Ellipsoids

Friederike Schmid and Nguyen H. Phuong

Theoretische Physik, Universität Bielefeld, D-33501 Bielefeld, Germany

Abstract. Computer simulations of simple model systems for liquid crystals are briefly reviewed, with special emphasis on systems of ellipsoids. First, we give an overview over some of the most commonly studied systems (ellipsoids, Gay-Berne particles, spherocylinders). Then we discuss the structure of the nematic phase in the bulk and at interfaces.

1 Introduction

Randomly distributed hard spheres in three dimensions form two types of structures, depending on their density: Fluid and crystalline. For randomly distributed anisotropic particles, the situation is different: Several phases exist at intermediate densities between the fluid and the solid state. These phases are called "mesophases". For example, one often observes a nematic phase, where the particles are oriented in one common preferred direction, but have no crystalline translational order. Other common phases are the smectic phases, in which the particles are arranged in layers of two dimensional fluids. Some of these structures are sketched in Fig. 1. Since the mesophases are neither crystalline nor truly liquid, they are commonly referred to as "liquid crystal phases".

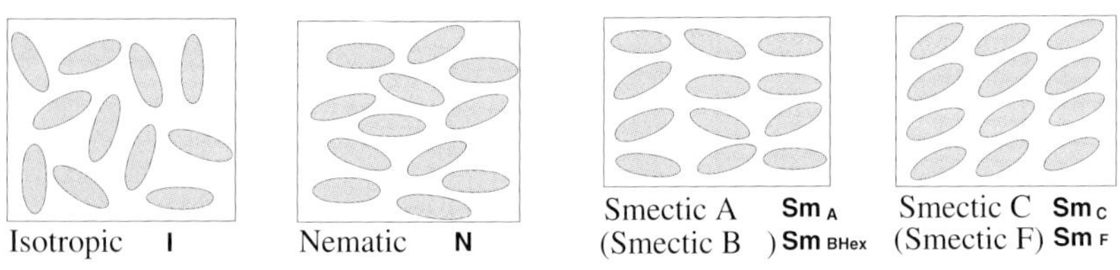

Fig. 1. Liquid crystalline phases. In the Smectic B_{Hex} and Smectic F phase, the structure within the layers is not entirely fluid, but has a type of order called hexatic. See [1, 2] for explanation.

Experimentally, such phases are observed in various systems of anisotropic particles: Small organic molecules, stiff polymers, self-associated wormlike micelles, or even whole virusses (one famous example being the tobacco mosaic virus). The phase transitions are triggered in some cases by the temperature (thermotropic liquid crystals), and in other cases by the concentration of the anisotropic particles (lyotropic liquid crystals). In reality, particles are of course not distributed "randomly", but according to a Boltzmann distribution $P \propto e^{-E/k_B T}$ which accounts for the energy E of their mutual interactions (T is the temperature and k_B the Boltzmann constant). Nevertheless, the mechanisms which drive the phase transitions in lyotropic liquid crystals are essentially the same as those in systems of hard anisotropic particles. Therefore, the latter are often employed to model phase transitions in liquid crystals. More generally, systems of anisotropic particles with simplified interactions have proven very useful to study generic properties of liquid crystals [1, 3, 4, 5, 6].

In the present contribution, we describe some of the computer simulation work that has been done in this direction. We do not intend to give a complete account of this active and rapidly progressing field of research. Space permits only a rather crude introduction and the discussion of a few examples. The reader who is interested in more exhaustive overviews is referred to, e.g. , the set of review articles in [6], or to Refs. [5, 7, 8].

The paper is organized as follows: First, we review some of the most commonly studied idealized models for particle based simulations. We will then focus on the nematic phase and discuss some of its bulk properties. Finally, we address specific issues of interfacial properties in nematic liquid crystals.

2 Model Systems

2.1 Ellipsoids

Perhaps the most obvious anisotropic generalization of hard spheres are hard ellipsoids of revolution with one symmetry axis of length L and transverse thickness D. The phase diagram in three dimensions has been established from computer simulations by Frenkel, Allen and coworkers [9, 10, 11, 12, 13, 14]. It is shown for a range of elongations $\kappa = L/D$ in Fig. 2. If the particles are sufficiently anisotropic, a nematic phase (N) intrudes between the isotropic (I) and the crystalline (X,PC) phase. No further liquid crystalline phases are present. In particular, smectic phases do not exist. This makes hard ellipsoids particularly suited to study the properties of nematic liquid crystals.

From a technical point of view, the study of ellipsoids is complicated by the fact that the contact distance $\sigma(\mathbf{u}_i, \mathbf{u}_j, \hat{\mathbf{r}}_{ij})$ between two particles i and j with orientations $\mathbf{u}_i$ and $\mathbf{u}_j$ in the center-center direction $\hat{\mathbf{r}}_{ij}$ ($|\mathbf{u}_{i,j}| = |\hat{\mathbf{r}}_{ij}| = 1$)) cannot be given in analytically closed form. Criteria to decide whether or not two particles overlap have been derived by Vieillard-Baron [16], and by Perram and Wertheim [17, 18]. Technical details on the implementation can be found in the review article [11]).

Both the simulation and the data analysis are simplified considerably if an explicit formula for $\sigma(\mathbf{u}_i, \mathbf{u}_j, \mathbf{r}_{ij})$ is available. For this reason, Berne and Pechukas [19] have proposed an approximate expression

$$\sigma(\mathbf{u}_i, \mathbf{u}_j, \hat{\mathbf{r}}_{ij}) = \sigma_0 \left\{ 1 - \frac{\chi}{2} \left[\frac{(\mathbf{u}_i \cdot \hat{\mathbf{r}}_{ij} + \mathbf{u}_j \cdot \hat{\mathbf{r}}_{ij})^2}{1 + \chi \mathbf{u}_i \cdot \mathbf{u}_j} + \frac{(\mathbf{u}_i \cdot \hat{\mathbf{r}}_{ij} - \mathbf{u}_j \cdot \hat{\mathbf{r}}_{ij})^2}{1 - \chi \mathbf{u}_i \cdot \mathbf{u}_j} \right] \right\}^{-1/2}, \tag{1}$$

with the anisotropy parameter $\chi = (\kappa^2 - 1)/(\kappa^2 + 1)$ This function gives the exact contact distance if the particles i and j are co-linear (parallel to each other), and overestimates it by a factor of at most $\sqrt{2}/\sqrt{1 + 2\kappa/(\kappa^2 + 1)} < \sqrt{2}$ otherwise. A few simulation studies have considered systems of "hard" particles with contact distances given by (1) ("hard gaussian overlap particles") [20, 21, 22, 23, 24, 25]. Much more commonly, however,

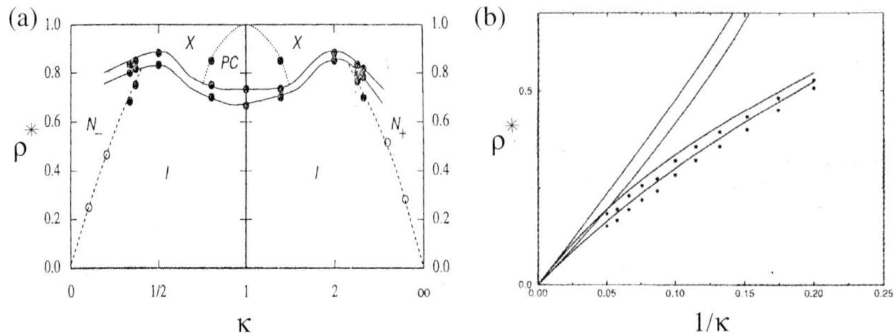

Fig. 2. Phase diagram of hard ellipsoids as a function of elongation $\kappa = L/D$ and reduced density ρ/ρ_{cp} (ρ_{cp} being the density at hcp packing). Phases are: I (isotropic phase), N (nematic phase), X (orientationally ordered crystalline phase), PC (orientationally disordered crystalline phase). (a) Full range of elongations: From [5] (Copyright 1995, Kluwer Publishers), with data from Refs. [9, 10, 12, 15]. (b) Detail from [14] (Copyright 1996, American Institute of Physics). Points denote simulation data, lines predictions from different theories.

the contact function (1) is used in conjunction with Lennard-Jones type interaction potentials, the Gay-Berne potential [26]. This class of models will be discussed in the next section. The present authors favor yet another potential, which is purely repulsive but soft.

$$V_{ij} = \begin{cases} 4\epsilon_0 \, (R_{ij}^{-12} - R_{ij}^{-6}) + \epsilon_0, & R_{ij}^{-6} < 2 \\ 0, & \text{otherwise} \end{cases}, \tag{2}$$

where R_{ij} is a reduced distance,

$$R_{ij} = (r_{ij} - \sigma(\mathbf{u}_i, \mathbf{u}_j, \hat{\mathbf{r}}_{ij}) + \sigma_0)/\sigma_0 \tag{3}$$

As far as we have seen so far, systems of such "soft ellipsoids" (at temperature $k_B T \sim \epsilon$) have almost the same structure as systems of hard ellipsoids.

2.2 Gay-Berne Particles

Despite their appealing simplicity, hard or soft repulsive ellipsoids are not in every respect best suited to study generic liquid crystal properties. For example, such models do not exhibit smectic phases. Furthermore, one cannot study the effect of attractive interactions, the interplay between liquid-crystalline order and liquid-vapor phase separation etc. Therefore, Gay and Berne [26] have introduced a class of ellipsoidal pair potentials with attractive interactions, which has become one of the standard liquid crystal potentials.

The functional form of these potentials is

$$V_{ij} = 4\,\epsilon_0\,\epsilon'(\mathbf{u}_i, \mathbf{u}_j, \hat{\mathbf{r}}_{ij})^{\mu}\,\epsilon''(\mathbf{u}_i, \mathbf{u}_j)^{\nu}\,(R_{ij}^{-12} - R_{ij}^{\;6}), \tag{4}$$

where the reduced distance R_{ij} defined is as above (eqn. (3) with σ taken from (1)), and the energy functions ϵ are given by

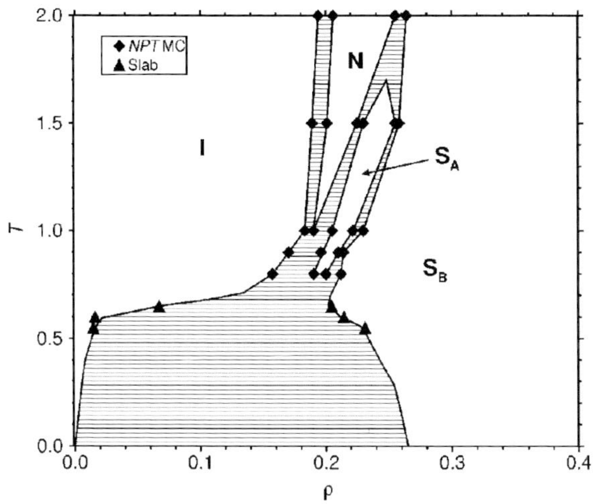

Fig. 3. Phase diagram of Gay-Berne ellipsoids with $\mu = 2, \nu = 1, \kappa = 4$, and $\kappa' = 5$. In addition to the phases observed for hard ellipsoids (Fig. 2), one also finds smectic phases S_A and S_B. The simulations did not allow to decide whether the smectic B phase is hexatic or solid, i.e., crystalline. In the smectic A phase, however, the smectic layers are undoubtedly fluid. From [32] (Copyright 1997, American Physical Society).

$$\epsilon'(\mathbf{u}_i, \mathbf{u}_j, \hat{\mathbf{r}}_{ij}) = 1 - \frac{\chi'}{2}\left[\frac{(\mathbf{u}_i \cdot \hat{\mathbf{r}}_{ij} + \mathbf{u}_j \cdot \hat{\mathbf{r}}_{ij})^2}{1 + \chi' \mathbf{u}_i \cdot \mathbf{u}_j} + \frac{(\mathbf{u}_i \cdot \hat{\mathbf{r}}_{ij} - \mathbf{u}_j \cdot \hat{\mathbf{r}}_{ij})^2}{1 - \chi' \mathbf{u}_i \cdot \mathbf{u}_j}\right] \tag{5}$$

$$\epsilon''(\mathbf{u}_i, \mathbf{u}_j) = \left[1 - \chi^2(\mathbf{u}_i \cdot \mathbf{u}_j)^2\right]^{-1/2} . \tag{6}$$

The new additional anisotropy parameter χ' accounts for the fact that the attractive energy should favor a side-by-side alignment of particles compared to an end-to-end alignment. It is related to the corresponding ratio of minimum energies (well depths) $\kappa' = \epsilon_{\text{side-side}}/\epsilon_{\text{end-end}}$ via $\chi' = (\kappa'^{1/\mu} - 1)/(\kappa'^{1/\mu} + 1)$. Gay and Berne originally suggested to choose the exponents $\mu = 2$ and $\nu = 1$. The phase behavior for this set of exponents has been studied in detail [27, 28, 29, 30, 31, 32], for various choices of κ and κ' [30, 31, 32]. An example is shown in Fig. 3. One observes a liquid-vapor coexistence region at low temperature. Furthermore, the attractive interaction stabilizes a smectic phase (Sm A), which consists of stacked layers of two dimensional fluids.

The Gay-Berne model is widely used in liquid crystal simulations [6]. Due to the option of varying not only κ and κ', but also the exponents μ and ν, it is very versatile [33, 34]. Luckhurst et al have demonstrated that it can even be adjusted to serve as a good model for a real thermotropic liquid crystal [35, 36].

2.3 Other Models

Ellipsoids are not the only particles that have been used to model liquid crystals on an idealized level. An almost equal amount of work has been devoted to systems of spherocylinders: Cylinders of length L and diameter D capped with hemispheres at

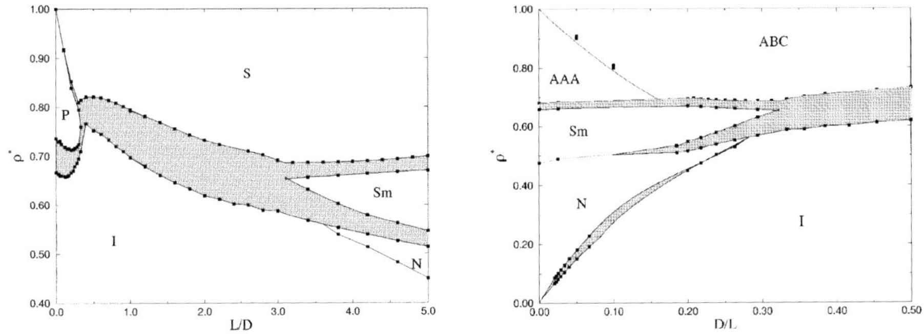

Fig. 4. Phase diagram of hard spherocylinders as a function of reduced density $\rho^* = \rho/\rho_{cp}$ and elongation L/D (a) or inverse elongation D/L. Phases are: I (isotropic phase), N (nematic phase), Sm (smectic A phase), AAA and S=ABC (two types of solid phases). From [46] (Copyright 1996, American Institute of Physics). The transition between the nematic and the smectic phase is first order at all elongations [47].

both ends. They have the advantage that the contact distance σ between two particles can be evaluated exactly [37]. Furthermore, they resemble quite closely certain real colloidal liquid crystal substances, e. g., rodlike virusses [38, 39, 40, 41].

The phase behavior of hard spherocylinders has been studied by Frenkel and others [42, 43, 44, 45, 46, 47]. The complete phase diagram for all elongations L/D as computed by Bolhuis and Frenkel [46] is shown in Fig. 4. It differs strikingly from the corresponding phase diagram for hard ellipsoids in that sufficiently elongated spherocylinders form smectic phases. This came as a surprise when it was first discovered in simulations [42]. Meanwhile, it has been reproduced by density functional theories [48, 49, 50, 51, 52].

Bolhuis et al [53] have studied the influence of attractive interactions on the phase behavior of spherocylinders. Like in systems of Gay-Berne particles (Fig. 3), a liquid-vapor coexistence region emerges at low temperatures (see also [54]). The influence of polydispersity – i. e., fluctuating rod length – on the phase diagram has been investigated by Bates et al [55]. Perhaps not surprisingly, the smectic phase is destabilized in favor of the nematic phase.

We shall only briefly mention some other particle based model systems for liquid crystals: In a number of studies, liquid crystalline molecules have been modeled by stiff chain molecules [56, 57, 58, 59, 60, 61, 62, 63]. with aspect ratio $\kappa < 1$ and spheres cut at both sides have been used to study discotic liquid crystals, i. e., fluids of disklike particles [64, 65, 66, 67, 68, 69]. Such fluids exhibit new, different types of liquid crystalline order. For example, the smectic phases are replaced by columnar phases: The discs assemble into hexagonal arrays of columns, but remain fluid in one dimension within the columns.

3 Properties of the Nematic Phase

A number of important properties of the nematic phase can already be deduced from very general considerations.

First, geometric arguments suggest that the transition between the isotropic and the nematic phase should be discontinuous or first order, i.e., there should be a density regime in which both phases coexist. This is a consequence of "Landau's rule" [1, 70], which relates the character of the phase transition to the symmetry groups of the two phases. Landau's rule is not rigorous, it does not exclude the possibility that the width of the coexistence region may happen to shrink to zero in certain points of the phase diagram. Moreover, order fluctuations may affect the nature of the phase transition. So far, however, the actually observed isotropic-nematic transitions have been discontinuous, albeit with rather narrow coexistence regions.

Second, the free energy cost of spatial modulations of the average particle orientation must vanish if the wavelength of the modulations tends to infinity. This is an example of Goldstone's theorem [71, 72]: If the order parameter in an ordered phase breaks a continuous symmetry, there exist soft massless fluctuation modes [73]. Modulations with finite wavelength are subject to elastic restoring forces [73, 74]. For symmetry reasons, these depend on only three material constants, the Frank elastic constants K_{11}, K_{22}, K_{33} [75, 76, 77]. Figure 5 illustrates the corresponding fundamental distortions, the "splay", "twist" and "bend" mode.

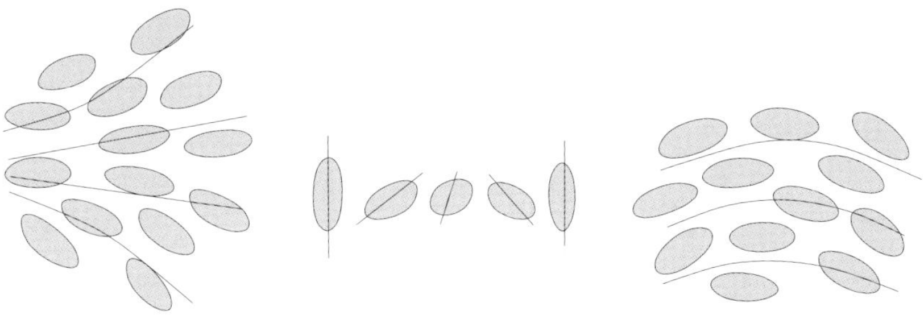

Fig. 5. Elastic modes in nematic liquid crystals.

We turn to give a more quantitative description. The nematic order is usually characterized by the 3×3 dimensional nematic order tensor [1]

$$\mathbf{Q} = \frac{1}{N}\sum_{i=1}^{N}(\frac{3}{2}\mathbf{u}_i \otimes \mathbf{u}_i - \frac{1}{2}\mathbf{1}), \tag{7}$$

where the sum runs over all particles i, $\mathbf{1}$ denotes the unit matrix and $\otimes$ the dyadic product. The largest eigenvalue of $\mathbf{Q}$ is the nematic order parameter per particle, and the corresponding eigenvector is the director $\mathbf{n}$ of the nematic liquid. It gives the direction of preferential alignment of the molecules. Long wavelength spatial modulations $\mathbf{n}(\mathbf{r})$ of the director are penalized by the elastic free energy [77]

$$\mathcal{F}\{\mathbf{n}(\mathbf{r})\} = \frac{1}{2}\int d\mathbf{r}\Big\{K_{11}[\nabla\cdot\mathbf{n}]^2 + K_{22}[\mathbf{n}\cdot(\nabla\times\mathbf{n})]^2 + K_{33}[\mathbf{n}\times(\nabla\times\mathbf{n})]^2\Big\}. \tag{8}$$

On macroscopic length scales, the structure of liquid crystals is controlled almost exclusively by the elastic energy and thus by the Frank elastic constants K_{11}, K_{22}, and K_{33}.

The quantitative characterization of the local, microscopic structure is much more involved. The most intensely studied quantities are usually the pair distributions or the pair correlation functions. On principle, a hierarchy of infinitely many N-particle distribution functions would be required to fully characterize a fluid. In systems with only pairwise interactions, however, the pair distributions stand out because they can be used to calculate several macroscopic quantities such as the pressure and the compressibility. Moreover, pair correlations are often accessible to experiments in real systems.

As before, we characterize the orientation of a particle by a unit vector $\mathbf{u}$. The one-particle distribution $\rho^{(1)}(\mathbf{u}, \mathbf{r})$ gives the probability density of finding a particle of orientation $\mathbf{u}$ at the position $\mathbf{r}$. Similarly, the two-particle distribution $\rho^{(2)}(\mathbf{u}_i, \mathbf{r}_i, \mathbf{u}_j, \mathbf{r}_j)$ is the probability density of finding simultaneously one particle of orientation $\mathbf{u}_i$ at the position $\mathbf{r}_i$, and another with orientation $\mathbf{u}_j$ at the position $\mathbf{r}_j$. In a homogeneous nematic phase, $\rho^{(1)}$ does not depend on the position $\mathbf{r}$, and $\rho^{(2)}$ depends only on the relative position $\mathbf{r}_{ij} = \mathbf{r}_i - \mathbf{r}_j$. Furthermore, particles at infinite distances are uncorrelated, hence

$$\rho^{(2)}(\mathbf{u}_i, \mathbf{u}_j, \mathbf{r}) \stackrel{r\to\infty}{\longrightarrow} \rho^{(1)}(\mathbf{u}_1)\rho^{(1)}(\mathbf{u}_2). \tag{9}$$

The correlations between particles at finite distance are characterized by the total correlation function [78, 79]

$$h(\mathbf{u}_1, \mathbf{u}_2, \mathbf{r}) = \frac{\rho^{(2)}(\mathbf{u}_1, \mathbf{u}_2, \mathbf{r})}{\rho^{(1)}(\mathbf{u}_1)\rho^{(1)}(\mathbf{u}_2)} - 1, \tag{10}$$

It subsumes the total effect of a particle i on a particle j. Due to the elastic interactions mentioned above, it decays only slowly at long distances with the algebraic power law $1/r$.

Of course, other definitions of correlation functions are possible. As an alternative to the total correlation function, one could consider the orientation correlation function defined by Penttinen and Stoyan [80]. It describes the pure orientation correlations between particles at a given distance, and should display the same asymptotic $1/r$ power law in the nematic fluid as h.

The long range correlations are mediated by the whole bulk of the surrounding nematic fluid. It seems desirable to separate these "indirect" effects from a more local "direct" effect of two particles on each other. This can be done by considering the "direct correlation function" $c(\mathbf{u}_i, \mathbf{u}_j, \mathbf{r}_{ij})$, which is defined through the Ornstein-Zernike equation [78, 79]

$$h(\mathbf{u}_i, \mathbf{u}_j, \mathbf{r}_{ij}) = c(\mathbf{u}_i, \mathbf{u}_j, \mathbf{r}_{ij}) + \int c(\mathbf{u}_i, \mathbf{u}_k, \mathbf{r}_{ik})\, \rho^{(1)}(\mathbf{u}_k)\, h(\mathbf{u}_k, \mathbf{u}_j, \mathbf{r}_{kj}) d\mathbf{u}_k d\mathbf{r}_k. \tag{11}$$

As it turns out, the direct correlation function is indeed short range even in the nematic phase.

We illustrate this with recent computer simulation results. We have performed computer simulations of 1000-8000 ellipsoids with elongation $\kappa = 3$ at the number density

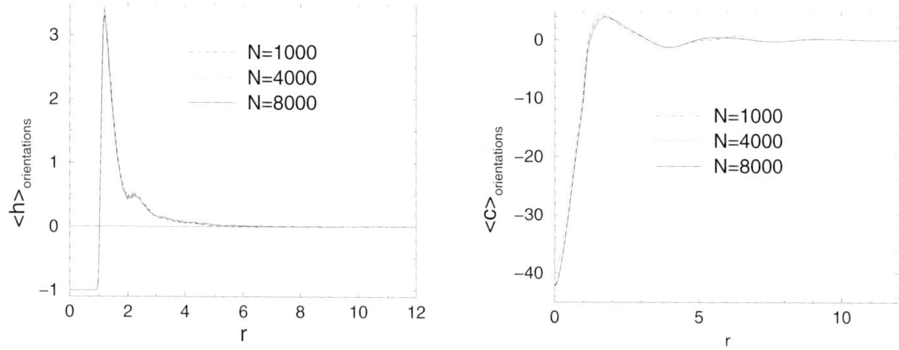

Fig. 6. Total correlation function h (left) and direct correlation function c (right) vs. molecular distance r, averaged over all orientations of $\mathbf{u}_i$,$\mathbf{u}_j$, and $\mathbf{r}_{ij}$, for different system sizes N.

$\rho = 0.3$, interacting with the potential (2) at temperature $T = 0.5$. Technical details can be found in Refs. [81, 83, 84].

The calculation of direct correlation functions from computer simulations is computationally demanding and involves a complicated data analysis. We were the first to determine exactly the direct correlation function in a nematic fluid. Direct correlation functions in isotropic fluids have been calculated earlier by Allen et al [85, 86], and approximate data for nematic fluids have been derived from simulations by Stelzer et al [87, 88].

Figure 6 shows the orientational average of the total correlation function and the direct correlation function. The average is carried out over the orientations $\mathbf{u}_i$, $\mathbf{u}_j$, and $\hat{\mathbf{r}}_{ij}$. Figure 6 demonstrates that long range algebraic correlations between particles are not apparent in this averaged curve. The elastic interactions affect only orientations, not total densities. Therefore, both h and c are short range.

In order to assess the effect of the elasticity, we need to study orientational dependent correlations. To this end, we expand all correlation functions in spherical harmonics $Y_{lm}(\mathbf{u})$.

$$F(\mathbf{u}_i, \mathbf{u}_j, \mathbf{r}) = \sum_{\substack{l_1,l_2,l \\ m_1,m_2,m}} F_{l_1 m_1 l_2 m_2 l m}(r)\, Y_{l_1 m_1}(\mathbf{u}_i)\, Y_{l_2 m_2}(\mathbf{u}_j)\, Y_{lm}(\hat{\mathbf{r}}), \qquad (12)$$

where F stands for $\rho^{(2)}$, h, or c. The z axis is chosen in the direction of the director. The orientation dependence of the correlations is reflected by the coefficients with nonzero indices. Figure 7 (left) shows a coefficient of the total correlation function with a particularly pronounced long range tail. As shown in Fig. 7 (right), the tail completely disappears in the direct correlation function.

The direct correlation function is a central quantity in theories of liquid matter [78]. It enters density functional theories which predict the structure of inhomogeneous fluids, e. g., fluids at interfaces. It allows to calculate a number of material constants of fluids. In nematic fluids, for example, it can be used to calculate the Frank elastic constants according to a set of equations first derived by Poniewierski and Stecki [89, 90]. We have

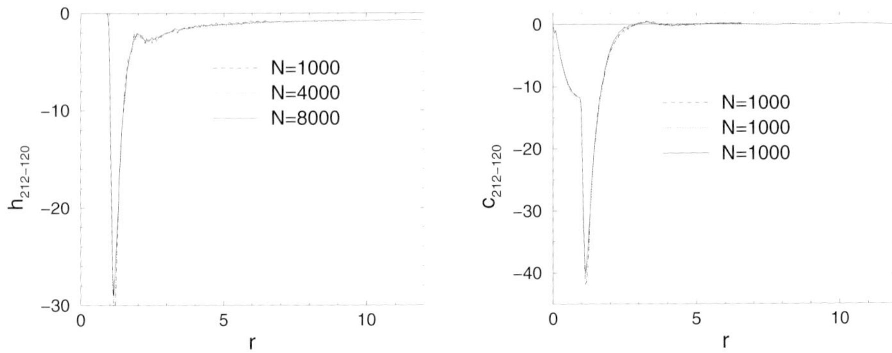

Fig. 7. Expansion coefficient with $l_1 = 2, l_2 = 2, m_1 = 1, m_2 = -1, l = 2$, and $m = 0$ of the total correlation function h (left), and the direct correlation function (right) vs. molecular distance r for different system sizes N. Left Figure: Dashed line indicates an extrapolation with a $1/r$ power law behavior. Inset shows the same data vs. $1/r$.

applied these equations to our data and obtained a set of values for K_{11}, K_{22}, and K_{33}, which is in agreement with results from a different, more direct method [81, 82]. Thus the direct correlation function establishes a bridge between the microscopic structure and the parameters which characterize the macroscopic structure.

4 Interfacial Properties

We turn to discuss briefly the interfacial properties in nematic liquid crystals. These are of great interest in the liquid crystal display technology [91]. Therefore, a growing number of simulations are devoted to the study of model nematics at surfaces [93, 94, 95, 96, 97, 98, 99, 100, 101, 102, 103, 104, 105, 106, 107] and interfaces [108, 109, 110, 111, 112] or in thin films [24, 25, 113, 114, 115, 116, 117, 118].

The microscopic structure at surfaces can be quite complex [93, 94, 95, 96]. From a macroscopic point of view, the presence of surfaces or interfaces introduces essentially three new effective parameters: the interfacial tension, the anchoring angle, and the anchoring energy. The anchoring angle is the angle with respect to the surface which the director preferably assumes close to the surface. The anchoring energy is related to the force needed to twist the director out of the anchoring angle.

The anchoring angle can be calculated from simulations in a straightforward way. Obtaining the anchoring energy is more difficult and time consuming. Allen and coworkers have devised different methods and applied them to systems of ellipsoids at hard, planar walls [106, 107]. The interfacial tension can be computed from the anisotropy of the pressure tensor close to the surface [111, 119], or in the case of fluid-fluid interfaces from their undulations (capillary waves) [112]. Recently, a number studies have been devoted to the interface between the nematic and the isotropic phase in Gay-Berne fluids [108] and fluids of spherocylinders [110] or ellipsoids [109, 111, 112]. They reveal among other a rather intriguing capillary wave spectrum. It reflects a complex interplay between the bare surface tension and the elastic interactions in the nematic phase [112].

5 Conclusions

In materials science, systems of ellipsoids serve as model systems which exhibit generic properties of nematic liquid crystals. They are studied for the general insight they can give into the physics of liquid crystals, and also because simulations of ellipsoids are computationally much cheaper than simulations of a comparable number of more realistic molecules. However, fluids of ellipsoids are also fascinating from a much more general point of view: They are simple systems where purely entropic effects generate an amazing variety of different structures, and lead to unusual long-range static and dynamic properties. Computer simulations are often the only way of investigating some of that behavior.

Acknowledgements

We have benefitted from discussions and interactions with N. Akino, M. P. Allen, K. Binder, G. Germano, and H. Lange. N. H. P. acknowledges financial support by the German Science Foundation (DFG). The computer simulations of our group are carried out mostly on the CRAY T3Es of the NIC in Jülich.

References

1. De Gennes P. G. (1995): *The Physics of Liquid Crystals*. (Oxford University Press, Oxford)
2. Chandrasekhar, S. (1992): *Liquid Crystals*. (Cambridge University Press, Cambridge)
3. Onsager L. (1949): 'The effects of shape on the interaction of colloidal particles'. Ann. N.Y. Acad. Sci. **51**, pp. 627–659
4. Maier W., A. Saupe (1958): 'Eine einfache molekulare Theorie des nematischen kristallinflüssigen Zustandes'. Z. Naturforschung **A13**, 564–566.
5. Allen M. P. (1995): 'Simulations and phase behaviour of liquid crystals'. In: *Observation, Prediction and Simulation of Phase Transitions in Complex Fluids*, ed. by M. Baus et al (Kluwer Academic Publ., Dordrecht), pp. 557–590.
6. Pasini P., C. Zannoni eds. (1998): *Advances in the Computer Simulations of Liquid Crystals*. (Kluwer Academic Publ., Dordrecht)
7. Crain J., A. V. Komolkin (1999): 'Simulating molecular properties of liquid crystals'. Adv. Chem. Phys. **109**, 39–113.
8. Singh S. (2000): 'Phase transitions in liquid crystals'. Phys. Rep. **324**, 108–269.
9. Frenkel D., B. M. Mulder (1985): 'The hard ellipsoid-of-revolution fluid. I. Monte Carlo simulations'. Molecular Physics **55**, 1171–92.
10. Allen M. P., M. R. Wilson (1989): 'Computer simulation of liquid crystals'. Journal of Computer Aided Molecular Design **3**, 335–353.
11. Allen M. P., G. T. Evans, D. Frenkel, B. Mulder (1993): 'Hard convex body fluids'. Adv. Chem. Phys. **86**, 1–166.

12. Samborski A., G. T. Evans, C. P. Mason, M. P. Allen (1994): 'The isotropic to nematic liquid crystal transition for hard ellipsoids: An Onsager-like theory and computer simulations'. Molecular Physics **81**, 263–76.
13. Allen M. P., C. P. Mason (1995): 'Stability of the nematic phase for the hard ellipsoid fluid' Molecular Physics **86**, 467–74.
14. Camp. P. J., C. P. Mason, M. P. Allen, A. A. Khare, D. A. Kofke (1996): 'The isotropic-nematic phase transition in uniaxial hard ellipsoid fluids: Coexistence data and the approach to the Onsager limit'. J. Chem. Phys. **105**, 2837–2949.
15. Frenkel D., A. J. C. Ladd (1984). 'New Monte Carlo method to compute the free energy of arbitrary solids. Application to the FCC and HCP phases of hard spheres'. J. Chem. Phys. **81**, 3188–93.
16. Vieillard-Baron J. (1972): 'Phase transitions of the classical hard-ellipse system'. J. Chem. Phys. **56**, 4729–4744.
17. Perram J. W., M. S. Wertheim, J. L. Lebowitz, G. O. Williams (1984): 'Monte Carlo simulation of hard spheroids'. Chem. Phys. Lett. **105**, 277–280.
18. Perram J. W., M. S. Wertheim (1985): 'Statistical mechanics of hard ellipsoids. I. Overlap algorithm and the contact function'. J. Comp. Phys. **58**, 409–416.
19. Berne J. B., P. Pechukas (1971): 'Gaussian model potentials for molecular interactions'. J. Chem. Phys. **56**, 4213–4216.
20. Rigby M. (1989): 'Hard gaussian overlap fluids'. Mol. Phys. **68**, 687–697.
21. Padilla P., E. Velasco (1997): 'The isotropic-nematic transition for the hard Gaussian overlap fluid: Testing the decoupling approximation'. J. Chem. Phys. **106**, 10299-10310.
22. Velasco E., P. Padilla (1998): 'Nematic virial coefficients of very long hard molecules and Onsager theory'. Mol. Phys. **94**, 335–339.
23. Huang S. L., V. R. Bhethanabotla (1999): 'Virial coefficients for the Hard Gaussian Overlap model'. Int. J. Mod. Phys. C **10**, 361–374 (1999).
24. Cleaver D. J., P. I. C. Teixeira (2001): 'Discontinuous structural transition in a thin hybrid liquid crystal film'. Chem. Phys. Lett. **338**, 1–6.
25. Chrzanowska A., P. I. C. Teixeira, H. Ehrentraut, D. J. Cleaver (2001): 'Ordering of hard particles between hard walls'. J. Phys. - Cond. Matt. **13**, 4715–4726 (2001).
26. Gay J. G, B. J. Berne (1980): 'Modification of the overlap potential to mimic a linear site-site potential'. J. Chem. Phys. **74**, 3316–3319.
27. Adams D. J., G. R. Luckhurst, R. W. Phippen (1987): 'Computer simulation studies of anisotropic systems. XVII. The Gay-Berne model nematogen'. Mol. Phys. **61** 1575–1580.
28. De Miguel E., L. F. Rull, M. K. Chalam, K. E. Gubbins, F. Van Swol (1991): 'Location of the isotropic-nematic transition in the Gay-Berne model'. Mol. Phys. **72** 593–605.
29. de Miguel E., L. F. Rull, M. K. Chalam, K. E. Gubbins (1991): 'Liquid crystal phase diagram of the Gay-Berne fluid'. Mol. Phys. **74**, 405–424.
30. Emsley J. W., G. R. Luckhurst, W. E. Palke, D. J. Tildesley (1992): 'Computer simulation studies of the dependence on density of the orientational order in nematic liquid crystals'. Liquid Crystals **11**, 519–530.
31. de Miguel E., E. Martin del Rio, J. T. Brown, M. P. Allen (1996): 'Effect of the attractive interactions on the phase behavior of the Gay-Berne liquid crystal model'. J. Chem. Phys. **105**, 4234–4249.
32. Brown J. T., M. P. Allen, E. Martin del Rio, E. de Miguel (1997): 'Effects of elongation on the phase behavior of the Gay-Berne fluid'. Phys. Rev. E **57**, 6685–6699.
33. Luckhurst G. R., R. A. Stephens, R. W. Phippen (1990): 'Computer simulation studies of anisotropic systems. XIX. Mesophases formed by the Gay-Berne model mesogen'. Liquid Crystals **8**, 451–464.
34. Berardi R., A. P. J. Emerson, C. Zannoni (1993): 'Monte-Carlo investigations of a Gay-Berne liquid crystal'. J. Chem. Soc. Faraday T. **89**, 4069–4078.

35. Luckhurst G. R., P. S. J. Simmonds (1993): 'Computer simulation studies of anisotropic systems. XXI. Parametrization of the Gay-Berne potential for model mesogens'. Mol. Phys. **80**, 233–252.
36. Bates M. A., G. R. Luckhust (1999): 'Computer simulation studies of anisotropic systems. XXX. The phase behavior and structure of a Gay-Berne mesogen'. J. Chem. Phys. **110**, 7087–7108.
37. Vieillard-Baron J. (1974): 'The equation of state of a system of hard spherocylinders'. Mol. Phys. **28**, 809–818.
38. Zasadzinski J. A. N., M. J. Sammon, R. B. Meyer, M. Cahoon, D. L. D. Caspar (1986): 'Freeze-fracture imaging of ordered phases of tobacco mosaic virus in water'. Mol. Cryst. Liqu. Cryst. **138**, 211–229.
39. Zasadzinski J. A. N., R. B. Meyer (1986): 'Molecular imaging of tobacco mosaic virus lyotropic nematic phases'. Phys. Rev. Lett. **56**, 636–6388.
40. Oldenbourg R., X. Wen, R. B. Meyer, D. L. D. Caspar (1988): 'Orientational distribution function in nematic tobacco-mosaic-virus liquid crystals measured by X-ray diffraction'. Phys. Rev. Lett. **61**, 1851–1854.
41. Dogic Z., S. Fraden (1997): 'Smectic phase in a colloidal suspension of semiflexible virus particles'. Phys. Rev. Lett. **78**, 2417–2420.
42. Frenkel D. (1987): 'Onsager's spherocylinders revisited'. J. Phys. Chem. **91**, 4912–4916.
43. Frenkel D., H. N. W. Lekkerkerker, A. Stroobants (1988): 'Thermodynamic stability of a smectic phase in a system of hard rods'. Nature **332**, 822–823.
44. Veerman J. A. C., D. Frenkel (1990): 'Phase diagram of a system of hard spherocylinders by computer simulation'. Phys. Rev. E **41**, 3237–3244.
45. McGrother S. C., D. C. Williamson, G. Jackson (1996): 'A re-examination of the phase diagram of hard spherocylinders'. J. Chem. Phys. **104**, 6755–6771.
46. Bolhuis P., D. Frenkel (1996): 'Tracing the phase boundaries of hard spherocylinders', J. Chem. Phys. **106**, 666–687.
47. Polson J. M., D. Frenkel (1997): 'First-order nematic-smectic phase transition for hard spherocylinders in the limit of infinite aspect ratio'. Phys. Rev. E **56**, R6260-R6263.
48. Poniewierski A., R. Holyst (1988): 'Density-functional theory for nematic and smectic-A ordering of hard spherocylinders'. Phys. Rev. Lett. **61**, 2461–2464.
49. Somoza A. M., P. Tarazona (1990): 'Nematic and smectic liquid crystals of hard sphero cylinders'. Phys. Rev. E **41**, 965–970.
50. Poniewierski A., T. J. Sluckin TJ (1991): 'Phase diagram for a system of hard spherocylinders'. Phys. Rev. E **43**, 6837–6842.
51. Graf H., H. Löwen (1999): 'Density functional theory for hard spherocylinders: phase transitions in the bulk and in the presence of external fields'. J. Phys.: Cond. Matt. **11**, 1435-1452.
52. Velasco E., L. Mederos, D. E. Sullivan (2000): 'Density-functional theory of inhomogeneous systems of hard spherocylinders'. Phys. Rev. D **62**, 3708-3718.
53. Bolhuis P. G., A. Stroobants, D. Frenkel, H. N. W. Lekkerkerker (1997): 'Numerical study of the phase behavior of rodlike colloids with attractive interactions'. J. Chem. Phys. **107**, 1551–1564.
54. Williamson D. C., F. del Rio (1998): 'The isotropic-nematic phase transition in a fluid of square well spherocylinders'. J. Chem. Phys. **109**, 4675–4686.
55. Bates M. A., D. Frenkel (1998): 'Influence of polydispersity on the phase behavior of colloidal liquid crystals: A Monte Carlo simulation study'. J. Chem. Phys. **109**, 6193–6198.
56. Wilson M. R. (1994): 'Molecular dynamics simulation of semiflexible mesogens'. Mol. Phys. **81**, 675–690.

57. Levesque D., M. Mazars, J. J. Weis (1995): 'Monte Carlo study of the thermodynamic stability of the nematic phase of a semiflexible liquid crystal model'. J. Chem. Phys. **103**, 3820–3831.
58. Mazars M., D. Levesque, J. J. Weis (1997): 'Monte Carlo study of a semiflexible liquid crystal model: The smectic phase'. J. Chem. Phys. **106**, 6107–6115.
59. Affouard F., M. Kröger, S. Hess (1996): 'Molecular dynamics of model liquid crystals composed of semiflexible molecules'. Phys. Rev. E **54**, 5178–5186.
60. Williamson D. C., G. Jackson (1998): 'Liquid crystalline phase behavior in systems of hard-sphere chains'. J. Chem. Phys. **108**, 10294–10302.
61. Faller R., A. Kolb, F. Müller-Plathe (1999): 'Local chain ordering in amorphous polymer melts: influence of chain stiffness'. PCCP Phys. Chem. Chem. Phys. **1**, 2071–2076.
62. Weber H., W. Paul, K. Binder (1999): 'Monte Carlo simulation of a lyotropic first-order isotropic-nematic phase transition in a lattice polymer model'. Phys. Rev. E **59**, 2168–2174.
63. McBride C., C. Vega, L. G. MacDowell (2001): 'Isotropic-nematic phase transition: Influence of intramolecular flexibility using a fused hard sphere model '. Phys. Rev. E **64**, 11703–11718.
64. Veerman J. A. C., D. Frenkel (1992): 'Phase behavior of disklike hard-core mesogens'. Phys. Rev. E **45**, 5632–5648.
65. Bates M. A., G. R. Luckhurst (1996): 'Computer simulation studies of anisotropic systems. XXVI Monte Carlo investigations of a Gay-Berne discotic at constant pressure'. J. Chem. Phys. **104**, 6696–6709.
66. De Luca M. D., M. K. Griffiths, C. M. Care, M. P. Neal (1994): 'Computer modelling of discotic liquid crystals'. Intnl. J. Electr. **77**, 907–917.
67. Emerson A. P. J., G. R. Luckhurst, S. G. Whatling (1994): 'Computer simulation studies of anisotropic systems. XXIII. The Gay-Berne discogen'. Mol. Phys. **82**, 113–124.
68. Zewdie H. (1998): 'Computer-simulation studies of diskotic liquid crystals'. Phys. Rev. E **57**, 1793–1805.
69. Bates M. A., D. Frenkel (1998): 'Infinitely thin disks exhibit a first order nematic-columnar phase transition'. Phys. Rev. E **57**, 4824–4826.
70. Landau L. D., E. M. Lifschitz (1969): *Course of theoretical physics, Vol.5 Statistical physics*. (Pergamon Press Ltd. Oxford)
71. J. Goldstone J. (1961): 'Field theories with 'superconductor' solutions'. Nuovo Cimento **19**, 154–164.
72. Goldstone J., A. Salam, S. Weinberg (1962). 'Broken Symmetries'. Phys. Rev. **127**, 965–970.
73. Forster D. (1975): *Hydrodynamic Fluctuations, Broken Symmetry and Correlation Functions*. Vol. 47 of *Frontiers in Physics* (Benjamin, Reading, MA)
74. Chaikin, P. M., Lubensky, T. C. (1995): *Principles of condensed matter physics*. (Cambridge University Press, Cambridge)
75. Oseen, C. (1933): 'The theory of liquid crystals'. Trans. Faraday Soc. **29**, 883–899.
76. Zöcher, H. (1933). Trans. Faraday Soc. **29**, 945.
77. Frank, F. C. (1958): 'On the theory of liquid crystals'. Discuss. Faraday Soc. **25**, 19–29.
78. Hansen, J. P., I. R. McDonald (1986): *Theory of Simple Liquids*. (Academic Press, London, 1986).
79. Gray C. G., K. E. Gubbins (1984): *Theory of Molecular Fluids*, Vol. 1. (Oxford University Press, New York)
80. Penttinen A. K., D. Stoyan (1989): 'Statistical analysis for a class of line segment processes'. Scand. J. Statist **16**, 153–168.
81. Phuong N. H., G. Germano, F. Schmid (2001): 'Elastic constants from direct correlation functions in nematic liquid crystals: A computer simulation study'. J. Chem. Phys. **115**, 7227–7234.

82. Allen M. P., M. A. Warren, M. R. Wilson, A. Sauron, W. Smith (1997): 'Molecular dynamics calculation of elastic constants in Gay-Berne nematic liquid crystals'. J. Chem. Phys. **105**, 2850–2858.
83. Phuong N. H., G. Germano, F. Schmid (2001): 'The direct correlation function in nematic liquid crystals from computer simulation'. Comp. Phys. Comm., in print.
84. Phuong N. N., F. Schmid (2001): 'Liquid structure of nematic and isotropic liquid crystals'. in preparation.
85. Allen M. P., C. P. Mason, E. de Miguel, J. Stelzer (1995): 'Structure of molecular liquids'. Phys. Rev. E **52**, R25–R52.
86. Allen M. P., J. T. Brown, M. A. Warren (1996): 'Computer simulation of liquid crystals'. J. Phys.: Cond. Matt. **8**, 9433–9437.
87. Stelzer J., L. Longa, H. R. Trebin (1995): 'Molecular dynamics simulations of a Gay-Berne nematic liquid crystal – elastic properties from direct correlation functions.' J. Chem. Phys. **103**, 3098–3107.
88. Stelzer J., L. Longa, H. R. Trebin (1995): 'Elastic constants of nematic liquid crystals from molecular dynamics simulations'. Mol. Crys. Liqu. Crys. A **262**, 455–461.
89. Poniewierski A., J. Stecki (1979): 'Statistical theory of the elastic constants of nematic liquid crystals'. Mol. Phys. **38**, 1931–1940.
90. Poniewierski A., J. Stecki (1982): 'Statistical theory of the Frank elastic constants'. Phys. Rev. A **25**, 2368–2370.
91. Bahadur B. (edt.) (1990): *Liquid crystals and uses* (World Scientific, Singapore)
92. Schadt M. (1997): 'Liquid crystal materials and liquid crystal displays'. Ann. Rev. Mater. Science, **27**, 305–379.
93. Stelzer J., P. Galatola, G. Barbero G, L. Longa (1997): 'Surface-induced order parameter profiles in a nematic liquid crystal from molecular dynamics simulations'. Mol. Cryst. Liqu. Cryst. A **229**, 61–64.
94. Stelzer J., P. Galatola, G. Barbero, L. Longa (1997): 'Molecular dynamics simulations of surface-induced ordering in a nematic liquid crystal'. Phys. Rev. E **55**, 477–480.
95. Stelzer J., L. Longa L, H. R. Trebin (1997): 'Homeotropic surface anchoring of a Gay-Berne nematic liquid crystal'. Phys. Rev. E **55**, 7085–7089.
96. Palermo V., F. Biscarini, C. Zannoni (1998): 'Abrupt orientational changes for liquid crystals adsorbed on a graphite surface'. Phys. Rev. E **57**, R2519–R2522.
97. Miyazaki T., K. Shigematsu, M. Yamashita (1998): ' Surface-stabilized smectic A phase in the Gay-Berne model'. J. Phys. Soc. Jpn **67**, 3477–3487.
98. Miyazaki T., H. Hayashi, M. Yamashita (1999): ' Surface-induced spatial ordering in nematic and smectic phases of Gay-Berne model'. Mol. Cryst. Liqu. Crys. A **330**, 1611–1618.
99. del Rio E. M., E. deMiguel (1997): 'Computer simulation study of the free surfaces of a liquid crystal model'. Phys. Rev. E **55**, 2916–2924.
100. De Miguel E., E. M. Del Rio (1999): 'Simulation of nematic free surfaces'. Int. J. Mod. Phys. C **10**, 431–433.
101. Doerr T. P., P. L. Taylor (1999): 'Molecular dynamics simulations of liquid crystal anchoring at an amorphous polymer surface'. Int. J. Mod. Phys. C **10**, 415–429.
102. Doerr T. P., P. L. Taylor (1999): 'Simulation of liquid crystal anchoring at an amorphous polymer surface from various initial configurations'. Mol. Cryst. Liqu. Cryst. A **330**, 1735–1740 (1999).
103. Emerson A. P. J., S. Faetti, C. Zannoni (1997): 'Monte Carlo simulation of the nematic-vapour interface for a Gay-Berne liquid crystal'. Chem. Phys. Lett. **271**, 241–246.
104. Binger D. R. , S. Hanna S (1999): 'Computer simulation of interactions between liquid crystal molecules and polymer surfaces - I. Alignment of nematic and smectic A phases'. Liqu. Cryst. **26**, 12050–1224.

105. Xu J. L., R. L. B. Selinger, J. V. Selinger, R. Shashidar (2001): 'Monte Carlo simulation of liquid-crystal alignment and chiral symmetry-breaking'. J. Chem. Phys. **115**, 4333–4338.
106. Allen M. P. (1999): 'Molecular simulation and theory of liquid crystal surface anchoring'. Mol. Phys. **96**, 1391–1397.
107. Andrienko D., G. Germano, M. P. Allen (2000): 'Liquid crystal director fluctuations and surface anchoring by molecular simulation'. Phys. Rev. E **62**, 6688–6693.
108. Bates M. A. , C. Zannoni (1997): 'A molecular dynamics simulation study of the nematic-isotropic interface of a Gay-Berne liquid crystal'. Chem. Phys. Lett. **280**, 40–45.
109. Allen M. P. (2000): 'Molecular simulation and theory of the isotropic-nematic interface'. J. Chem. Phys. **112**, 5447–5453.
110. Al-Barwani M. S., M. P. Allen (2000) 'Isotropic-nematic interface of soft spherocylinders'. Phys. Rev. E **62**, 6706–6710.
111. McDonald A. J., M. P. Allen, F. Schmid (2001): 'Surface tension of the isotropic-nematic interface'. Phys. Rev. E **63**, 10701-10704.
112. Akino N., F. Schmid, M. P. Allen (2001): 'Molecular-dynamics study of the nematic-isotropic interface'. Phys. Rev. E **63**, 41706–41713.
113. Wall G. D. , D. J. Cleaver (1997): 'Computer simulation studies of confined liquid-crystal films'. Phys. Rev. E **56**, 4306–4316.
114. Mills S. J., C. M. Care, M. P. Neal, D. J. Cleaver (1998): 'Computer simulation of an unconfined liquid crystal film'. Phys. Rev. E **58**, 3284–3294.
115. Gruhn T., M. Schön (1998): 'Substrate-induced order in confined nematic liquid-crystal films'. J. Chem. Phys. **108**, 9124–9136. J CHEM PHYS 108: (21) 9124-9136 JUN 1 1998
116. Gruhn T., M. Schön (1998): 'Grand canonical ensemble Monte Carlo simulations of confined 'nematic' Gay-Berne films'. Thin solid films **330**, 46–58.
117. van Roij R., M. Dijkstra, R. Evans (2000): 'Orientational wetting and capillary nematization of hard-rod fluids'. Europhys. Lett. **49**, 350–356.
118. Dijkstra M., R. van Roij, R. Evans (2001): 'Wetting and capillary nematization of a hard-rod fluid: A simulation study' Phys. Rev. E **63**, 051703–051710.
119. Allen M. P. (2000): 'Pressure tensor profiles at the isotropic-nematic interface'. Chem. Phys. Lett. **331**, 513–518.

Two-Dimensional Fluid Foams at Equilibrium

François Graner

Spectrométrie Physique, BP 87, F-38402 St Martin d'Hères Cedex, France;
CNRS UMR 5588 et Université Grenoble I
e-mail: graner@ujf-grenoble.fr.

Abstract. This is an introduction to fluid foams for non-specialists. After presenting some applications and modeling of foams in general, we focus on ideal two-dimensional fluid foams at equilibrium. We discuss : the interplay of topology, geometry and forces; the central role of pressure and energy; an analogy with 2D electrostatics; the role of disorder; the minimal perimeter problem.

1 Introduction

There is a place where chemists, engineers, mathematicians and physicists routinely meet: congresses about foams [26, 40, 36]. In fact, fluid foams stimulate truly interdisciplinary fundamental research with a wide range of industrial applications. They have even fascinated some artists. We will try to explain here why foams are at the crossroads of different approaches, and why they are so attractive.

Section 2 presents their main properties and the corresponding applications. Section 3 tries to catch the essence of what a foam is, and how far we should simplifiy it to model it. Section 4 then proceeds by presenting the position of the problem for physicists and mathematicians. In the remaining of the paper, we mostly focus on our own research, which concerns the statistical physics of two-dimensional (2D) foams at equilibrium. Section 5 tries to simplify the problem and reduce it to known results of electrostatics; Sects. 6 and 7 apply it to determinations of the foam's energy and pressure field, respectively; Sect. 8 addresses the physical aspects of the mathematical "minimal perimeter problem". For 3D aspects, which we do not treat here, we refer to excellent existing books: for the mathematical approach, see [20]; for the physical and historical point of view see [19, 35].

The branch of mathematics which deals with the minimal perimeter problem in two or more dimensions is achieving remarkable progresses [20, 17, 18]. Readers interested in the mathematical aspects of foams, especially their specific interplay between topology, geometry and pressure, might read Sects. 4, 5, 6, and 8. These readers can skip Sects. 3 and 7, which are more intended for foam physicists.

However, the point of view we want to present here is completely different. It is the approach of statistical physics: to start from exact detailed properties and deduce average global properties. We have tried to mention in the text which formulas are approximate: this is the case for Sects. 5.2, 5.3, 6.1, 6.2, 7.2, 8.2, 8.3.

The morphology of foams, *i.e.* the shape of the cellular domains, may be quantified by the techniques presented in the contributions to this volume by Arns *et al.* (1.2) and by

Fig. 1. Example of fluid foam, here the surface of a soap froth. Picture courtesy of S. Courty.

Beisbart *et al.* (3.2) applying Minkowski functionals on spatially disordered materials. Orientational and tensorial features of patterns are characterised by methods described by P. Soille (3.1) and Beisbart *et al.* (4.2) in this volume, also useful in the context of foams.

2 A Wide Range of Properties and Applications

"Fluid foams" consist of *gas bubbles separated by a continuous film of liquid* (Fig. 1). Their applications usually combine simultaneously several of their properties, related to their structure, mechanics, chemistry, and surface tension. For reviews, see for instance [19, 39, 3, 12, 13, 25, 34, 11, 14, 33].

2.1 Structure

Since the gas occupies most of the volume, foams have a low density, and a high specific surface per unit mass. They can uniformly cover large surfaces. They help in fire-fighting: for oil, which they cover and insulate from air (while water would only fall down, and possibly later boil explosively), and in confined media; foams are much less destructive than water, and the air they contain allows endangered persons to breathe.

Since they can fill volumes with convoluted shapes, they are easy to mould, hence an intermediate step in most processes to prepare solid foams [36]: for cushion stuffing, building materials, bumpers, filters, insulators, catalysts, heat exchangers. Fluid foams fill and can clean pipes or pores, using a small amount of mass; this is essential in nuclear decontamination or chemical pollution where it is necessary to reduce the amounts of cleaning residues. They considerably reduce the cost of all fiber treatments: brushing hair, protecting crops; but also waterproofing or dying textiles, being easier and cheaper to remove by drying than bulk water.

2.2 Mechanics

Fill a bottle with water, add some dishwashing product, shut it with your thumb, and shake it: the foam damps the shock and your thumb can barely feel it. Such efficient energy dissipation has shock damping applications from demining to gun silencers.

Other police and military applications use foams for confinement: as anti-riots agents, or for preparation and separate storage of H-bombs sub-components. Fluid foams are also useful as temporary insulators: spermicides, thermal insulation of greenhouses during the night, protection of wounds against toxic chemicals, trap for dust in coal mines, prevention of beer oxydation during bottling, prevention of polluant dispersion in soils.

Resistance to pressure is crucial in several applications of foams to oil drilling: as pressurising fluid carrying sand to fracture the rock; to carry away forage residues in under-pressurised regions; to eliminate water; or, by filling empty pores, to prevent vapor, water or gas from leaking through porous rocks.

Note that a foam has a shape, to which it returns after a small mechanical perturbation: it is elastic. Under higher stress, it deforms plastically; under high shear rate, it flows. This triple, elasto-plasto-viscous behaviour, although not specific to foams, is one of their most attractive features. For instance, in shaving or in airplane windshield de-icing, they can be deposed in large amounts on vertical walls thanks to their elasticity, and are later evacuated through the shear produced by the blade or the wind, respectively.

2.3 Chemistry

Materials with high affinity for water, especially aqueous solutions, are called "hydrophilic". "Hydrophobic" materials are not wet by water; for instance oil and water do not spontaneously mix. Mixing oil with water, so that rinsing can evacuate the oil, requires cleaning agents, usually "amphiphilic" molecules with a hydrophilic head and a long oily chain made of carbons. Amphiphilic molecules are often also foaming agents, so that foams appear as by-products in cleaning. Neither the presence, nor the stability of the foam seems to improve the quality of cleaning [6]. In that case their presence is a marker of the dose injected, and their destruction marks the presence of impurities.

The amphiphilicity of the foams implies they have an affinity for most materials. Fine tuning their chemical composition leads to selective affinity. This is used in mines to separate ore and useless materials in huge flotation tanks. Foams also can separate colorants by fractional distillation.

2.4 Model Systems

In fundamental research, foams are used as model systems. Since they minimise their surface energy by decreasing the area of the bubbles, they are examples of the mathematical "minimum perimeter" problem which we discuss below, and of minimal surfaces used in architecture. They are often compared with other cellular patterns : solid foams, biological epithelia or aggregates, ordered pattern like the retina or honeycombs, fracture pattern like in basalts or salt lakes. However, the most useful analogy is with cellular patterns which do have a surface energy and minimise it: grain boundaries in crystals, magnetic domains. Since their characteristics are entirely measurable from images, without any hidden variables, foams are also models for more complex elasto-plasto-viscous materials.

2.5 Other

Although the above list of applications is long and could be even longer, foams are often associated with an image of futility. This might originate from their consistence and short life-expectancy. But there might be another reason: foams which are visible in our daily life have less essential applications; or often no applications at all, when they are by-products or polluants.

As light and attractive materials, they are essential to the image of foods and especially drinks like sparkling wines, sodas, or beers, which they protect from oxydation. Without attempting at listing all representation of bubbles and foams in poetry or pictures [19, 9], let us mention "Soap bubbles", by Edouard Manet, or a delicate *azulejo* in the north-east corner of the garden, in Palacio de la Frontera near Lisbon. They appear in movies, in foam parties, or in both together: "The Party", by Blake Edwards, ends with an unforgettable invasion by a foam provoked by Peter Sellers.

3 A Generic Model for Fluid Foams

3.1 Common Characteristics

Foams exist if and only if two generic conditions are met:
(i) *within each bubble, the matter is conserved*;
and:
(ii) *the energy decreases when the total surface of the walls decreases.*
This deserves some discussion and examples.

In a foam the surface energy is an increasing, but not necessarily a linear function of the surface. For instance, the surface tension can be non-uniform due to temperature gradient: Marangoni effect. Or, a crystal usually consists in many small monodomain crystals; the boundaries between these grains actually meet both above conditions: here the surface energy is not uniform nor anisotropic. To the contrary, condition (ii) excludes aggregates of vesicles, although they look like soap bubbles [27]: since their walls do not communicate together, each vesicle has a different tension; the total energy is a function of each vesicle's surface separately, not of their sum.

In addition to this surface energy, the total energy of the foam always contains additional, specific terms which vary from foam to foam. For instance, the repulsion between the electrostatic dipoles of amphiphilic molecules on either side of the film provides a microscopic energy term to stabilise the walls against breakage.

Minimisation and equilibration of energy requires both phases to be fluid. This includes emulsions, which are fluid-in-fluid foams, like oil in water. This specifically excludes solid foams, granular systems like dry or wet sand.

Due to condition (i), an equation of state relates the pressure and volume of each bubble separately. The surface of the walls decrease and the pressure within each bubble increases until they exactly balance each other. This ensures the mechanical stability of the foam. Two simple cases are often studied: the perfect gas, where the pressure and volume are inversely proportional; and the incompressible emulsion, where the volume is constant and only the pressure varies.

To observe a foam, conditions (i) and (ii) need only be valid during the time required to reach the mechanical equilibrium: typically tens of milliseconds for a soap froth. Almost all real foams violate these conditions at longer time scales. A main cause is wall breakage: foams produced without amphiphilic molecules by violent mixing gas and liquid (froth on sea waves) or chemical reactions (metallic foams) are transient and out-of-equilibrium, and only last a few seconds. A slower phenomenon is drainage of the water under gravity, like in mine flotation tanks; in a beer glass it only takes seconds or minutes. Finally, depending on the solubility of the gas in the liquid, bubbles can lose gas into their neighbours and disappear, resulting in the foam losing its bubbles and coarsening over minutes or hours.

3.2 The Simplest Foam

Ideal foam: To catch the essence of the physics common to all foams, we focus on a idealised foam defined as follows.

• *Dry foam:* the gas occupies most of the foam's volume, while the liquid occupies a very small fraction. In physical experiments this dryness is limited by the stability of the film: typically, in soap froths it can go down to one or two percents. In the ideal foam we take the limit of fluid fraction going to zero.

• *Incompressible foam:* Although foams are full of gas, we can usually neglect their compressibility: of order $P_{atm}^{-1} = 10^{-5}$ m^3.J^{-1}, that of a perfect gas. This is much smaller than the effect of the surface tension, characterised by the Laplace overpressure (2) of order of the bubble wall curvature, typically a centimeter, divided by the surface energy, typically 10^{-2}J.m^{-2}, hence 1 m^3.J^{-1}. Emulsions, which are full of liquid, are of course even less compressible.

• *Energy proportional to the surface:* this means that surface tension is uniform and anisotropic, and dominates all other contributions to energy. A soap froth with bubbles much larger than molecular sizes often provides a correct example.

• *No temperature effect:* room temperature does not play any significant role in the structure of a foam, see Sect. 8.1 for order of magnitudes.

Two-dimensional (2D) foams: We chose to focus on two dimensional (2D) foams, although specialists of actual (3D) experiments often object to this choice.

• The first usual objection to 2D foams is their lack of practical applications by themselves. The answer is that they turn particularly useful via their use in fundamental research. 2D experiments and simulations are easier to realise and, what is more important, to analyse both qualitatively and quantitatively. Theories are also easier to develop in 2D, before generalisation to 3D, especially since there are more exact results in 2D; this will be addressed in Sect. 4.

• The second, more radical objection is that experimental foams are never truly 2D. The answer is that it does not really matter. To understand the physics, we perform truly 2D simulations and theories, and approximately 2D experiments. An experiment is a correct model of 2D foam if the energy is proportional to the total perimeter. For instance, in a soap froth between two parallel plates of glass, each line has an energy proportional to its length, 3D effects such as the curvature in the vertical plane appearing only in the value of the proportionality constant. Some 2D foams are stabilised by dipolar interactions without amphiphilic molecules [29]: in Langmuir monolayers [4], magnetic garnets [37], oil/ferrofluid emulsions [7]; they are acceptable model of 2D foams if and only if the dipolar energy is small enough not to affect their behaviour.

4 Position of the Problem

From now on, we consider only ideal, dry 2D foams at equilibrium. This section deals with its structure: first the position of the problem, then some directions of research. Most results below come from [15], to which we refer for details.

4.1 Energy

An ideal 2D foam consists in N bubbles with fixed areas $\{A_i,\ i = 1, \dots, N\}$, Fig. 2. Its energy [41] is simply the sum of wall lengths:

$$H = \gamma \sum_{0 \le i < j \le N} \ell_{ij} = \lambda \sum_{i,j=0}^{N} \ell_{ij} = \lambda \sum_{i=0}^{N} \mathcal{L}_i, \tag{1}$$

where ℓ_{ij} is the length of the wall between bubbles i and j, $\mathcal{L}_i = \sum_j \ell_{ij}$ is the perimeter of bubble i, $i = 0$ denotes the medium that surrounds the foam, $\ell_{ii} = 0$. Here λ is the line energy of the interface between water and gas, $\gamma = 2\lambda$ is the line energy of a wall consisting of two interfaces, both in J.m^{-1}.

If $\langle A \rangle$ is the average bubble area, note that $H/\lambda \langle A \rangle^{1/2}$ is a dimensionless function of the bubbles' shapes only. In this sense, minimising the energy is a purely geometrical problem, universal in the sense that it is the same for *any* perimeter-minimising system.

4.2 Local Rules: Plateau

At equilibrium, the ideal foam obeys the following local rules, stated by Plateau [24].

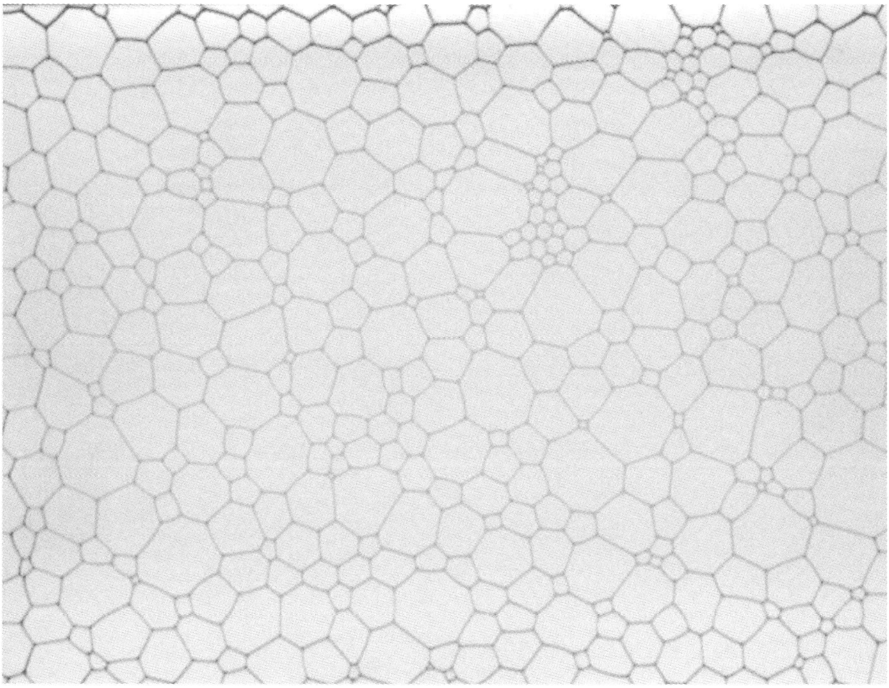

Fig. 2. Example of a 2D foam, here a soap froth between two plates of glass. The line tension λ is here twice the usual surface tension between water and gas (here 27.3 mN/m) multiplied by the distance between plates (1 mm). Picture courtesy of O. Lordereau.

Walls: According to Laplace's law, the wall between bubbles i, j has a curvature which balances the 2D pressure difference:

$$\kappa_{ij} = \frac{P_i - P_j}{\gamma}. \tag{2}$$

where $\kappa_{ij} = -\kappa_{ji}$ is the algebraic curvature, counted positive when bubble i is convex compared with bubble j. As a consequence, along the wall, the curvature is uniform: each wall is an arc of circle.

Vertices: Bubble edges meet in triples at $2\pi/3$ angles (Fig. 3). Foams have the topology of three-fold connected networks; four-fold vertices cost more energy and are unstable (Fig. 7).

As a consequence of (2), algebraic curvatures of the three edges that meet at the same vertex must add to zero:

$$\kappa_{ij} + \kappa_{jk} + \kappa_{ki} = \frac{P_i - P_j}{\gamma} + \frac{P_j - P_k}{\gamma} + \frac{P_k - P_i}{\gamma} = 0. \tag{3}$$

Equation (3) holds for any closed contour crossing more edges.

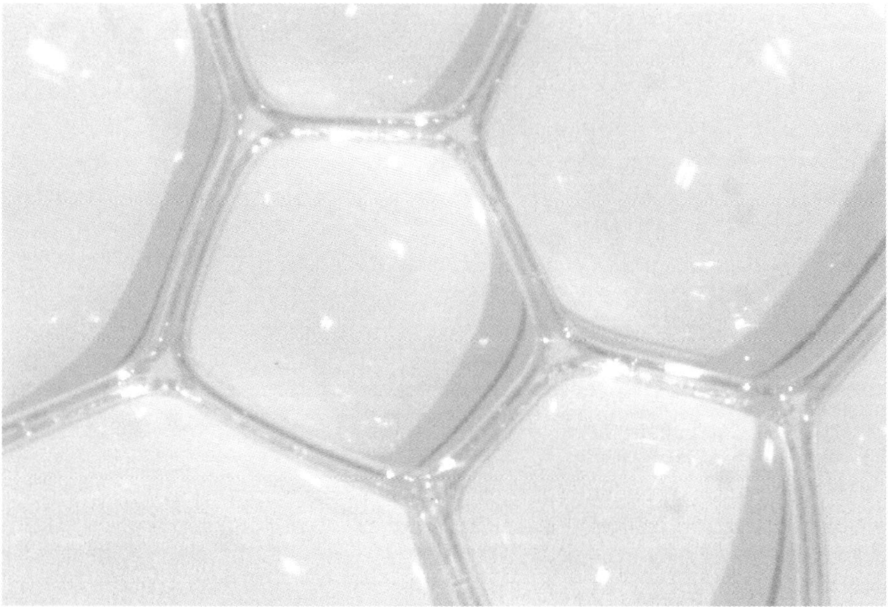

Fig. 3. Illustration of Plateau rules, on a 2D soap froth. Bubble walls are arcs of circles meeting by three at (approximately, since the foam is not ideally dry) 120° angles. A pentagon is convex and at high pressure, with respect to its neighbours. Picture courtesy of S. Courty [5].

4.3 What Is the Problem?

At first sight, the ideal foam appears truly simple. The expression of its energy, (1), is one of the simplest one can dream of. It immediately yields very intuitive consequences, the above Plateau rules, which considerably reduce the number of degrees of freedom necessary to describe the structure: for each vertex, an angle, modulo $2\pi/3$; for each side, a length and a curvature.

Now, we give ourselves the area of each of the N bubbles, and the topology of the foam, that is the list of which bubbles share a common edge, and we wonder: what can we say about the structure of the foam, the bubble shape, or the total perimeter? The answer is that there is nothing simple! The disorder inherent to the foam, especially due to the area distribution, but also due to the unavoidable foam boundaries, prevents us from any intuition. There is an intricate interplay between topology, geometry, and pressures, specific to foams, which make them both fascinating and difficult to solve, as the next sections will discuss.

5 Electrostatic Analogy

5.1 Pressure, Geometry, Topology

One bubble: Consider a bubble of a foam; let n be its number of sides. It is characterised by a topological quantity which quantifies the deviation from a hexagonal lattice: its topological charge $6 - n$. Equivalently, we can say that an n-sided bubble is a disinclination bearing a geometrical charge:

$$q = (6 - n)\,\frac{\pi}{3} = 2\pi - \frac{n\pi}{3}. \tag{4}$$

Now, circle once counterclockwise around this bubble. What the Gauss theorem states, is here particularly simple: the tangent vector $\hat{t}$ along the bubble edge, which travels around the bubble once counterclockwise, rotates by 2π. Note i the number of this bubble, j its neighbours. Each side ij is an arc of length ℓ_{ij} and of curvature κ_{ij}, thus contributes $\kappa_{ij}\ell_{ij}$ to this rotation, while each vertex rotates the tangent vector $\hat{t}$ by $\pi/3$. The Gauss theorem hence links the average curvature of a bubble to its number of sides [23]:

$$\sum_j \frac{1}{\gamma}\,(P_i - P_j)\,\ell_{ij} = \sum_j \kappa_{ij}\ell_{ij} = (6 - n_i)\,\frac{\pi}{3} = q_i. \tag{5}$$

This classical but essential equation is what makes 2D foams unique. It relates the topology, the geometry (curvatures, lengths), and the pressure (hence forces) in a single bubble. This means for instance that 4-sided or 5-sided bubbles are convex, and their pressure is higher than their neighbours'; see the hatched pentagon in Fig. 4. On the opposite, 7-sided or 8-sided bubbles are concave and have a low pressure. If this relation was not satisfied, the bubble would not be at equilibrium: the walls would move, and the pressure adapt, until the condition (5) is fulfilled.

Several bubbles: Extension to more than one bubble uses a topological theorem [30, 2]. Consider a contour C which follows only bubble edges and encloses a few bubbles, see the black contour in Fig. 4. Following the notation of [2], we define the number of vertices which originate an outward (inward) pointing edge as v^+ (v^-) (Fig. 4); *e.g.* $v^+ = 0$ if C is the boundary of the total foam, $v^- = 0$ if C encloses a single bubble. The *topological* Gauss theorem [30, 2] states that C encloses a total topological charge $\sum_{k\in C}(6 - n_k)$ equal to:

$$\sum_{k\in C}(6 - n_k) = (6 - v^+ + v^-), \tag{6}$$

where k labels the bubbles enclosed by C. For instance, if C encloses a single bubble, $v^- = 0$, $v^+ = n$, and one recovers (5).

We can write the Gauss theorem, exactly as we wrote it above for one bubble, by stating that the tangent vector $\hat{t}$ along the contour C, which travels around the bubble once counterclockwise, rotates by 2π. We thus find that:

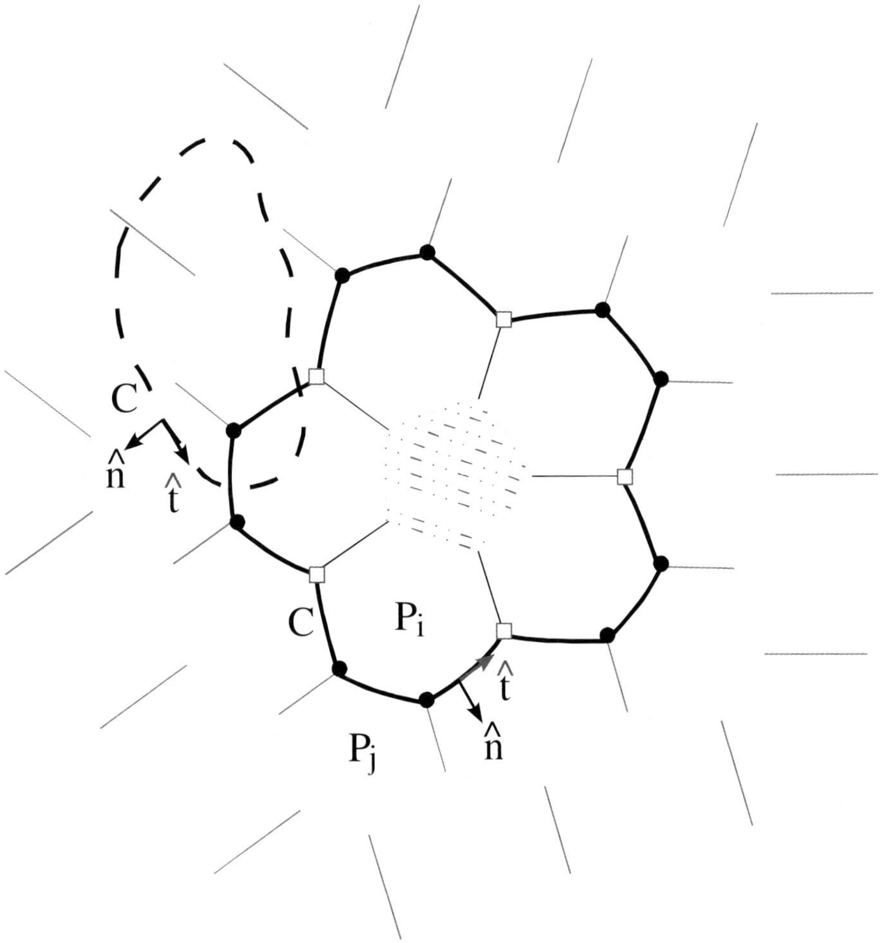

Fig. 4. A schematic of a 2D foam. The black contour C follows bubble edges and encloses a geometrical charge Q ($Q = \pi/3$ due to the pentagon). Vertices with squares point inwards ($v^- = 5$) and vertices with circular discs point outwards ($v^+ = 10$). The dashed contour C crosses bubble edges transversely. Along the contours tangent $\hat{t}$ and normal $\hat{n}$ vectors (arrows) are defined by the right-hand rule. Drawing by Y. Jiang [15].

$$\sum_C \kappa_{ij}\ell_{ij} = (6 - v^+ + v^-)\frac{\pi}{3}. \tag{7}$$

Combining (6) and (7), we have expressed the geometrical Gauss theorem for a closed contour C which follows the bubble edges ij and encloses a total geometrical charge Q [15]:

$$\sum_j \frac{P_i - P_j}{\gamma}\ell_{ij} = \sum_C \kappa_{ij}\ell_{ij} = Q(C) = \sum_{k\in C} q_k. \tag{8}$$

This theorem establishes at a global level the essential relation between topology, geometry and pressure in 2D foams.

Finally, (3) can also be generalised for a contour C crossing the bubble edges perpendicularly, see the dashed contour in Fig (4). The same argument that applied to a contour around a single vertex now yields, for the whole contour, that the curvatures of all edges encountered add to zero:

$$\sum_C \kappa = 0. \tag{9}$$

5.2 Continuous Limit

We now rewrite the above equations as an analogy with 2D electrostatics (Table 1), and discuss its advantages [15].

Table 1. Proposed analogy between foams and electrostatics in two dimensions.

	potential	field	charge
$2D$ electrostatics	potential V	electric field $-\boldsymbol{\nabla} V$	electric charge e
$2D$ foams	pressure P	curvature $\kappa \propto -\boldsymbol{\nabla} P$	geometrical charge $(6-n)\frac{\pi}{3}$

In this analogy, each bubble represents a conducting platelet bearing a charge $q = (6-n)\pi/3$ distributed along its edges, which (along each link) is uniform at equilibrium because the charge density is proportional to the curvature. The walls are a thin layer of insulating material, which bear a large gradient of pressure (almost a discontinuity). Since we are not interested in the actual pressure within them, we have thus chosen to interpolate the pressure in such a way that $\Delta P = 0$ within walls. We introduce the notation $\boldsymbol{E} \equiv -\boldsymbol{\nabla} P$:

$$\begin{aligned} \boldsymbol{E} &\propto \kappa\, \hat{n} \qquad &&\text{(across walls)}, \\ \boldsymbol{E} &= 0 \qquad &&\text{(within bubbles)}, \end{aligned} \tag{10}$$

where κ is the curvature of the walls and $\hat{n}$ its outwards normal.

The pressure thus appears as a potential, the curvature as its gradient (Table 1). Let the unit vectors $\hat{n}$, $\hat{t}$ be the normal and the tangent to C, respectively, as drawn in Fig. 4. In the dry foam limit (see [15] for details), (3,8) can be rewritten under a more familiar continous form:

$$\oint_C \boldsymbol{E} \cdot \hat{n}\, d\ell \propto Q(C),$$
$$\oint_C \boldsymbol{E} \cdot \hat{t}\, d\ell = 0. \tag{11}$$

These approximate relations hold for any contour C (not necessarily parallel or perpendicular to edges), although it is easier to visualise (11a) on the black contour, (11b) on the dashed one. Although both (11a,b) look similar, they are physically different, describing respectively an outwards flux through C (which, in 2D, is a line integral instead of a surface integral) and a circulation along C. They make the analogy to electrostatics obvious.

5.3 Examples

Pressure and curvature fields: A positive q corresponds to a local high pressure. In turn the pressure gradient correlates with the concavity of the bubbles. For illustration, we consider a foam with all bubbles having the same area A, that is a honeycomb of side L. Take one single "defect" bubble with a geometrical charge $q = (6 - n)\pi/3$ as a germ bubble (disinclination), for instance a regular pentagon, hatched in Fig. 4. The foam consists of concentric shells around this germ. When the foam is sufficiently large, in the s shell, $s \gg 1$, the asymptotic limits of both κ and P are approximately :

$$\kappa(s) \propto \frac{q}{Ls},$$
$$P = \gamma \sum_s \kappa(s) \propto P_0 - \frac{\gamma q}{L} \ln s. \tag{12}$$

Here P_0 is a constant *e.g.* equal to the pressure P_b at the boundary of the foam, and P grows logarithmically with the foam size. Equations (12) are characteristic of 2D electrostatics.

Dipoles and quadrupoles: Foams thus enter the well-known class of universality of 2D Coulomb interactions, which encompasses e.g. electrostatics, lines of disinclinations, or arrays of vortices. Two opposite charges $-q+q$ (*e.g.* a pentagon-heptagon pair, Fig. 5a) a distance $d \sim L$ apart constitute a dislocation, that is a dipole of moment $\boldsymbol{p} = q\boldsymbol{d}$, which deforms neighboring bubbles and induces curved edges in the hexagons around it. At a distance $s \gg d$ away from this dipole, the pressure goes approximately as $1/s$ and the curvature as $1/s^2$: similar to the dipolar potential in 2D. A dipole can pair with another dipole (Fig. 5a) to form a topological quadrupole (Fig. 5b), which affects the honeycomb lattice over a much shorter range, the pressure going as $1/s^2$ and the curvature as $1/s^3$.

6 The Energy of the Foam

6.1 Perimeter Increase due to a Defect

The analogy is deeper and extends to the expression of energy itself. The energy, or sum of perimeters, of a 2D foam is similar to the energy in 2D electrostatics. We can even

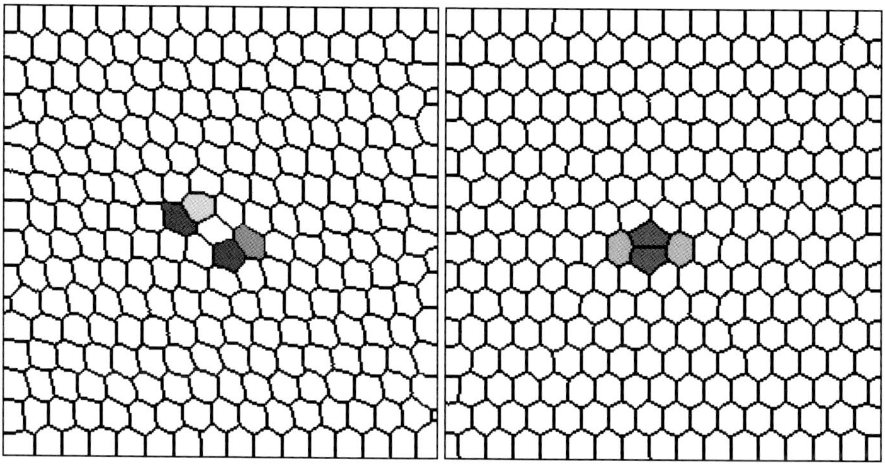

Fig. 5. (a) Simulation of two dipoles in a regular foam with equal areas ($\delta A / \langle A \rangle = 0.4\%$) and periodic boundaries. Two pentagon-heptagon pairs result in a curvature field in the hexagons around them. (b) An artificially constructed pentagon-heptagon-pentagon-heptagon cluster forms a topological quadrupole, with the rest of the honeycomb lattice undisturbed. Simulation by Y. Jiang [15]

define an energy density for 2D foams. The demonstration applies to honeycomb lattices with only few topological defects, and proceeds as follows.

Energy density: In a honeycomb, when a topological defect induces a curvature in the neighboring edges, the total foam perimeter increases. To estimate this increase, consider two vertices separated by a distance a. An arc with curvature κ connecting them has a length $\ell = 2\alpha/\kappa$, where $\alpha = \arcsin(a\kappa/2)$ is half of the subtended angle. The difference between a and ℓ is the length increase due to the curvature:

$$\ell - a \approx \ell \left[\frac{\kappa^2 \ell^2}{24} + O\left(\kappa^4 \ell^4\right) \right]. \tag{13}$$

As expected, $\ell - a$ is positive and quadratic in κ. This is the expression of the energy density, proportional to the square of the field.

Energy: By summing over all edges of the foam, and keeping only the leading order when $\kappa\ell \ll 1$, we obtain the increase in energy due to curvature:

$$H_t \approx \gamma \sum_{i<j} \ell_{ij} \times \frac{\kappa_{ij}^2 \ell_{ij}^2}{24}. \tag{14}$$

For instance, around the single topological charge, $\kappa(s)$ decreases approximately as q/s (12). The summation over all curved edges yields the energy cost due to this single topological defect:

$$H_t \propto \gamma \sum_s sL\kappa^2(s)$$
$$\propto \gamma L q^2 \sum_s \frac{1}{s}. \tag{15}$$

It grows logarithmically with the size of the foam, as expected in analogy with the self-energy of a 2D electrostatic charge. For a large foam, a single topological charge costs so much energy that it never occurs in real foams.

6.2 Several Topological Charges

Consider two charges q, q' separated by a (topological) distance $s \gg L$, the interaction energy goes approximately as:

$$H_t \approx -qq'\gamma L \ \ln\left(\frac{s}{L}\right). \tag{16}$$

It decreases when charges of same sign separate, or charges of opposite sign aggregate. Heptagon-pentagon pairs are more likely than isolated heptagons. This "effective interaction" of topological origin, which previous work has assumed [28] or derived empirically [25], has important consequences. For instance, this interaction could explain the origin of the correlations between bubbles: the side number distribution and the Aboav-Weaire law [28]. As an illustration, in Fig. 6, the heptagon (indicated by a number 7) has two pentagonal neighbors.

This description remains valid even for many charges, as long as the curvature fields they induce are small and can be added by linear superposition. The pressure fields of multiple charges are added linearly too. However, the present calculations assume equal areas; they become more difficult when the area disorder couples with the topological disorder.

6.3 Relation Between Pressure and Energy

For a free foam, in any equilibrium state, there exists a relation between the energy and the pressure field. Surprisingly, it is strictly exact:

$$H = 2\sum_{i=1}^{N}(P_i - P_b)A_i, \tag{17}$$

where P_b is the pressure of the outer fluid at the boundary of the foam. This generalises in D dimensions:

$$H = \frac{D}{D-1}\sum_{i=1}^{N}(P_i - P_b)A_i. \tag{18}$$

There exist different demonstrations [1]; the shortest is probably the following, based on a Legendre transformation, using P_b as a Lagrange multiplier [15]. Consider a free foam with fixed pressures, not areas. At equilibrium, the enthalpy, $\mathcal{H} \equiv H - \sum_{i=1}^{N}(P_i - P_b)A_i$, is extremal. Thus, under dilation, $L \to \lambda^{D-1}L$, $A \to \lambda^D A$, the function:

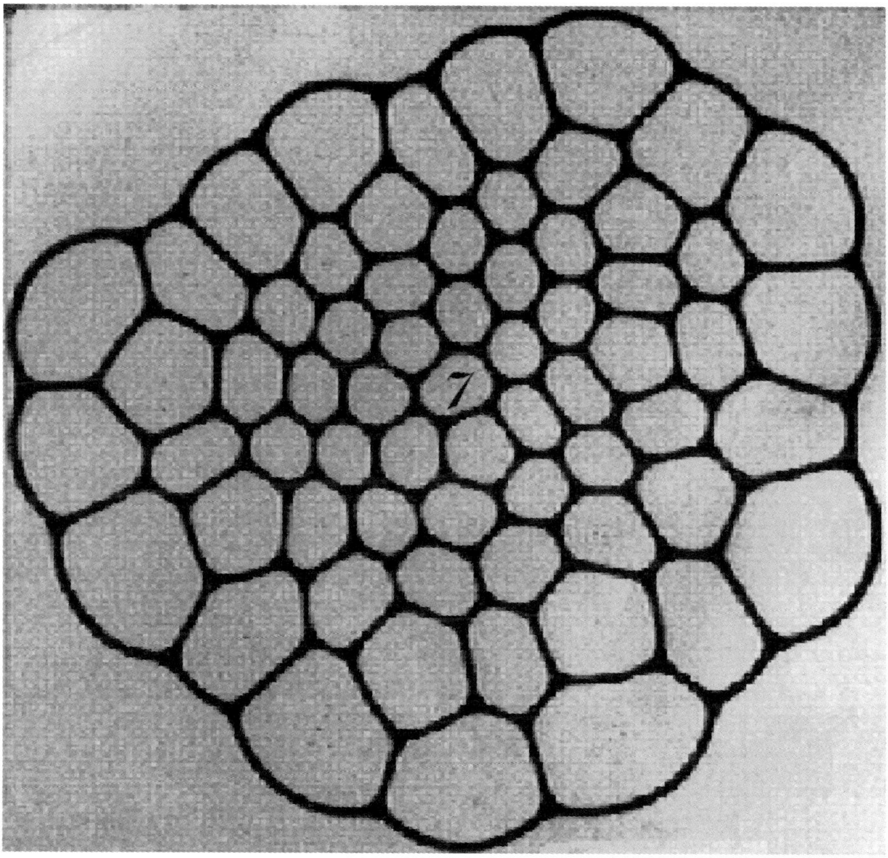

Fig. 6. Annealed 2D ferro-fluid foam. *Foam preparation:* An ionic magnetic fluid (aqueous magnetic liquid, black) and oil (white-spirit, grey) do not mix, their surface tension is 15 mN/m. They are trapped together in the 1 mm space between two parallel Plexiglas plates. A homogeneous magnetic field of 9 kA/m perpendicular to the plates induces the cellular structure and fixes the bubble areas and wall thickness. The same oil as that fills the bubbles surrounds them. For details see [7]. *Annealing procedure:* We tilt the Plexiglas plates from the horizontal plane to an angle of $0.1°$, inducing a low effective gravity field. Large bubbles drift upwards, small bubbles downwards, resulting in vertical sorting according to size. We then bring the plates back to horizontal, and the bubbles slowly drift and settle. This procedure allows the bubbles to rearrange and explore the energy space to find a lower energy configuration. The picture displays the final stable pattern. Picture size is 10 cm. Picture by E. Janiaud, for details see [15].

$$\mathcal{H}(\lambda) = \lambda^{D-1} H - \lambda^{D} \sum_{i=1}^{N} (P_i - P_b) A_i, \tag{19}$$

is extremal at $\lambda = 1$; this proves (17,18).

7 The Pressure Field within the Foam

The electrostatic analogy emphasises the central role played by the pressure. How can we determine experimentally the pressure field within a 2D foam? We summarise and compare different methods which could be tried.

7.1 Measurements

Direct measurements: An advantage of 2D foams is that each bubble is accessible from outside. It is possible to replace one of the glass plates by an array of pressure sensors [F. Elias, private communication]. This method would present no fundamental difficulty, and would yield an unambiguous result, requiring only a compromise between the price of the equipment and the desired precision. The following methods do not use any specific equipment, only image analysis.

Through areas: The measurement of bubble areas yields access to the pressure if stringent conditions are met. The bubble must be compressible and its equation of state known (e.g. a perfect gas). The area variations must result from pressure variations, not from variation of the amount of matter enclosed in the bubble. The pressure of the bubble must be known at least at one time (for instance at equilibrium with atmospheric pressure). Each bubble area must be tracked individually in time. This method has been successfully applied to a 2D soap froth (M. Asipauskas, to be published).

Through coarsening: Conversely, the measurement of bubble area variations applies under a different set of conditions. The area variation is due only to diffusion of matter from one bubble to its neighbour across their wall, resulting in the coarsening of the foam. The pressure difference is related with the flow rate across the wall, itself related with the normal velocity of the foam if each bubble is incompressible.

Through curvatures: By measuring the curvatures from a picture, and using iteratively eqs (2), the pressure P_i in each bubble can be determined up to an additive constant P_0, the pressure in a reference place (a given bubble, or the outer medium). We only need to chose a path which links bubble i with the reference place, and sum the algebraic curvatures of the walls crossed. Two different paths would give the same result: this is readily seen by joining both path to close a circuit, and apply (9) or its continuous equivalent (10b). Thus the choice of the path is arbitrary, and the result is unambiguous.

In practice, this method is imprecise and suited only to very dry foams with good image contrast, due to the notorious difficulty of curvature measurements [O. Cardoso, private communication]. In the example of Fig. 6 the error is around 15 % [15]. An improvement could be to measure all curvatures, typically $3N$ measurements, and try to find the N pressures which best fit the $3N$ (2).

Through vertex positions: M.-L. Chabanol obtained a significant progress by noting that vertex positions yield as much information as curvatures. In fact, the vertex positions yield the length of the chord which joins them, which approximates the value of the wall's length. We thus rewrite the essential (5):

$$\sum_j \frac{P_i - P_j}{\gamma} \ell_{ij} = (6 - n_i) \frac{\pi}{3}.$$

It is easy to solve these N linear equations with N unknown P_i, $i = 1, ..., N$. We then obtain approximate values of the P_is, from which we deduce approximate κ_{ij} and ℓ_{ij}. Reinjecting these values of ℓ_{ij} in the $N \times N$ equations systems enables an iteration until stable values for pressures, curvatures and lengths are found.

Preliminary tests with C. Monnereau-Pittet and N. Pittet have yielded a quick convergence to surprisingly good results. The method is correct even on images with low contrast, thick walls or bad focus, because vertex positions and topologies are well measured. The method seems operational, and it is easy to include boundaries. It seems to converge to a unique set of solutions. Care is required: since the method does not enforce the rule of 120° angles, the solution corresponds to a real foam only if the positions fed in the system are the positions of a foam's actual vertices. Applications are in progress: they could concern extremely precise determinations of curvatures, for instance for coarsening studies.

7.2 Laplacian of the Pressure

The analogy with electrostatics indicates that an $N \times N$ equations system can be brought back to a single equation. In fact, (10) imply that the Laplacian of the pressure P is proportional to the topological charge. We were thus tempted to conjecture that the topology, along with boundary conditions, uniquely fixes the pressure field [15]. This is not strictly true, since the bubble shapes are not rigidly fixed; in fact, D. Weaire and S. Cox found a counter-example [38]: two patterns, with exactly the same topology and bubble areas, but a different set of pressures, curvatures and lengths.

However, the Laplacian point of view becomes very interesting on larger scales. On a "mesoscopic" scale much larger than one bubble, but smaller than the foam size, the net topological charge of the foam is zero [16]. This implies that:

$$(\text{mesoscopic}) \qquad \Delta P \approx 0. \tag{20}$$

The mesoscopic average of P thus has all properties of zero-laplacian fields. For instance, it is a very "smooth" field; minima or maxima of pressure can exist only on the boundary of the foam (like Earnshaw theorem in electrostatics [22]). If the foam is free, i.e. surrounded by an outer medium with uniform pressure (Fig. 6), its pressure is thus roughly uniform. If the foam is in a closed box, its maxima are imposed by the shape of the box; hence, as N. Pittet noticed, the pressure field and hence the coarsening should differ in round, square and hexagonal boxes. If the foam is flowing in a channel, there is a pressure gradient fom the entrance to the exit. This mesoscopic point of view helps considering the foam as a continuous medium [21].

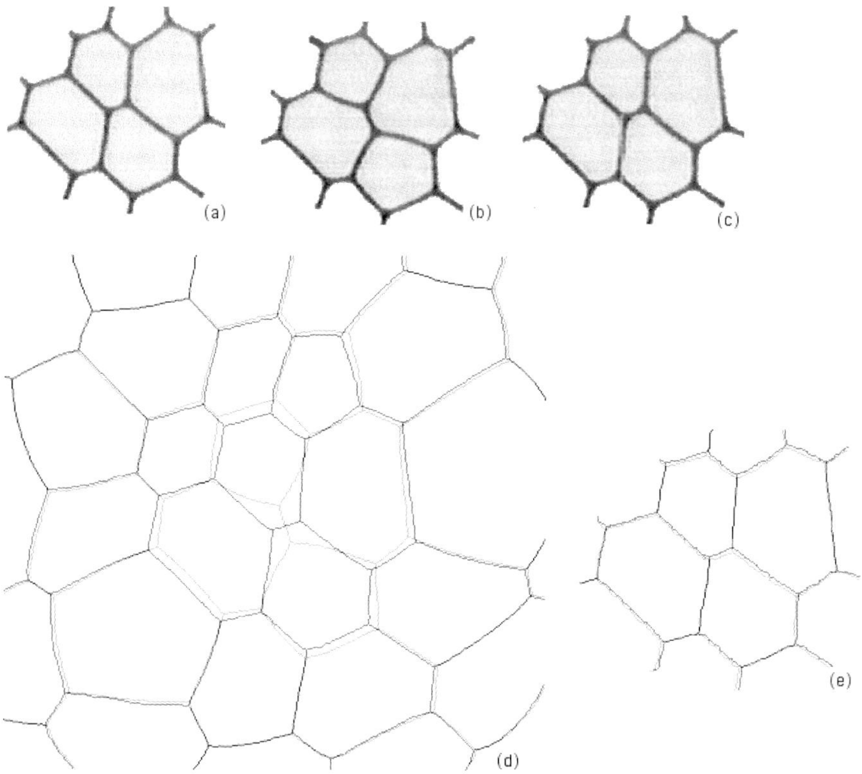

Fig. 7. Artificial side-swapping, or T1 process. We start from an initial equilibrium state (a). A metallic pin placed above the ferro-fluid foam locally channels the magnetic field lines and attracts some ferro-fluid. This enables us to locally move a vertex and bring it closer to the second one, until they merge and form an unstable 4-fold vertex. The foam spontaneously relaxes towards a new equilibrium state with a different topology (b): bubbles have swapped a side and changed their neighbours. Although the energy difference between (a) and (b) was much smaller than our sensitivity, (b) has visibly more regular bubbles, hence likely a lower energy than (a). The skelettised image (d) is the difference between (a) and (b), stressing the range of modifications induced by this T1: typically 3 bubble sizes. Inducing the reverse T1 was possible (c), and we fall back in the same state, see the comparison between (a) and (c) displayed in (e): both (a) and (b) are lasting equilibrium states. However, the reverse T1 was much more difficult: since the 4-fold vertex systematically decayed into configuration (b), we had to nucleate the side we wanted to appear. Hence the energy barrier between (a) and (b) is asymmetric, again suggesting that (b) has a lower energy than (a). Picture by F. Elias. For details see [8].

8 The Minimum Perimeter Problem

8.1 Different Equilibrium States

Topological changes: When the foam is at equilibrium, it remains there. However, the equilibrium can be modified, either by mechanical perturbation such as shear, or by variation of control parameters such as the content of each bubble, through diffusion. If one wall length happens to vanish, four bubbles come in contact. Such four-fold vertex is unstable and decays into one of either possible configuration. The foam spontaneously choses one, likely the one with the lowest energy, and eventually reaches a new equilibrium (Fig. 7). If bubbles have exchanged their neighbours, this elementary topological change is called a side-swapping, or T1 process. A T1 process conserves not only the total charge but also the total dipolar moment, altering only the quadrupolar moment of the foam: it is a current of dipoles.

The energy difference between both equilibrium states is usually much smaller than the barrier which separates them (Fig. 7). In fact, creating a four-fold vertex costs of order the line tension times a wall length, typically 10^{-7} J. The foams with the lowest barrier, namely in Langmuir monolayers, have a cost larger than a pN $\times$ μm, i.e. at least 3 orders of magnitude larger than k_BT (k_BT being reached at the scale of the nm, which is the realm of micro-emulsions rather than foams). Thus a T1 never occurs spontaneously under thermal fluctuations, and local energy minima are metastable.

Position of the problem: This raises the question: if we could vary freely the topology, keeping the area distribution fixed, what would be the global energy minimum, i.e. the lowest accessible value of the energy? And what would be the ground state(s), i.e. the corresponding configuration(s)?

This question, called "the minimal perimeter problem", is extremely challenging [20]. Mathematicians are progressing at a quick pace, but exact results still mainly concern small numbers of bubbles, typically $N = 2, 3$ [18]. However, studying a few bubbles does not help understanding foams with large N. Since a topological change can modify the shape of bubbles over a distance of order three bubble sizes (Fig. 7d), the problem is non-local.

In fact, the mathematical result which is the most useful for physicists, is so useful and natural that physicists believed it was obvious: if all bubbles have the same area A, the best pattern is a regular hexagonal lattice! The global energy minimum is then $2^{3/2}3^{1/4}\lambda\sqrt{A}$. This famous "honeycomb conjecture" is now a theorem, recently proved by Hales [17]. We use this strong result as a starting point for the following approximate, large scale approach [15].

8.2 Global Energy Minimum (Minimal Perimeter at Free Topology)

An estimate: If bubbles have different areas, not all bubbles need be hexagons. However, it surprisingly happens that regular bubbles obeying Plateau rules, with n curved edges meeting at $2\pi/3$ angles, all have almost the same ratio of perimeter to square root of area: of order $2^{3/2}3^{1/4} \approx 3.722...$, and always larger than $2(\pi)^{3/2}/3 \approx 3.712...$ [15].

We thus suggest that H is close to the energy H_h it would have if all bubbles were regular hexagons:

$$H \approx H_h = 3.72\,\lambda \sum_{i=1}^{N} \sqrt{A_i} = 3.72\,\lambda N \left\langle \sqrt{A} \right\rangle , \tag{21}$$

so that H_h is a good approximation of the foam's global energy minimum. This estimate depends only on the area distribution of the bubbles. It depends on $\left\langle \sqrt{A} \right\rangle$, while our intuition rather often uses $\langle A \rangle$. If the foam has a boundary, there is an additional term of order $\sqrt{N \langle A \rangle}$.

Tests: We have checked this estimate by examining different examples: simulations and experiments on disordered foams [15], analytical and numerical calculations on periodic foams [10], small clusters with large boundary contributions [32]. The main result which emerges is that the estimate (21) seems consistently correct, its precision being typically the percent. The foam energy is usually slightly higher than H_h, but surprisingly, we could find ordered foams with a value slightly lower than H_h, see Sect. 8.3.

In addition, most configurations tend to have an energy close to the global minimum, and it is very difficult to prepare a foam with an energy a few percent higher (Fig. 8). This explains for instance why we could not measure the energy variation after a T1 process (Fig. 7). This means that the energy estimate is an acceptable approximation. It is operational, since it depends on the area distribution, which is an intrinsic characteristic of the foam, but not on any particular topology.

8.3 Ground State Configuration (Perimeter-Minimizing Pattern)

Difficulty: Since most patterns have almost the same energy, the problem of determining the ground state configuration(s) is subtle. Physical foams come always close to the global energy minimum, not close to a ground state configuration.

Equation (17) indicates that the energy reaches its minimum when the largest bubbles have as low a pressure as possible. It is tempting to use it to determine the ground state configuration. However, this research direction has not yet been successful, to our knowledge [M. Fortes, private communication]. The reason is probably that the pressure field is "smooth", as mentioned above (Sect. 7.2); it might differ from bubble to bubble but these spatial fluctuations remain local and invisible at large scales, hence difficult to analyse in details.

We thus use another, approximate approach, to find the ground state configuration(s), in two opposite extreme cases.

Size-sorting: Consider two bubbles inside a foam, with areas A_1, A_2 respectively. If A_1/A_2 is close to 1, the total perimeter should be minimal if both bubbles are close to regular hexagons. Both bubbles share an edge, which length is a compromise between $\frac{3.722}{6}\sqrt{A_1}$ and $\frac{3.722}{6}\sqrt{A_2}$, where $\frac{3.722}{6} \approx \frac{2^{1/2}}{3^{3/4}} \approx 0.62$ is the side of a regular hexagon of

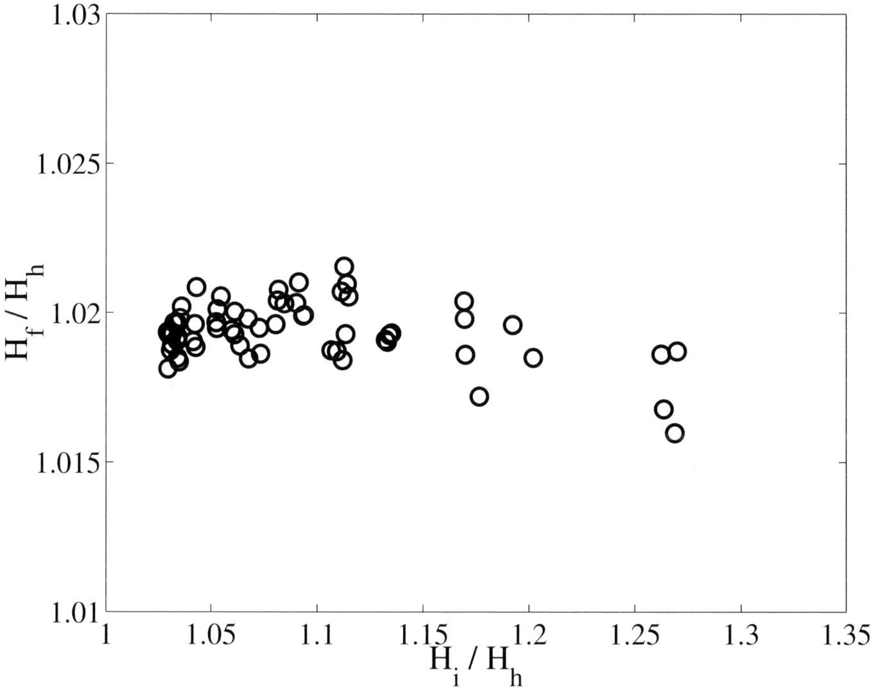

Fig. 8. Test of the estimate for the minimum energy. We simulated out-of-equilibrium, highly deformed foams using the Potts model [15], with a high initial energy H_i. The foam spontaneously relaxes towards an equilibrium state, with a final energy H_f. Note the difference between vertical and horizontal scales: although we could reach initial energies H_i up to 30% higher than H_h, the final energy is systematically the same, barely higher than H_h. Each symbol corresponds to one foam. Simulations by Y. Jiang [15].

area 1. We have suggested and checked that, in average, the common edge has a length of order:

$$L(A_1, A_2) \approx \frac{3.722}{6}\sqrt{A_1^{1/2}A_2^{1/2}} \approx 0.62(A_1A_2)^{1/4}. \tag{22}$$

Thus the cost in total perimeter is lower when bubbles have neighbours of almost the same area. This leads to size-sorting, with bubbles segregating according to their sizes.

We have tried to observe this effect by producing an annealed foam, i.e. a foam as close as possible to the ground state configuration. This is difficult, since the energy barrier between different equilibrium states is larger than the energy differences (Fig. 7). We have succeeded by using ferrofluid-foams, easier to manipulate thanks to their high viscosity / low relaxation time. This size-sorting is favoured by a large area distribution, which is here continuous, so that each bubble can find neighbours of almost the same area and be as regular as possible. Such size-sorting is visible on Fig. 6.

Size-mixing: On the opposite, when the area distribution is bidisperse, and A_1/A_2 is very different from 1, bubbles do not remain 6-sided; they rather change their topology.

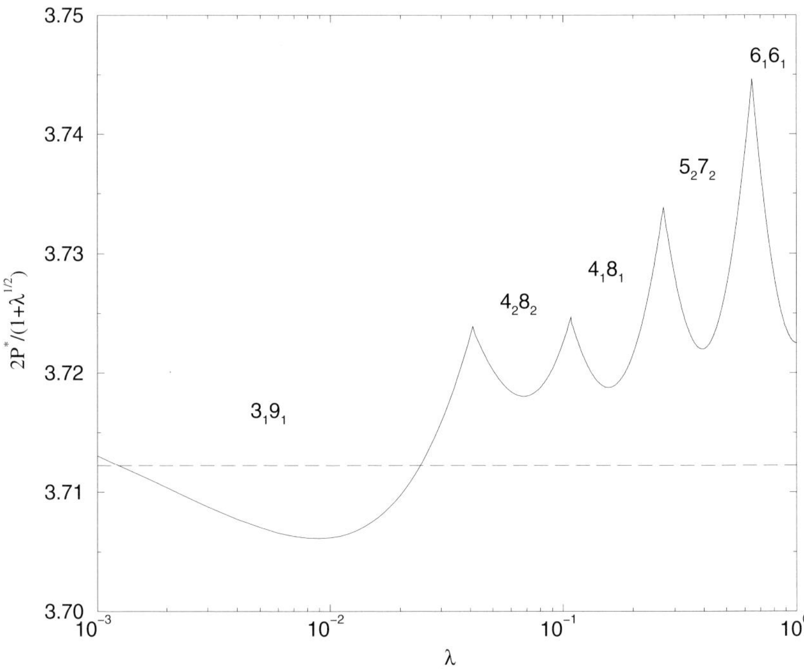

Fig. 9. The best periodic patterns found by Fortes and Teixeira [10]. Two types of cells, of areas $A_1 = \lambda$ and $A_2 = 1$, are mixed in equal numbers. The vertical axis is twice the perimeter, divided by $(1 + \lambda^{1/2})$ to enable comparison with H_h (21). For each value of λ, only the best pattern found is plotted. The figures indicate the numbers of bubble sides, the indices indicate the number of bubbles per unit cell. For instance $6_1 6_1$ indicates alternate hexagons; $3_1 9_1$ indicates an array of large bubbles paving a regular honeycomb, but decorated on each other corner by a small three-sided bubble: surprisingly, this beats the value of regular bubbles, $2\pi^{3/2}/3$ (dashed line). Graph by P. Teixeira.

Large bubbles ignore small ones, small bubbles decorate the vertices between large ones. Fortes and Teixeira [10] have calculated the energies of different periodic patterns, and looked for the most favorable one, for each given value of A_1/A_2. Their results appear in Fig. 9. Starting from $A_1/A_2 = 1$, and decreasing A_1/A_2, they found that the best pattern is successively: alternate hexagons, then mixture of 5- and 7- sided bubbles, then mixtures of 4 and 8 (two types of periodic tilings), and finally of 3 and 9.

Recently, we have found [31] that the periodic tiling of 9-sided and 3-sided bubbles (large bubbles placed in honeycomb lattice, with tiny 3-sided bubbles decorating half of their vertices) beats H_h, and even slightly beats the value 3.712 $\lambda \sum_{i=1}^{N} \sqrt{A_i}$ (Fig. 9). We are currently investigating this surprising result.

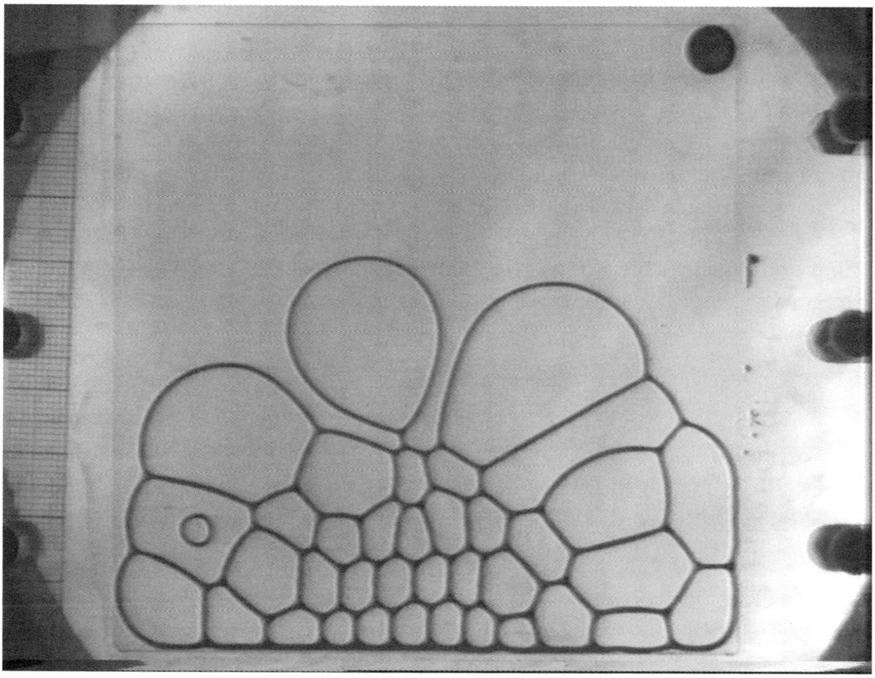

Fig. 10. A rather exotic 2D foam! This is a true experiment, actually a ferro-fluid foam. Do you think the 0-angle (i.e. round) bubble can be at equilibrium? And the 1-angle bubble? Picture courtesy of C. Flament.

9 Conclusion

We have tried here to show how foams link mathematical properties, such as topology and curvature, with physical quantities such as energy and pressure (Sect. 5.1). We have adopted the point of view of statistical physics, linking the apparently simple properties of each bubble (Sect. 4) with the global behaviour of a disordered foam.

We have considered here an ideal 2D foam (Sect. 3.2) at equilibrium, where each bubble has a given constant area. We have tried to show the central role of the energy and pressure field. We can estimate both on large scales without knowing microscopic details. The energy is approximately a function of the area distribution only, and more precisely on $\left\langle \sqrt{A} \right\rangle$ (21); it is almost equal to the perimeter it would have if each bubble had a regular shape, and can even be slightly lower (Sect. 8.3). The pressure resembles a field with zero laplacian (Sect. 7.2), as we could understand using an analogy with 2D electrostatics (Sect. 5). Both quantities are also related together (Sect. 6.3).

We currently try to open perspectives into three directions: actual 3D foams; mechanical properties of foams out of equilibrium; and more complex fluids such as emulsions and elastic-plastic-viscous materials. We hope to have shown the triple interest of foam research: as a truly fascinating interdisciplinary field, as a carefully-chosen model system, and for their own applications.

Acknowledgments

The results presented here have been derived together with M. Aubouy, S. Courty, F. Elias, J.A. Glazier, Y. Jiang. I would also like to acknowledge the collaboration with M. Asipauskas, O. Cardoso, M.-L. Chabanol, S.J. Cox, R. Delannay, B. Dollet, C. Flament, M. Fortes, S. Ifadir, E. Janiaud, O. Lordereau, C. Monnereau-Pittet, N. Pittet, P. Teixeira, M.F. Vaz. Thanks are also due to G. Debregeas, K. Kassner, N. Kern, J. Lajzerowicz, K. Mecke, F. Morgan, D. Weaire for useful discussions. D. Weaire and F. Elias critically read the manuscript. L. Mousson, from the Association des Buveurs d'Orges (www.space.ch/abo), was an endless source of information about beer foams.

References

1. Aref, H., D. Vainchtein (2000): 'The equation of state of a foam', *Phys. Fluids*, **12**, pp. 23-28 contains two demonstrations. Two simpler demonstrations were then successively proposed by us, in [15], and by Fortes, M. (2001): 'The surface energy of finite clusters of soap bubbles', *Phys. Fluids*, **13**, pp. 3542-3546. Note that it was presented as an already old conjecture in Ross, S. (1969): 'Bubbles and Foam, *Ind. Eng. Chem.*, **61**, p. 48 sqq.
2. Aste, T., D. Boosé, N. Rivier (1996): 'From one bubble to the whole froth: A dynamical map', *Phys. Rev. E* **53**, 6181-6191.
3. Aubert, J., A. Kraynik, P. Rand (1986): 'Les mousses aqueuses', *Pour la Science* Juillet 1986 pp. 62-71
4. Berge, B., A. Simon, A. Libchaber (1990): 'Dynamics of gas bubbles in monolayers', *Phys. Rev. A* **41** pp. 6893-6900.
5. Courty, S. (2001): *Solides bidimensionnels à la surface de l'eau: étude mécanique et optique*, PhD dissertation, University of Grenoble, unpublished.
6. (2001): 'Décapants pour four', *Que Choisir* mars 2001, **380**, pp. 53-57.
7. Elias, F., C. Flament, J.-C. Bacri, O. Cardoso, F. Graner (1997): 'Two-dimensional magnetic froth: Coarsening and topological correlations', *Phys. Rev. E* , **56**, pp. 3310-3318 .
8. Elias, F., C. Flament, J. A. Glazier, F. Graner, Y. Jiang (1999): 'Foams out of stable equilibrium: cell elongation and side-swapping', *Phil. Mag. B* **79**, pp. 729-751.
9. Emmer, M. (1991): *Bolle di saponi: un viaggio tra arte, scienzia a fantasia* (La Nuova Italia Editrice, Scandicci, Firenze).
10. Fortes, M., P. Teixeira (2001): 'Minimum perimeter partitions of the plane into equal numbers of regions of two different areas', *Eur. Phys. J. E* **6**, 131-138.
11. de Gennes, P.-G. (2000): *Les mousses*, Cours au Collège de France, Paris, unpublished.
12. Glazier, J.A (1989): *Dynamics of Cellular Patterns*, Ph.D. dissertation, University of Chicago, unpublished.
13. Glazier, J.A., D. Weaire (1992): 'The kinetics of cellular patterns', *J. Phys.: Cond. Matt.* **4**, pp. 1867-1894.
14. Graner, F. (2001): 'La mousse', *La Recherche*, **345**, pp. 46-49.
15. Graner F, Y. Jiang, E. Janiaud, C. Flament (2001): 'Equilibrium states and ground state of two-dimensional fluid foams'. *Phys. Rev. E*, **63** pp. 011402/1-13
16. Graustein, W.C. (1931): 'On the average number of sizes of polygons in a net', *Annals of Math.* **32**, pp. 149-153. The demonstration is based on Euler theorem, applied here to a non simply- convex foam (e.g. on a torus) [12, 25].
17. Hales T. C. (2001): 'The honeycomb conjecture'*D*iscrete Comput. Geom. **25**, pp. 1-25; see also *Science News*, July 24, 1999; Klarreich E., *American Scientist* **88**, March-April 2000, 152.

18. Hutchings, M., F. Morgan, M. Ritoré, A. Ros (2000): 'Proof of the double-bubble conjecture', *Electron. Res. Announc. Amer. Math. Soc.*, **6** pp. 45-49, www.ugr.es/~ritore/bubble/bubble.htm. See also Morgan, F. (2001): 'Proof of the double bubble conjecture' *American Mathematical Monthly*, **108** (March 2001), pp. 193-205.
19. Hutzler, S., D. Weaire (1999): *Physics of Foams* (Oxford University Press, Oxford).
20. Morgan F. (2000): *Geometric Measure Theory: a Beginner's Guide*, revised 3rd ed. (Academic Press, Boston).
21. Jiang, Y., M. Asipauskas, J.A. Glazier, M. Aubouy, F. Graner (2000): '*Ab initio* derivation of stress and strain in fluid foams', in [40], pp. 297-304.
22. Landau, L.D., E.M. Lifschitz (1973): *Teoriya Polia* (Nauka, Moscow); english translation: *The classical theory of fields*, fourth revised edition, 2000 (Reed Elsevier, Oxford), chapter 5, p. 95.
23. von Neumann J. (1952): 'Discussion', in: *Metal Interfaces*, ed. C. Herring (American society for metals, Cleveland Ohio) pp. 108-110, quoted *e.g.* in [12, 13].
24. Plateau, J (1873): *Statique Expérimentale et Théorique des Liquides Soumis aux Seules Forces Moléculaires* (Gauthier-Villars, Paris).
25. Rivier, N. (1993): 'Order and disorder in packings and froths'. In: *Disorder and granular media*, eds. D. Bideau, A. Hansen, (Elsevier Science Publ., Amsterdam), pp. 55-102.
26. Sadoc J.-F., N. Rivier eds. (1999): *Foams and Emulsions*, Proceedings of 1997 summer school, Cargèse, France (Nato ASI series E, Kluwer, Dordrecht).
27. Sandre, O. (1999): unpublished pictures.
28. Schliecker G., S. Klapp (1999): *Europhys. Lett.*, **48**, p. 122.
29. Seul, M., D. Andelman (1995): 'Domain Shapes and Patterns: The Phenomenology of Modulated Phases', *Science* **267** pp. 476- 483.
30. Smith, C.S (1952): 'Grain shapes and other metallurgical applications of topology', in: *Metal Interfaces*, ed. C. Herring (American society for metals, Cleveland Ohio) pp. 65-108. Smith's notations E_0, E_b correspond respectively to v^+ and $v^+ + v^-$, respectively.
31. Teixeira P., F. Graner, M. Fortes (2002): 'Mixing and sorting of bidisperse two-dimensional bubbles', preprint.
32. Vaz M.F., F. Graner, M. Fortes (2002): 'Energy of free clusters of bubbles", preprint.
33. Vignes-Adler, M., F. Graner (2002): 'La vie éphémère des mousses', *Pour la Science*, **293**, pp. 48-55.
34. Weaire, D. (1995): De la bulle à la mousse', *La Recherche* **273** (mars 1995) pp. 246-252.
35. Weaire, D. ed. (1997): *The Kelvin problem* (Taylor and Francis, London).
36. Weaire, D., J. Banhart (1999): *Foams and Films*, Proceedings of 1999 international workshop on foams and films, Leuven, Belgium (MIT Verlag, Bremen).
37. Weaire, D., F. Bolton, P. Molho, J.A. Glazier (1991): 'Investigation of an elementary model for magnetic froth', *J. Phys.: Cond. Matt.* **3**, pp. 2101-2114.
38. Weaire, D., S.J. Cox, F. Graner (2002): 'On the non-uniqueness of foam equilibrium geometry', *Eur. Phys. J. E*, to appear.
39. Weaire, D., N. Rivier (1984): 'Soap, Cells and Statistics - Random Patterns in Two Dimensions'. *Contemp. Phys.* **25**, pp. 59 -99.
40. Zitha, P., J. Banhart, G. Verbist (2000): *Foams, Emulsions and their Applications*, Proceeding of the Eurofoam 2000 conference, Delft, Netherlands (MIT Verlag, Bremen).
41. We use the word "energy" throughout the paper. We do not distinguish here between equations such as (1), where H is a functional of the coordinates, hence strictly speaking a hamiltonian; and such as (17), where it is a value, hence an energy. Moreover, since the temperature does not play any significant role (Sect. 8.1), we do not specify the fact that it is actually a free energy, as are λ and γ.

Part II

Spatial Statistics and Morphology

Morphological Texture Analysis: An Introduction

Pierre Soille

EC Joint Research Centre, Institute for Environment and Sustainability,
I-21020 Ispra (Va), Italy
e-mail: Pierre.Soille@jrc.it

Abstract. Texture is commonly defined as the characteristic physical structure given to an object by the size, shape, arrangement, and proportions of its parts. Texture characterisation is therefore a difficult problem that cannot be solved by a single measurement. Mathematical morphology, being a theory for the analysis of spatial structures, naturally offers a variety of techniques useful for texture characterisation. We show that the shape, size, orientation, and periodicity of ordered textures as well as some features of disordered textures can be revealed by appropriate morphological transformations.

1 Introduction

An informal definition of the notion of texture is *'the characteristic physical structure given to an object by the size, shape, arrangement, and proportions of its parts'* [2]. The goal of texture analysis in image processing is to map the image of a textured object into a set of quantitative measurements revealing its very nature. The success of this mapping can be assessed by determining whether the resulting vectors are discriminant: measurement vectors of similar textures should form a cluster in the associated feature space and should be well separated from measurement vectors corresponding to different textures. In addition, the dimensionality of the vectors should be as small as possible for efficiency considerations. In this sense, texture analysis can be considered as a pattern recognition/classification problem [11].

Similarly to shape measurements, texture measurements may require invariance to image transformations such as translations, rotations, scalings of the image definition domain, and monotone transformations of the pixel intensity values. For global texture measurements, a unique vector of features is produced for the whole image frame. In the case of local measurements, one vector is available for each pixel by performing calculations within a moving window centred on each pixel. Local measurements are necessary when an image contains more than one texture since they enable the partitioning of the image into connected regions whose texture features are homogeneous. This process is referred to as texture segmentation.

Following the boundless variety of textures, a tremendous amount of texture features have been proposed since the early days of digital image processing [12, 38, 34, 22, 36]. In this paper, we concentrate on features revealed by mathematical morphology transformations. The suitability of a transformation for generating discriminant features depends on the input texture class. We will therefore refer to the taxonomy for texture description detailed in [21] and based on the distinction between ordered and disordered textures, as well as their composition.

This introduction to morphological texture analysis is organised as follows. Necessary background notions about mathematical morphology are presented in Sect. 2. In Sect. 3, we show that granulometries are suited to the extraction of shape and size characteristics of a texture. Technical issues related to the design of granulometric families of discrete line segments and disks are also addressed. The extraction of global and local directional information of oriented textures is detailed in Sect. 4. Several additional methods such as those based on the morphological covariance and the watershed transformation are briefly reviewed in Sect. 5. Concluding remarks and a summary table are presented in Sect. 6.

2 Basic Principles of Mathematical Morphology

In mathematical morphology, 2-dimensional grey tone images are seen as 3-dimensional sets by associating each image pixel with an elevation proportional to its intensity level. Another set of known shape and size, called *structuring element* (SE), is then used to investigate the morphology of the input set. This is achieved by positioning the origin of the structuring element to every possible position of the space and testing, for each position, whether the structuring element either is included or has a non-empty intersection with the studied set. By keeping those points for which the inclusion test is positive, we obtain an *erosion* of the input set while the non-empty intersection test leads to a *dilation* of this set. Erosions and dilations are the letters of the morphological alphabets since all other transformations are based on these two elementary transformations. The shape and size of the structuring element must be selected according to the morphology of the targeted image structures. For instance, line segments are suited to the analysis of elongated structures such as fission tracks or fibrous materials while disks are more appropriate for granular structures.

In mathematical terms, the erosion ε of a set X by a structuring element B is the set of points $\mathbf{x}$ such that when the origin of B coincides with $\mathbf{x}$, B is included in X: $\varepsilon_B(X) = \{\mathbf{x} \mid B_\mathbf{x} \subseteq X\}$. Similarly, the dilation δ is the set of points having a non-empty intersection with X when placing the structuring element at these points: $\delta_B(X) = \{\mathbf{x} \mid B_\mathbf{x} \cap X \neq \emptyset\}$. For example, the erosion of a shape by itself preserves only the origin of the shape while the dilation of horizontal line segment by a vertical line segment generates a rectangle. Erosions and dilations are dual operations with respect to set complementation $\complement$ in the sense that the erosion of the complement of a set is equivalent to the complement of the dilation of this set: $\varepsilon_B[\complement(X)] = \complement[\delta_B(X)]$.

When processing a discrete grey tone image f, the erosion (resp. dilation) by a grey tone structuring element g at a given pixel $\mathbf{x}$ is equivalent to the following minimum (resp. maximum) value: $[\varepsilon_g(f)](\mathbf{x}) = \min_{\mathbf{b} \in \mathcal{D}_g}\{f(\mathbf{x}+\mathbf{b}) - g(\mathbf{b})\}$ and $[\delta_g(f)](\mathbf{x}) = \max_{\mathbf{b} \in \mathcal{D}_g}\{f(\mathbf{x}+\mathbf{b}) + g(\mathbf{b})\}$, where $\mathcal{D}_g$ denotes the definition domain of g. Unless pixel values actually represent heights above a reference surface, the set representation of grey tone images has heterogeneous units: spatial along the axis of the image definition domain and non-spatial along the grey tone axis. To avoid mixing up these units, structuring elements having zero heights are considered: $\forall\, \mathbf{b},\ g(\mathbf{b}) = 0$. Owing to their morphology, such structuring elements are called *flat*. They are referred to by capital letters such as B to emphasise that they have only spatial extent such as disks for 2-D images or spheres for 3-D images. Erosions (resp. dilations) by flat structuring elements come down to calculating, for each pixel, the minimum (resp. maximum) image value within the neighbourhood defined by the structuring element. The erosion and dilation of an image by a 3×3 square whose origin is at its centre are illustrated in Figs. 1a-c.

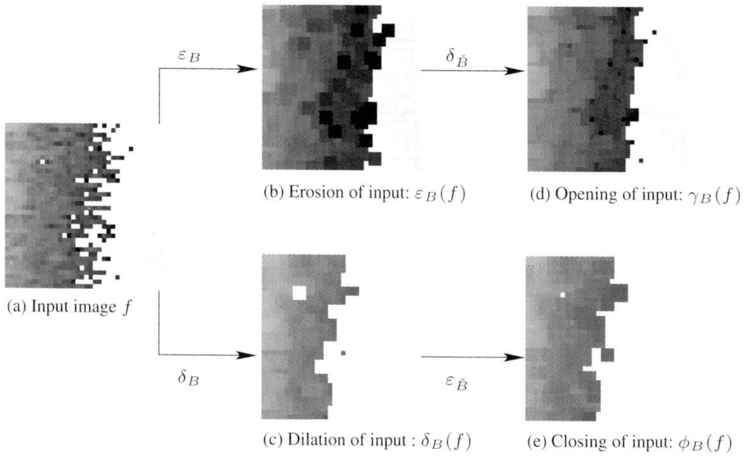

Fig. 1. Erosion, dilation, opening, and closing of a wrapped phase map associated with a fringe pattern. The considered structuring element B is a 3×3 square. Note that the opening filters out the bright 'noise' while the closing filters out dark 'noise' (dual behaviour).

The search for an operator recovering as much as possible structures shrinked by the erosion leads to the notion of *opening*. This is achieved by dilating the eroded image. In general, the original structures are only partly recovered. For example, connected components of pixels completely destroyed by the erosion are not recovered at all. This behaviour is at the very basis of the nonlinear filtering properties of the opening operator: foreground image structures are selectively filtered out, the selection depending on the shape and size of the structuring element.

The geometric formulation of the morphological opening of a set in terms of structuring element fits is as follows: *"Does the structuring element fit the set?"* Each time the answer to this question is affirmative, the *whole* structuring element must be kept (for the erosion, it is the origin of the structuring element that is kept). For example, the opening of a rectangle by a disk rounds off its corners. Denoting by γ the opening transformation, we have the following formulation for the opening of a set X by a structuring element B: $\gamma_B(X) = \cup_{\mathbf{x}}\{B_{\mathbf{x}} \mid B_{\mathbf{x}} \subseteq X\}$. It can be shown that the same result is achieved by performing an erosion by B followed by a dilation with the reflection of B, denoted by $\check{B}$: $\gamma_B(X) = \delta_{\check{B}}[\varepsilon_B(X)]$. This latter equation extends directly to the processing of grey tone images: $\gamma_B(f) = \delta_{\check{B}}[\varepsilon_B(f)]$. By omitting the operand, we obtain the following compact definition for the opening: $\gamma_B = \delta_{\check{B}}\varepsilon_B$. By construction, the opening of an image can only decrease the intensity value of its pixels: $\left[\gamma_B(f)\right](\mathbf{x}) \leq f(\mathbf{x})$. In other words, the opening γ is ordered with the identity transformation id: $\gamma_B \leq \text{id}$. This property is called the anti-extensivity. Openings satisfy two further properties. First, they preserve ordering relations between images: $f \leq g$ implies that $\gamma_B(f) \leq \gamma_B(g)$. This property is called the increasingness. Second, openings are idempotent transformations in the sense that the opening of an image that has already been opened by this opening has no effect: $\gamma_B \gamma_B = \gamma_B$. More generally, it can be shown that any image transformation that is anti-extensive, increasing, and idempotent can be represented as the union of openings by a family of structuring elements. For example, the operation which removes all connected objects whose area is less than a given threshold is equivalent to the union of the openings by the connected structuring elements whose area equals this threshold value. This operation is known as the *area opening*. Note that the union (resp. intersection) between a series of grey tone images is defined as the pointwise maximum $\vee$ (resp. minimum $\wedge$) between these images: $\left[\vee_i (f_i)\right](\mathbf{x}) = \max_i \left[f_i(\mathbf{x})\right]$.

The morphological *closing* ϕ is the dual transformation of the opening with respect to set complementation: $\phi_B = \varepsilon_{\check{B}} \delta_B$. Closings are therefore extensive, increasing, and idempotent transformations. The opening and closing by a square structuring element are illustrated on a sampled region of a wrapped phase map in Fig. 1. Note that, in this example, $B = \check{B}$ since B represents a 3×3 square with the origin located at the centre pixel.

Beyond these basic concepts, a thorough presentation of the principles and applications of morphological image analysis is given in [26, 27]. A more mathematical exposition is proposed in [13] while key seminal reference books are [18, 23, 24].

3 Granulometry

We first define the concept of granulometry and show its application for texture classification and segmentation. We then address technical issues related to the generation of discrete line segments and disks satisfying the properties of a granulometric analysis.

3.1 Principle

When analysing granular materials, a granulometry is performed by sieving a sample through sieves of increasing mesh size while measuring the mass retained by each sieve. A granulometric curve is a decreasing curve plotting the measured mass for each sieve size. Interestingly, the sieving of materials through a sieve shares all properties of an opening: (i) what is left in the sieve can only be a subsample of the input sample (anti-extensivity), (ii) when sieving a subsample of a larger sample, what remains in the sieve is a subsample of what would remain when sieving the whole sample (increasingness), and (iii) sieving a sample twice through the same sieve does not sieve further this sample (idempotence). However, because the granulometry process involves a series of sieves, it satisfies one further property called the *absorption property*: what remains after sieving a sample through two sieves of arbitrary size is only driven by the size of the largest sieve. Consequently, the concept of granulometry can be transposed to image data by opening the image with a family of openings of increasing size λ while making sure that the absorption property is satisfied. That is, the composition of any two openings of the family must come down to applying the opening with the largest size: $\gamma_{\lambda_i}\gamma_{\lambda_j} = \gamma_{\lambda_j}\gamma_{\lambda_i} = \gamma_{\max(\lambda_i,\lambda_j)}$ or, equivalently, the openings must satisfy the following ordering: $\gamma_{\lambda_i} \leq \gamma_{\lambda_j}$ if $\lambda_i \geq \lambda_j$. The granulometric curve plots the sum of the pixel values (called *volume* V) of the opened image versus the size of the opening, i.e., $V(\gamma_\lambda)$ versus λ or $V(\gamma_\lambda)/V(\mathrm{id})$ versus λ for a normalised granulometric curve. In practice, granulometric curves are often interpreted by computing their discrete derivative, i.e., loss of volume between two successive openings versus the size of the opening: $V(\gamma_{\lambda_i} - \gamma_{\lambda_{i-1}})$ versus λ_i. The resulting curve is called *size distribution* or *pattern spectrum* because its peaks indicate the prevailing sizes of the image structures. More generally, volume measurements can be substituted by measurements revealing other texture features. For example, one may consider the evolution of the Euler characteristic (also called connectivity number) of the image opened by a disk of increasing size.

Note that a granulometry by opening produces information concerning image structures brighter than their neighbourhood. Indeed, when performing an opening, we test only whether the structuring element fits the foreground structures. Therefore, information about the arrangement of structures which appear darker than their neighbourhood is obtained by closing the image with the same family of structuring elements. Because closings are extensive rather than anti-extensive, this process is referred as an *anti-granulometry* or simply a *granulometry by closing*. In addition, *pseudo-granulometries* (resp. anti-pseudo-granulometries) are sometimes preferred to actual granulometries [25]. They are obtained by performing erosions (resp. dilations) of increasing size and measuring for each size the volume of the eroded (resp. dilated) image. The corresponding pseudo-granulometric curves are also referred to as the *erosion and dilation curves* [33, 19].

3.2 Texture Classification Using Global Granulometries

Granulometries provide us with useful texture features because they reflect information regarding the shape and size of the patterns defining ordered textures as well as the degree of granularity of disordered texture. For example, Fig. 2 shows two types of texture occurring in satellite images of forest stands together with their granulometries by opening with squares of increasing size. These granulometries can be used for discriminating the input textures because they highlight that the upper texture is coarser than the lower texture. Indeed, the size at which the maximum of the derivative of the granulometry occurs is significantly larger for the coarser than the finer texture.

Granulometries have already been used in several fields for classifying textured images. In [35], granulometric curves with disks and line segments are utilised for discriminating textures recorded by Brodatz [6]. The methodology is then applied to the classification of opacities found in chest radiographic images of lungs suffering from pneumoconiosis. In [7], the classification of grey scale textures using simultaneously opening and closing granulometric curves is investigated. It has also been shown in [39] that granulometries with line segments and disc-shaped structuring elements allow for the discrimination of tree species from aerial photographs. In [4], pavement defects are classified using a Gaussian model of the particle size distributions to eliminate the effects of the normal texture so that the variations caused by the defects are emphasised for subsequent analysis. Features derived from the residues of granulometries based on

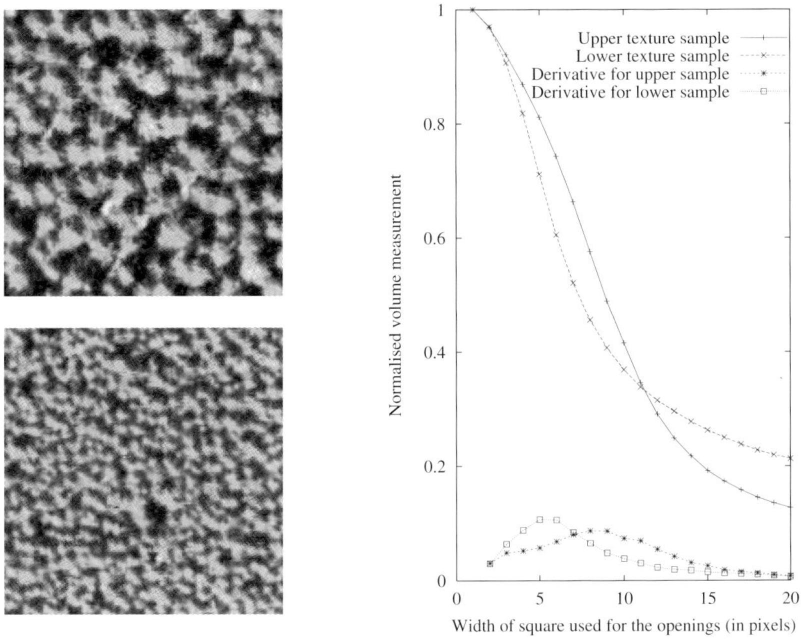

Fig. 2. Left: Two samples of a 1 m resolution panchromatic IKONOS satellite image showing two forest stands of different age (crown size). Right: Their granulometries by opening using squares of increasing width.

openings and closings by reconstruction with squares are used in [16] for classifying a variety of Brodatz textures. The quantitative assessment of stone decay is assessed in [20] by taking into account both the maximum of the size distribution (revealing the dominant pattern in the texture) and the area below the granulometric curve. Finally, it is pointed out in [3] that the discrete derivative of the erosion and dilation curves are more discriminant than granulometries for classifying the texture of plastering mortars.

3.3 Texture Segmentation Using Local Granulometries

A segmentation of a binary image into regions of homogeneous textures using local granulometries is proposed in [10]. This is achieved by computing a granulometry for each pixel within a centred window. A new image, whose grey values are statistics resulting from the local size distributions, is created. This image is then segmented assuming that regions of homogeneous texture have similar statistics (e.g., similar pattern spectrum variance). This technique has been applied to the analysis of electrophotographic images [9] and to the classification of bone structures to detect osteoporosis in magnetic resonance images [8]. An application to the segmentation of forest stands of different age is presented in Fig. 3.

The decomposition procedure developed in [37] is identical to morphological granulometries. Nevertheless, the authors propose an interesting and novel approach for se-

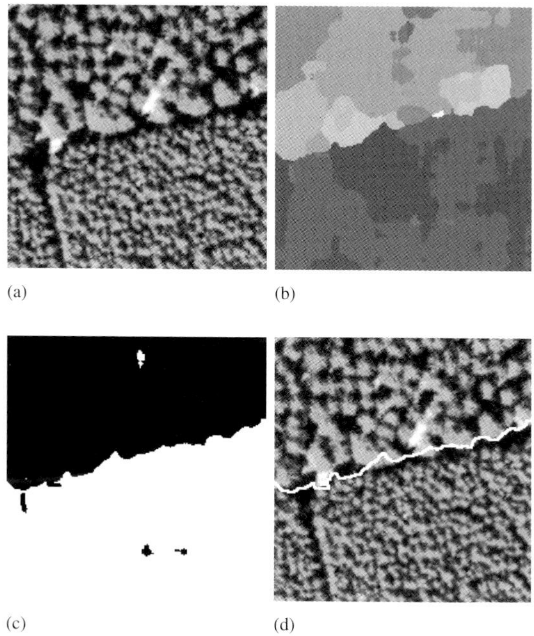

Fig. 3. Local granulometries for texture segmentation. (a) Input 1 m resolution panchromatic IKONOS satellite image of two forest stands of different age (crown size). (b) Image of the size at which the maximum of the derivative occurs for each pixel (the local granulometries have been computed within a 60×60 window centred at each pixel). (c) Global threshold: small crowns for size ≤ 5 pixels appear in white. (d) Boundary of filtered threshold superimposed on input satellite image. The granulometric curves of the two types of textures occurring in the input image are displayed in Fig. 2.

lecting the window position at a given pixel for texture segmentation purposes. Indeed, a common problem of texture segmentation is that centred windows do not provide good results when they contain more than one texture. Therefore, the authors propose to define the output class at a given pixel as the class given by the window containing this pixel and having the highest a posteriori probability (in the context of supervised classification). Note that an optimal window size must be chosen according to the size of the textured regions: small windows decrease the regions where mixed textures occur but at the same time they tend to decrease the accuracy of the estimation of the local size distribution.

3.4 Discrete Line Segments and Disks

Structuring elements useful for practical applications and satisfying the property of a granulometry are disks and line segments of increasing size. Granulometries with disks give information about the maximal width of the structures occurring in the image while line segments give information about the length distribution of the image structures (either in a given orientation or for all possible orientations). Therefore, granulometries with line segments are more suited to the analysis of anisotropic (directional) textures. In practice, computations are performed with *discrete* approximations of Euclidean line segments and disks. However, these approximations must be chosen so as to make sure that generated families of structuring elements comply with the absorption property.

Discrete line segments. A first discrete approximation of Euclidean line segment consists in considering only those points of the segment matching the centre of image pixels (i.e., the nodes of a square grid). Starting from a node of a square grid, only Euclidean lines with a rational slope intersect other grid nodes. This idea is at the very basis of the notion of periodic line [14]. Starting from an arbitrary grid node, a *periodic line* P containing n pixels is defined by considering the grid nodes which are defined by a multiple of the elementary constant vector $\boldsymbol{v} = (a, b)$ where $a, b \in \mathbb{Z}$: $P_{n,\boldsymbol{v}} = \bigcup_{i=0}^{i=n-1} i\boldsymbol{v}$. We assume in the sequel that the slope b/a is in the form of an irreducible fraction including the forms $0/1$ and $1/0$ for referring to horizontal and vertical lines respectively.

By construction, periodic lines are not 8-connected unless $\boldsymbol{v}$ matches one of 4 principal directions of the square grid. Connected line segments of arbitrary direction in the square grid are usually constructed using the Bresenham algorithm [5] illustrated in Fig. 4. We denote by $L_{\lambda_i,(a,b)}$ the connected line segment obtained by considering λ

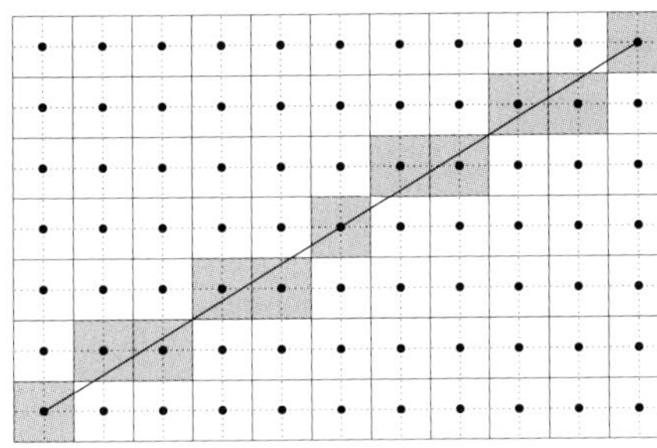

Fig. 4. Euclidean line of slope $3/5$ and corresponding Bresenham discrete line. Each time the Euclidean line cuts the vertical lines linking the centres of the pixels (i.e., the vertical dotted lines), the centre of the pixel which is closest to the Euclidean line defines a pixel of the discrete line. Had the slope been larger than 1, the horizontal dotted lines linking the centres of the pixels should have been considered instead of the vertical ones.

successive pixels of a Bresenham line of slope b/a, starting from the ith pixel of the line. Note that the same connected line segment appears periodically along the Bresenham line, the periodicity k being equal to $\max(|a|, |b|)$, i.e., $L_{\lambda_{i+k},(a,b)} = L_{\lambda_i,(a,b)}$ for all i. A family of line segments satisfying the absorption property required by granulometries can be generated by extending a given line segment by k pixels at a time. In practice, we start with one of the elementary connected patterns $L_{k_i,(a,b)}$ occurring along the line and extend it by $(n-1)k$ pixels by dilating it with the periodic line of the same slope and containing n pixels: $L_{nk_i,(a,b)} = L_{k_i,(a,b)} \oplus P_{n,(a,b)}$, where $\oplus$ denotes Minkowski addition, i.e., $X \oplus B = \delta_{\check{B}}(X)$. An example of this dilation cascade is displayed in Fig. 5. Consequently, the number of pixels nk_i of the connected line segment is a multiple of the periodicity k. However, by considering the union of openings with all k possible structuring elements of slope b/a and length λ pixels, we can increase the length of the structuring element one pixel at a time while satisfying the absorption property. The resulting directional granulometry for a given slope b/a and size parameter λ is then defined as follows: $\gamma_{\lambda,(a,b)} = \bigvee_{i=1}^{i=k} \gamma_{L_{\lambda_i,(a,b)}}$.

Using the raw implementation of erosions and dilations necessary for computing an opening, $2k(\lambda - 1)$ comparisons per pixel are required for each size λ. We show hereafter that an implementation requiring a fixed number of comparisons (namely $6k$) whatever the length λ is possible. Indeed, there exists an algorithm for erosions (resp. dilations) along Bresenham lines requiring only 3 minimum (resp. maximum) comparisons whatever the length of the structuring element [30]. However, because this algorithm operates along Bresenham lines, the shape of the connected line segment of λ pixels varies with a periodicity k. The actual output at a given image pixel is therefore not

(a) (b) (c)

Fig. 5. Generation of a connected line segment by dilating an elementary connected pattern with a periodic line of the same slope. (a) The second pattern occurring on a Bresenham line of slope 5/3 and considering a length of 5 pixels: $L_{5_2,(5,3)}$. (b) The periodic line of slope 5/3 with 2 pixels: $P_{2,(5,3)}$. (c) The resulting dilation cascade: $L_{10_2,(5,3)} = L_{5_2,(5,3)} \oplus P_{2,(5,3)}$.

translation-invariant (non-TI): it depends on the position of the pixel along the Bresenham intersecting this pixel when sweeping the image definition domain with a Bresenham line (there are k possibilities). We denote by $\Psi_{BL_{\lambda_i},(a,b)}$, where $i \in \{1, \ldots, k\}$ and $k = \max(|a|, |b|)$, the k possible outputs of an image transformation Ψ by a line segment of length λ applied along Bresenham lines sweeping the image definition domain. It can be shown [32] that the union of all k non-TI possible openings $\bigvee_{i=1}^{i=k} \gamma_{BL_{\lambda_i},(a,b)}$ is identical to the union of the k possible TI openings $\bigvee_{i=1}^{i=k} \gamma_{L_{\lambda_i,(a,b)}}$ but the former requires only $6k$ comparisons per pixel whatever the length λ of the line segment.

The discrepancies between directional non-TI and TI granulometries are exemplified in Fig. 6. Non-TI granulometries should be avoided because, beyond the fact that they not translation invariant, they miss many structures as revealed by significantly lower volume measurements. Indeed, assuming that the input image would just contain a randomly placed discrete line segment of slope b/a and length greater than that of the considered structuring element, the non-TI opening along Bresenham lines of slope b/a has a probability of 1 out of k to preserve it. The TI opening by one of the k possible line segments would always preserve it but may remove some points at the extremity of the line segment depending on whether its extremities can be covered by the selected structuring element. By contrast, the TI opening by a union of directional openings would preserve all pixels of an input image containing discrete line segments of slope b/a and length greater than that of the considered structuring element.

Rather than computing a granulometry for a given direction, one may be interested in the size distribution for all possible directions simultaneously so as to estimate the distribution of the length of elongated structures independently of their orientation. This can be achieved by considering for each size a union of directional openings. However, the selection of a family of line segments ensuring the absorption property is not trivial because, in a discrete grid, the angular resolution increases with the line segment length as illustrated in Fig. 7 [28]. The first two families of line segments leading to a granulometry by union of directional openings are illustrated in Fig. 8.

Discrete disks. In digital image analysis, Euclidean disks are often approximated by either squares, diamonds, or octagons of increasing size. Octagons of increasing size are obtained by starting with the elementary diamond (a pixel and its 4 direct neighbours) and then alternating dilations by the elementary 3×3 square and the elementary diamond.

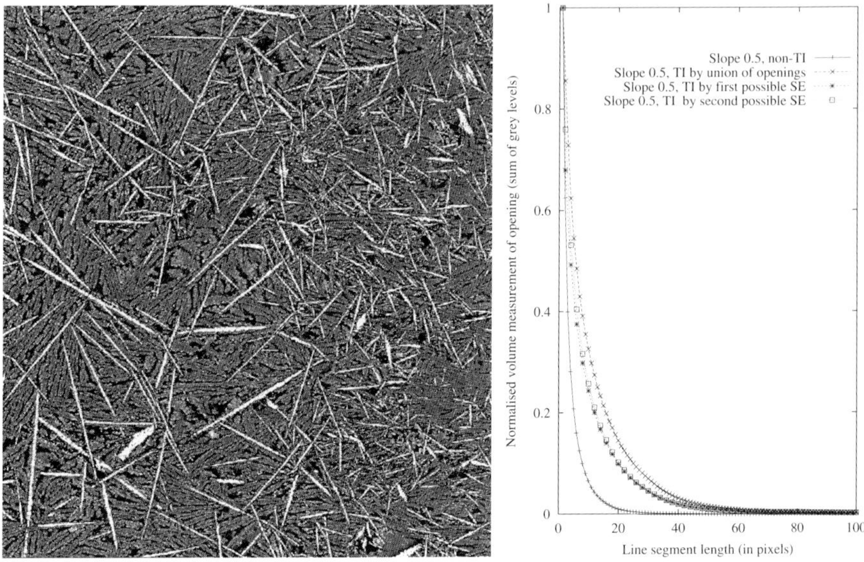

Fig. 6. Effect of the choice of the structuring element for a directional granulometry of slope 0.5 on an electron microprobe image of fibres. The first curve is for non-TI openings along Bresenham lines $\gamma_{BL_{\lambda_1,(2,1)}}$, the second is by considering the TI union of the two possible structuring elements occurring along these lines $\bigvee_{i=1}^{i=k} \gamma_{BL_{\lambda_i,(2,1)}} = \bigvee_{i=1}^{i=k} \gamma_{L_{\lambda_i,(2,1)}}$, while the third and fourth curves show the TI results when using either the first or second pattern. i.e., $\gamma_{L_{\lambda_1,(2,1)}}$ or $\gamma_{L_{\lambda_2,(2,1)}}$ respectively.

Fig. 7. Graphical representation of all possible discrete slopes in the range $[0, 1]$ and for a discrete line segment length less than or equal to 32 pixels. Each slope is obtained by linking the origin (centre of lower left pixel) to the centre of a given dark pixel. Note that this diagram corresponds to a graphical representation of the Farey sequence.

Whether squares, diamonds, or octagons are considered, the corresponding families of structuring elements satisfy the absorption property. However, octagons should be preferred because they are less anisotropic then squares and diamonds.

The design of better discrete approximations of Euclidean disks satisfying the absorption property is not straightforward. For example, discrete disks defined by all grid points falling within a given Euclidean distance calculated from a central point do not from a granulometric family. Indeed, the first approximation is the elementary diamond while the second is the 3×3 square which already contradicts the absorption property because the latter is not opened by the former.

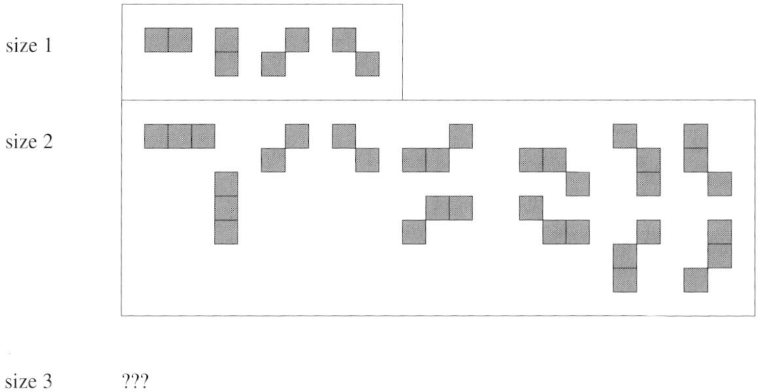

Fig. 8. The construction of a granulometry by considering for each size the union of directional openings is not trivial because one has to make sure that the absorption property is satisfied.

Another approach is to cascade dilations by line segments. In the plane $\mathbb{R}^2$, we have the following result due to Matheron [18]: for $\theta_1, \ldots, \theta_n$ distinct in $[0, \pi)$, $\lambda_1, \ldots, \lambda_n > 0$, then $L_{\lambda_1,\theta_1} \oplus \cdots \oplus L_{\lambda_n,\theta_n}$ is a convex polygon of $2n$ sides whose opposite edges are of length λ_i and have orientation given by θ_i for some i. This idea can be transposed to the discrete space $\mathbb{Z}^2$ for generating granulometric approximations of disks by cascading dilations of carefully selected periodic lines [14]. More precisely, suppose we look for the discrete approximation of an Euclidean disk using cascades of line segments whose Euclidean length equals l. We then consider Fig. 7 and look for all slopes defined for this length and use the corresponding periodic lines with the maximal number of pixels so that the length of the corresponding Euclidean line segment does not exceed l. We denote by B_l the corresponding approximation. For example, we obtain the following cascades for the eight first approximations depicted in Fig. 9:

$$\begin{aligned}
B_1 &= P_{2,(1,0)} \oplus P_{2,(0,1)}, \\
B_{\sqrt{2}} &= P_{2,(1,0)} \oplus P_{2,(0,1)} \oplus P_{2,(1,1)} \oplus P_{2,(1,-1)}, \\
B_2 &= P_{3,(1,0)} \oplus P_{3,(0,1)} \oplus P_{2,(1,1)} \oplus P_{2,(1,-1)}, \\
B_{\sqrt{5}} &= P_{3,(1,0)} \oplus P_{3,(0,1)} \oplus P_{2,(1,1)} \oplus P_{2,(1,-1)} \\
&\quad \oplus P_{2,(2,1)} \oplus P_{2,(2,-1)} \oplus P_{2,(1,2)} \oplus P_{2,(1,-2)}, \\
B_{2\sqrt{2}} &= P_{3,(1,0)} \oplus P_{3,(0,1)} \oplus P_{3,(1,1)} \oplus P_{3,(1,-1)} \\
&\quad \oplus P_{2,(2,1)} \oplus P_{2,(2,-1)} \oplus P_{2,(1,2)} \oplus P_{2,(1,-2)}, \\
B_3 &= P_{4,(1,0)} \oplus P_{4,(0,1)} \oplus P_{3,(1,1)} \oplus P_{3,(1,-1)} \\
&\quad \oplus P_{2,(2,1)} \oplus P_{2,(2,-1)} \oplus P_{2,(1,2)} \oplus P_{2,(1,-2)}, \\
B_{\sqrt{10}} &= P_{4,(1,0)} \oplus P_{4,(0,1)} \oplus P_{3,(1,1)} \oplus P_{3,(1,-1)} \\
&\quad \oplus P_{2,(2,1)} \oplus P_{2,(2,-1)} \oplus P_{2,(1,2)} \oplus P_{2,(1,-2)} \\
&\quad \oplus P_{2,(3,1)} \oplus P_{2,(3,-1)} \oplus P_{2,(1,3)} \oplus P_{2,(1,-3)}
\end{aligned}$$

$$\begin{aligned} B_{\sqrt{13}} = & P_{4,(1,0)} \oplus P_{4,(0,1)} \oplus P_{3,(1,1)} \oplus P_{3,(1,-1)} \\ & \oplus P_{2,(2,1)} \oplus P_{2,(2,-1)} \oplus P_{2,(1,2)} \oplus P_{2,(1,-2)} \\ & \oplus P_{2,(3,1)} \oplus P_{2,(3,-1)} \oplus P_{2,(1,3)} \oplus P_{2,(1,-3)} \\ & \oplus P_{2,(3,2)} \oplus P_{2,(3,-2)} \oplus P_{2,(2,3)} \oplus P_{2,(2,-3)} . \end{aligned}$$

B_1 $B_{\sqrt{2}}$ B_2 $B_{\sqrt{5}}$ $B_{2\sqrt{2}}$ B_3 $B_{\sqrt{10}}$ $B_{\sqrt{13}}$

Fig. 9. Discrete disks of increasing size using cascades of dilations with periodic lines. Each disk is symmetric, discrete convex, and is opened for all disks of smaller size: $\gamma_{B_i}(B_j) = B_j$ for all $i \leq j$ (see text for the definition of each B_i).

Note that, by construction, these disks form a granulometry with size parameter given by l. In contrast, radial decompositions using Bresenham lines [1] cannot be used to generate a granulometry. However, due to the discrete nature of $\mathbb{Z}^2$, it is impossible to generate disks of arbitrary diameter by a cascade of periodic lines. A similar conclusion was reached in the context of disks generated by dilation cascades of the extremity points of line segments defined by the successive radii of a discrete disk [40].

4 Orientation of Directional Structures

Natural textures such as eutectic lamellars, fingerprints, or copolymers imaged by scanning force microscopy devices (see contribution by Magerle [17] in this volume) display directional textures. A characterisation of the directional information of these textures can be achieved by applying a series of directional filters to the input texture and measuring, for each direction, the sum of the grey levels of the filtered image. This approach is *global* in the sense that it tells us whether there are some predominant orientations without indicating the position of the corresponding oriented image structures. This latter task is the goal of *local* orientation measurements whereby each image point is given a value indicating the orientation of the image structure it belongs to. Both approaches are detailed hereafter.

4.1 Global Orientation Information

While granulometries provide information about the size (or length) of predominant image structures, global orientation procedures tell us about the predominant orientation of oriented image structures. The amount of directional information in a given direction can be estimated by computing the volume (i.e., sum of grey levels) of the image transformed by a directional structuring element in this direction. Predominant orientations appear as peaks in the curve displaying the measured volume for a varying orientation. The corresponding polar diagram is referred to as a *rose of directions*.

Two types of directional structuring elements are considered in practice. The first is simply a pair of points separated by a fixed distance in a given direction. For example, Fig. 10a shows a binary texture with parallel lines along two principal directions. Area measurements of the image eroded by the pair of points versus the direction of these pairs of points are shown in Fig. 10b. When the pair of points is in the same direction as the grid lines, the image is not as much eroded as for other directions. Therefore, the directions corresponding to the two peaks of the diagram clearly indicate the directions of the lines.

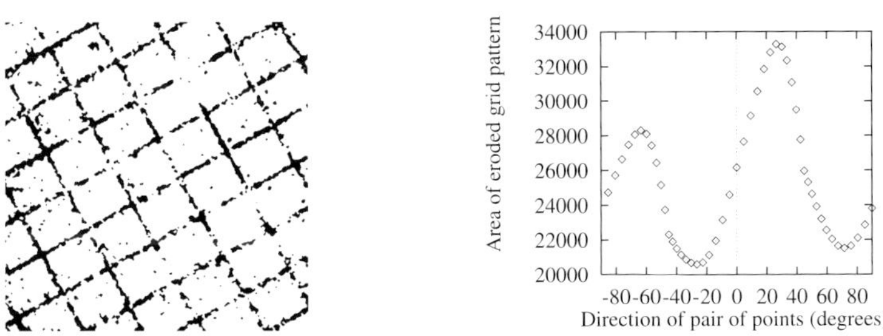

(a) A binary image with lines in two directions.

(b) Measure of the erosion of *(a)* using a pair of points at fixed distance as SE.

Fig. 10. The measure of the area of the original image eroded by a pair of points in various directions can be used for determining the two main directions of the grid.

Note that dilations by a pair of points should be performed instead of erosions in situations where the directional structures are darker than their background neighbourhood.

The second type of structuring element consists of line segments. They are better suited to textures containing connected directional patterns such as those appearing in the scanning force microscopy image of a block copolymer thin film displayed in Fig. 11a. Directional openings in the principal directions of the square grid are shown in Figs. 11b–e. The rose of directions showing the distribution of the normalised volume measurement of the directional openings is displayed in Fig. 12 together with those obtained for longer line segments. This figure reveals the anisotropy of the structures appearing in the input image (predominant horizontal orientation).

The rose of directions can also be used for determining whether two oriented textures contain the same distribution of oriented structures. Indeed, their rose of directions should be identical. Similarly, rose of directions computed within local windows can be helpful for segmenting images containing textures with varying orientation distributions.

4.2 Local Orientation Information (Orientation Field)

Principle. In contrast to the rose of directions, the local orientation information is assessed at the level of each individual pixel. Similarly to the rose of directions by

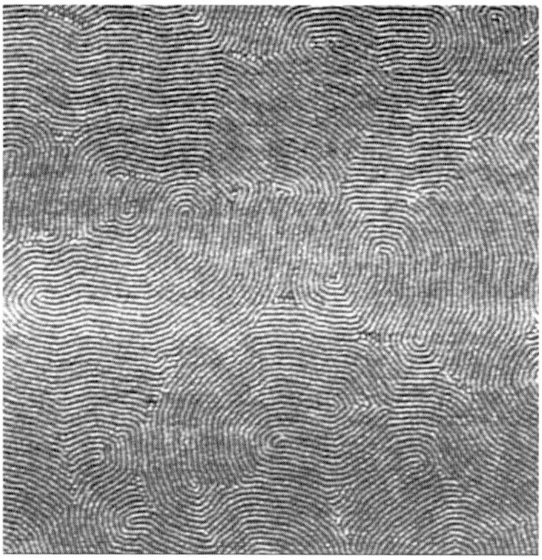

(a) Input image

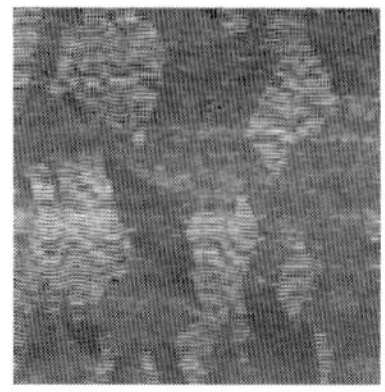

(b) Horizontal opening

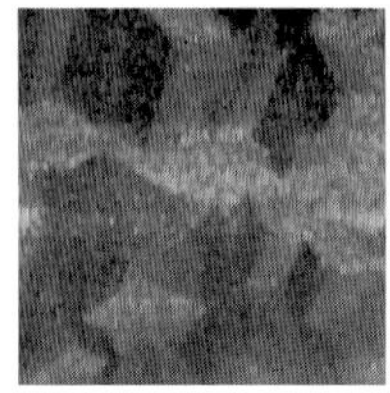

(c) Vertical opening

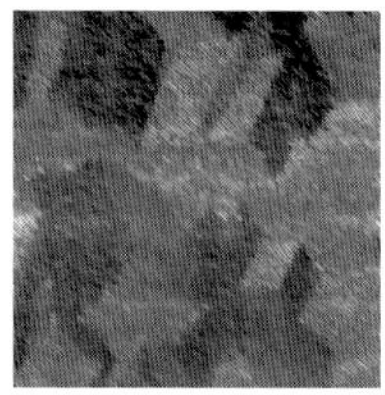

(d) Opening at 45°

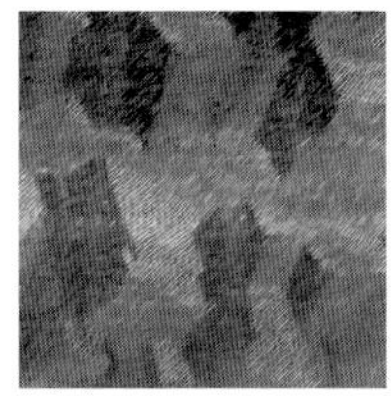

(e) Opening at -45°

Fig. 11. (a) Tapping mode scanning force microscopy image of a block copolymer thin film [Courtesy of A. Knoll, see also Chap. 4 in this volume [17]]. (b)–(e) Directional openings with 7 pixel long line segments.

Fig. 12. Roses of direction by opening representing the polar distribution of the normalised volume measurement of the image displayed in Fig. 11a and opened by line segments. The outermost distribution is for a length of 7 pixels, the middle one for a length of 11 pixels, and the innermost for a length of 15 pixels.

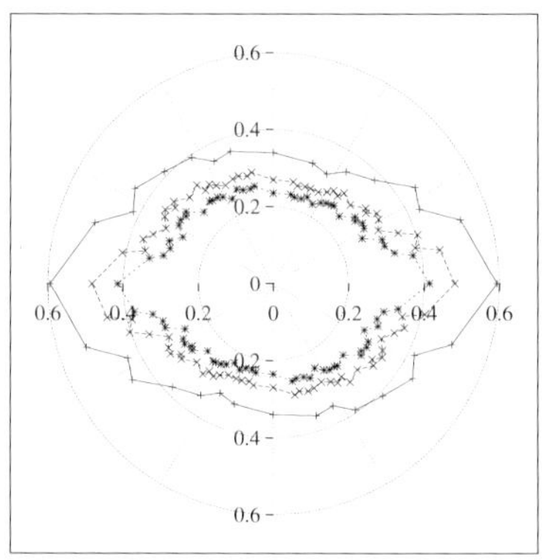

opening, the image is first transformed by a series of directional openings. However, rather than measuring the volume of the output images, we analyse the evolution of the grey scale value of the transformed images for each individual pixel [32]. Indeed, if a pixel belongs to a directional structure, it will be preserved for a directional opening in the direction of this structure but removed when considering other directions. The directional signature of a pixel displays the evolution of the grey scale values of the pixel for a series of directional filters (see example in Fig. 13). That is, the highest value of the directional signature of a pixel reveals the strongest direction occurring at this pixel, the strength of the direction being defined as the difference between the maximum and minimum value of the directional signature.

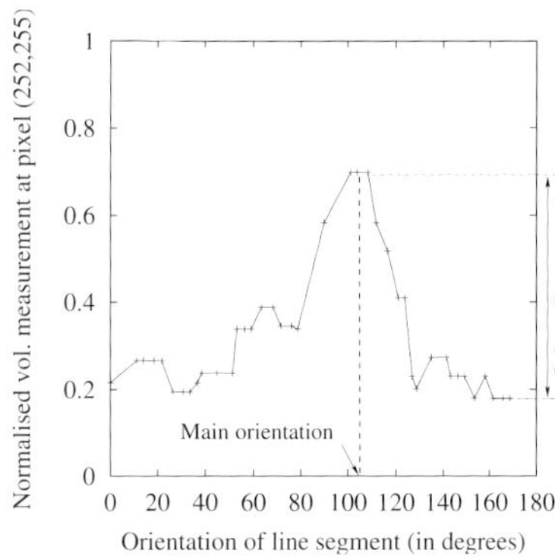

Fig. 13. Directional signature of a pixel located in a bright directional structure at the centre of the image shown in Fig. 11a. The normalised grey scale values occurring at this pixel have been calculated for a series of directional openings by a line segments of 11 pixels. The peak indicates that the main orientation at this pixel is about 104°. The contrast between the maximum and minimum value of the signature defines the orientation strength of this pixel.

In practice, directional openings are suited to the determination of the directional signature of bright structures while closings should be used for dark structures. Let us now translate all these concepts into formal definitions. We define the orientation by opening at a given image pixel $\mathbf{x}$ and for a given scale λ as the orientation θ of the directional morphological opening of length λ which modifies the least the original image value at position $\mathbf{x}$. We denote the orientation by opening dir^γ, the orientation by closing dir^ϕ being defined by duality:

$$\left[\mathrm{dir}^\gamma_\lambda(f)\right](\mathbf{x}) = \{\theta_i \mid \left[\gamma_{L_{\lambda,\theta_i}}(f)\right](\mathbf{x}) \geq \left[\gamma_{L_{\lambda,\theta_j}}(f)\right](\mathbf{x}),\ \forall\,\theta_i \neq \theta_j\}, \tag{1}$$

$$\left[\mathrm{dir}^\phi_\lambda(f)\right](\mathbf{x}) = \{\theta_i \mid \left[\phi_{L_{\lambda,\theta_i}}(f)\right](\mathbf{x}) \leq \left[\phi_{L_{\lambda,\theta_j}}(f)\right](\mathbf{x}),\ \forall\,\theta_i \neq \theta_j\}, \tag{2}$$

where the possible discrete orientations θ depend on the line segment length as already discussed in Sect. 3.4. If all orientations output the same value, it means that there is no orientation for the chosen structuring length. An arbitrary orientation can then be selected because the strength of the orientation (see below) will be zero. However, for pixels belonging to wide (with respect to the length λ of the structuring element) oriented objects of constant grey tone, several distinct orientations may output the maximum value. Assuming that the range of orientations at the maximum value is connected, we break the tie by selecting the middle orientation. If the range is not connected, it means that there are oriented structures crossing each other. The *strength* gdir^γ (resp. gdir^ϕ) of the calculated orientation at position $\mathbf{x}$ is defined as the difference between the maximum and minimum output of the directional openings (resp. closings) at this position:

$$\left[\mathrm{gdir}^\gamma_\lambda(f)\right](\mathbf{x}) = \Big[\bigvee_{\theta_i} \gamma_{L_{\lambda,\theta_i}}(f)\Big](\mathbf{x}) - \Big[\bigwedge_{\theta_i} \gamma_{L_{\lambda,\theta_i}}(f)\Big](\mathbf{x}), \tag{3}$$

$$\left[\mathrm{gdir}^\phi_\lambda(f)\right](\mathbf{x}) = \Big[\bigvee_{\theta_i} \phi_{L_{\lambda,\theta_i}}(f)\Big](\mathbf{x}) - \Big[\bigwedge_{\theta_i} \phi_{L_{\lambda,\theta_i}}(f)\Big](\mathbf{x}). \tag{4}$$

When comparing the values of gdir^γ and gdir^ϕ of a given pixel, it is possible to detect whether it belongs to a positive or negative image structure: positive, if $\mathrm{gdir}^\gamma > \mathrm{gdir}^\phi$, negative, otherwise. We denote by gdir the pointwise maximum between the images gdir^γ and gdir^ϕ: $\mathrm{gdir} = \mathrm{gdir}^\gamma \vee \mathrm{gdir}^\phi$. The image of directions dir is then defined as follows:

$$[\mathrm{dir}_\lambda(f)](\mathbf{x}) = \begin{cases} [\mathrm{dir}_\lambda^\gamma(f)](\mathbf{x}), \text{ if } [\mathrm{gdir}_\lambda(f)](\mathbf{x}) = [\mathrm{gdir}_\lambda^\gamma(f)](\mathbf{x}), \\ [\mathrm{dir}_\lambda^\phi(f)](\mathbf{x}), \text{ otherwise.} \end{cases} \tag{5}$$

A colour representation of the local orientation information is then simply achieved by equating (i) the orientation information (i.e., either dir^γ, dir^ϕ, or dir) to the hue and (ii) the strength of the orientation information (i.e., either gdir^γ, gdir^ϕ, or gdir) to the lightness component of the image, the colours being fully saturated in all cases. The three local orientation images (also called orientation fields) of the copolymers displayed in Fig. 11 are shown in Fig. 14. A filtering of the resulting directional fields is required for

(a) Orientation field by opening: dir^γ. (b) Orientation field by closing: dir^ϕ. (c) Combined orientation field: dir.

Fig. 14. Local orientation fields of the image of block copolymer thin film displayed in Fig. 11a. A structuring element length of 11 pixels has been considered in this experiment.

suppressing noisy pixels occurring within regions of homogeneous orientation and due to irregularities of the boundary of the input patterns.

Applications. Local orientation maps can be used to segment oriented textures into zones of homogeneous orientation. At its simplest level, this idea can be used for discriminating directional from non-directional textures. Note that the contrast of the oriented structure is taken into account: the higher the contrast of an oriented structure, the higher the value of the intensity component in the colour representation of the orientation map. By analyzing the orientation content within a moving window, boundaries between different textures can be enhanced.

5 Some Other Techniques

In this section, we briefly survey additional morphological techniques useful for texture analysis. A detailed introduction to the discussed morphological transformations can be found in [27].

5.1 Morphological Covariance

Principle. The morphological *covariance* K of an image f consists in measuring the volume of the image eroded by a pair of points $P_{2,\boldsymbol{v}}$, i.e., two points separated by a vector $\boldsymbol{v}$:

$$K(f; P_{2,\boldsymbol{v}}) = V[\varepsilon_{P_{2,\boldsymbol{v}}}(f)]. \tag{6}$$

The covariance is interpreted by looking at the evolution of the volume measurement for an increasing vector length. In the case of binary images, the covariance is nothing but the area A of the intersection of the image X with its translation by the vector $\boldsymbol{v}$: $K(X; P_{2,\boldsymbol{v}}) = A(X \cap X_{\boldsymbol{v}})$. Therefore, in the binary case, the morphological covariance is equivalent to the autocorrelation function. The proposed definition for grey scale images comes down to the binary covariance of the subgraph of the image (i.e., sum of the area of the erosion by a pair of point of each cross-section of the image). In contrast to the definition adopted here, the covariance for grey scale images is often defined as the autocorrelation function, e.g., [23, p. 273].

Applications. Depending on whether a texture is coarse or thin, its covariance at the origin will drop off slowly or quickly. In addition, the covariance can be used for automatically determining the period of periodic textures. Indeed, peaks of the covariance are located at a multiple of the distance separating the texture elements (in the direction defined by the pair of points). Information about the size of the periodic texture elements is proportional to the width of the peaks: sharp peaks indicate thin texture patterns and vice versa. For example, the sum of the grey levels for erosions by points separated by an increasing distance along the horizontal direction are shown in Fig. 15. The anisotropy of directional textures can be highlighted by the directional dependence of the covariance, i.e., by plotting the "Volume of erosion by a pair of equidistant points versus the orientation of these points". This latter principle has already been detailed when introducing the concept of rose of directions in Sect. 4.1 (see example in Fig. 10). In [20], the average of vertical and horizontal covariances is used for characterising the texture of stone surfaces in historic buildings.

Rather than displaying in a 1-D diagram volume measurements when varying either the orientation or the distance between the pair of points, a 2-D diagram rendering all possible combinations can be generated as illustrated in Fig. 16.

5.2 Multiscale Surface Area Measurements

Principle. Minkowski suggested to measure the length L of a curve $\mathcal{C}$ of $\mathbb{R}^2$ by replacing each point on the curve by a disc B of diameter λ and dividing the area of the resulting ribbon by λ. Replacing each point on the curve by a disc B_λ comes down to performing the dilation of the curve with this disc: $\mathsf{L}_\lambda(\mathcal{C}) = A(\mathcal{C} \oplus B_\lambda)/\lambda$. This technique can be extended to the graph (intensity surface) of a 2-D image by dilating it with a sphere of increasing size. In practice however, the intensity surface of an image has luminance units along the grey scale axis and spatial units along the image plane. To avoid mixing up these units, dilations by disks rather than spheres must be considered. The surface area at scale λ of the graph G of a 2-D grey scale image f is then defined as the volume

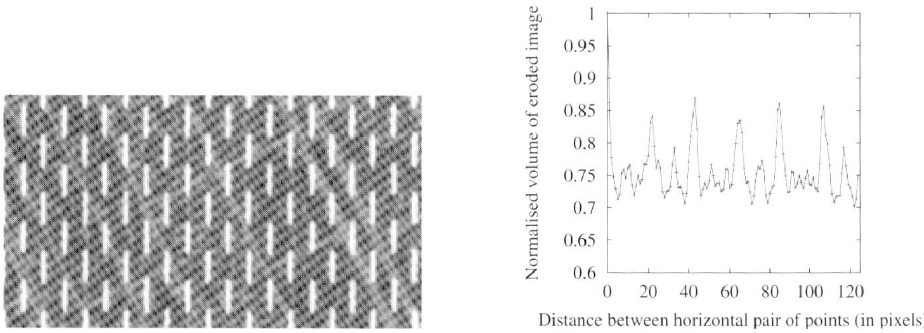

Fig. 15. Use of covariance for determining the periodicity of periodic texture elements. Left: input image showing a textured region. Right: normalised sum of grey levels of erosions by the horizontal pair of points $P_{2,(i,0)}$ versus i (i denotes the distance separating the pair of points).

(a) For the texture shown in Fig. 15.

(b) For the copolymer film shown in Fig. 11a.

Fig. 16. Two-dimensional diagrams of the morphological covariance rendering the normalised volume measurements of the erosion by an arbitrary pair of point.

of the of the pointwise difference between the dilation δ and the erosion ε with a disk B of diameter λ as SE, this latter operation being referred to as the morphological gradient operation ρ, i.e., $\rho = \delta - \varepsilon$:

$$A_\lambda[G(f)] = \frac{V[\rho_{B_\lambda}(f)]}{\lambda}. \tag{7}$$

Applications. The measurements "surface area vs. scale" is usually plotted in a log-log diagram. If the texture under study is statistically self-similar, the plotted values should follow a line whose slope equals $2 - D$, D denoting the fractal dimension. However, in most practical applications, a linear relationship is not observed. This indicates that the texture does not satisfy the self-similarity assumption. In this latter case, the whole diagram should be used for featuring the texture [31].

5.3 Extrema Analysis and Watershed Segmentation

A characteristic of the human visual system is that important texture information is contained in the relative frequency of local extrema as well as the shape of their influence zones. By viewing a grey tone image as a topographic surface, it becomes apparent that the influence zone of an image minimum corresponds to the concept of *catchment basin* known in hydrology. Therefore, the set of all catchment basins of a grey tone image provides us with a partition of its definition domain. The morphological *watershed transformation* is precisely extracting the boundaries of the influence zones (catchment basins) of all image minima. The shape and size of these influence zones are sometimes discriminant texture features that can be used for segmenting an image into regions of homogeneous texture. For example, in [3], the size distribution in area and volume of the catchment basins of an image and its complement is suggested as a potential signature for texture characterisation.

Texture segmentation based on watershed transformation is illustrated in Fig. 17 on a tapping mode scanning force microscopy image of a block copolymer thin film [15] (see also the contribution by Magerle [17] in this volume).

From a visual inspection of this image, three types of microdomains are clearly visible, each with a different texture. We propose to extract the microdomains corresponding to hexagonally ordered dark and bright spots using the watershed paradigm. Because the watershed transformation produces one catchment basin per image minimum, it is of paramount importance to remove irrelevant image minima before applying this transformation (otherwise the watershed transformation produces an oversegmentation). Let us first restrict our attention to the *dark* hexagonally ordered spots. Since they have a relatively high contrast, we can filter them using a depth parameter. In mathematical morphology, this is achieved by the so-called h-minima transformation. The input image filtered by this transformation is shown in Fig. 17b. The watershed transformation of this filtered image is displayed in Fig. 17c. We then use an area parameter for discriminating those catchment corresponding to the influence zones of dark spots from all other catchment basins. The selected catchment basins are shown in Fig. 17d. A closing by a 2×2 is then used to connect neighbour catchment basins, as depicted Fig. 17e. Isolated catchment basins are then removed using another area threshold. The resulting mask of the microdomains corresponding to dark hexagonally ordered spots is displayed in Fig. 17f. Because the watershed transformation extracts the influence zones of the image minima, we need to consider the complement (also called negative) of the input image for extracting the microdomains corresponding to *bright* hexagonally ordered spots appearing in the input image. After complementation, we apply the same procedure as that described for dark spots. This procedure leads to the mask shown in Fig. 17g. The resulting segmentation of the input image in regions of homogeneous texture appears in Fig. 17h.

6 Concluding Remarks and Summary Table

Mathematical morphology provides us with a wide variety of transformations for featuring textured objects. They are summarised in Table 1. All techniques can be used for both classification and segmentation purposes (by applying them within local windows). The criteria shape, size, orientation, and periodicity concern ordered textures. The second last column indicates whether the method is useful for processing disordered textures. The last column concerns features related to the connectivity of the texture elements

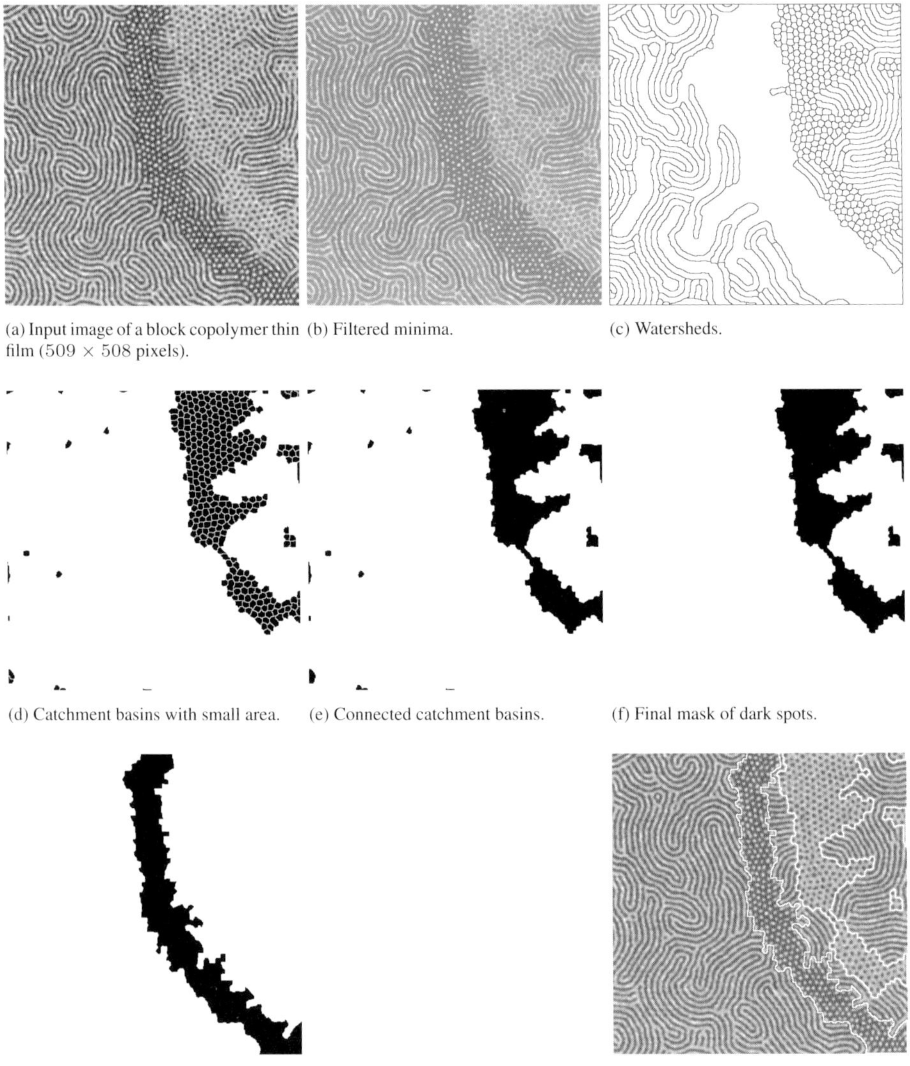

(a) Input image of a block copolymer thin film (509 × 508 pixels). (b) Filtered minima. (c) Watersheds.

(d) Catchment basins with small area. (e) Connected catchment basins. (f) Final mask of dark spots.

(g) Final mask of bright spots. (h) Final segmentation.

Fig. 17. Texture segmentation based on the analysis of the shape of the catchment basins extracted by the watershed transformation. The input image is from [15], see also Chap. 4 of the contribution by R. Magerle in in this volume [17]. [Full explanation in text].

(local measurements of the connectivity number computed with the morphological grey scale hit-or-miss transform [29] could also be used).

In general, morphological texture characterisation techniques are not sufficient for discriminating all possible textures. Therefore, real applications dealing with the analysis of texture images may require other techniques to be used in conjunction with, or even in replacement of, morphological techniques. The choice between one technique or another

Table 1. Synoptic table indicating the suitability of a given method for extracting morphological features of a texture.

Method	Shape	Size	Orientation	Periodicity	Disorder (granularity)	Connectivity
Granulometry	yes	yes	yes	no	yes	no
Rose of directions	no	yes[a]	yes	no	no	no
Orientation map	no	yes[a]	yes	no	no	no
Morphological covariance	yes	yes	yes	yes	yes	no
Surface area	no	no	no	no	yes	no
Watersheds	yes	yes	yes	no	yes	yes

[a] Length

should be guided by an analysis of the discriminant efficiency of each computed texture feature. A general table indicating what feature is best suited to what class of texture is out of the scope of this introduction to morphological texture analysis.

References

1. R. Adams. Radial decomposition of discs and spheres. *Computer Vision, Graphics, and Image Processing: Graphical Models and Image Processing*, 55(5):325–332, September 1993.
2. Anonymous. *Webster's Encyclopedic Unabridged Dictionary*. Gramercy Books, New York, 1994.
3. A. Aubert, D. Jeulin, and R. Hashimoto. Surface texture classification from morphological transformations. In J. Goutsias, L. Vincent, and D. Bloomberg, editors, *Mathematical Morphology and its Applications to Image and Signal Processing*, pages 253–262, Boston, 2000. Kluwer Academic Publishers.
4. C. Bhagvati, D. Grivas, and M. Skolnick. Gaussian normalization of morphological opening distributions to increase their sensitivity to texture variations and its application in pavement distress assessment. In *International Conference on Computer Vision and Pattern Recognition*, pages 800–804, Seattle, Washington, USA, June 1994.
5. J. Bresenham. Algorithm for computer control of digital plotter. *IBM System Journal*, 4: 25–30, 1965.
6. P. Brodatz. *Textures: A photographic album for artists and designers*. Dover Publications, New York, 1966.
7. Y. Chen and E. Dougherty. Gray-scale morphological granulometric texture classification. *Optical Engineering*, 33(8):2713–2722, August 1994.
8. Y. Chen, E. Dougherty, S. Totterman, and J. Hornak. Classification of trabecular structure in magnetic resonance images based on morphological granulometries. *Magnetic Resonance in Medicine*, 29:358–370, 1993.
9. E. Dougherty and J. Pelz. Morphological granulometric analysis of electrophotographic images – size distribution statistics for process control. *Optical Engineering*, 30(4):438–445, 1991.
10. E. Dougherty, J. Pelz, F. Sand, and A. Lent. Morphological image segmentation by local granulometric size distributions. *Journal of Electronic Imaging*, 1(1):40–60, January 1992.

11. R. Duda and P. Hart. *Pattern classification and scene analysis*. John Wiley & Sons, New York, 1973.
12. R. Haralick. Statistical and structural approaches to texture. *Proceedings of the IEEE*, 67 (5):768–804, May 1979.
13. H. Heijmans. *Morphological image operators*. Advances in Electronics and Electron Physics Series. Academic Press, Boston, 1994.
14. R. Jones and P. Soille. Periodic lines: Definition, cascades, and application to granulometries. *Pattern Recognition Letters*, 17(10):1057–1063, September 1996. `http://www.elsevier.nl/gej-ng/10/35/61/13/12/18/abstract.html`
15. A. Knoll, A. Horvat, K. Lyakhova, G. Krausch, G. Sevink, A. Zvelindovsky, and R. Magerle. Phase behavior in thin films of cylinder forming block copolymers. *Physical Review Letters*, 2002. Submitted.
16. W. Li, V. Haese-Coat, and J. Ronsin. Residues of morphological filtering by reconstruction for texture classification. *Pattern Recognition*, 30(7):1081–1093, 1997.
17. R. Magerle. Nanotomography: Real-space volume imaging with scanning probe microscopy. *Lecture Notes in Physics*, This volume, 2002.
18. G. Matheron. *Random sets and integral geometry*. Wiley, New York, 1975.
19. J. Mattioli and M. Schmitt. On information contained in the erosion curve. In Y.-L. O, A. Toet, D. Foster, H. Heijmans, and P. Meer, editors, *Shape in Picture: Mathematical Description of Shape in Grey-level Images*, pages 177–195. Springer-Verlag, 1994.
20. A. Mauricio and C. Figueiredo. Texture analysis of grey-tone images by mathematical morphology: A nondestructive tool for the quantitative assessment of stone decay. *Mathematical Geology*, 32(5):619–642, 2000.
21. A. Rao. *A taxonomy for texture description and identification*. Springer-Verlag, New York, 1990.
22. T. Reed and J. du Buf. A review of recent texture segmentation and feature extraction techniques. *Computer Vision and Image Understanding*, 57(3):359–372, May 1993.
23. J. Serra. *Image analysis and mathematical morphology*. Academic Press, London, 1982.
24. J. Serra, editor. *Image analysis and mathematical morphology. Volume 2: Theoretical advances*. Academic Press, London, 1988.
25. K. Sivakumar and J. Goutsias. Discrete morphological size distributions and densities: estimation techniques and applications. *Journal of Electronic Imaging*, 6(1):31–53, January 1997.
26. P. Soille. *Morphologische Bildverarbeitung*. Springer-Verlag, Berlin, Heidelberg, 1998. ISBN 3-540-64323-0. URL `http://web.ukonline.co.uk/soille/buch1st`.
27. P. Soille. *Morphological image analysis*. Springer-Verlag, Berlin, New York, 1999. URL `http://ams.egeo.sai.jrc.it/soille/book1st`.
28. P. Soille. Morphological operators with discrete line segments. *Lecture Notes in Computer Science*, 1953:78–98, 2000.
29. P. Soille. Advances in the analysis of topographic features on discrete images. *Lecture Notes in Computer Science*, 2301:175–186, March 2002.
30. P. Soille, E. Breen, and R. Jones. Recursive implementation of erosions and dilations along discrete lines at arbitrary angles. *IEEE Transactions on Pattern Analysis and Machine Intelligence*, 18(5):562–567, May 1996.
31. P. Soille and J.-F. Rivest. On the validity of fractal dimension measurements in image analysis. *Journal of Visual Communication and Image Representation*, 7(3):217–229, September 1996. URL `http://www.idealibrary.com/links/artid/jvci.1996.0020`.
32. P. Soille and H. Talbot. Directional morphological filtering. *IEEE Transactions on Pattern Analysis and Machine Intelligence*, 23(11):1313–1329, November 2001. URL `http://computer.org/tpami/tp2001/i1313abs.htm`.

33. D. Stoyan and H. Stoyan. *Fractals, random shapes, and point fields*. John Wiley & Sons, Chichester, 1994.
34. L. Van Gool, P. Dewaele, and A. Oosterlinck. Texture analysis anno 1983. *Computer Vision, Graphics, and Image Processing*, 29:336–357, 1985.
35. M. Vanrell and J. Vitrià. Mathematical morphology, granulometries, and texture perception. In E. Dougherty, P. Gader, and J. Serra, editors, *Image algebra and morphological image processing IV*, volume SPIE-2030, pages 152–161, July 1993.
36. P. Wagner. Texture analysis. In B. Jähne, H. Haußecker, and P. Geißler, editors, *Handbook of Computer Vision and Applications*, volume 2, chapter 12, pages 275–308. Academic Press, San Diego, 1999.
37. D. Wang, V. Haese-Coat, A. Bruno, and J. Ronsin. Texture classification and segmentation based on iterative morphological decomposition. *Journal of Visual Communication and Image Representation*, 4(3):197–214, September 1993.
38. H. Wechsler. Texture analysis: A survey. *Signal Processing*, 2:271–282, 1980.
39. X. Zheng, P. Gong, and M. Strome. Characterizing spatial structure of tree canopy using colour photographs and mathematical morphology. *Canadian Journal of Remote Sensing*, 21(4):420–428, 1995.
40. Z. Zhou and A. Venetsanopoulos. Analysis and implementation of morphological skeleton transforms. *Circuits Systems Signal Process*, 11(1):253–280, 1992.

Vector- and Tensor-Valued Descriptors for Spatial Patterns

Claus Beisbart[1,2], Robert Dahlke[2], Klaus Mecke[3,4], and Herbert Wagner[2]

[1] University of Oxford, Nuclear & Astrophysics Laboratory, Keble Road, Oxford OX1 3RH, U.K.
[2] Ludwig-Maximilians-Universität, Sektion Physik, Theresienstraße 37, D-80333 München, Germany
[3] Max-Planck-Institut für Metallforschung, Heisenbergstr. 1, D-70569 Stuttgart, Germany
[4] Institut für Theoretische und Angewandte Physik, Fakultät für Physik, Universität Stuttgart, Pfaffenwaldring 57, D-70569 Stuttgart, Germany
e-mails: beisbart/dahlke/wagner@theorie.physik.uni-muenchen.de, mecke@fluids.mpi-stuttgart.mpg.de

Abstract. Higher-rank Minkowski valuations are efficient means of describing the geometry and connectivity of spatial patterns. We show how to extend the framework of the scalar Minkowski functionals to vector- and tensor-valued measures. The versatility of these measures is demonstrated by using simple toy models as well as real data.

1 Introduction

The spatial patterns originating from the polymorphic aggregation of matter in Nature occur on vastly different length scales and with unlimited variety. For instance, the pattern shown in Fig. 1 could represent a biological structure obtained with X-ray tomography, or a colloidal cluster in a milky emulsion. In fact, Fig. 1 displays an isodensity contour of cosmic matter as traced by the galaxies surrounding the Milky way, which lies in the center of the plot.

The visual impression from such pictures may already convey valuable insight into the dynamic evolution or – in the context of biology – into the functionality of such structures. However, for unbiased inferences as well as for the comparison with model simulations, one needs objective and quantitative measures to characterize the geometry and topology of typical structural motifs.

In this article we deal with a family of functions which provide a unique description of the morphology displayed by spatial patterns. We call these functions the Minkowski valuations; they are extensions of the well-known Minkowski functionals often mentioned within this volume (see the contributions by C. Arns et al. and M. Lösche and P. Krüger in this volume as well as [25] in the previous volume LNP 554). Therefore they share some of the simplicity and beauty which the Minkowski functionals own. In particular, they are characterized by simple axioms which set intuitively reasonable standards for a morphological description. In the same way as the Minkowski functionals generalize the notion of the volume, the Minkowski valuations extend the notion of the mass vector (or center of mass) and of the inertia tensor. Like their scalar relatives the majority of the Minkowski valuations are curvature measures integrated over the surface of a pattern.

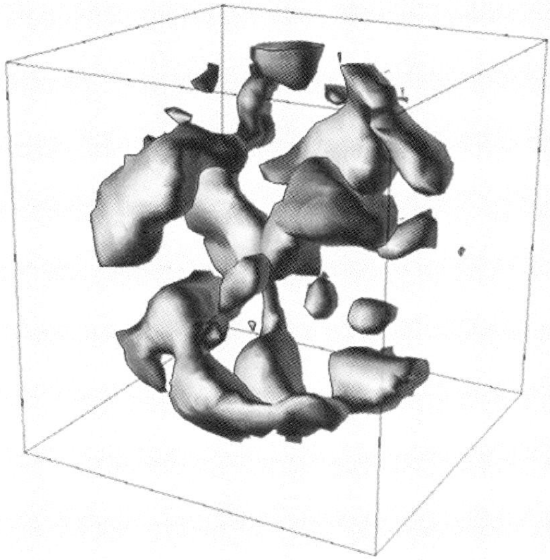

Fig. 1. The matter distribution around our galaxy as traced by the PSCz survey. The galaxy number density is smoothed; one isodensity contour is shown (from the PSCz homepage at http://star-www.dur.ac.uk/cosmology/theory/pscz.html).

We start with a short motivation in Sect. 2 by showing that there are good reasons to go beyond the scalar Minkowski functionals. After having outlined ways to generalize the Minkowski functionals (Sect. 3), we review some of the extensions' mathematical properties in Sect. 4. To illustrate how useful the Minkowski valuations are, we discuss concrete applications covering submolecular up to cosmological scales (Sects. 5 and 6).

Before, two cautionary remarks are appropriate: first, in order to distinguish conceptually between the scalar Minkowski functionals and their higher-rank extensions, we shall address the whole entirety of Minkowski functionals and their extensions via the term "Minkowski valuations" (MVs), whereas "Minkowski functionals" (MFs) is reserved to the scalar Minkowski functionals as before. Secondly, there is a vivid theoretical development of higher-rank Minkowski valuations these days [44, 45, 9]. Therefore, this overview can not provide a complete survey on Minkowski valuations; on the contrary, substantial progress is to be expected both from the mathematical side as well as from applications. For instance, morphological analysis based on Minkowski valuations may be useful to quantify molecular orientations (see the contributions by F. Schmid and N. H. Phuong, U. S. Schwarz and G. Gompper, D. Jeulin, C. Arns et al., and H.-J. Vogel in this volume). Our vector-valued descriptors may also be used to refine the morphological analysis of Langmuir monolayer phases (presented by M. Lösche and P. Krüger in this volume) which has so far been based merely on scalar MFs. Another application may cover the geometric characterization of liquid foams described by F. Graner in this volume, where stress tensors need to be related to a complex spatial structure.

2 Going Beyond Scalar Minkowski Functionals – A Physical Motivation

The following problem is typical for challenges occurring in physics: the formation of galaxy clusters depends in a complicated way on the background cosmology usually described in terms of a cosmological model. Such a model is defined through the values of the cosmological parameters and the power spectrum of the initial density perturbations [31, 6]. In order to investigate the impact of the background cosmology on galaxy clusters one carries out gravitational N-body simulations; typical results comprise cluster images such as those shown in Fig. 2. To compare the models and to check which model fits best present-day cluster observations, one has to use quantitative and discriminative morphological measures.

Fig. 2. The projected matter distribution of two simulated galaxy clusters (from the GIF-project). Their different morphologies are due to different values of the cosmological parameters and to different initial power spectra imposed (left panel: τ Cold Dark Matter model; right panel open Cold Dark Matter model, see [6] for more details). Apart from that, however, the clusters stem from comparable initial conditions. In order to distinguish between both models we need a thorough morphological description.

The Minkowski functionals certainly place practical tools at the physicist's disposal. Defined on the convex ring $\mathcal{K}_d$, i.e. for finite unions of compact, convex sets in d-dimensional Euclidean space, they are the only motion-invariant, convex-continuous, and additive descriptors. Despite this very general characterization there are only a finite number of Minkowski functionals, namely $(d+1)$ in d dimensions (characterization theorem, [16, 21]). They have simple geometrical meanings derived from their representation as integrals. These integrals either extend over the pattern itself to yield the volume, or cover its surface which is weighted with combinations of the local curvatures in order to give – in two dimensions, e.g. – the perimeter and the integrated curvature (which equals the Euler characteristic). The Minkowski functionals also arise quite naturally as expansion coefficients in the Steiner formula [43] specifying the volume of the parallel body of a convex body. Indeed, the Minkowski functionals have been extensively applied to physical structures at vastly different length scales, see [25] for an overview and [20] for astrophysical applications. On the mathematical side, numerous useful results such as the principal kinematic formulae or the Crofton formulae (see [16]) have been proven with integral geometric methods.

However, the morphometry with Minkowski functionals is by no means comprehensive. Combining two of their defining principles, additivity and homogeneity, one can easily see, that, e.g., moving a connected part of a pattern results in significant visible changes which can not be detected by the Minkowski functionals at all. This is illustrated in the first panel of Fig. 3 representing schematic clusters. It is surely of interest, however, in which way the subclumps of a system are arranged, and how they are aligned with respect to each other (see the second panel of Fig. 3). Therefore, we have to move beyond the scalar Minkowski functionals. It turns out that one can generalize the Minkowski functionals without leaving the mathematical framework of integral geometry.

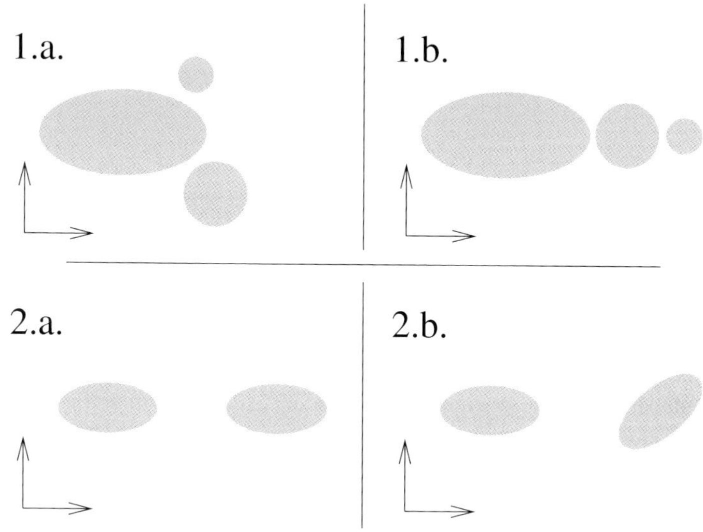

Fig. 3. Neither the schematic clusters displayed in panel 1.a. and 1.b. nor both patterns in panel 2.a. and 2.b. can be distinguished from each other using only the scalar Minkowski functionals. Additivity and motion-invariance prevent the MFs from discriminating between the different positions (upper panels) or different orientations (lower panels) of the subcomponents, respectively.

3 The Hierarchy of Minkowski Valuations – Extending the Framework

There are several ways to extend the framework of the Minkowski functionals. For instance, one may start with Steiner's formula.

The Steiner formula. It is a natural choice to employ the volume $V(K) = \int_K \mathrm{d}V$ of a body K for its very coarse description. If the parallel body K_ϵ is considered, which contains all points closer to K than $\epsilon > 0$, and if K is convex, then one finds from Steiner's formula an expansion for $V(K_\epsilon)$ in powers of ϵ,

$$V\left(K_{\epsilon}\right)=\sum_{\nu=0}^{d}\binom{d}{\nu}\epsilon^{\nu}W_{\nu}(K) \quad , \tag{1}$$

where the Minkowski functionals $W_\nu(K)$ arise as expansion coefficients. The latter functionals are useful morphometric concepts and can be extended to non-convex patterns. In this sense, the volume quite generally leads to a family of related measures.

Since we saw that in some applications (recall the upper panels of Fig. 3) the position of a body may be worth considering, we specify the location of K with its *mass vector*

$$V^{1}\left(K\right)\equiv\int_{K}\mathrm{d}V\mathbf{x} \quad . \tag{2}$$

Alternatively, we could introduce its center of mass $\mathbf{p}_0 \equiv V^1/V$.

Moreover, as the lower panels of Fig. 3 indicate, the body's orientation may be relevant. In this case we have to move on to even higher moments; for instance, the tensor

$$V^{2}\left(K\right)\equiv\int_{K}\mathrm{d}V\mathbf{x}\mathbf{x} \tag{3}$$

(where $\mathbf{xx}$ denotes the symmetric tensor product[5]) contains information on main directions of the body K. Note, that both the mass vector of a body and its inertia tensor (the traceless part of (3): $V^2(K) - \mathrm{Tr}\left(V^2(K)\right)E_d$, where E_d is the d-dimensional second-rank unit tensor) are of major interest for the physicist: the mass center of a rigid body moves in space as if the whole force acting on the body was concentrated onto a point particle at its center of mass. The inertia tensor determines how a body is rotating, given a fixed angular momentum. – Higher-rank moments such as symmetric r-rank tensors of the form $V^r\left(K\right) \equiv \int \mathrm{d}V\mathbf{x}^r$ with $r > 2$, may be considered, too.

Now it is again possible to write down a Steiner-type formula for the transition to the parallel body of a convex set K [41, 44]: the scalar expansion coefficients have to be replaced by vector- or tensor-valued coefficients. The following simple argument shows that not only the character, but also the number of coefficients has to change: for a unit ball centered at the origin, $B_R(0) \equiv \left\{\mathbf{x}\middle||\mathbf{x}| \leq R\right\}$, the construction of the parallel body results in $B_{(\epsilon+R)}(0)$. On the other hand, as a simple calculation shows, $V^r\left(B_{(\epsilon+R)}(\mathbf{x})\right) \propto (\epsilon+R)^{d+r}$. Thus, powers up to $(d+r)$ arise in the generalization of Steiner's formula, which we write formally as

$$V^{r}\left(K_{\epsilon}\right)=\sum_{\nu=0}^{d+r}\epsilon^{\nu}M_{r,\nu} \quad . \tag{4}$$

As for the scalar Minkowski functionals, one can obtain an integral representation of the Steiner coefficients by considering a smooth and convex body K. From the integral

[5] As [9] shows, there are no non-trivial *anti*symmetric Minkowski valuations of rank $r \geq 1$ at all, i.e. no motion-covariant, additive, and conditionally continuous tensor functions with rank $r \geq 1$, see below Sect. 3 for the explanation of these concepts.

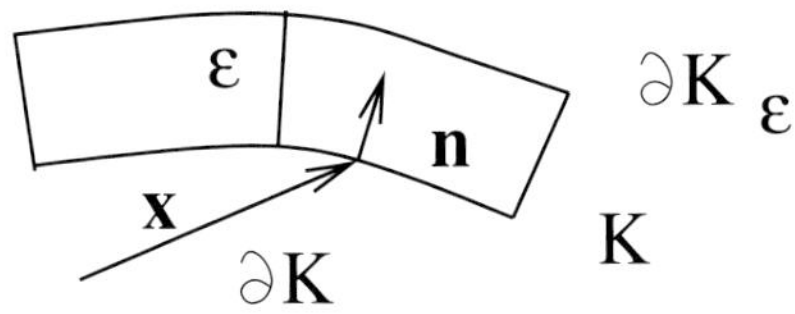

Fig. 4. The fact that each volume element in K is weighted with the position vector in order to give $V^r(K)$ for $r > 0$, makes the local normal $\mathbf{n}$ relevant for the parallel body K_ϵ.

representation of the scalar MFs and from the form of the moments V^r one might guess that the coefficients $M_{r,\nu}$ are built by functions of the form:

$$\int_{\partial K} \mathrm{d}S^{d-1} \mathbf{x}^r s_\nu\left(\kappa_1, .., \kappa_{d-1}\right) \tag{5}$$

with the elementary symmetric functions s_ν for $\nu = 0, .., (d-1)$ of the local curvatures $\kappa_1, .., \kappa_{d-1}$ (for a definition, see [44]). But this will not suffice, since we need $(d+r)$ coefficients. A closer look shows (see Fig. 4), that indeed the local normal vector comes into play; for, if $\mathbf{x} \subset \partial K$, then $\mathbf{x} + \epsilon\mathbf{n}$ lies on the surface of K_ϵ; here $\mathbf{n}$ denotes the local normal vector to ∂K at the point $\mathbf{x}$. The weighting with the local position in the definition of V^r therefore leads to contributions from local normal vectors in the Steiner formula. In [44] it is shown that in general the expansion coefficients $M_{r,\nu}$ can be written as linear combinations of the quantities:

$$W_\nu^{p,q} \equiv \frac{1}{\nu\binom{d}{\nu}} \int_{\partial K} \mathrm{d}S^{d-1} s_{\nu-1}\left(\kappa_1, .., \kappa_{d-1}\right) \mathbf{x}^p \mathbf{n}^q \tag{6}$$

for $\nu = 1, .., d$, and $p, q \in \mathbb{N}_0$. Now the $W_\nu^{p,q}$ look more basic than the coefficients arising in the Steiner formula for the volume tensor. It seems therefore, as if, in contrast to the case of the scalars, the Steiner coefficients are no longer the basic quantities. In order to generalize the Minkowski functionals we therefore have to move to an alternative route.

Spatial moments of the Minkowski functionals. Let us take the moments V^r together with the $W_\nu^{r,s}$ $(\nu = 0, .., (d-1), r, s = 0, 1, ..)$ as basic quantities. In order to arrive at a coherent notation, we set $W_0^{r,0} \equiv V^r$ and $W_0^{r,j} = 0$ for $j > 0$. Higher-rank Minkowski valuations therefore are higher moments of the scalar Minkowski functionals where the local surface is weighted with both position and surface normal. If the body K is shifted with a translation vector $\mathbf{t}$, then the $W_\nu^{r,s}$ transform like

$$W_\nu^{r,s}\left(K+\mathbf{t}\right) = \sum_{p=0}^{r} W_\nu^{p,s} \mathbf{t}^{r-p} \binom{r}{p} \quad , \tag{7}$$

as follows immediately from their definition. For rotations R, we have

$$W_\nu^{r,s}\left(RK\right) = R W_\nu^{r,s}\left(K\right) \quad , \tag{8}$$

where we use the fact that the d-dimensional orthogonal group $O(d)$ has a natural representation in each tensor space.

Recalling that the Minkowski functionals can be described axiomatically, one would be inclined at this point to look for a characterization theorem for vectors and tensors. It turns out that one has to include even more valuations in order to achieve that purpose.

Axiomatic characterization. A rigorous way to extend the Minkowski functionals is therefore to relax one of their defining axioms. We keep the *additivity*

$$\Phi(A \cup B) = \Phi(A) + \Phi(B) - \Phi(A \cap B) \tag{9}$$

for a tensor valuation Φ; we also maintain the *conditional continuity*, i.e. continuity for the subclass of convex bodies[6]; but, in the spirit of our previous findings, we replace the motion invariance by *motion covariance* defined as follows: for translations we require the existence of a number $n \in \mathbb{N}_0$ and of functions Φ_i $(i = 0, .., n)$ such that for all K

$$\Phi(K + \mathbf{t}) = \sum_{i=0}^{n} \Phi_i(K)\mathbf{t}^i \quad . \tag{10}$$

Note, that products such as $\Phi_i(K)\mathbf{t}^i$ always denote symmetric tensor products. For rotations we use the natural representation of $O(d)$ in the space of tensors as before, and require for $R \in O(d)$:

$$\Phi(RK) = R\Phi(K) \quad . \tag{11}$$

In a seminal work, Alesker [4] proved the following *characterization theorem*: every motion-covariant, convex-continuous, and additive mapping Φ can be written as a linear combination of the functionals

$$E_d^k W_\nu^{r,s} \tag{12}$$

with $i, j, k \in \mathbb{N}_0$ and $\nu = 0, .., d$. E_d denotes the d-dimensional unit matrix[7].

Equation (12) gives all of the Minkowski valuations and finally answers the question how to generalize the MFs in a suitable manner. It contains more valuations than those given in (6). However, some caution has to be exercised: Alesker's theorem [4] only bounds the dimension of the tensor functional spaces from above. It is not obvious that all of the above functionals are linearly independent. Indeed, [24] proved some linear relationships holding among the tensors. In order to give one example, Gauß' theorem yields

$$W_1^{1,1} = \int_{\partial K} \mathrm{d}S^{d-1}\mathbf{x}\mathbf{n} = W_0^{0,0}(K)E_d \quad . \tag{13}$$

One therefore has to explore the different tensor ranks further.

[6] On the space of convex bodies one uses the Hausdorff metric, see [43], p. 48 for a definition.

[7] For higher than two-dimensional spaces, it does not make a difference whether we require $O(d)$- or $SO(d)$-covariance. In the two-dimensional case, however, it does [4]. For the one-dimensional case see also [3].

4 Exploring Higher-Rank Minkowski Valuations

4.1 Some Mathematical Results

Minkowski vectors. Minkowski vectors, i.e. Minkowski valuations of rank one, have been known for some years [18, 41, 42, 17]. There is a strict one-to-one correspondence between the vectors and the scalars: for each MF $W_\nu \propto \int_{\partial K} s_{\nu-1}(\kappa_1, .., \kappa_{d-1})$ there is a related vector $\int_{\partial K} s_{\nu-1}(\kappa_1, .., \kappa_{d-1})\,\mathbf{x}$, such that $(d+1)$ vectors exist altogether. The reason why no normals enter at this level is basically because for compact convex bodies $\int_{\partial K} s_{\nu-1}(\kappa_1, .., \kappa_{d-1})\,\mathbf{n} = 0$ for $\nu = 0, .., (d-1)$ (for an elementary proof see [17]). The integral geometry of vectors was extensively scrutinized, see, e.g., [18]. For physical applications of vectors (see [8, 10]) the curvature centroids

$$\mathbf{p}_i \equiv W_\nu^{1,0} / W_\nu^{0,0} \qquad (\nu = 0, .., d) \tag{14}$$

are a useful concept. Their meaning is straight forward: they locate morphological information. We shall see below that, if the curvature centroids do not coincide within one point, strong evidence for asymmetry is given.

Second-rank tensors. The situation becomes more complicated for higher-rank tensors. It can be shown, that at the second rank, the following $(3d+1)$ tensors provide a base [9]:

$$\begin{aligned} &\int_K \mathrm{d}V \mathbf{x}\mathbf{x} \;, \\ &\int_{\partial K} \mathrm{d}S^{d-1} s_\nu \mathbf{x}\mathbf{x} \;, \quad \int_{\partial K} \mathrm{d}S^{d-1} s_\nu \mathbf{x}\mathbf{n} \;, \quad \int_{\partial K} \mathrm{d}S^{d-1} s_\nu \mathbf{n}\mathbf{n} \end{aligned} \tag{15}$$

for $\nu = 0, .., (d-1)$. Not all of them are interesting from a physical point of view, since some of them are related to Minkowski functionals times the unit tensor and thus do not contain any new information. Principal kinematic formulae and Crofton formulae are known for second-rank tensors [44, 9, 45].

For a geometrical interpretation, we have to bear in mind, that the values of the tensors depend on the origin adopted. For $W_\nu^{2,0}$ $(\nu = 0, .., d)$, however, the corresponding curvature centroid $\mathbf{p}_\nu$ is a natural reference point; the other tensors listed in (15) are translation-invariant. Obviously, with such a choice the tensors are sensitive to morphological anisotropies, in the case of symmetry they align along the symmetry axes, otherwise they distinguish morphologically interesting directions. It is therefore useful to consider their eigenvalues and eigendirections.

Beyond second rank. Although higher-rank tensors may be interesting as well, we shall not consider them further at this point.

4.2 Examples

In order to give a first example of how the Minkowski valuations discriminate spatial structures, we consider a few simple patterns as shown in Fig. 5. In the case of a circle (panel 1) the curvature centroids coincide within the circle's center; the tensors – here

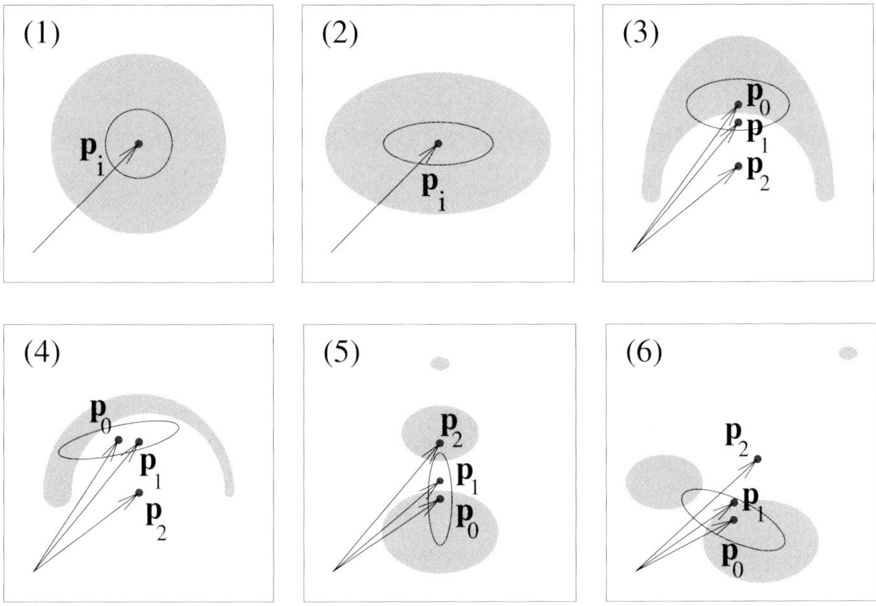

Fig. 5. A number of simple patterns together with their curvature centroids $\mathbf{p}_i$ ($i = 0, .., 2$). The ellipse represents the mass tensor $W_0^{2,0}/W_0^{0,0}$; it is calculated with the center of mass $\mathbf{p}_0$ as reference point. The ellipses indicate the ratio of the eigenvalues and the eigendirections as well.

represented as an ellipse featuring their eigenvalues and the corresponding directions – are isotropic. It is useful to divide the tensors $W_\nu^{r,s}$ by the corresponding Minkowski functionals $W_\nu^{0,0}$. The trace of the normalized tensor (i.e. the sum of the eigenvalues) quantifies to which extent the morphological information is concentrated around the center, how large the morphological fluctuations around the center are.

The transition to an ellipse of the same volume as the circle (panel 2) can be traced either with the tensors becoming anisotropic or with the ratio of the squared surface to the volume (isoperimetric ratio, see [39]). Furthermore, as the morphological concentration is decreasing, the traces of most of the normalized tensors increase, especially of $W_2^{2,0}$: the more elongated the ellipse becomes, the higher is its curvature at its very fringes along the horizontal direction. The normalized trace of $W_1^{1,1}$, on the contrary, becomes smaller, since the normal and the local position vector cease to be parallel to each other. The curvature centroids still coincide with each other because of the point symmetry. – If we reduce the symmetry further to a merely axially symmetric configuration (panel 3), the curvature centroids fan out along the symmetry axis. In the case of no symmetry at all (panel 4) they may constitute a triangle.

Finally, suppose one is given three convex bodies (or $d + 1$ ones in d dimensions) with known Minkowski functionals in order to combine them into a pattern such as indicated in panels 5 and 6. Given the scalar measures of the whole pattern, especially its Euler characteristic, one can constrain the number of intersections or subclumps in a pattern, which does not make any difference in our examples. For disentangling the positions of the grains, however, one has to know the curvature centroids. The centroids weight the subclusters according to their Minkowski functionals. Therefore, whereas from the volume's point of view, only the larger components contribute significantly, all subclumps have the same weight (Euler characteristic $= 1$) for the curvature centroid $\mathbf{p}_2$. If the subclumps do not overlap, one can indeed completely recover their positions.

In Fig. 6, we show all tensors for the last example in more detail. Obviously, there are not only anisotropies; rather different morphological directions are relevant. The volume concentration around the center of mass is relatively high, since the volume is dominated by the two largest subclusters, being close to each other. It is quite the opposite with $W_2^{2,0}$: $\mathbf{p}_2$ is relatively far away from all the clumps, the curvature around $\mathbf{p}_2$ is dispersed leading to a high value of tensor's trace. The main relevant direction connects the pair of large clumps with the third subcluster. The tensor $W_1^{1,1}$ is always isotropic, as shown in (13), its trace is connected to the volume. Obviously, $W_2^{1,1}$ and $W_1^{0,2}$ complement each other; indeed they add up to a diagonal tensor (recall the linear relationships described in [44]). Their eigenvectors' alignment with the coordinate axes is due to the fact that all the subclumps are aligned parallel to the axes as well. Otherwise, $W_2^{1,1}$ and $W_1^{0,2}$ would average over the orientations of the subclumps. Finally, $W_2^{0,2}$ is isotropic, its trace equals the Euler characteristic.

The detailed description of patterns given by the Minkowski valuations can be used in practical applications, as we shall see in the next section.

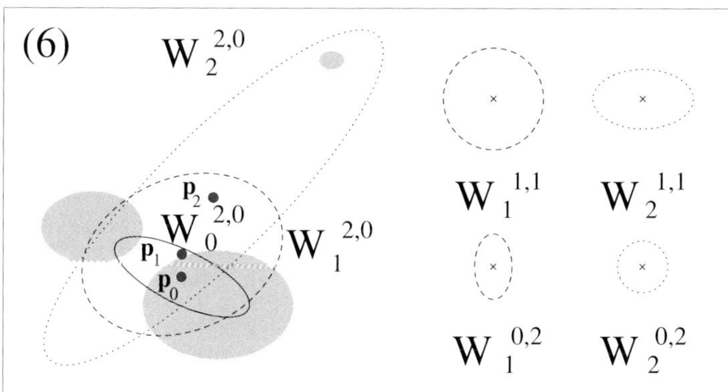

Fig. 6. We show all of the tensors $W_0^{2,0}$, $W_\nu^{2,0}$, $W_\nu^{1,1}$, and $W_\nu^{0,2}$ for $\nu = 1, 2$ – they span a base of the second-rank tensor space. The tensors $W_\nu^{r,s}$ were divided by $W_\nu^{0,0}$ ($i = 0, .., 2$); the tenors with a twofold position weighting ($W_\nu^{2,0}$) are calculated around the corresponding centroid. All of the other tensors are translation-invariant. Therefore we depict them outside the pattern on the right-hand-side. Their eigenvalues are multiplied by a factor of 6 in order to yield a more convenient representation. $W_0^{2,0}$: solid line; $W_1^{2,0}$, $W_1^{1,1}$, $W_1^{0,2}$: dashed line; $W_2^{2,0}$, $W_2^{1,1}$, $W_2^{0,2}$: dotted line.

5 Physical Applications of Higher-Rank Minkowski Valuations

5.1 The Inner Structure of Spiral Galaxies

The classification of spiral or disc galaxies into specific types (the Hubble types, e.g., [19]) is an important step towards their physical understanding; there are competing theoretical pictures trying to explain the formation of spiral arms [23, 15]. In Fig. 7 we display a real galaxy observed face-on together with a simulation according to a percolation model [15]. So far one has mainly relied on an assignment by eye in order to classify observed disc galaxies and to compare observational data with theoretical models [29]. One of the reasons for this is that spiral arms are easily detected by the human eye, but are rather hard to quantify. Here we use Minkowski valuations to reveal aspects of the inner structure of spiral galaxies (see [37, 11, 22] for some discussion of how to quantify morphometric features of spiral galaxies, and especially [1, 2]).

In order to render an optical galaxy image accessible to a Minkowski analysis, it is helpful to understand it as a two-dimensional surface brightness field $u(\mathbf{x})$. We may then estimate the Minkowski valuations of the excursion sets $M_v = \{\mathbf{x}|u(\mathbf{x}) > v\}$. Contrary to large scale cosmological data (e.g. the distribution of galaxies or clusters of galaxies), such a density field is neither homogeneous nor isotropic; rather it reveals a point of considerable symmetry: the center of mass. We expect the curvature centroids to coincide within this point and therefore concentrate on second-rank Minkowski tensors.

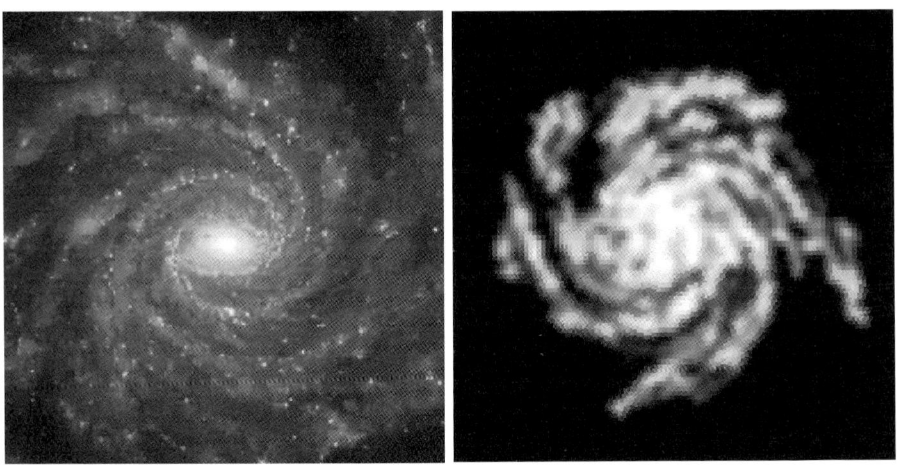

Fig. 7. The spiral galaxy NGC 1232 (as observed by the VLT/ESO) and a simulated galaxy following the model by [46].

In order to detect a possible overall anisotropy, we divide the smaller eigenvalue of $W_0^{2,0}$ by the larger one; this yields the parameter X. Since $X \leq 1$ by definition, $\epsilon \equiv 1 - X$ can serve as a measure of the anisotropy. Such an anisotropy may reflect a physical anisotropy, induced, for instance, by the presence of two massive spiral arms, but may also indicate that the galaxy is not observed face on, but rather under an inclination angle.

As the tensors $W_\nu^{2,0}$ ($\nu = 0, .., 2$) weight the volume or the surface, respectively, with $\mathbf{xx}$, their eigenvalues are increased, if the configuration is more widespread (see Fig. 8). This fact can be used to quantify the concentration of a density field around its center of mass, by calculating the trace of $W_\nu^{2,0}$. Therefore another characteristic feature

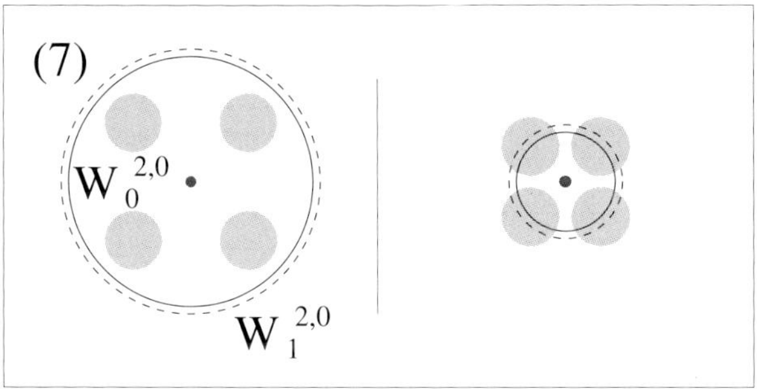

Fig. 8. The tensors $W_0^{2,0}$ (solid line) and $W_1^{2,0}$ (dashed line) for a wide spread (right panel) and a compact (left panel) configuration of circles. The eigenvalues depend quadratically on the separation of the circles.

of disc galaxies, namely the way in which the radial surface brightness distribution declines (with increasing radius), can be analyzed by scanning through various threshold values ν. We applied (suitably normalized) concentration parameters P_ν (built from $\mathrm{Tr}(W_\nu^{2,0})$) to the images of both real galaxies (C. Moellenhoff, private communication) as well as to galaxies produced by a percolation model ([46]; see [13] for details). These morphological descriptors reveal strong luminosity concentration for real galaxies, which could not be reproduced by any of the model galaxies (see Fig.9) with reasonable choices for the model parameters. Using the morphology of observed disc galaxies, one can therefore rule out the percolation model (at least in its simple form) as a viable attempt to explain the formation of spiral arms.

5.2 The Morphology of Galaxy Clusters

As mentioned in Sect. 2, the morphology of clusters may constrain the values of the cosmological parameters. Intuitively, one may expect clusters in cosmological low-density models to have less substructure than in high-density models [33]. Using Minkowski valuations, we check whether this effect can be observed for the cosmological GIF simulations [6]. We analyze the clusters' projected mass distributions such as shown in Fig. 2. Observations from the gravitational lensing through clusters [40, 14] trace approximately the same cluster information.

In order to make the simulation data, which consist of "particles tracing matter", accessible to the Minkowski valuations, we construct a pixelized density field $u(\mathbf{x})$ and smooth it with a Gaussian kernel of width[8] $\lambda = 0.3h^{-1}$Mpc. The resulting field $u_\lambda(\mathbf{x})$

[8] Length scales are given in units of h^{-1}Mpc. One Megaparsec (Mpc) equals about 3.26 million light years, the constant $h \sim 0.65$ accounts of the uncertainty in the measurements of the Hubble constant.

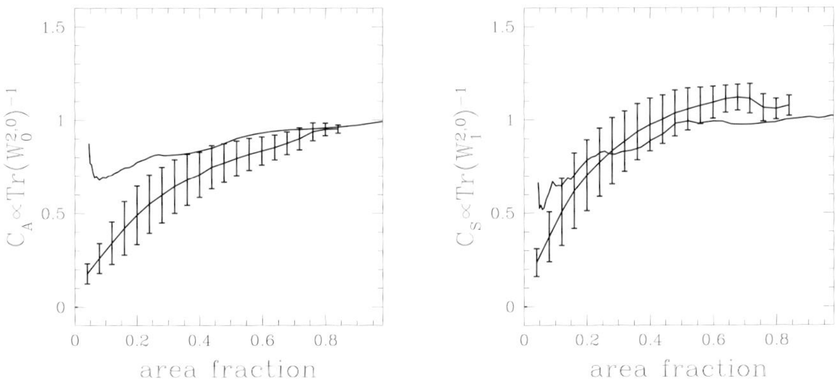

Fig. 9. Results of a quantitative morphometry for spiral galaxies using Minkowski tensors. We consider the galaxy NGC 1232 (shown in the left panel of Fig. 7) and a sample of model galaxies following a percolation model (an example of a model prediction is displayed in the right panel of Fig. 7). We plot two concentration parameters (as described in the main text) versus decreasing surface brightness threshold values, i.e. increasing surface area. The error bars indicate 1σ fluctuations around the expectation value of the model, whereas the single line refers to NGC 1232. The parameter C_A reveals significant differences in the concentration between the surface brightness profiles of real and model galaxies. The particular percolation model used thus is incompatible with real data.

contains the overall shape of the cluster. Again we estimate the Minkowski valuations of the excursion sets, which contain all pixels above a given density threshold v. Our method to estimate the Minkowski functionals from the grid is based upon Crofton's formulae [47, 38, 7]. For each cluster we average a function of the Minkowski valuations f over different density thresholds:

$$\frac{1}{(u_\lambda)_2 - (u_\lambda)_1} \langle f \rangle \equiv \int_{(u_\lambda)_1}^{(u_\lambda)_2} du_\lambda f \tag{16}$$

in order to obtain a compact set of parameters. Here $(u_\lambda)_1$ and $(u_\lambda)_2$ denote appropriately chosen integration limits. We consider five types of substructure.

1. The *clumpiness* $\mathrm{C} \equiv \sqrt{\langle (\frac{1}{\pi} W_2 - 1)^2 \rangle}$ counts the number of subclumps in quantifying how much the Euler characteristic W_2/π deviates from one.
2. The construction of the *asymmetry* parameter A is based on the fact that curvature centroids which do not coincide with each other, indicate asymmetry. Therefore, we calculate the perimeter of the triangle constituted by the curvature centroids, $2W_1(\Delta(\mathbf{p}_i))$, $\mathrm{A} \equiv \langle W_1(\Delta(\mathbf{p}_i)) \rangle$.
3. The ratio of the tensors' eigenvalues, $\gamma_1(W_\nu^{r,s})$ and $\gamma_2(W_\nu^{r,s})$, tells us to which extent *anisotropies* are present. Here we consider the tensor $W_1^{2,0}$ and define $\mathrm{X} \equiv \left\langle 2\left(\gamma_1\left(W_1^{2,0}\right) - \gamma_2\left(W_1^{2,0}\right)\right) / \left(\gamma_1\left(W_1^{2,0}\right) + \gamma_2\left(W_1^{2,0}\right)\right)\right\rangle$ (compare [27, 28] for a related approach).

4. In general the curvature centroids $\mathbf{p}_\nu$ wander in space if the density threshold is varied; their fluctuations therefore measure the *shift of morphological properties*. Focusing on $\mathbf{p}_1$ we define S$\equiv$ $\langle(\mathbf{p}_1 - \langle\mathbf{p}_1\rangle)^2\rangle$. This parameter is related to the centroid variation used in cluster investigations before [27, 28].
5. Likewise, the eigendirections (parametrized by an angle α) of the tensors are changing; this generates the *twist of morphological* properties. Restricting ourselves to $W_1^{2,0}$ we employ the parameter T$\equiv$ $\langle(\alpha - \langle\alpha\rangle)^2\rangle$ (this generalizes a method used by [27, 28]).

For more technical details, see [7]. For each time period (or cosmological redshift) we estimate the order parameters for three projections per cluster and average over all clusters in one cosmological model in order to extract a typical morphology. In Fig. 10, we display the evolution of the averaged morphological parameters vs. redshift. One recognizes immediately a relaxation process, leading to less substructure (and lower values of the morphological order parameters) as time goes by (from right to left). Further-

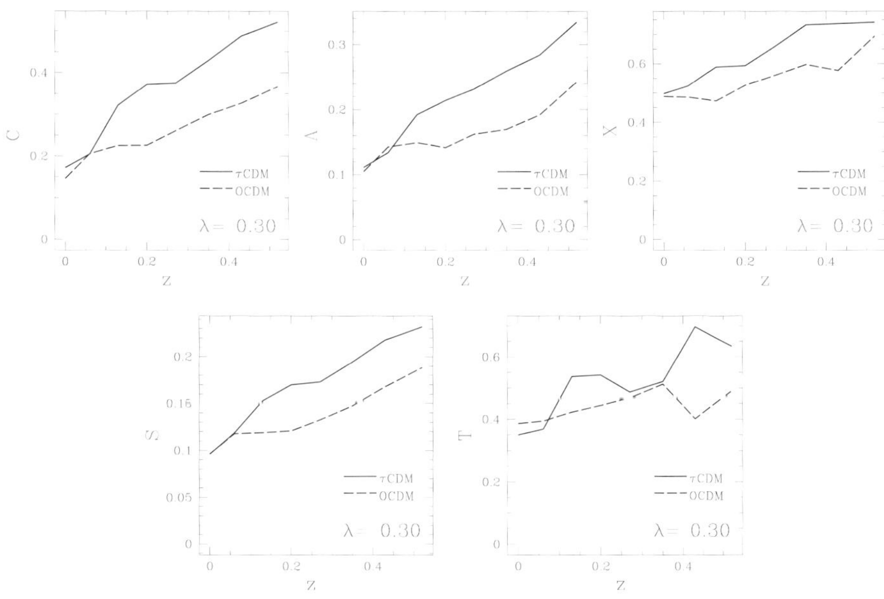

Fig. 10. The average morphological evolution of galaxy clusters from the GIF simulations. Our morphological order parameters clumpiness C, asymmetry A, anisotropy X, shift of morphology S and twist of morphology T are shown versus time. The time is given in terms of the observable cosmological redshift. Redshift $z = 0$ means the present day, higher redshifts lead back into past. In order to get typical results for two cosmological models (low-density model with an open geometry, called open Cold Dark Matter, OCDM: dashed line; high-density model with a massive τ neutrino, named τCDM: solid line) we average over about 15 clusters and 3 orthogonal projections per cluster within each model. As the results show, our morphological order parameters quantitatively feature the differences between the models which are in accordance with the theoretical predictions. For details on the cosmological Cold Dark Matter models see [30].

more, clear differences between the cosmological background models arise. Indeed the theoretical predictions according to which the low-density clusters (OCDM) exhibit less substructure, are confirmed for the past; for redshift $z = 0$, the differences disappear.

All of our order parameters show a similar behavior. The reason probably is that we average over different types of relaxation processes. An investigation of single clusters should reveal differences between, e.g., a bimodal merger dominated by two colliding subclumps and a more isotropic mass infall.

5.3 The Geometry of the Electric Charge Distribution in Molecules

An important implication of the atom hypothesis is that the structures of molecules are to be explained in terms of their constituent parts, the atoms. Heuristic principles based upon this assumption work quite well in chemistry. However, if one tries to justify this hypothesis in the framework of quantum mechanics, the notion of an individual atom seems to disappear, since the negative charge distributions of the atoms merge, blurring the electron clouds which define the single atoms.

Bader [5] proposed an elegant method to recover the notion of atoms in molecules. Since, from the quantum mechanical point of view, the charge density profile $\varrho(\mathbf{r})$ is the only relevant quantity, its geometry and topology must contain the notion of atoms. Using the critical points of the density distribution ϱ, where $\nabla\varrho = 0$, one can define atoms as local density maxima together with a surrounding domain enclosed by zero-flux surfaces. This geometrical approach can be extended to chemical bonds and has been justified on physical grounds. Moreover, it provides us with means to visualize molecules in a proper way using the charge density profile, see Fig. 11 for an illustration. A distinguished role is played by the surface $\nabla^2\varrho = 0$ which separates the zone where chemical reactions are likely to happen from the rest of space.

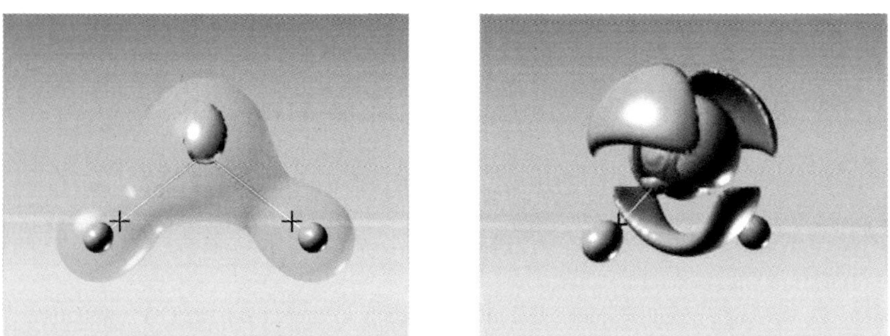

Fig. 11. The water molecule H_2O. The bright lines denote the bonds connecting the H-atom with both of the O-atoms. The crosses mark the points where the zero-flux surfaces separating the atoms intersect with the bonds. Moreover, some iso-surfaces of the Laplacian of the charge density, $\triangle\varrho$, are shown. The images are from http://www.nas.nasa.gov/~creon/papers/mgms96/ (courtesy C. Levit). The panels show different projections and focus on different iso-surfaces.

Obviously, this is an area of research where physics and geometry strongly interact, and the applications of higher-rank Minkowski valuations may yield useful information on the structure of the chemical bonding. First, the morphology of the whole molecules can be described in terms of the Minkowski valuations of the density field. The atoms, which Bader defines [5], may not be compact in every case, so the single atoms can not be understood in terms of the Minkowski valuations. However, the fact that the atoms' positions are specified in Bader's theory gives us the chance of probing the molecular structure with the Boolean grain method where the atoms' positions are decorated using spheres (either monodisperse or with radii mirroring physical quantities). Note, that at this point one really needs tensors, since in this case the locations and the orientations of subsystems are of crucial importance; and this local information lies beyond the scope of the scalar measures. Bader's theory constructs also a graph of bonds, that may be analyzed with Minkowski valuations straightforwardly.

But given the fact that both the molecules and the constituting atoms are defined in terms of geometry, one may go one step further and try to discover phenomenological principles allowing one to predict the shape of molecules given their constituents.

These endeavors clearly are beyond the scope of this article. However, in order to illustrate to some extent how useful the Minkowski valuations may prove in this area, we consider briefly some of the Minkowski tensors for the H_2^+ ion.

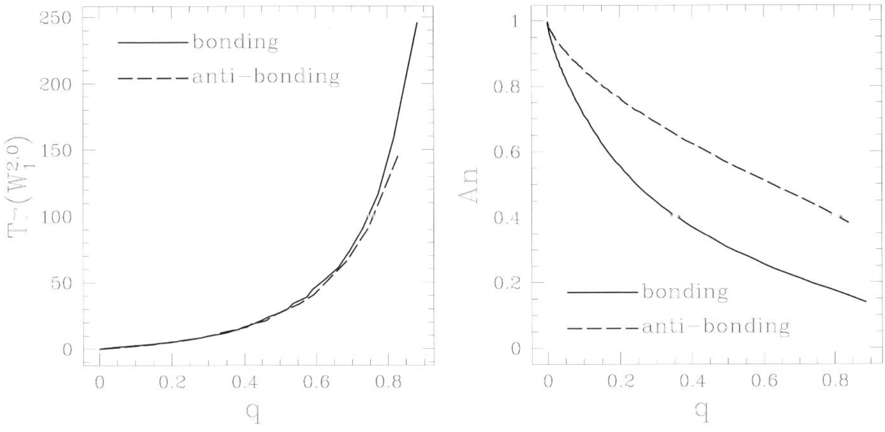

Fig. 12. The morphology of the bonding and the anti-bonding states of the electron within the H_2^+ ion. We explore the charge distribution of the electron via the Minkowski tensors of excursion sets from the density field. In the left panel, we show the trace of $W_1^{2,0}$ as a function of the charge fraction q enclosed within the density contour. In the right panel, we consider the anisotropy of the charge distribution. It is quantified via $An \equiv \left(\left(W_1^{2,0}\right)_{xx} - \left(W_1^{2,0}\right)_{yy}\right) / \mathrm{Tr}\left(W_1^{2,0}\right)$. The anti-bonding state, shown as the dashed line is much more anisotropic than the bonding state which wraps both atoms with a smooth electron hull.

In order to treat the H_2^+ ion consisting of two protons and an electron in the framework of quantum mechanics, one often considers the protons as fixed in space (Born-Oppenheimer approximation) thereby reducing the problem to an effective one-particle problem of an electron moving in the potential of two positively charged nuclei. A reasonable ansatz for the quantal wave function of the electron – at least suitable to illustrate what is basically happening – is a superposition of quantum states where the electron is bound to either of the protons. To be more precise, let us consider the protons as being separated by their equilibrium distance of approximately two Bohr radii[9]. Let us label the protons with 1 and 2 and let ψ_i denote the unperturbed (well-known) hydrogen ground state wave function of an electron bound to proton i for $i = 1, 2$. For symmetry reasons, only the linear combinations

$$\psi_a = \frac{1}{\sqrt{2}} (\psi_1 - \psi_2) \quad \text{and} \quad \psi_b = \frac{1}{\sqrt{2}} (\psi_1 + \psi_2) \tag{17}$$

are of interest. Since ψ_b is favored on energetic grounds, it is a bonding state; consequently ψ_a is called antibonding state.

The morphology of the electron charge distribution is expected to distinguish between both states. Let us consider excursion sets of the three-dimensional electron charge density profile $\varrho = |\psi|^2$ for both states. In Fig. 12 we display the surface tensor $W_1^{2,0}$ as a function of the charge fraction enclosed within the excursion set. We start from the very peaks of the charge distribution (high charge density thresholds) and move on to lower charge density thresholds enclosing a significant part of the whole electron charge. The trace of the tensor $W_1^{2,0}$ shows that the iso-surfaces of the bonding state are slightly more extended than those of the anti-bonding state. But on the whole, the differences are small: the concentration of the charge distribution is basically determined by the distance of the protons. If we ask for morphologically relevant directions, the differences between both states are much more perspicuous. The anisotropy of the azimuthal symmetric quantum states can be quantified via

$$An \equiv \left(\left(W_1^{2,0} \right)_{xx} - \left(W_1^{2,0} \right)_{yy} \right) / \mathrm{Tr} \left(W_1^{2,0} \right) \ , \tag{18}$$

the protons lying at $(x, y, z) = (-1, 0, 0)$ and $(1, 0, 0)$, respectively. For small charge fractions, the tensor $W_1^{2,0}$ is dominated by the peaks around the positively charged protons yielding a maximum anisotropy, but no essential difference between both states. However, as we move to lower density thresholds and higher charge fractions, the bonding state isotropises much more quickly than the anti-bonding one. In the bonding state, therefore, the electron forms a cloud spreading out over both atoms in a quite isotropic way, whereas in the anti-bonding state, the signature of the two distinct protons remains still visible. The chemical bond therefore is mirrored by the geometry of the electron density distribution, as quantified via Minkowski tensors.

This example is but an elementary first illustration of the nature of the chemical bond. In the spirit of Bader's theory, one can possibly go much further in order to consider single atoms in molecules. Additivity can serve as a guiding principle, since Bader found that there are stable atom configurations which persist through different types of molecules, presenting a sort of rigid module movable as a whole; this fact is well-known from chemistry, of course.

[9] In the following we consider spatial distances in units of Bohr radii. The Bohr radius a_0 marks the typical scale of the hydrogen ground state wave function. For an introduction into the physical treatment of the H_2^+ ion one may consult, e.g. Complement G_{XI} in [12].

6 Density Functional Theory Based on Minkowski Valuations

Having shown the morphometric versatility of MVs, we conclude our discussion of applications by considering systems where integral geometric quantities enter a physical approximation. The quantities that come into play at this point generalize the Minkowski valuations further and may indicate a new direction of interest.

Density functional theory became the standard method to study inhomogeneous classical fluids in the last 20 years. The main feature of such physical systems is the spatial variation of the equilibrium density $\rho_{eq}(\mathbf{x})$ which minimizes the free energy Ω. Density functional theory is based on the exact result that the free energy of the inhomogeneous fluid can be expressed as a functional $\Omega[\rho(\mathbf{x})]$ of the density $\rho(\mathbf{x})$. Thus, all relevant thermodynamic properties such as surface tensions or phase transitions of confined fluids can be calculated from $\Omega[\rho_{eq}(\mathbf{x})]$; even equilibrium correlation functions that describe the microscopic structure of inhomogeneous fluids can be determined by derivatives of the functional.

In general, the form of the functional $\Omega[\rho(\mathbf{x})]$ is not known; deriving the exact functional would be equivalent to solving the statistical mechanics for fluid systems exactly which seems to be an impossible task. Therefore, one has to restrict oneself to reasonable approximations for $\Omega[\rho(\mathbf{x})]$. For convenience, one separates an excess intrinsic free energy functional $\mathrm{f}_{ex}[\{\rho_i\}]$ from exactly known external contributions

$$\Omega[\rho(\mathbf{x})] = \int \mathrm{d}\mathbf{x} \ \left(\mathrm{f}_{ex}[\rho(\mathbf{x})] \ + \ k_B T \rho(\mathbf{x}) \left(\ln \Lambda^3 \rho(\mathbf{x}) - 1\right) \ + \ \rho(\mathbf{x}) \left(V(\mathbf{x}) - \mu\right)\right) \tag{19}$$

where $V(\mathbf{x})$ is an external potential, μ the chemical potential, $\beta = 1/(k_B T)$ the inverse temperature T, and Λ the thermal wavelength of the fluid. k_B denotes the Boltzmann constant.

The most elaborated and reliable functional for hard-sphere fluids is that developed by Rosenfeld [34] based on the local scalar and vector-valued Minkowski functionals

$$W_\nu^{p,q}[\rho(\mathbf{x})] \equiv \frac{1}{\nu\binom{d}{\nu}} \int_{\partial K} \mathrm{d}S^{d-1} s_{\nu-1}(\kappa_1, .., \kappa_{d-1})\, \mathbf{x}^p \mathbf{n}^q \ \rho(\mathbf{x}) \, . \tag{20}$$

In contrast to the definition (6) and the applications considered in the previous sections, the functionals here are locally weighted with the density $\rho(\mathbf{x})$. Notice, that for inhomogeneous densities the vector valuations $W_\nu^{0,1}[\rho(\mathbf{x})]$ in general do not vanish for a convex body K and that the functionals do inherently depend on the location of K. The following ansatz for the excess free energy functional proves to be useful:

$$\begin{aligned} \beta \mathrm{f}_{ex}[\rho(\mathbf{x})] = \ & f_1(W_0^{0,0}) \ W_3^{0,0} + f_2(W_0^{0,0}) \ W_1^{0,0} W_2^{0,0} + f_3(W_0^{0,0}) \ (W_2^{0,0})^3 \\ & + f_4(W_0^{0,0}) \ W_1^{0,1} W_2^{0,1} + f_5(W_0^{0,0}) \ W_2^{0,0} (W_2^{0,1} W_2^{0,1}) \ , \end{aligned} \tag{21}$$

where we drop the argument $[\rho]$ for convenience. This ansatz is inspired by the observation that the physical dimensions are $[W_3^{0,0}] = [W_1^{0,q} W_2^{0,q}] = [(W_2^{0,q})^3] = [\beta f_{ex}] = [\text{length}]^{-3}$, and that $[W_0^{0,0}] = 1$ is a dimensionless weighted density.

To obtain the functions $f_i(W_0^{0,0})$ one may apply an argument from scaled particle theory for a homogeneous fluid [32], namely, that the excess chemical potential for big spheres of radius $\sigma/2 \to \infty$ is $\lim \mu/V \to p$ with the volume $V = \frac{\pi}{6}\sigma^3$, the chemical potential $\mu = \frac{\partial f_{ex}}{\partial \rho}$ and the pressure p. Within the present ansatz (21) one finds in this limit the reversible work to create a cavity of volume V, $\mu = \sum_\nu \frac{\partial f_{ex}}{\partial W_\nu^{0,0}} \frac{\partial W_\nu^{0,0}}{\partial \rho}$, and therefore in the limit $\sigma/2 \to \infty$ the chemical potential $\mu/V \to \frac{\partial f_{ex}}{\partial W_0^{0,0}}$. Using the equation of state $\Omega = -pV$, i.e., $p = -f_{ex} + \sum_\nu \frac{\partial f_{ex}}{\partial W_\nu} W_\nu + W_3^{0,0}$ one gets the well known 'scaled particle' differential equation

$$\frac{\partial f_{ex}}{\partial W_0^{0,0}} = -f_{ex} + \sum_{\nu=0}^{3} \frac{\partial f_{ex}}{\partial W_\nu^{0,0}} W_\nu^{0,0} + \frac{\partial f_{ex}}{\partial W_1^{0,1}} W_1^{0,1} + \frac{\partial f_{ex}}{\partial W_2^{0,1}} W_2^{0,1} + W_3^{0,0} . \tag{22}$$

Within the ansatz (21) the solution of the differential equation reads

$$\begin{aligned}\beta f_{ex}[\rho(\mathbf{x})] &= -\frac{3W_3^{0,0}}{4\pi} \ln(1 - W_0^{0,0}) + \frac{9}{4\pi} \frac{W_1^{0,0}W_2^{0,0} - W_1^{0,1}W_2^{0,1}}{1 - W_0^{0,0}} \\ &+ \frac{9W_1^{0,0}}{8\pi} \frac{(W_1^{0,0})^2 - 3W_1^{0,1}W_1^{0,1}}{(1 - W_0^{0,0})^2} .\end{aligned} \tag{23}$$

This functional yields excellent results for thermodynamic quantities and for structures in the fluid phase of hard spherical molecules [36]. For instance, in Fig. 13 the inhomogeneous density profile $\rho(z)$ close to a hard wall is shown as compared to simulation results.

Unfortunately, the functional in (23) was derived only for hard spheres in three dimensions; however, a proposal has been made for general convex bodies [35] based on an approximative decomposition of f_{ex} in the low density limit. Here, we show how Rosenfeld's idea of decomposing f_{ex} for spheres [34] may be extended exactly to non-spherical convex bodies by applying integral geometric techniques such as kinematic formulae and an explicit local expression for the Euler characteristic $\chi(K \cap K')$ of two overlapping bodies [26]. The main idea is to decompose the exact expression for the local excess free energy $f_{ex}[\rho(\mathbf{x})]$ at low densities in terms of the vector-valued Minkowski functionals. To this end one has to evaluate the configurational integral $\int \mathrm{d}K = \int \mathrm{d}R \int \mathrm{d}\mathbf{x}$ over all orientations R and positions $\mathbf{x}$ of a body $K = RK_{\mathbf{x}}$, namely,

$$\begin{aligned}\frac{4\pi}{3} \int \mathrm{d}\mathbf{x}\, \beta f_{ex}[\rho(\mathbf{x})] &\to \frac{2\pi}{3} \int \mathrm{d}K \int \mathrm{d}K'\, \chi(K_{\mathbf{x}} \cap K_{\mathbf{x}'})\ \rho(\mathbf{x})\rho(\mathbf{x}') \\ = &\int_{\mathbb{R}^d} \mathrm{d}\mathbf{x}\ W_3^{0,0}[\rho(\mathbf{x})] W_0^{0,0}[\rho(\mathbf{x})] \\ &+ \int_{\mathbb{R}^d} \mathrm{d}\mathbf{x}\, 3W_1^{0,0}[\rho(\mathbf{x})]W_2^{0,0}[\rho(\mathbf{x})] - 3W_1^{0,1}[\rho(\mathbf{x})]W_2^{0,1}[\rho(\mathbf{x})] \\ &+ \int_{\mathbb{R}^d} \mathrm{d}\mathbf{x} \int_{\mathbf{r}} \mathrm{d}K_{\mathbf{x}}\rho(\mathbf{r}) \int_{\mathbf{r}'} \mathrm{d}K'_{\mathbf{x}}\rho(\mathbf{r}') \frac{\kappa_1 - \kappa_2}{2} \frac{(\mathbf{v}_1\mathbf{v}_1 - \mathbf{v}_2\mathbf{v}_2)_{ij}\,(\mathbf{n}'\mathbf{n}')_{ij}}{1 + \mathbf{n}\mathbf{n}'}\end{aligned} \tag{24}$$

where $\mathbf{v}_i$ denotes the principal direction of the curvature κ_i at $\mathbf{r}$ and $\mathbf{n}$ ($\mathbf{n}'$) are the normal vectors of $RK_{\mathbf{x}}$ ($R'K_{\mathbf{x}}$) at $\mathbf{r}$ ($\mathbf{r}'$), respectively. The integral over rotations R and over the

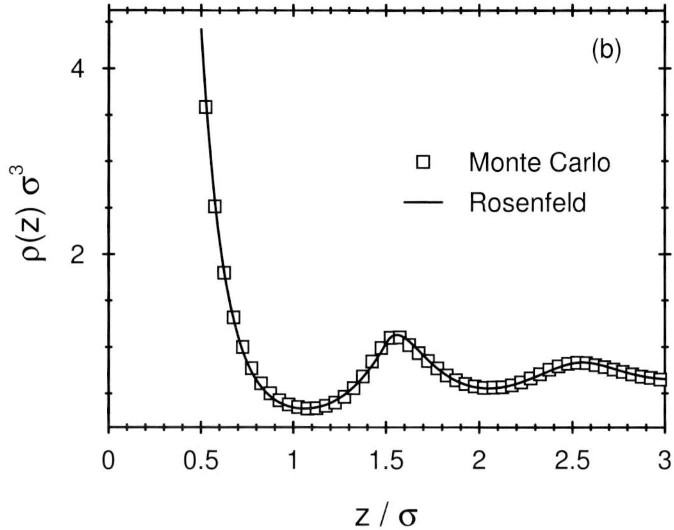

Fig. 13. The density profile of a hard sphere fluid close to a hard wall for a bulk density $\rho(2\sigma)^3 = 0.715$ calculated with the Rosenfeld functional (solid line) as compared to simulation results (squares, Figure from R. Roth, 1999, see [36]). σ denotes the diameter of the spheres.

surface of the body $K_\mathbf{x}$ at $\mathbf{x}$ is written as $\int_\mathbf{r} \mathrm{d}K_\mathbf{x} = \int \mathrm{d}R \int_{\mathbf{r}\in\partial RK_\mathbf{x}} \mathrm{d}S^{d-1}$. For spheres with $\kappa_1 = \kappa_2$ one recovers immediately the low-density limit of the intrinsic free energy given by (23). But unfortunately the last term on the right hand side of (24) vanishes only for spheres, so that only for hard spheres a finite number of Minkowski functionals is, by accident, sufficient to decompose the excess free energy into contributions stemming alone from the individual bodies, whereas for non-spherical shapes in arbitrary dimensions an infinite number of fundamental curvature measures is required. Nevertheless, based on the most general analytical expression (24), several approximations by a finite number of tensor-valued Minkowski functionals (including the proposal in [35]) may be proposed and numerically tested which make Rosenfeld's density functional applicable to general convex bodies, in particular, to spherocylinders and ellipsoids [26].

7 Conclusions

Concluding, we argue that a new perspective emerges as regards the Minkowski functionals. Instead of marking a small number of distinguished measures (which, of course, they do in a certain sense), the Minkowski functionals are embedded into a more extensive class of Minkowski valuations. These measures generalize the Minkowski functionals without leaving the framework of integral geometry. The essential step to extend the scalar functionals is to replace the motion invariance by motion covariance. This step yields a hierarchy of tensor-valued measures with the Minkowski functionals at the bottom. The higher-rank valuations are moments of the Minkowski functionals where local geometry is weighted with either the position or the normal vector. The full, comprehensive description of the tensor-valued measures is still a mathematical challenge,

although some important results concerning the integral geometry of tensor measures have already been achieved, especially in low dimensions (see [45]).
The extensions of the Minkowski functionals are not only mathematically interesting; rather they can be motivated on physical grounds. Since the Minkowski functionals proved to be useful for a number of applications, we are justified to expect that also their generalizations may be suitable tools. Using physical applications at various length scales we exemplified how higher-rank Minkowski valuations complement the information featured by the scalar functionals. From this point of view we hope that the higher-rank Minkowski valuations will become a standard tool for morphometry and some other fields of research concerned with pattern formation.

Acknowledgement

We thank C. Möllenhoff for providing us with recent observations of spiral galaxies. Furthermore, we thank C. Levit for the permission to show the images of the molecules and J. Colberg for the GIF cluster data. We thankfully acknowledge comments on the manuscript by M. Kerscher and R. Schneider. Last but not least we thank the participants of the second Wuppertal conference for vivid discussions.

References

1. Abraham, R. G., F. Valdes, H. K. C. Yee, S. van den Bergh (1994): 'The Morphologies of Distant Galaxies. I. An Automated Classification System'. *Ap. J.* **432**, pp. 75-90.
2. Abraham, R. G., S. van den Bergh, K. Glazebrook, R. S. Ellis, B. X. Santiago, P. Surma, R. E. Griffiths (1996): 'The Morphologies of Distant Galaxies. II. Classifications from the Hubble Space Telescope Medium Deep Survey'. *Ap. J. Suppl.* **107**, pp. 1.
3. Alesker, S. (1999): 'Continuous Rotation Invariant Valuations on Convex Sets'. *Ann. of Math.* **149 (3)**, p. 977.
4. Alesker, S. (1999): 'Description of Continuous Isometry Covariant Valuations on Convex Sets'. *Geom. Dedicata* **74**, p. 241.
5. Bader, R. F. W. (1990): *Atoms in Molecules – A Quantum Theory*. (Oxford University Press, Oxford).
6. Bartelmann, M., A. Huss, J. M. Colberg, A. Jenkins, F. R. Pearce (1998): 'Arc Statistics with Realistic Cluster Potentials. IV. Clusters in Different Cosmologies'. *Astron. Astrophys.* **330**, pp. 1-9.
7. Beisbart, C. (2001): *Measuring Cosmic Structure. Minkowski Valuations and Mark Correlations for Cosmological Morphometry*. Ph.D. thesis, Ludwig-Maximilians-Universität München.
8. Beisbart, C., T. Buchert, H. Wagner (2001): 'Morphometry of Spatial Patterns'. *Physica A* **293/3-4**, p. 592.
9. Beisbart, C., K. Mecke (2002): 'Tensor Valuations'. In preparation.
10. Beisbart, C., R. Valdarnini, T. Buchert (2001): 'The Morphological and Dynamical Evolution of Simulated Galaxy Clusters'. *Astron. Astrophys.* **379**, pp. 412–425.
11. Block, D. L., I. Puerari, R. J. Buta, R. Abraham, M. Takamiya, A. Stockton (2001): 'The Duality of Spiral Structure, and a Quantitative Dust Penetrated Morphological Tuning Fork at Low and High Redshift'. In 'ASP Conf. Ser. 230: Galaxy Disks and Disk Galaxies'.
12. Cohen-Tannoudji, C., B. Diu, F. Laloë (1977): *Quantum Mechanics: Volume II*. (John Wiley & Sons, Chichester), 2nd edition.

13. Dahlke, R. (2000), 'Dynamische Perkolationsmodelle zur Morphologie von Galaxien', Diploma thesis, Ludwig-Maximilians-Universität München.
14. Fischer, P., G. Bernstein, G. Rhee, J. A. Tyson (1997): 'The Mass Distribution of the Cluster 0957+561 from Gravitational Lensing'. *A. J.* **113**, p. 521.
15. Gerola, H., P. E. Seiden (1978): 'Stochastic Star Formation and Spiral Structure of Galaxies'. *Ap. J.* **223**, pp. 129-135.
16. Hadwiger, H. (1957): *Vorlesungen über Inhalt, Oberfläche und Isoperimetrie*. (Springer Verlag, Berlin).
17. Hadwiger, H., C. Meier (1974): 'Studien zur vektoriellen Integralgeometrie'. *Math. Nachr.* **56**, p. 361.
18. Hadwiger, H., R. Schneider (1971): 'Vektorielle Integralgeometrie'. *Elemente der Mathematik* **26**, p. 49.
19. Hubble, E. (1926): 'Extra-galactic Nebulae'. *Ap. J.* **64**, p. 321.
20. Kerscher, M. (2000): 'Statistical analysis of large-scale structure in the Universe'. In: *Statistical Physics and Spatial Statistics: The Art of Analyzing and Modeling Spatial Structures and Pattern Formation*, ed. by K. R. Mecke, D. Stoyan, Number 554 in Lecture Notes in Physics (Springer, Berlin), astro-ph/9912329.
21. Klain, D. A. (1995): 'A Short Proof of Hadwiger's Characterization Theorem'. *Mathematika* **42**, pp. 329-339.
22. Kornreich, D. A., M. P. Haynes, R. V. E. Lovelace, L. van Zee (July 2000): 'Departures From Axisymmetric Morphology and Dynamics in Spiral Galaxies'. *A. J.* **120**, pp. 139-164.
23. Lin, C. C., F. H. Shu (1964): 'On the Spiral structure of Disk Galaxies'. *Ap. J.* **140**, p. 646.
24. McMullen, P. (1997): 'Isometry Covariant Valuations on Convex Bodies'. *Rend. Circ. Mat. Palermo (2)* **50**, pp. 259–271.
25. Mecke, K. (2000): 'Additivity, Convexity, and beyond: Application of Minkowski Functionals in Statistical Physics'. In: *Statistical Physics and Spatial Statistics: The Art of Analyzing and Modeling Spatial Structures and Pattern Formation*, ed. by K. R. Mecke, D. Stoyan, Number 554 in Lecture Notes in Physics (Springer, Berlin).
26. Mecke, K. R. (2002): 'Fundamental Measure Density Functional for Mixtures of Non-spherical Hard Particles'. In preparation.
27. Mohr, J. J., D. G. Fabricant, M. J. Geller (1993): 'An x-ray-method for detecting substructure in galaxy clusters'. *Ap. J.* **413**, p. 492.
28. Mohr, J. J., A. E. Evrard, D. E. Fabricant, M. J. Geller (1995): 'Cosmological Constraints from X-ray Cluster Morphologies'. *Ap. J.* **447**, p. 8.
29. Naim, A., O. Lahav, R. J. Buta, H. G. Corwin, G. de Vaucouleurs, A. Dressler, J. P. Huchra, S. van den Bergh, S. Raychaudhury, L. Sodre, M. C. Storrie-Lombardi (1995): 'A Comparative Study of Morphological Classifications of APM Galaxies'. *Mon. Not. R. Astron. Soc.* **274**, pp. 1107-1125.
30. Peacock, J. (1999): *Cosmological Physics*. (Cambridge University Press, Cambridge).
31. Peebles, P. J. E. (1993): *Principles of Physical Cosmology*. (Princeton University Press, Princeton, New Jersey).
32. Reiss, H., H. Frisch, J. L. Lebowitz (1959): 'Statistical Mechanics of Rigid Spheres'. *J. Chem. Phys.* **31**, pp. 369-380.
33. Richstone, D., A. Loeb, E. L. Turner (July 1992): 'A Lower Limit of the Cosmic Mean Density from the Ages of Clusters of Galaxies'. *Ap. J.* **393**, p. 477.
34. Rosenfeld, Y. (1989): 'Free-energy Model for the Inhomogeneous Hard-sphere Fluid Mixture and Density-functional Theory of Freezing', *Phys. Rev. Lett.* **63**, pp. 980-983.
35. Rosenfeld, Y. (1994): 'Density Functional Theory of Molecular Fluids: Free-energy Model for the Inhomogeneous Hard-body Fluid'. *Phys. Rev. E* **50**, pp. R3318-3321.
36. Roth, R. (1999): *Depletion Forces in Hard Sphere Mixtures*. Ph.D. thesis, University Wuppertal, (WUB-DIS 99-19).

37. Russell, W. S., W. W. Roberts (September 1993): 'Analysis of the Distribution of Pitch Angles in Model Galactic Disks – Numerical Methods and Algorithms'. *Ap. J.* **414**, pp. 86-97.
38. Schmalzing, J., T. Buchert (1997): 'Beyond Genus Statistics: a Unifying Approach to the Morphology of Cosmic Structure'. *Ap. J. Lett.* **482**, p. L1.
39. Schmalzing, J., T. Buchert, A. L. Melott, V. Sahni, B. S. Sathyaprakash, S. F. Shandarin (1999): 'Disentangling the Cosmic Web. I. Morphology of Isodensity Contours'. *Ap. J.* **526**, p. 568.
40. Schneider, P., J. . Ehlers, E. E. Falco (1992): *Gravitational Lenses.* (Springer-Verlag, Berlin. Also Astronomy and Astrophysics Library).
41. Schneider, R. (1972): 'Krümmungsschwerpunkte konvexer Körper (I)'. *Abh. Math. Sem. Univ. Hamburg* **37**, p. 112.
42. Schneider, R. (1972): 'Krümmungsschwerpunkte konvexer Körper (II)'. *Abh. Math. Sem. Univ. Hamburg* **37**, p. 202.
43. Schneider, R. (1993): *Convex Bodies: the Brunn-Minkowski Theory.* (Cambridge University Press, Cambridge).
44. Schneider, R. (2000): 'Tensor Valuations on Convex Bodies and Integral Geometry'. *Rend. Circ. Mat. Palermo, Ser. II, Suppl.* **65**, p. 295.
45. Schneider, R., R. Schuster (2001): 'Tensor Valuations on Convex Bodies and Integral Geometry, II'. In preparation.
46. Seiden, P. E., H. Gerola (1982): 'Propagating Star Formation and the Structure and Evolution of Galaxies'. *Fundamentals of Cosmic Physics* **7**, pp. 241-4311.
47. Serra, J. (ed.) (1994): *Mathematical Morphology and its Applications to Image Processing* (Kluwer, Dordrecht).

Computational Topology for Point Data: Betti Numbers of α-Shapes

Vanessa Robins

Department of Applied Mathematics, Research School of Physical Sciences and Engineering, Australian National University, Canberra ACT 0200.
e-mail: vanessa.robins@anu.edu.au

Abstract. The problem considered below is that of determining information about the topology of a subset $X \subset \mathbb{R}^n$ given only a finite point approximation to X. The basic approach is to compute topological properties – such as the number of components and number of holes – at a sequence of resolutions, and then to extrapolate. Theoretical foundations for taking this limit come from the inverse limit systems of shape theory and Čech homology. Computer implementations involve constructions from discrete geometry such as alpha shapes and the minimal spanning tree.

1 Introduction

Two objects have the same topology if they are homeomorphic, i.e. when there is a continuous transformation from one to the other, with a continuous inverse. This means topological properties give a fundamental description of structure, and one that is independent of geometry. It is clear that objects can have the same topology and completely different geometry (a coffee cup and a donut are archetypal examples). However, the converse is also true: objects can have very similar geometric properties but a vastly different topology – the fractals in Sect. 5.2 are one example. Thus, in order to characterize spatial structures, we need both geometric and topological information.

Much attention has been given to the computation of geometric quantities from data, for example the Minkowski functionals [21, 22] and fractal dimensions [3, 26], but the field of 'computational topology' is relatively new [6]. The earliest work on extracting topological information from data targeted digital images. For black and white pixel or voxel images there are algorithms for labelling connected components [16], and for computing the Euler characteristic [19, 25] (a measure of connectivity that also appears as the zero-dimensional Minkowski functional). The technique of erosion and dilation is used in conjunction with these algorithms to probe structure at different length scales in digital images [2].

In this paper we discuss topological properties such as the number of connected components and number of k-dimensional holes, i.e., the Betti numbers, β_k. The Betti numbers are related to the Euler characteristic, χ, via the Euler-Poincaré formula:

$$\chi = \beta_0 - \beta_1 + \beta_2 - \ldots + \beta_d .$$

It follows that the Betti numbers give a more detailed description of topological structure than the Euler characteristic alone. Formal definitions of the Betti numbers are given in Sect. 2.

We assume the data are given as a finite point pattern, $S \subset \mathbb{R}^d$. To give this finite set of points some non-trivial topological structure, we must first fatten or coarse-grain the set. This can be achieved, for example, by overlaying a digital mesh and colouring pixels if they contain data points, or by attaching spheres of radius α at each point. We focus on the latter objects, called alpha neighbourhoods or alpha parallel bodies, because they form a more flexible framework.

Once we choose the coarsening method, we have to set an appropriate level of coarse-graining. This might be physically motivated, or we may have complete freedom to choose. Without any a priori information, it is helpful to coarse-grain at a sequence of resolutions. This has the additional advantages of allowing us to detect fractal scaling, and if the data, S, approximate an underlying object, $X \subset \mathbb{R}^d$, we can extrapolate information about the topology of X. Theoretical underpinnings of this coarse-graining and extrapolation are described in Sect. 3.

To compute Betti numbers, we need a discrete complex (e.g., a triangulation) that has the same topology as the alpha neighbourhood. One solution comes from Edelsbrunner's alpha shapes [8] and their construction is described in Sect. 4. We use the publicly available alpha shape software [1] to compute Betti numbers of some example data in 2D and present the results in Sect. 5.

2 Homology Groups and Betti Numbers

Homology theory is a branch of topology that attempts to distinguish between objects by constructing algebraic invariants that reflect their connectivity properties. Readers should be warned that the homology groups do not completely determine the topology of an object. If two objects have different homology groups, then they certainly have different topologies. The converse, however, does not hold. For example, the set X shown in Fig. 3 has the same homology groups as a circle, but it is not homeomorphic to a circle because of the branch-point where the 'O' joins the 'C'.

In this section, we give a brief outline of the definitions for simplicial homology, following Munkres [24]. The basic building block is an *oriented k-simplex*, σ^k, the convex hull of $k+1$ geometrically independent points, $\sigma^k = [x_0, x_1, \ldots, x_k]$. Its orientation is defined by an arbitrary but fixed ordering of the vertices. Even permutations of this ordering give the same orientation and odd permutations reverse it. For example, a 0-simplex is just a point, a 1-simplex is a line segment, a 2-simplex a triangle, and a 3-simplex is a tetrahedron. A *simplicial complex*, $\mathcal{C}$, is a collection of oriented simplices with the property that the non-empty intersection of two simplices in $\mathcal{C}$ must itself be a simplex in $\mathcal{C}$. The set-theoretic union of all simplices from $\mathcal{C}$, when viewed as a subset of $\mathbb{R}^d$, is called a *polytope*. If a subset X of $\mathbb{R}^d$ is homeomorphic to a polytope, we say X is *triangulated* by $\mathcal{C}$.

The simplicial complex, $\mathcal{C}$, is given a group structure by defining the addition of k-simplices in a similar manner to addition in a vector space. The resulting free group is called the *chain group*, C_k. Its elements consist of *k-chains*, the sum of a finite number of oriented k-simplices: $c_k = \sum_i a_i \sigma_i^k$. The coefficients, a_i, are typically integers, but in general they can be elements of any abelian group, G. In this paper, we use the rational or real numbers for coefficients since this simplifies the homology group structure (we

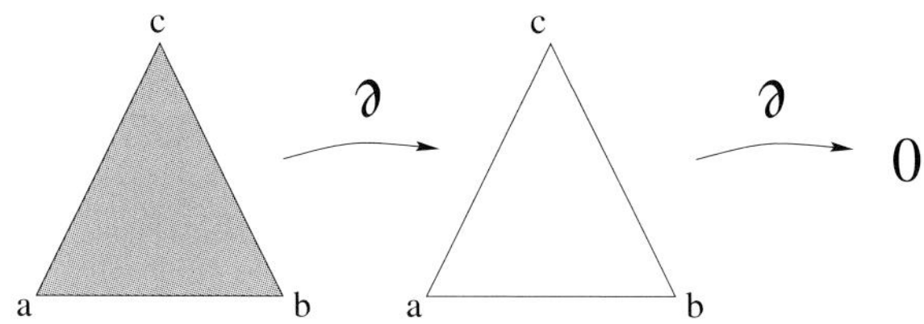

Fig. 1. The boundary operator, ∂, maps a 2-simplex onto the sum of three 1-simplices $\partial[a, b, c] = [b, c] - [a, c] + [a, b]$. The boundary of this 1-chain is empty. This property holds for any k-chain: $\partial\partial(c_k) = 0$.

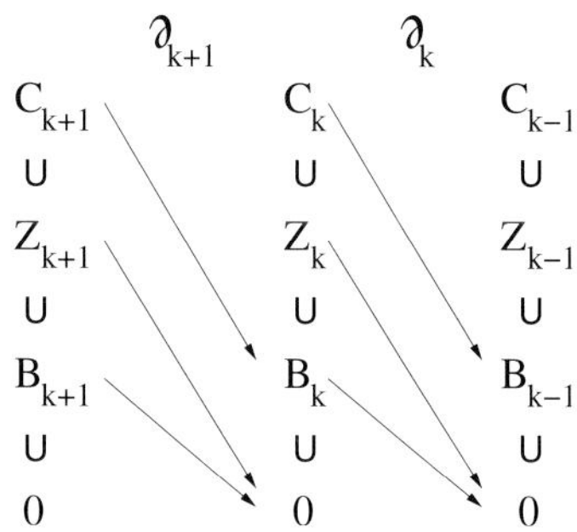

Fig. 2. The boundary operator, ∂_k, maps all k-chains onto the $k-1$-boundaries and the k-cycles to zero.

can ignore torsion). A negative coefficient is interpreted as reversing the orientation of the simplex: $-[x_0, x_1, \ldots, x_k] = [x_1, x_0, \ldots, x_k]$.

The next step towards defining the homology groups is to look at how the k-chains are related to the $k-1$-chains. This is done using the *boundary operator*, $\partial_k : C_k \to C_{k-1}$, a linear operator that maps a k-simplex onto the oriented sum of all $(k-1)$-simplices in its boundary

$$\partial_k[x_0, x_1, \ldots, x_k] = [x_1, x_2, \ldots, x_k] - [x_0, x_2, \ldots, x_k] + \ldots \\ +(-1)^k[x_0, x_1, \ldots, x_{k-1}].$$

For example, in Fig. 1 the boundary of the triangle is its three edges.

The action of the boundary operator on the chain groups leads to the definition of three more groups. Firstly, the image of ∂_k is a subgroup of C_{k-1} called the boundary group, and is denoted B_{k-1}. Secondly, the set of all k-chains that have empty boundary forms the group of k-cycles, Z_k (i.e. Z_k is the kernel or null space of ∂_k). These two groups are related by the fact that the boundary of a boundary is empty. This is the fundamental property of the boundary operator, $\partial_k\partial_{k+1} = 0$. It implies that B_k is a subgroup of Z_k. The relationships between the chain, cycle, and boundary groups are illustrated in Fig. 2.

The structures we are really interested in are the k-cycles that do not bound anything, since these indicate the presence of a k-dimensional 'hole'. It is this idea that leads to the definition of the homology groups as the quotient group $H_k = Z_k/B_k$. This means

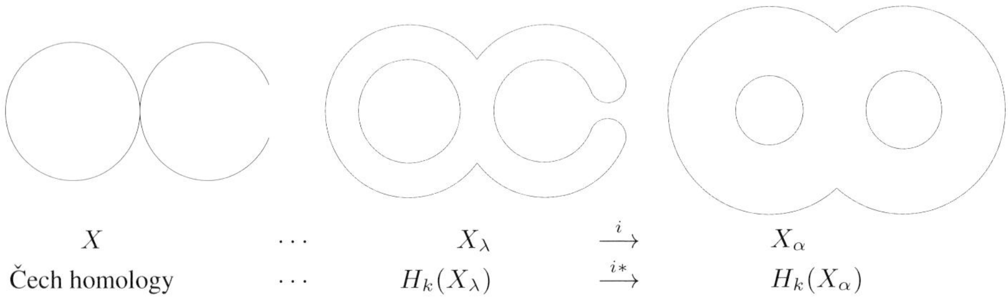

Fig. 3. The inclusion maps, $i : X_\lambda \to X_\alpha$, allow us to detect those holes which persist in the limit. In this example $\beta_1(\alpha) = 2$, but $\beta_1^\lambda(\alpha) = 1$.

that two k-cycles w_k and z_k belong to the same homology class if the k-chain formed by their difference is the boundary of some $(k+1)$-chain, $z_k - w_k = \partial v_{k+1}$.

The number of distinct equivalence classes of H_k is the kth Betti number β_k. The kth Betti number effectively counts the number of k-dimensional holes in X, so is exactly the type of information we seek. When $k = 0$, the Betti number counts the number of path-connected components of X. For subsets of $\mathbb{R}^3$, we can interpret β_1 as the number of independent tunnels, and β_2 as the number of enclosed voids. For example, the solid torus has $\beta_0 = 1$, $\beta_1 = 1$, and $\beta_2 = 0$ (the same as a simple circle), whereas the surface of a torus has $\beta_0 = 1$, $\beta_1 = 2$, and $\beta_2 = 1$.

Given a finite simplicial complex, the computability of Betti numbers using linear algebra techniques is well established [24]. However, these algorithms have run times at best cubic in the number of simplices. The development of fast algorithms for computational homology is thus an active area of research [5, 7, 13, 18]. We will describe one of these algorithms in Sect. 4.2.

3 A Theoretical Framework for Computational Homology

As mentioned in the introduction, we are interested in the situation where a finite number of points, S, approximate a subset $X \subset \mathbb{R}^d$. Our hope is to obtain accurate information about the topological structure of X by coarse-graining the data S at a sequence of resolutions, then extrapolating the limiting behaviour. There are two problems to be addressed here. First, is the extrapolation valid, even supposing we have perfect information about X? Second, what happens when we only have the finite approximation, S? The mathematical theory that helps answer these questions comes from Čech homology [15] and shape theory [20]. We give a brief overview of the concepts here – details can be found in [30, 31].

We can coarse-grain a set by taking its intersection with a digital mesh, or by applying opening and closing operations. In this paper we use the closed α-neighbourhood of a set (also known as the parallel set)

$$X_\alpha = \{y \text{ such that } \inf_{x \in X} |y - x| \leq \alpha\}$$

where $|.|$ is the Euclidean metric on $\mathbb{R}^d$. The following theory is easily adapted to the case of digital meshes and other standard methods of coarse-graining.

Our question now is: for what spaces do the Betti numbers of the α-neighbourhoods, X_α, converge to those of X? That is, does $\beta_k(X_\alpha) \to \beta_k(X)$ as $\alpha \to 0$? For the number of components, $\beta_0(X_\alpha)$, this holds for any compact set (i.e. closed and bounded subsets of $\mathbb{R}^d$) [15, 28]. This restriction to compact sets is not unreasonable, since we are primarily interested in objects that are well approximated on a computer by a finite number of points.

Higher-order Betti numbers present a more subtle problem, because fattening a set to its α-neighbourhood can introduce holes, as illustrated in Fig. 3, and also remove holes by filling them in. Mathematically, this problem is resolved by incorporating information about how a smaller neighbourhood maps inside a larger one. This leads to the definition of *persistent Betti number*. Suppose $0 \leq \lambda < \alpha$, so that $X_\lambda \subset X_\alpha$. The inclusion map $i : X_\lambda \to X_\alpha$ induces a homomorphism on the homology groups, $i* : H_k(\lambda) \to H_k(\alpha)$. We define the λ-persistent Betti number, $\beta_k^\lambda(\alpha)$, to be the rank of $i*$, i.e. the number of non-equivalent, non-bounding k-cycles in $H_k(\alpha)$ that are the image of a k-cycle from $H_k(\lambda)$. This leads to the formula

$$\beta_k^\lambda(\alpha) = \text{rank}[Z_k(\lambda)] - \text{rank}[i_*(Z_k(\lambda)) \cap B_k(\alpha)] \tag{1}$$

Geometrically, $\beta_k^\lambda(\alpha)$ is the number of holes in X_λ that do not get filled in by taking a fatter neighbourhood, X_α.

Cycles in X_α that genuinely come from a cycle in X are the 0-persistent Betti numbers, $\beta_k^0(\alpha) \leq \beta_k^\lambda(\alpha)$ for all $0 \leq \lambda \leq \alpha$. From the continuity of Čech homology, we know that the 0-persistent Betti numbers of the α-neighbourhoods converge to those of the original space: $\beta_k^0(\alpha) \to \beta_k(X)$, when X is compact. Thus, to find the regular Betti number of X, $\beta_k(X)$, from the persistent Betti numbers $\beta_k^\lambda(\alpha)$, we must first fix a sequence of α-values and let $\lambda \to 0$ to get the limits $\beta_k^\lambda(\alpha_n) \to \beta_k^0(\alpha_n)$. Then we can find $\beta_k(X)$ as the limit of this sequence: for $\alpha_n \to 0$, $\beta_k^0(\alpha_n) \to \beta_k(X)$.

The second problem we must address concerns the relationship between the neighbourhoods of the data, S_α, and those of the set they approximate, X_α. Intuitively speaking, the idea is that if the data and the underlying set are "close" then their α-neighbourhoods should have similar topological properties, provided α is sufficiently large. We measure the distance between the data, S, and the underlying space, X, using the Hausdorff metric on compact sets. This is defined as

$$d_H(S, X) = \min\{\alpha \mid X \subset S_\alpha \text{ and } S \subset X_\alpha\}. \tag{2}$$

If $\rho = d_H(S, X)$, it follows that $S_\alpha \subset X_{\alpha+\rho}$ and $X_\alpha \subset S_{\alpha+\rho}$ for all $\alpha > 0$. These inclusions allow us to derive a number of inequalities relating the persistent Betti numbers of X to those of S [31]. For example, we have that for $\alpha > \rho$, $\beta_k^0(X_{\alpha+\rho}) \leq \beta_k^\rho(S_\alpha)$ and $\beta_k^{\alpha-\rho}(S_{\alpha+\rho}) \leq \beta_k(X_\alpha)$. For the number of connected components, the persistent Betti number is the same as the regular Betti number and we find that $\beta_0(S_{\alpha-\rho}) \geq \beta_0(X_\alpha) \geq \beta_0(S_{\alpha+\rho})$. We refer to ρ as the *cutoff resolution*, since we cannot hope to get valid topological information about X from S_α unless $\alpha > \rho$.

4 Computer Implementation

We now describe some algorithms for computing Betti numbers from scattered point data at a sequence of α-values. We begin with the simple problem of counting the connected components of S_α.

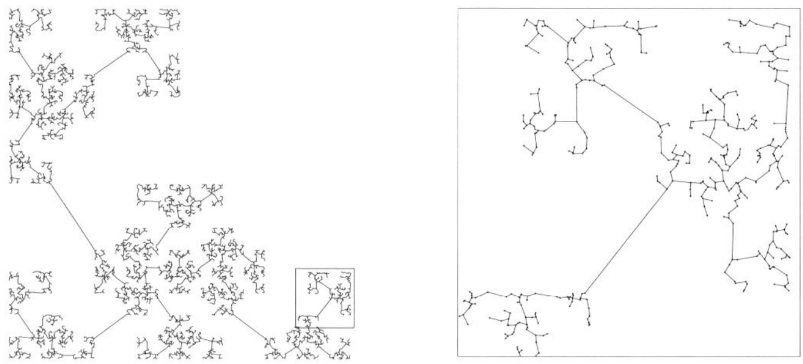

Fig. 4. The MST of 10^4 points on a Cantor set, and a close up of a small region.

4.1 Connected Components

For scattered point data, $S \subset \mathbb{R}^d$, the Euclidean minimal spanning tree (MST) encodes all the relevant information about connected components at all resolutions [29]. Recall that the MST is a tree (i.e. a graph with no cycles) that passes through every point in such a way as to minimize the total length. There are a number of algorithms for constructing the MST; the most easily implemented one is due Prim [4, 27]. This starts with any point and joins its nearest neighbour to create an initial subtree. The algorithm proceeds incrementally by adding the point which is closest to the existing subtree, finishing when all points have been added. The computational complexity is quadratic in the number of data points.

The property of connecting closest points gives the MST a natural correspondence with the connected components of the α-neighbourhood, S_α, [29]. In fact the number of components, $\beta_0(\alpha)$, is just one more than the number of MST edges that are longer than 2α:

$$\beta_0(\alpha) = 1 + \#\{\text{MST edges} > 2\alpha\}. \tag{3}$$

Note that we need only build the MST once to get component information at all scales.

4.2 Alpha Shapes and Homology

We now want to compute higher-order Betti numbers of the α-neighbourhoods S_α for a sequence of α values. This requires a sequence of simplicial complexes that reflect the topology of the α-neighbourhoods. An elegant solution to this problem comes from alpha shapes - a construction based on the Voronoi diagram and its dual, the Delaunay triangulation. Alpha shapes were defined by Edelsbrunner *et al.* for finite sets of points

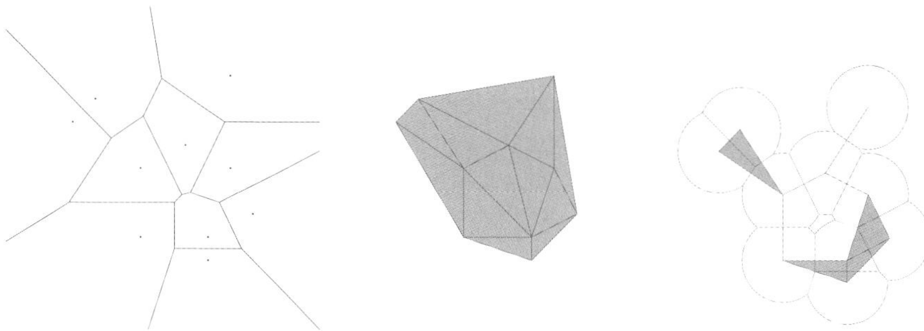

Fig. 5. Left: the Voronoi diagram of ten points. Middle: the corresponding Delaunay triangulation. Right: The alpha diagram is the intersection of the Voronoi diagram with the union of balls of radius α centred on the data points; it is shown together with its geometric dual, the alpha complex.

in the plane [9], then generalised to higher dimensions [11], and to points with weights [8]. Here we describe unweighted alpha shapes, and an algorithm to compute their Betti numbers.

Let $S = \{x_1, \ldots, x_n\}$ be a set of n points in $\mathbb{R}^d$ which satisfy the general position assumption that no four points simultaneously lie on the boundary of a circle. We write $V(x_i)$ for the Voronoi cell of x_i, i.e. the region that contains all points in $\mathbb{R}^d$ closer to x_i than to any other point in S. The dual tessellation of the Voronoi diagram is the Delaunay complex, $\mathcal{D}(S)$. This is obtained by associating a point with each Voronoi cell, an edge with two cells that share a $d-1$ face, a triangle with any triple of intersecting cells, and so on. See Fig. 5 for an example in the plane.

Edelsbrunner introduces a resolution parameter by taking the intersection of the Voronoi cell of each x_i with a ball of radius α centered at x_i, $B_\alpha(x_i)$. This gives us the *alpha diagram*:

$$\mathcal{Q}_\alpha(S) = \bigcup_{i=1}^{n} B_\alpha(x_i) \cap V(x_i). \tag{4}$$

The *alpha complex*, $\mathcal{C}_\alpha(S)$, is the dual of the alpha diagram. The *alpha shape* is the subset of $\mathbb{R}^d$ covered by the alpha complex. Figure 5 shows, for a simple example, the alpha diagram and its dual alpha complex. From the above definitions we have that $\mathcal{C}_\alpha(S) \subset \mathcal{D}(S)$ for all $\alpha > 0$. Also, if $\alpha_1 < \alpha_2$, then $\mathcal{Q}_{\alpha_1}(S) \subset \mathcal{Q}_{\alpha_2}(S)$, and this implies that $\mathcal{C}_{\alpha_1}(S) \subset \mathcal{C}_{\alpha_2}(S)$. Note that for sufficiently small α, the alpha complex consists of the vertices, $x_1, \ldots, x_n$, while at the other extreme, for large enough alpha, we get the Delaunay complex, $\mathcal{C}_\alpha(S) = \mathcal{D}(S)$.

The power of this construction comes from transforming the "continuous space" alpha diagrams into the discrete geometry of alpha complexes. The nerve theorem in topology [8, 15] guarantees that the alpha diagram is homotopy equivalent to the alpha shape. Thus, for a given alpha, each of these objects has the same Euler characteristic and Betti numbers.

Delfinado and Edelsbrunner's algorithm [5] for computing the Betti numbers of a simplicial complex relies on a sequential ordering of simplices, termed a filtration. The subset property $\mathcal{C}_{\alpha_1} \subset \mathcal{C}_{\alpha_2}$, for $\alpha_1 < \alpha_2$, means that every simplex σ has a threshold,

$\alpha_c(\sigma)$, such that $\sigma \in \mathcal{C}_\alpha$ for all $\alpha > \alpha_c(\sigma)$. The filtration orders the simplices according to their thresholds. Note that two simplices may have the same threshold, and in this situation they are ordered by adding the lower dimensional simplices first. The Betti numbers are computed incrementally as each simplex is added to the complex. This process depends on a test to determine whether the new simplex belongs to a k-cycle of the new complex. There are efficient algorithms for testing 1-cycles, and homology-cohomology duality theorems transform the $(d-1)$-cycles into 1-cocycles that are equally easy to test for. However, there is no fast test for other k-cycles, so Delfinado and Edelsbrunner's algorithm applies only to subcomplexes of $\mathbb{R}^2$ or $\mathbb{R}^3$.

The computational complexity involved in building the Delaunay triangulation in $\mathbb{R}^3$ is quadratic in the number of points, and the incremental algorithm for computing Betti numbers is barely super-linear in the number of simplices. The NCSA ftp site provides software that implements the above alpha shape constructions in $\mathbb{R}^2$ and $\mathbb{R}^3$ [1]. The Betti number data for the examples in the following section were generated using this software.

The numerical computation of persistent Betti numbers is possible using linear algebra techniques but is not yet implemented. Recently, Edelsbrunner [10] has made a similar definition of persistence specific to alpha shapes and has developed an incremental algorithm for this quantity.

5 Example Data

The random point clouds we study tend to have three regimes determined by the cutoff resolution ρ discussed in Sect. 3. For $\alpha \gg \rho$ we recover the coarse-scale topology of the approximated region X. At the other extreme, $\alpha \ll \rho$ we see purely the statistical effects of a point process. When α is on the same order as ρ, there is transitional behaviour as the finite data effects begin to dominate the underlying topological structure.

In the following two subsections we analyse the Betti numbers for data from a binomial point process and a family of self-similar fractals. For a finite domain, the coarse scale topology of the binomial point pattern is extremely simple, so the interesting Betti number behaviour occurs only for $\alpha \leq \rho$. The fractal examples have identical fractal dimension, but greatly differing topological structure. The Betti number data at coarse resolutions distinguishes between the different topological types. For $\alpha < \rho$, however, the fractals show essentially the same behaviour, presumably dictated by the iterated function system technique of generating the data points.

5.1 Binomial Point Process

We begin by analysing N points that are randomly distributed in the unit square according to a binomial point process. The coordinates of each point are assigned using the uniform distribution on $[0, 1]$ and the point pattern therefore constitutes a finite-domain approximation to a Poisson process with intensity $\lambda = N/1$. We describe the behaviour of a typical single realisation of N points, and then present results from a numerical investigation of the mean values of β_0 and β_1 as functions of the radius α.

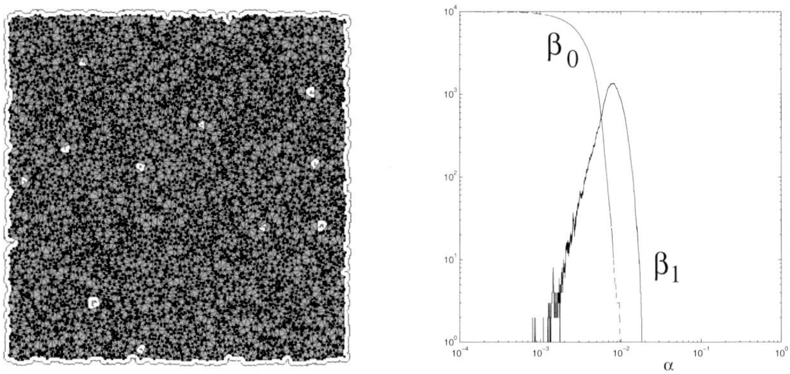

Fig. 6. Left: An example alpha neighbourhood (faint outline) and alpha shape (shaded region) of 10^4 points sampled from the uniform distribution on the unit square. On the right we plot the Betti number data on logarithmic axes; — — the number of components β_0, and – number of holes β_1.

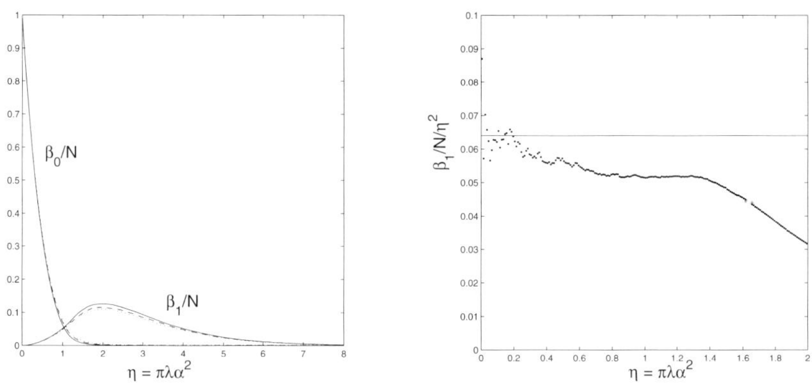

Fig. 7. On the left we show that the Betti number data rescale to functions of $\eta = \pi\lambda\alpha^2$. We compute sample mean values of β_0/N and β_1/N for $N = 500$ $(\cdots)$, $N = 1000$ (- -), and $N = 5000$ (—).

Figure 6 shows for an example the α-shape of $N = 10^4$ points. The points form an approximation to the unit square, a connected region that has extremely simple topology and geometry. For α above the cutoff resolution we expect to see a single connected component with no holes, and this is reflected by the data presented in Fig. 6 for $\alpha > 0.04$. For sufficiently small values of α, each point is a single component, so there are no holes and $\beta_0(\alpha) = N$, the number of points. As α increases, edges are added to the alpha complex, decreasing β_0 and possibly increasing β_1. When a triangle is added to the complex, β_1 may decrease; this is the reason for the noisy non-monotonic behaviour in the number of holes. As alpha approaches the cutoff resolution there is a sharp drop in the number of components until the point when the alpha neighbourhood forms a single connected component. At around the same α-value, the number of holes reaches a maximum then starts to decrease with increasing α as the holes are filled in.

Readers familiar with the field of stochastic geometry will recognize that the α-neighbourhoods of a random point process are analogous to a Boolean model with circular grains of radius α. Most geometric quantities associated with such a circular-grain Boolean model are really functions of $\eta = \pi\lambda\alpha^2$ [32]. For example, the mean Euler characteristic per unit area is $e = \lambda(1-\eta)e^{-\eta}$ [21]. Unsurprisingly, the expected value of the mean Betti numbers per unit area (written b_0 and b_1) are also functions of this quantity. The graphs in Fig. 7 show the mean specific Betti number curves from 500, 1000, and 5000 points in the unit square.

Hopes for deriving analytic formulas for the mean Betti numbers are limited by their connection with percolation. (The percolation transition is signalled by a singularity in the second moment of the cluster size distribution and β_0 is the total number of clusters.) However, an expression for the expected number of components per unit area in the low-intensity limit, $\lambda \to 0$, is known [14]:

$$b_0(\eta) = \lambda(1 - 2\eta + \tfrac{2}{3}(4 - 3\sqrt{3}/\pi)\eta^2) + O(\eta^3).$$

Assuming $e = b_0 - b_1$ [23], we can find the limiting behaviour of $b_1(\eta)$ as $\eta \to 0$. Specifically

$$b_1(\eta) = \lambda(\tfrac{7}{6} - 2\sqrt{3}/\pi)\eta^2 + O(\eta^3) \approx 0.0640\lambda\eta^2 + O(\eta^3).$$

Data from the alpha shapes closely match these expressions, see Fig. 7.

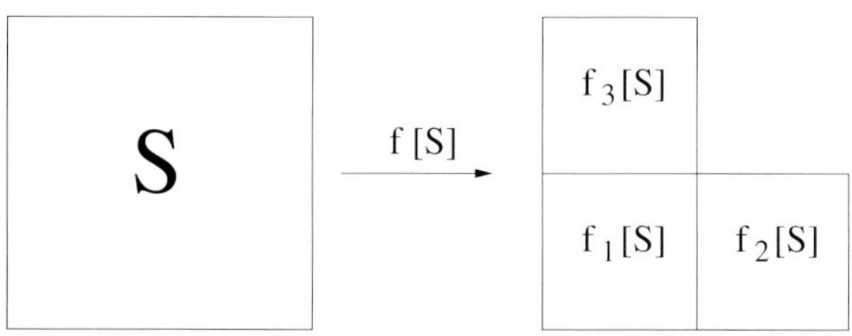

Fig. 8. Template for the iterated function system that generates the Sierpinski triangle relatives.

5.2 Sierpinski Triangle Relatives

The fractals we study here are a family of iterated function systems [12], illustrated by the template in Fig. 8:

$$S = f[S] = f_1[S] \cup f_2[S] \cup f_3[S].$$

The functions f_i are similarity transformations of the unit square with contraction ratio $\frac{1}{2}$, i.e. $|f_i(x) - f_i(y)| = \frac{1}{2}|x - y|$. Those that generate the Sierpinski triangle, shown in Fig. 9, are simple contractions followed by a translation; the generators of the other examples involve additional rotation or reflection symmetries of the square. There are 232 different fractals in this family [26]. Their topology ranges from simply connected

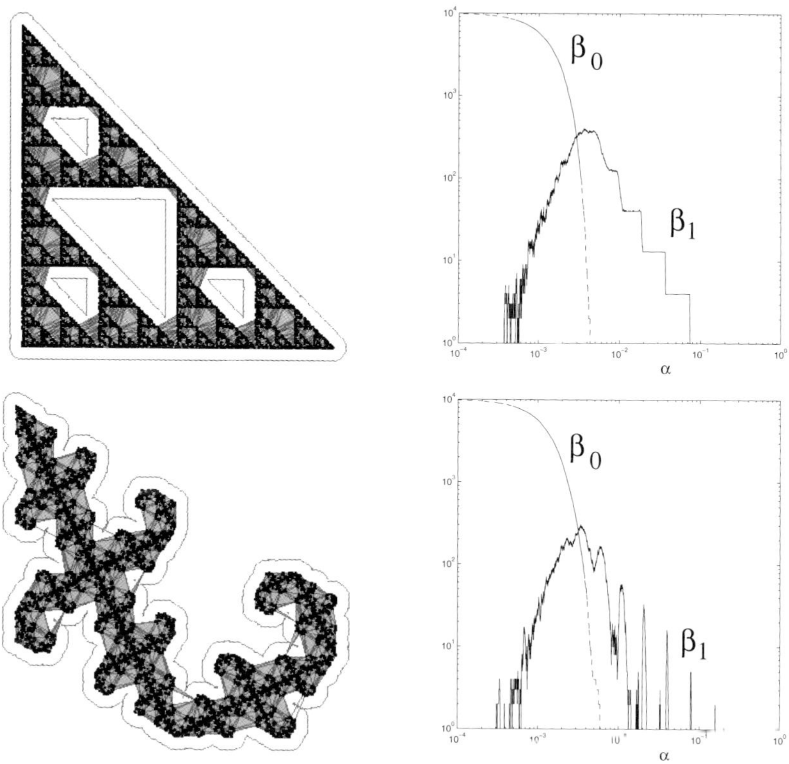

Fig. 9. Left column: an example alpha neighbourhood (outline) and alpha shape (shaded) of 10^4 points on two fractals. Right column: Betti number data on logarithmic axes. The Sierpinski triangle (top row) is a connected set with infinitely many holes. The Betti numbers of its alpha neighbourhoods are $\beta_0(\alpha) = 1$ for $\alpha \geq 0$, and $\beta_1(\alpha_1/2^n) = \frac{1}{2}(3^{n+1} - 1)$, where $n = 0, 1, 2, \ldots$ and α_1 is the radius of the circle inside the largest triangular hole. This behaviour is accurately reflected by data from the finite point approximation for $\alpha \geq 0.005$. The other fractal is simply connected, so in principle $\beta_0 = 1$ and $\beta_1 = 0$. There are spikes in the hole data however, and these are due to the geometry of the fractal – the way it hooks around creates non-persistent holes in the α-neighbourhoods.

(Fig. 9 bottom) to connected (Fig. 9 top) to totally disconnected (Fig. 10 top) to a class of examples with infinitely many connected components of non-zero diameter (Fig. 10 bottom). This range of topological structure makes them ideal test cases for our techniques [28, 29, 31].

We generate a finite number of points on a fractal using Barnsley's chaos game technique [3]. This records the trajectory of an initial point under the iteration $x_{n+1} = f_{i_n}(x_n)$, setting $i_n = 1, 2$ or 3 with equal probability at each step.

We present data for the number of components, $\beta_0(\alpha)$, and the number of holes, $\beta_1(\alpha)$, for a fractal of each topological type in Fig. 9 and Fig. 10. The Betti numbers are plotted on a logarithmic scale to show the self-similar fractal scaling.

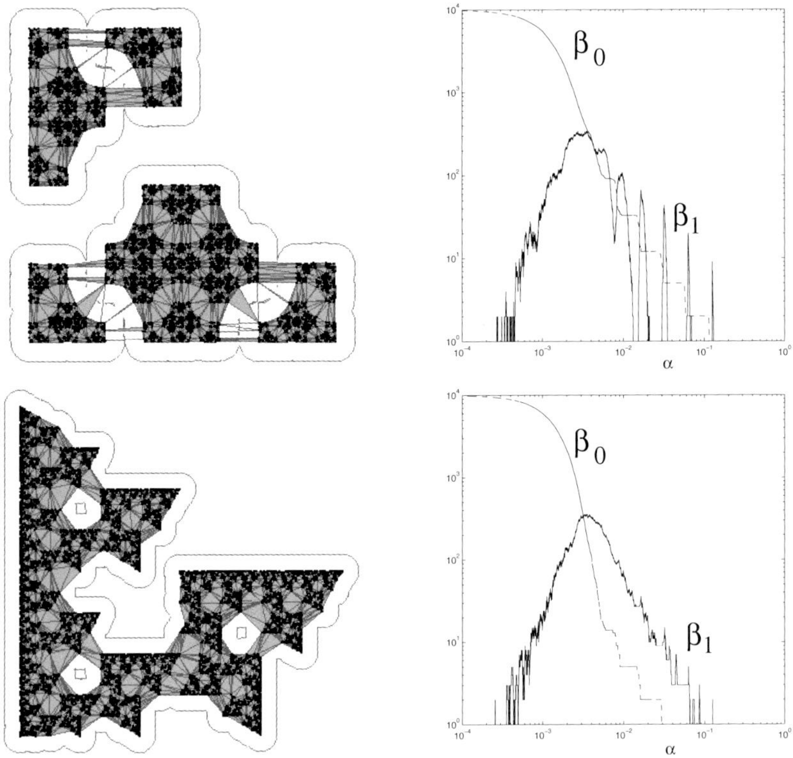

Fig. 10. Left column: an example alpha neighbourhood (outline) and alpha shape (shaded) of 10^4 points on two fractals. Right column: Betti number data on logarithmic axes. The upper fractal is a Cantor set so it is totally disconnected and perfect. We find [28] $\beta_0(\alpha_0/2^n) \to 3^n$ as $n \to \infty$ (where α_0 is the width of the largest gap) and $\beta_1(\alpha) = 0$. Although topologically the Cantor set does not contain any holes, the graph of $\beta_1(\alpha)$ has regularly spaced spikes. The reason for these is seen in the α-neighbourhood picture. The edges bridging the gaps cause holes in the triangulation that disappear for slightly smaller values of α when the edges are removed from the alpha shape. The lower fractal is disconnected, and consists of infinitely many line segments. Again, there are no topological holes in this fractal. It is the geometry of the set, i.e., the arrangement of the line segments in the plane, that creates holes in the α-neighbourhoods. Again, $\beta_0(\alpha_0/2^n) \to 3^n$. The regular Betti number shows fractal growth $\beta_1(\alpha_1/2^n) = 3^n$, while the 0-persistent Betti numbers are $\beta_1^0(\alpha) = 0$.

Each of these fractals has a cutoff resolution $\rho \approx 0.005$. For $\alpha < \rho$, the behaviour of the Betti number data is quantitatively similar for each fractal and qualitatively similar to that of the point process in Sect. 5.1. Preliminary analysis suggests that for each of these fractals, the Betti numbers are really functions of the rescaled quantity $\eta = \pi N \alpha^{d_f}$, where $d_f = \log 3/\log 2$ is the fractal dimension. In particular we find $\beta_0(\eta)/N \to 1 - \eta$ and $\beta_1(\eta) = O(\eta^{d_f})$ as $\eta \to 0$.

It is in the region $\alpha > \rho$ that we see the effects of the different topology of each fractal. The difference between the connected and disconnected fractals shows up clearly in the number of components. The connected sets in Fig. 9 have $\beta_0(\alpha) = 1$ for $\alpha > \rho$. The difference between the Sierpinski triangle and the simply connected fractal is seen the behaviour of β_1. The Sierpinski triangle has regular staircase growth in the number of holes, as expected, while the simply connected fractal has no persistent holes.

The disconnected fractals show a stair-step decrease in the number of components due to their self-similar structure. The slope is equal to the fractal dimension, $\log 3/\log 2$, in both cases. In order to distinguish between the totally disconnected Cantor set, and the other disconnected fractal, we need information about the size of the connected components [28, 29]. (For a totally disconnected set, the diameters go to zero as $\alpha \to 0$.) The disconnected fractal at the bottom of Fig. 10 illustrates the need to compute the persistent Betti numbers to obtain an accurate diagnosis of the underlying topology.

Acknowledgements

This paper is based on part of my dissertation [30]. I would like to thank my thesis advisors Professors J.D. Meiss and E. Bradley for many stimulating discussions in relation to this work. I am also grateful to Dr Klaus Mecke for inviting me to attend the Second Conference on Spatial Statistics and Statistical Physics.

References

1. Software available from NCSA via anonymous ftp from ftp.ncsa.uiuc.edu/Visualization/Alpha-shape/.
2. Arns, C.H., M.A. Knackstedt, W.V. Pinczewski, and K. Mecke (2001): 'Characterisation of irregular spatial structures by parallel sets and integral geometric measures'. *Applied Mathematics*, preprint. (RSPhysSE, ANU, Canberra)
3. Barnsley, M.F. (1993): *Fractals Everywhere*, second edition. (Academic Press, Boston)
4. Cormen, T.H., C.E. Leiserson, and R.L. Rivest (1990): *Introduction to Algorithms*. (MIT Press, Cambridge, MA)
5. Delfinado, C.J.A., and H. Edelsbrunner (1995): 'An incremental algorithm for Betti numbers of simplicial complexes on the 3-sphere'. *Computer Aided Geometric Design* **12**, pp. 771–784.
6. Dey, T.K., H. Edelsbrunner, and S. Guha. (1999): 'Computational topology'. In: *Advances in Discrete and Computational Geometry*, volume 223 of *Contemporary Mathematics*. ed. by B. Chazelle, J.E. Goodman, and R. Pollack (American Mathematical Society)
7. Dey, T.K., and S. Guha. (1998): 'Computing homology groups of simplicial complexes in $\mathbf{R}^3$'. *Journal of the ACM* **45** pp. 266–287.
8. Edelsbrunner, H. (1995): 'The union of balls and its dual shape'. *Discrete and Computational Geometry* **13**, pp. 415–440.
9. Edelsbrunner, H., D.G. Kirkpatrick, and R. Seidel (1983): 'On the shape of a set of points in the plane'. *IEEE Transactions on Information Theory* **29**, pp. 551–559.
10. Edelsbrunner, H., D. Letscher, and A. Zomorodian (2000): 'Topological persistence and simplification'. *41st Annual Symposium on Foundations of Computer Science*.
11. Edelsbrunner, H., and E.P. Mücke (1994): 'Three-dimensional alpha shapes'. *ACM Transactions on Graphics* **13**, pp. 43–72.
12. Falconer, K. (1990): *Fractal Geometry: Mathematical Foundations and Applications*. (Wiley, Chichester)

13. Friedman, J. (1998): 'Computing Betti numbers via combinatorial Laplacians'. *Algorithmica* **21**, pp. 331–346.
14. Hall, P. (1988): *Introduction to the Theory of Coverage Processes*. (Wiley, Chichester)
15. Hocking, J.G., and G.S. Young (1961): *Topology*. (Addison-Wesley)
16. Hoshen, J. and Kopelman, R. (1976): 'Percolation and cluster distribution. I. Cluster multiple labeling technique and critical concentration algorithm'. *Physical Review B* **14**, pp. 3438–3445.
17. Hutchinson, J.E. (1981): 'Fractals and self similarity'. *Indiana University Mathematics Journal* **30**, pp. 713–747.
18. Kalies, W.D., K. Mischaikow, and G. Watson (1999): 'Cubical approximation and computation of homology'. *Banach Center Publications* **47**, pp. 115–131.
19. Lee, C.-N., T. Posten, and A. Rosenfeld (1991): 'Winding and Euler numbers for 2D and 3D digital images'. *CVGIP: Graphical Models and Image Processing* **53**, pp. 522-537.
20. Mardešić, S. and J. Segal (1982): *Shape Theory*. (North-Holland)
21. Mecke, K. (1998): 'Integral geometry and statistical physics'. *International Journal of Modern Physics B* **12**, pp. 861–899.
22. Mecke, K. and D. Stoyan (Eds.) (2000): *Statistical Physics and Spatial Statistics – The Art of Analyzing and Modeling Spatial Structures and Pattern Formation* Lecture Notes in Physics **554** (Springer-Verlag, Berlin)
23. Mecke, J., and Stoyan, D. (2001): 'The specific connectivity number of random networks'. *Advances in Applied Probability* **33**, pp. 576–583.
24. Munkres, J.R. (1984): *Elements of Algebraic Topology*. (Benjamin Cummings)
25. Nagel, W., J. Ohser, and K. Pischang (2000): 'An integral-geometric approach for the Euler-Poincaré characteristic of spatial images'. *Journal of Microscopy* **198**, pp. 54–62.
26. Peitgen, H.-O., S. Jürgens, and D. Saupe. (1992): *Chaos and Fractals: New Frontiers of Science*. (Springer-Verlag, Berlin)
27. Prim, R.S. (1957): 'Shortest connection networks and some generalizations'. *The Bell System Technical Journal* November 1957, pp. 1389–1401.
28. Robins, V., J.D. Meiss, and E. Bradley (1998): 'Computing connectedness: An exercise in computational topology'. *Nonlinearity* **11**, pp. 913–922.
29. Robins, V., J.D. Meiss, and E. Bradley (2000): 'Computing connectedness: Disconnectedness and discreteness'. *Physica D* **139**, pp. 276–300.
30. Robins, V. (2000): *Computational topology at multiple resolutions*. PhD Thesis, Department of Applied Mathematics, University of Colorado, Boulder.
31. Robins, V. (1999): 'Towards computing homology from finite approximations'. *Topology Proceedings* **24**, pp. 503–532.
32. Stoyan, D., W.S. Kendall, and J. Mecke (1987): *Stochastic Geometry and its Applications*. (Wiley, Chichester)

The Euler Number of Discretized Sets – On the Choice of Adjacency in Homogeneous Lattices

Joachim Ohser[1], Werner Nagel[2], and Katja Schladitz[1]

[1] Fraunhofer Institut für Techno- und Wirtschaftsmathematik, Gottlieb-Daimler-Straße, 67663 Kaiserslautern, Germany
[2] Friedrich-Schiller-Universität Jena, Fakultät für Mathematik und Informatik, 07740 Jena, Germany

Abstract. Two approaches for determining the Euler-Poincaré characteristic of a set observed on lattice points are considered in the context of image analysis – the integral geometric and the polyhedral approach. Information about the set is assumed to be available on lattice points only. In order to retain properties of the Euler number and to provide a good approximation of the true Euler number of the original set in the Euclidean space, the appropriate choice of adjacency in the lattice for the set and its background is crucial. Adjacencies are defined using tessellations of the whole space into polyhedrons. In $\mathbb{R}^3$, two new 14 adjacencies are introduced additionally to the well known 6 and 26 adjacencies. For the Euler number of a set and its complement, a consistency relation holds. Each of the pairs of adjacencies $(14.1, 14.1)$, $(14.2, 14.2)$, $(6, 26)$, and $(26, 6)$ is shown to be a pair of complementary adjacencies with respect to this relation. That is, the approximations of the Euler numbers are consistent if the set and its background (complement) are equipped with this pair of adjacencies. Furthermore, sufficient conditions for the correctness of the approximations of the Euler number are given. The analysis of selected microstructures and a simulation study illustrate how the estimated Euler number depends on the chosen adjacency. It also shows that there is not a uniquely best pair of adjacencies with respect to the estimation of the Euler number of a set in Euclidean space.

1 Introduction

In statistical physics and in materials' science the Euler number (or its density) is used as a characteristic describing the connectivity of the components (constituents) of a composite material or the pore space of a porous medium, see [19, 20, 1] and [24].

From the mathematical point of view the Euler-Poincaré characteristic $\chi^n(X)$, or Euler number for short, of an n-dimensional set X in the Euclidean space $\mathbb{R}^n$ is a basic quantity of integral geometry. By means of Crofton's intersection formulae, the quermassintegrals (Minkowski functionals or intrinsic volumes) can be expressed in terms of the Euler numbers χ^k defined on sections of X with k-dimensional planes, $k = 0, \ldots, n-1$, see [29]. This is a basis of the measurement of the quermassintegrals in image analysis. Definitions and properties of the Euler number are recalled in Sect. 2.

In the context of image analysis it is usually assumed that the set X is observed on a point lattice $\mathbb{L}^n$. The intersection $X \cap \mathbb{L}^n$ is the mathematical expression for the observable information about X. In practical applications we consider the lattice $\mathbb{L}^n$ restricted to a bounded window $W \subset \mathbb{R}^n$. Let 1_X denote the characteristic function of X. The set $\{(x, 1_X(x)) : x \in W \cap \mathbb{L}^n\}$ is said to be the (binary) image of X observed in W, and the elements of the image are called pixels.

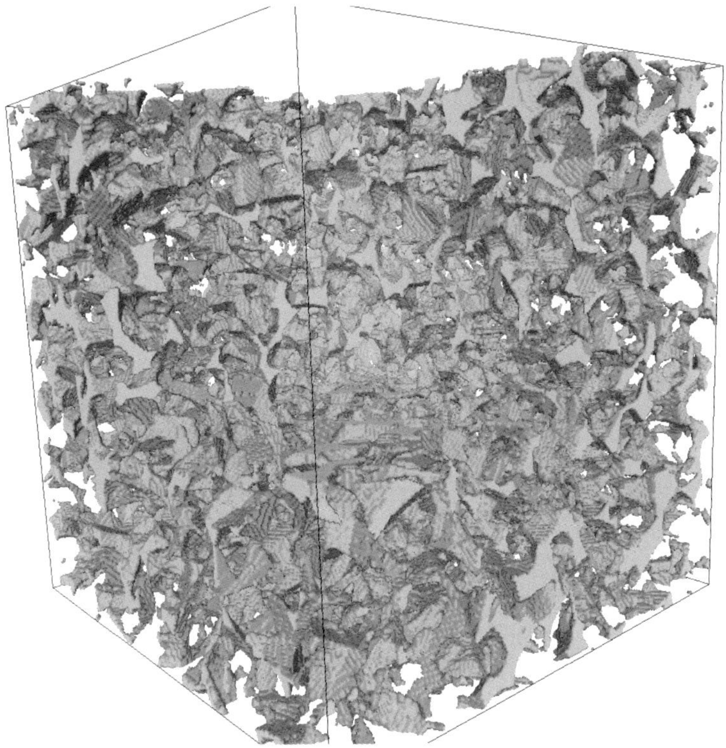

Fig. 1. The microstructure of Fontainebleau sandstone. The XCT scan consists of $299 \times 300 \times 300$ pixels arranged in a simple cubic lattice of spacing $7.5\ \mu\mathrm{m}$. This visualization of the 3d image shows the pore space.

There are two ways of considering such discretized images. One way is based on theoretical foundation and development of formulae of discrete (lattice) geometry, in particular integral geometry on the lattice and introduction of discrete Minkowski functionals; see [32], [19] (or [20] for a short summary). In the present paper the alternative view is taken, namely that the analysis of the discretized image aims of a good estimation of parameters or features of the original (i.e. non-discretized) set in the Euclidean space. This aspect was introduced systematically in [31].

In order to reconstruct or approximate the connectivity of the set or to measure the Euler number, the crucial problem is the choice of the neighborhood graph which links lattice points of $\mathbb{L}^n$ in an appropriate way. Consistency of the pair of estimators for the Euler numbers of X and of its complement X^c can only be achieved if the neighborhoods for object (i.e. foreground) and background 'fit' according to Theorem 1 in Sect. 2. We call a pair of adjacencies complementary if it provides a pair of consistent estimators.

In 2d, there are the $(4,8)$, $(8,4)$ and the $(6,6)$ adjacencies as pairs of complementary adjacencies. In this notation, the first component of a pair indicates the number of neighbors a lattice point is connected to if this lattice point belongs to X (the object) and the second component denotes the number of used connections to neighbors if the lattice point belongs to the complement X^c (the background). The $(6, 6)$ adjacency is of particular importance since it is self-complementary in the sense that both object and background are treated the same way. For an introduction of these 2d adjacencies see [31], [26], [24].

In 3d, the variety of reasonable pairs of adjacencies is much larger than in 2d and not easy to survey. In image analysis, object and background are usually endowed with different neighborhoods in order to ensure a digital Jordan surface theorem. Typical pairs are $(6, 26)$ and $(6, 18)$, see e.g. [12, 15, 17]. In [13], the 18 neighborhood is modified to yield a 14 neighborhood. An exception to the asymmetry is [10], where both object and background are endowed with the same neighborhood but the neighborhood graph loses its invariance with respect to the translation group of the lattice.

Most algorithms for determining the Euler number, see [12, Sect. 8] and [17, Sect. 3.1.2] for surveys, work for 6 connectivity of the object and 26 connectivity of the background or vice versa. In [15], Lee, Poston, and Rosenfeld use algebraic topology to determine the contribution to the Euler number of all possible $2 \times 2 \times 2$ configurations ('marching cube algorithm'). The idea of this algorithm was published first by Serra, see [30]. Imiya and Eckhardt's [8] method is similar although with different theoretical background. Bieri and Nef [2] deduce a recursive algorithm from Hadwiger's recursive definition. This definition is the starting point for our integral geometric approach, too, see Sect. 3.2. Our polyhedral approach in Sect. 3.4 follows the setup suggested in [12, Sect. 8].

For $n > 2$, the neighborhood graph does not describe uniquely the connectivity within the lattice cells. Thus, instead of neighborhood graphs, we use adjacency systems consisting of k-faces, $k = 0, \ldots, n$. In Sect. 3.3, we reconsider the well known 6 and 26 adjacencies and introduce two new 14 adjacencies in 3d using the new method of polyhedral tessellation of the lattice cells. The 14 adjacencies are self-complementary which is proved in 3.5. In Sect. 4 we show how the choice of adjacency influences the measurement results. Simulation studies show that using the 14 adjacencies can yield better results than $(6, 26)$ or $(26, 6)$ but that there are also models where they behave worse. This phenomenon is due to tiny structures in X or in X^c, respectively, in the range of the lattice constant, i.e. in the range of the resolution of the equipment used for discretization.

In this paper we focus on proving results in the 3d case. Basically, the approach for n-dimensional lattices is the same.

2 The Euler Number of a Set and of Its Complement

Assume that $X \subset \mathbb{R}^n$ is bounded and topologically closed, i.e. compact and that its Euler number $\chi^n(X)$ is defined. As an important subclass of such sets the reader may have in mind finite unions of compact convex sets (polyconvex sets). The set of all finite unions of compact convex sets is called the convex ring.

A set $X \subset \mathbb{R}^n$ is called *topologically regular* if $X = \overline{(X^{int})}$, where X^{int} and $\overline{X}$ denote the set of all interior points of X and the closure of X, respectively. Topological regularity means that there are no isolated points or lower-dimensional parts (like fibers) in X. Notice that also the empty set $\emptyset$ is regular.

The complement $X^c = \mathbb{R}^n \setminus X$ of a bounded and topologically closed set X is unbounded and topologically open. Furthermore, if X belongs to the convex ring, neither the complement X^c nor the topological closure of the complement $\overline{X^c}$ do belong to the convex ring. Hence it is not obvious how to introduce the Euler number of the complement and to establish relationships between $\chi^n(X)$ and $\chi^n(\overline{X^c})$.

Finally, we recall the terms 'morphologically closed' and 'morphologically open' (see e.g. [31]): A set X is said to be morphologically closed with respect to a compact set B if $X \bullet B = X$ where $X \bullet B = (X \oplus \check{B}) \ominus B$ denotes the morphological closure of X, $X \oplus B$ and $X \ominus B$ are the (Minkowski) sum and the (Minkowski) difference, respectively, and $\check{B}$ is the reflection of B. The set X is said to be morphologically open with respect to B if $X \circ B = X$ where $X \circ B = (X \ominus \check{B}) \oplus B$ is called the morphological opening. We call a set X to be *morphologically regular with respect to* B if it is morphologically open as well as morphologically closed with respect to B. Thus a morphologically regular set has no small (compared to B) particles, no thin spikes or lamellae, no small holes.

2.1 Definition

There are several equivalent ways for the definition of the Euler number of a set, see the books of Hadwiger [4], Matheron [18], Schneider [28], and Serra [31], cf. also [34].

In the case when X belongs to the convex ring, i.e. it can be represented as a finite union $X = \cup_{i=1}^m X_i$ of compact convex sets X_i, the Euler number can be defined using the initial settings $\chi^n(\emptyset) = 0$ and $\chi^n(X_0) = 1$ for every compact convex set $X_0 \neq \emptyset$ and the inclusion-exclusion principle, i.e.

$$\chi^n(X) = \sum_{i=1}^{m} \chi^n(X_i) - \sum_{i=1}^{m-1} \sum_{j=i+1}^{m} \chi^n(X_i \cap X_j) + \cdots + (-1)^{m+1} \chi^n\left(\bigcap_{i=1}^{m} X_i\right). \tag{1}$$

The inclusion-exclusion principle is equivalent to the additivity of the functional χ^n, see [28].

For the Euler number of a polytope, i.e. the convex hull of a finite set of points, there is an alternative definition in terms of the numbers of its lower-dimensional faces. Let $P \subset \mathbb{R}^n$ be a polytope. For $k = 0, \ldots, n$ denote by $\mathcal{F}^k(P)$ the set of all k-faces of P. In particular, $\mathcal{F}^0(P)$ is the set of vertices, $\mathcal{F}^1(P)$ the set of edges, $\mathcal{F}^{n-1}(P)$ the set of facets, and $\mathcal{F}^n(P)$ is the polytope itself, $\mathcal{F}^n(P) = P$. Furthermore, let $\#\mathcal{F}^k(P)$ be the cardinal number of $\mathcal{F}^k(P)$. Define the Euler number by

$$\chi^n(P) = \sum_{k=0}^{n} (-1)^k \#\mathcal{F}^k(P). \tag{2}$$

For a polytope, the Euler-Poincaré formula yields that the right hand side of (2) equals 1, see [33, Theorem 3.5.1]. Using (1), formula (2) can be additively extended to the case when P is a finite union of polytopes, i.e. when P is a polyhedron, see also [9, Sect. 4].

In the general case, when X belongs to the convex ring, $\chi^n(X)$ can be equivalently introduced using Hadwiger's recursive definition. Let $\mathcal{L}_k^n$ be the set of all k-dimensional subspaces of $\mathbb{R}^n$. For $L \in \mathcal{L}_k^n$ denote by $L^\perp$ the orthogonal space. Consider a pair $(L_x, L_{x+\varepsilon u})$ of parallel hyperplanes with distance ε where $L \in \mathcal{L}_{n-1}^n$, $x \in L^\perp, u \in L^\perp \setminus \{o\}, \varepsilon > 0$, and $L_x = L + x$. Starting with the initial setting $\chi^0(\emptyset) = 0$ and $\chi^0(\{o\}) = 1$ in the zero-dimensional case, the Euler number can be recursively defined for any dimension $n > 0$ by

$$\chi^n(X) = \sum_{x \in L^\perp} \lim_{\varepsilon \to 0} \left[\chi^{n-1}(X \cap L_x) - \chi^{n-1}(X \cap L_{x+\varepsilon u})\right]. \tag{3}$$

In (3), only finitely many items different from 0 occur. The limit differs from 0 when L_x is an 'upper' tangent hyperplane to X or a 'lower' tangent hyperplane to X^c. Notice that the value of the sum does not depend on the direction of the hyperplane. Here, the assumption that X is a topologically closed set is crucial. Note that (3) can also be applied to unbounded sets whose complement is bounded.

If X belongs to the convex ring and both X and X^c are morphologically closed (i.e. X is morphologically regular) with respect to a segment $[o, u]$ with $u \in L^\perp \setminus \{o\}$, then (3) can be rewritten as

$$\chi^n(X) = \frac{1}{|u|} \int_{L^\perp} \left[\chi^{n-1}(X_{(x)} \cup X_{(x+u)}) - \chi^{n-1}(X_{(x)})\right] dx \tag{4}$$

$$= \frac{1}{|u|} \int_{L^\perp} \left[\chi^{n-1}(X_{(x)}) - \chi^{n-1}(X_{(x)} \cap X_{(x+u)})\right] dx \tag{5}$$

with $X_{(x)} = (X \cap L_x)_{-x} = X_{-x} \cap L \subset L$, see [25]. Roughly, the assumptions make sure that the details of the structure in X and in X^c are not too tiny compared with the structuring element $[o, u]$, i.e. with respect to the distance of the pair of sweeping hyperplanes used in (4) and (5). This guarantees that there is no doubt concerning the correct connectivities in the set X. Equations (4) and (5) can be used to prove the correctness of approximations of the Euler number based on local observations of discretized sets, see [23].

2.2 Consistency Relation

In the 2d case, a geometric interpretation of the Euler number of a set is $\chi^2 = n_0 - n_1$, where n_0 denotes the number of connected subsets ('components') and n_1 is the 'number of holes'. For the 3d case it is $\chi^3 = n_0 + n_1 - g$, where n_0 is again the number of connected components, and n_1 is the number of 'cavities'. Here g is the genus of X. Heuristically, the genus is the maximal number of non-separating cuts through the set not increasing the number of connected components. Here a 'cut' must fully penetrate the set, see [17]. Hence, the genus of a set is the same as the genus of the complement. We remark that the terms '1st Betti number', 'number of handles', and 'number of tunnels' are used synonymously for 'genus'. For a more thorough discussion of Betti numbers see [16] and the contribution by Robins in this volume.

It is easy to see that for 2d sets the Euler number is equal to the negative Euler number of the complement, and for the 3d case the Euler numbers of a set and its complement are equal. In the following theorem, we prove this fact for arbitrary dimensions.

Theorem 1. (Consistency relation for the Euler number)
Assume that $X \subset \mathbb{R}^n$, $n \geq 1$, *permits a representation* $X = \cup_{i=1}^m X_i$, *with compact convex sets* X_i *such that all* X_i *and all intersections* $X_{i_1} \cap ... \cap X_{i_k}$ *for* $1 \leq i_1 < ... < i_k \leq m$ *and* $k = 2, ..., m$, *are topologically regular. Then*

$$\chi^n(\overline{X^c}) = (-1)^{n+1}\chi^n(X). \tag{6}$$

Sketch of the proof: By induction over the dimension n it can be shown easily that (6) is true for all compact convex sets X with inner points (which implies topological regularity). Although $\overline{X^c}$ is an unbounded set which does not belong to the convex ring, the sum in (3) contains exactly one item different from zero. If $X = X_1 \cup X_2$ where X_1, X_2 are compact convex sets and $X_1, X_2, X_1 \cap X_2$ fulfill the regularity condition, the additivity of χ^n (the inclusion-exclusion principle) (1) provides

$$\begin{aligned}
\chi^n(X_1 \cup X_2) &= \chi^n(X_1) + \chi^n(X_2) - \chi^n(X_1 \cap X_2) \\
&= (-1)^{n+1}\left[\chi^n(\overline{X_1^c}) + \chi^n(\overline{X_2^c}) - \chi^n(\overline{(X_1 \cap X_2)^c})\right] \\
&= (-1)^{n+1}\left[\chi^n(\overline{X_1^c}) + \chi^n(\overline{X_2^c}) - \chi^n(\overline{X_1^c} \cup \overline{X_2^c})\right] \\
&= (-1)^{n+1}\chi^n(\overline{X_1^c} \cap \overline{X_2^c}) \\
&= (-1)^{n+1}\chi^n(\overline{(X_1 \cup X_2)^c}).
\end{aligned}$$

Induction over the number m of sets generating X completes the proof.

□

Remark 1. The regularity condition for all intersections $X_{i_1} \cap ... \cap X_{i_k}$ means that these sets are either empty or have inner points, i.e., the sets $X_1, ..., X_m$ may not 'touch' each other in only one point or a set with a dimension less than n. It is obvious that without regularity conditions there is no chance to establish a relation between $\chi^n(X)$ and $\chi^n(\overline{X^c})$. As an example let X be a finite set of m isolated points. Then $\chi^n(X) = m$ and $\chi^n(\overline{X^c}) = \chi^n(\emptyset) = 0$.

An alternative proof of the consistency relation (6) can be given for those bounded sets X for which both $X \cap B$ and $\overline{X^c} \cap B$ belong to the convex ring for all convex bodies B, see [14, Appendix]. In this case the space $\mathbb{R}^n$ can be tessellated by convex bodies (e.g. cubes or parallelohedrons), and the additivity (1) can be applied. In particular, this approach works for bounded polyhedrons. We will use it for discretized sets.

3 Approximation of the Euler Number of a Set

3.1 General

The problem considered here is to find an appropriate approximation for the Euler number $\chi^n(X)$ based on the set $X \cap \mathbb{L}^n$ of lattice points. Formally, for every set X find an approximation $\tilde{\chi}^n(X \cap \mathbb{L}^n)$ being close to the true value $\chi^n(X)$.

In order to describe approximation, we provide some concepts regarding homogeneous lattices. A subset $\mathbb{L}^n$ of the linear space $\mathbb{R}^n$ is called a homogeneous lattice if there exists a base $\{u_1, \dots, u_n\}$ of $\mathbb{R}^n$ such that $\mathbb{L}^n = \{x \in \mathbb{R}^n : x = \sum_{i=1}^n \lambda_i u_i,\ \lambda_i \in \mathbb{Z}\} = U\mathbb{Z}^n$ with the matrix $U = (u_1, \dots, u_n)$, $\det(U) \neq 0$, where $\mathbb{Z}$ denotes the set of integers. The $|u_i|$ are the lattice spacings. Typical examples of homogeneous lattices are the *cubic primitive* lattice $\mathbb{L}^n = c_0\mathbb{Z}^n$, $c_0 > 0$, and the *orthorhombic primitive* lattice $\mathbb{L}^n = \text{diag}\mathbb{Z}^n$ where diag is a diagonal matrix with non-zero entries. We remark that the homogeneous lattices – in their general setting – are known as the *triclinic* lattices. The unit cell C of $\mathbb{L}^n$ is the Minkowski sum of the half-open segments $[o, u_i) = \{pu_n : 0 \leq p < 1\}$ between the origin o and the lattice points u_i; $C = [o, u_1) \oplus \dots \oplus [o, u_n)$. For more facts concerning lattices, see e.g. [3].

In this section, we present two approaches for the approximation of the Euler number. First, in Sect. 3.2, we recall the integral geometric approach based on a discretization of the recursive formulae (4) and (5).

Subsections 3.3 to 3.5 are devoted to the polyhedral approach. In image processing the standard solution for measuring the Euler number is to fix an adjacency relation for the lattice points and to determine the Euler number with respect to the corresponding polyhedron. Usually, the Euler number is computed by means of the Euler-Poincaré formula (2). This requires a clear and consistent definition of the k-faces to be counted. This definition is not unique for all neighborhood graphs or adjacency relations. Therefore, it is more precise to extend the neighborhood graph to the description of the whole adjacency system corresponding to a polyhedral approximation of the set X.

Very generally, an adjacency is a system of k-faces, $k = 0, \dots, n$. One way of obtaining an adjacency system is to use a tessellation $\mathcal{G}$ of $\mathbb{R}^n$ having the same translation group as the lattice. Then the adjacency system with respect to $\mathcal{G}$ – in the following denoted by $\mathbb{F} = \mathbb{F}(\mathcal{G})$ – is the set of all k-faces, $k = 0, \dots, n$, of polytopes belonging to $\mathcal{G}$.

Denote by $\mathbb{F} \sqcap X$ the union of all k-faces of $\mathcal{G}$ 'restricted' to X, $k = 0, \dots, n$ (i.e. the set of all k-faces of $\mathbb{F}$ for which all vertices belong to X; see also Sect. 3.4). This will be understood as a discretization of X. In Sect. 3.4 the following scheme is suggested:

$$X \to X \cap \mathbb{L}^n \xrightarrow{\mathbb{F}} \mathbb{F} \sqcap X \xrightarrow{(2)} \tilde{\chi}^n(\mathbb{F} \sqcap X) \approx \chi^n(X).$$

For the complement X^c we chose another adjacency system $\mathbb{F}_c$, and an approximation of the Euler number of X^c can be obtained analogously to the above scheme from

$$X^c \to X^c \cap \mathbb{L}^n \xrightarrow{\mathbb{F}_c} \mathbb{F}_c \sqcap X^c \xrightarrow{(2)} \tilde{\chi}^n(\mathbb{F}_c \sqcap X^c) \approx \chi^n(\overline{X^c}).$$

It is natural to use those pairs $(\mathbb{F}, \mathbb{F}_c)$ of adjacency systems for which the consistency relation (6) in the form $\tilde{\chi}^n(\mathbb{F}_c \sqcap X^c) = (-1)^{n+1}\tilde{\chi}^n(\mathbb{F} \sqcap X)$ is fulfilled for all X. Examples of such pairs are given in Sect. 3.5. Sufficient conditions for the correctness of the approximations are given in Sect. 3.6.

3.2 The Integral Geometric Approach

In [23] an approximation of the Euler number is given together with morphological regularity conditions on X and X^c which are a consequence of the regularity conditions for (4) and (5). To describe this approach more detailed, we introduce intersections of the lattice $\mathbb{L}^n$ with j-dimensional subspaces $L \in \mathcal{L}_j^n$. We consider only those subspaces L for which the intersection $\mathbb{L}^n \cap L$ forms a j-dimensional sublattice $\mathbb{L}^j$. Let $\mathcal{L}_j^n(\mathbb{L}^n)$ denote the set of all homogeneous j-dimensional sublattices of $\mathbb{L}^n$. The lattice $(\mathbb{L}^j)^\perp$ is the set of all points $y \in L^\perp$ with the property that $L_y \cap \mathbb{L}^n$ is a lattice on L_y. We call $(\mathbb{L}^j)^\perp$ the lattice orthogonal to $\mathbb{L}^j$. Note that $(\mathbb{L}^j)^\perp$ is not necessarily a sublattice of $\mathbb{L}^n$.

Let $\mathbb{L}^{n-1} \in \mathcal{L}_{n-1}^n(\mathbb{L}^n)$ and let v_n be the base of the orthogonal lattice $(\mathbb{L}^{n-1})^\perp$. Define $X_{(i)} = (X - iv_n) \cap \mathbb{L}^{n-1}$, i.e. $X_{(i)}$ is a discretization of the intersection $(X - iv_n) \cap L$. Then from (4) and (5) we obtain

$$\tilde{\chi}^n(X \cap \mathbb{L}^n) := \sum_{i\in\mathbb{Z}} \left[\tilde{\chi}^{n-1}(X_{(i)} \cup X_{(i+1)}) - \tilde{\chi}^{n-1}(X_{(i)})\right] \tag{7}$$

$$= \sum_{i\in\mathbb{Z}} \left[\tilde{\chi}^{n-1}(X_{(i)}) - \tilde{\chi}^{n-1}(X_{(i)} \cap X_{(i+1)})\right]. \tag{8}$$

Recursive application of one of these formulae for decreasing dimension of the considered spaces leads to an approximation $\tilde{\chi}^n(X \cap \mathbb{L}^n)$ of $\chi(X)$. Bieri and Nef's [2] recursive algorithm is deduced directly from (7).

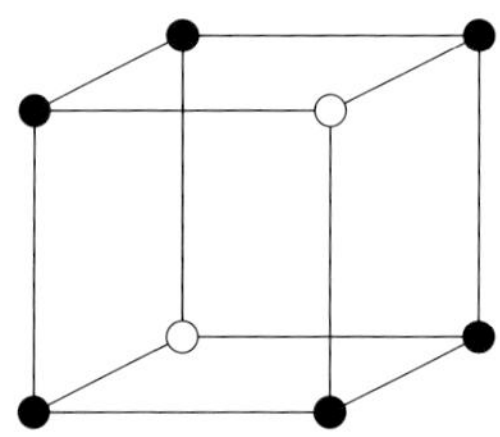

Fig. 2. Pixel configuration in a 3d image with a simple cubic lattice. The solid discs • are assigned to the pixels covered by X whereas the circles ◦ are the pixels that hit the complement X^c. If X is morphologically closed with respect to the edges and diagonals of the lattice cell, then $\chi(X) = 1$ whereas (7) provides $\chi^n(X \cap \mathbb{L}^n) = 0$. This difference is due to the fact that X is not morphologically open with respect to space diagonals.

We recall from [23] the sufficient conditions for the correctness of the approximation. Let C_v denote the unit cell of the lattice $\mathbb{L}_v^n$ with base $\{v_1, \ldots, v_n\}$ where $v_n, \ldots, v_1$ are the bases of the orthogonal lattices $(\mathbb{L}^{n-1})^\perp, \ldots, (\mathbb{L}^0)^\perp$ used in the recursion.

Theorem 2. *Let $\mathcal{S}_v$ be the set of all closed segments that connect the vertices of the unit cell C_v; $\mathcal{S}_v = \{[x,y] : x, y \in \mathcal{F}^0(\overline{C_v}),\ x \neq y,\ (x,y) \subset C_v\}$. If X is morphologically regular with respect to all segments of $\mathcal{S}_v$ then $\tilde{\chi}^n(X \cap \mathbb{L}^n)$ provides the exact value of the Euler number, i.e. $\tilde{\chi}^n(X \cap \mathbb{L}^n) = \chi^n(X)$.*

The proof follows from [23, Theorem 2], since $\mathbb{L}_v^n$ is by definition an orthorhombic primitive lattice.

If the regularity assumptions for X are relaxed, then there may occur pixel configurations where the true value $\chi^n(X)$ differs from $\tilde{\chi}^n(X \cap \mathbb{L}^n)$ provided by (7) and (8), respectively, see Fig. 2.

3.3 Adjacencies, Tessellations and Graphs

For sets X having a too tiny structure compared to the spacings of the observation lattice $\mathbb{L}^n$, i.e. sets not being morphologically regular with respect to the edges and diagonals of the cells of the lattice, one has to make guesses concerning the connectivity structure between the isolated lattice points. In image analysis, this guess is usually based on neighborhood graphs attached to the lattice.

We consider polyhedral approximations of the sets X and X^c rather than neighborhood graphs. These approximations are based on adjacencies. We first give a very general definition of adjacency:

Definition 1. *An* adjacency system $\mathbb{F}$ *is a set of k-faces,* $\mathbb{F} = \cup_{k=0}^{n} \mathcal{F}^k$ *with*

(i) each k-face $F \in \mathbb{F}$ is a subset of a (topologically closed) lattice cell and the vertices are lattice points
(ii) $\mathbb{F}$ is invariant with respect to lattice translations, $\mathbb{F} + x = \mathbb{F}$ for all $x \in \mathbb{L}^n$, and
(iii) the union of all polytopes of $\mathbb{F}$ is $\mathbb{R}^n$.

Adjacency systems can be constructed in various ways, for example using tessellations, collections or superpositions of tessellations, or forming the convex hull. Meaningful adjacencies are identified in Sect. 3.5 via the consistency relation (6). In this section we consider adjacency based on a special class of tessellations – so-called admissible tessellations.

A convex tessellation $\mathcal{G}$ of $\mathbb{R}^n$ is defined as a family $\{P_1, P_2, \dots\}$ of polytopes (with inner points) satisfying $P_i^{int} \cap P_j^{int} = \emptyset$ for $i \neq j$, $\#\{P \in \mathcal{G} : P \cap B \neq \emptyset\} < \infty$ for all bounded subsets $B \subset \mathbb{R}^n$, and $\cup_i P_i = \mathbb{R}^n$. In the following, we consider admissible tessellations only :

Definition 2. *The tessellation* $\mathcal{G} = \{P_1, P_2, \dots\}$ *is called an* admissible tessellation *with respect to the lattice $\mathbb{L}^n$ if*

(i) the vertices of each tessellation cell are lattice points, i.e. $\mathcal{F}^0(P) \subset \mathbb{L}^n,\ P \in \mathcal{G}$,
(ii) the tessellation is invariant with respect to lattice translations, i.e. $\mathcal{G} + x = \mathcal{G},\ x \in \mathbb{L}^n$,
(iii) the cells of the tessellation are subsets of the lattice cells, i.e., for every $P \in \mathcal{G}$ there exists a point $x \in \mathbb{L}^n$ such that $P \subseteq \overline{C} + x$, and
(iv) the intersection of two tessellation cells is either empty or a face of these cells, i.e., $P_i \cap P_j$ is an element of $\cup_{k=0}^{n-1} (\mathcal{F}^k(P_i) \cap \mathcal{F}^k(P_j)) \cup \{\emptyset\},\ P_i, P_j \in \mathcal{G},\ i \neq j$.

To be more flexible, we consider also the superposition $\mathcal{G}$ of admissible tessellations $\mathcal{G}_1, \dots, \mathcal{G}_m$ defined as

$$\mathcal{G} = \{P : P = Q_1 \cap \dots \cap Q_m : Q_1 \in \mathcal{G}_1, \dots, Q_m \in \mathcal{G}_m\},\ m \geq 1,$$

see [21, Sect. 3.4]. Note that for $m > 1$, the superposition $\mathcal{G}$ is not necessarily an admissible tessellation. In particular, there can exist vertices of cells of $\mathcal{G}$ that are not lattice points, $\mathcal{F}^0(\mathcal{G}) \supseteq \mathbb{L}^n$.

Now we introduce the systems $\mathcal{F}^k(\mathcal{G})$ of the k-faces of the superposition $\mathcal{G} = \{P_1, P_2, ...\}$ in the following way: $\mathcal{F}^k(\mathcal{G}) = \cup_{i=1}^{\infty} \mathcal{F}^k(P_i), k = 0, \dots, n$. Obviously, admissible tessellations and their superpositions yield adjacencies.

Definition 3. *Let $\mathcal{G}$ be the superposition of admissible tessellations. Then the set* $\mathbb{F}(\mathcal{G}) = \cup_{k=0}^{n} \mathcal{F}^k(\mathcal{G})$ *is called the* adjacency system with respect to $\mathcal{G}$.

In order to link our polyhedral approach and the characterization of neighborhood usually used in image processing, we introduce the neighborhood graph Γ consisting of non-oriented edges $\langle x, y\rangle$ that link lattice points $x, y \in \mathbb{L}^n$. Define the neighborhood relation for $\mathbb{F}$ as

$$\gamma = \left\{ \langle x, y\rangle : x, y \in \mathcal{F}^0(\overline{C}),\ (x, y) \subset C,\ [x, y] \subset \bigcup \{F \in \mathcal{F}^1(\mathbb{F})\} \right\}.$$

Then the neighborhood graph with respect to $\mathbb{F}$ can be described by γ; $\Gamma = \cup_{x \in \mathbb{L}^n}(\gamma + x)$. The connectivity $c = c(\Gamma)$ of the neighborhood graph is the order of the nodes; $c = \#\{\langle x, y\rangle \in \Gamma : y \in \mathbb{L}^n\}$ for every lattice point x. It is easy to see that $c = 2\#\gamma$. In the literature, adjacencies are usually called c neighborhoods.

Consider the admissible tessellation $\mathcal{G}_0 = \{\overline{C} + x : x \in \mathbb{L}^n\}$ consisting of the lattice cells. Then $\gamma_0 = \{\langle o, u_1\rangle, \ldots, \langle o, u_n\rangle\}$ is the generator of the neighborhood graph Γ_0 of the adjacency $\mathbb{F}(\mathcal{G}_0)$, and the connectivity is $c(\Gamma_0) = 2n$. Since $c(\Gamma_0) \le c(\Gamma)$ for all adjacencies $\mathbb{F}(\mathcal{G})$ with respect to an admissible tessellation $\mathcal{G}$ and with graph Γ, the adjacency $\mathbb{F}(\mathcal{G}_0)$ is said to be the *minimal adjacency*. The adjacency $\mathbb{F}(\mathcal{G}_{\max})$ with respect to the superposition $\mathcal{G}_{\max}$ of the set of all admissible tessellations is called the *maximum adjacency*. Its neighborhood graph $\Gamma_{\max}$ is of order $c(\Gamma_{\max}) = 3^n - 1$.

Examples of 3d adjacencies. We consider the adjacencies in 3d which correspond to the well known 6 and 26 neighborhoods. (For the 6 neighborhood each pixel is connected only by the edges of the lattice cells with its neighbors, and in the case of 26 neighborhood the pixels are connected by all edges, face diagonals and space diagonals of the cells.) Furthermore, we introduce two – apparently new – 14 neighborhoods. In order to describe the corresponding tessellations we introduce the vertices $x_0, \ldots, x_7$ of C given by $x_0 = o$, $x_1 = u_1$, $x_2 = u_2$, $x_3 = u_1 + u_2$, $x_4 = u_3$, $x_5 = u_1 + u_3$, $x_6 = u_2 + u_3$, and $x_7 = u_1 + u_2 + u_3$, where $\{u_1, u_2, u_3\}$ is the base of $\mathbb{L}^3$.

1. Clearly, for the 3d case the minimal adjacency $\mathbb{F}(\mathcal{G}_0)$ corresponds to the 6 neighborhood; $c(\Gamma_0) = 6$. The adjacency $\mathbb{F}(\mathcal{G}_0)$ is invariant with respect to rotation in the sense that $UMU^{-1}\mathbb{F}(\mathcal{G}_0) = \mathbb{F}(\mathcal{G}_0)$ for every rotation M from the rotation group $\Omega(\mathbb{Z}^3)$ of the discrete space $\mathbb{Z}^3$. ($\Omega(\mathbb{Z}^3)$ is called the octaeder group.)

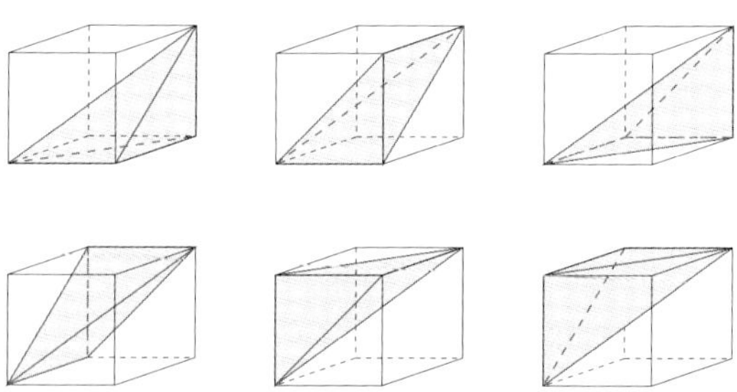

Fig. 3. The tessellation of the unit cell defining the 14.1 neighborhood.

2. Consider the tessellation of the unit cell into the 6 tetrahedrons $G_1, \dots, G_6$ which are convex hulls of vertices of C:

$$
\begin{aligned}
G_1 &= \operatorname{conv}(\{x_0, x_1, x_3, x_7\}),\ G_2 = \operatorname{conv}(\{x_0, x_1, x_5, x_7\}),\\
G_3 &= \operatorname{conv}(\{x_0, x_2, x_3, x_7\}),\ G_4 = \operatorname{conv}(\{x_0, x_4, x_5, x_7\}),\\
G_5 &= \operatorname{conv}(\{x_0, x_4, x_6, x_7\}),\ G_6 = \operatorname{conv}(\{x_0, x_2, x_6, x_7\}),
\end{aligned}
$$

   see Fig. 3. Let $\mathcal{G}_1$ consist of all translations of the G_i, $\mathcal{G}_1 = \{G_i + x : i = 1, \dots, 6,\ x \in \mathbb{L}^3\}$. Then $\mathcal{G}_1$ is an admissible tessellation of $\mathbb{R}^3$ with respect to $\mathbb{L}^3$. For the corresponding adjacency $\mathbb{F}(\mathcal{G}_1)$, the set γ consists of three edges of C, three face diagonals, and one space diagonal, i.e. $c(\Gamma) = 14$. Hence $\mathbb{F}(\mathcal{G}_1)$ has a 14 neighborhood. Note that $\mathbb{F}(\mathcal{G}_1)$ is not invariant under rotation. But even besides its rotations there is a further adjacency based on a tessellation of C into 6 tetrahedrons. To indicate the difference we call $\mathbb{F}_{14.1} = \mathbb{F}(\mathcal{G}_1)$ the 14.1 neighborhood.

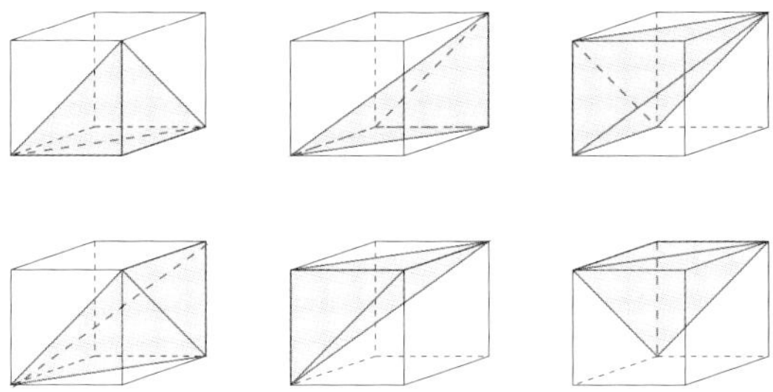

Fig. 4. The tessellation of the unit cell defining the 14.2 neighborhood.

3. Consider now the tetrahedrons

$$
\begin{aligned}
G_1 &= \operatorname{conv}(\{x_0, x_1, x_3, x_5\}),\ G_2 = \operatorname{conv}(\{x_0, x_2, x_3, x_7\}),\\
G_3 &= \operatorname{conv}(\{x_0, x_2, x_4, x_7\}),\ G_4 = \operatorname{conv}(\{x_0, x_3, x_5, x_7\}),\\
G_5 &= \operatorname{conv}(\{x_0, x_4, x_5, x_7\}),\ G_6 = \operatorname{conv}(\{x_2, x_4, x_6, x_7\})
\end{aligned}
$$

   which tessellate the unit cell as shown in Fig. 4. Again, the translations of the G_i form an admissible tessellation $\mathcal{G}_2$, and the adjacency $\mathbb{F}_{14.2} = \mathbb{F}(\mathcal{G}_2)$ is a 14 neighborhood, too, in the following called the 14.2 neighborhood.
4. The neighborhood graph of the maximum adjacency is of order 26. Hence we write $\mathbb{F}_{26} = \mathbb{F}(\mathcal{G}_{\max})$. The maximum adjacency $\mathbb{F}_{26}$ is invariant with respect to rotation (as $\mathbb{F}_6$).

3.4 The Euler Number with Respect to Adjacency

The elements of $\mathbb{F}(\mathcal{G})$ are the geometric objects to be used for the discretization of the set X and for the computation of its Euler number.

For simplicity we first consider the case when $\mathbb{F}(\mathcal{G})$ is an adjacency with respect to an admissible tessellation. Then the set of the k-faces for which all vertices belong to X can be formally introduced as $\mathcal{F}^k(\mathcal{G}) \sqcap X = \{F \in \mathcal{F}^k(\mathcal{G}) : \mathcal{F}^0(F) \subseteq X\}$.

If $\mathcal{G}$ is a superposition of admissible tessellations, there can occur difficulties with the extra vertices $\mathcal{F}^0(\mathcal{G}) \setminus \mathbb{L}^n$ in which the set X is not observed. To overcome this problem, we assume that a vertex $z \in \mathcal{F}^0(\mathcal{G}) \setminus \mathbb{L}^n$ belongs to X if there is an edge $\langle x, y \rangle$ of Γ with $x, y \in X$ and $z \in [x, y]$. That is, z lies on a segment between two neighboring nodes of the neighborhood graph Γ. In order to describe rigorously the discretization of X with respect to $\mathbb{F}(\mathcal{G})$, we augment X by the additional vertices:

$$X^* = X \cup \{z \in \mathcal{F}^0(\mathcal{G}) \setminus \mathbb{L}^n : \text{there are } x, y \in X \text{with } \langle x, y \rangle \in \Gamma \text{and } z \in [x, y]\}.$$

Now the discretization of the set X with respect to $\mathbb{F}(\mathcal{G})$ can be defined as follows:

Definition 4. *Let $\mathcal{F}^k(\mathcal{G}) \sqcap X$ denote the set of those k-faces for which all vertices belong to X^* and*

$$\mathcal{F}^k(\mathcal{G}) \sqcap X = \{F \in \mathcal{F}^k(\mathcal{G}) : \mathcal{F}^0(F) \subseteq X^*\}.$$

For a superposition $\mathcal{G}$ of admissible tessellations we call

$$\mathbb{F}(\mathcal{G}) \sqcap X = \cup_{k=0}^{n} \mathcal{F}^k(\mathcal{G}) \sqcap X$$

the discretization of X *with respect to the adjacency* $\mathbb{F}(\mathcal{G})$.

Notice that if X is a bounded subset of $\mathbb{R}^n$, its discretization is a polyhedron, i.e. a finite union of polytopes. Thus the Euler-Poincaré formula (3) yields an easy formula for the computation of the Euler number,

$$\tilde{\chi}^n(\mathbb{F}(\mathcal{G}) \sqcap X) = \sum_{k=0}^{n} (-1)^k \#(\mathcal{F}^k(\mathcal{G}) \sqcap X), \tag{9}$$

see also [9, (4)].

For practical applications it is helpful to give also a 'local version' of this formula based on 'local versions' of the $\mathcal{F}^k(\mathcal{G}) \sqcap X$. The approximation $\tilde{\chi}^n(\mathbb{F}(\mathcal{G}) \sqcap X)$ can be based on estimates in each of the cells and their combination according to the inclusion-exclusion principle. In order to ensure that each k-face is counted once only, we have to apply some kind of edge correction. Here we choose weighting the k-faces with the reciprocal of the number of closed lattice cells they intersect.

For $\ell \geq k$ let $\mathcal{F}_0^{k\ell} \sqcap X$ be the set of all k-faces $F \in \mathcal{F}^k(\mathcal{G}) \sqcap X$ with:

(i) There is an l-face $F_\ell \in \mathcal{F}^\ell(\overline{C})$ such that $F \subseteq F_\ell$ and
(ii) there is no j-face $F_j \in \mathcal{F}^j(\overline{C})$, $j < \ell$, such that $F \subseteq F_j$.

Then (9) can be rewritten as

$$\tilde{\chi}^n(\mathbb{F}(\mathcal{G}) \sqcap X) = \sum_{x \in \mathbb{L}^n} \sum_{k=0}^{n} (-1)^k \sum_{\ell=k}^{n} 2^{\ell-n} \#(\mathcal{F}_0^{k\ell} \sqcap X_{-x}). \tag{10}$$

For the 3d case the interpretation of the weights is as follows: All vertices $\mathcal{F}_0^{0,0} \sqcap X$ get weight $1/8$. The edges $\mathcal{F}_0^{1,1} \sqcap X$, $\mathcal{F}_0^{1,2} \sqcap X$, and $\mathcal{F}_0^{1,3} \sqcap X$ get $1/4$, $1/2$, and 1, respectively. The faces $\mathcal{F}_0^{2,2} \sqcap X$ and $\mathcal{F}_0^{2,3} \sqcap X$ are equipped with $1/2$ and 1, respectively, and all cells $\mathcal{F}_0^{3,3} \sqcap X$ with 1.

Remark 2. The local version (10) of (9) is easy to apply for $\mathbb{F}_6, \mathbb{F}_{14.1}$, and $\mathbb{F}_{14.2}$, but tedious for $\mathbb{F}_{26}$. For $\mathbb{F}_{26}$, the set $\mathcal{F}_0^{k\ell} \sqcap X_{-x}$ can be obtained from the convex hull of the vertices of $\overline{C}$ belonging to X.

Remark 3. Algorithmically, using the half open lattice cell C is easier. Let $\mathcal{F}_0^k \sqcap X = \{F \in \mathcal{F}^k(\mathcal{G}) \sqcap X : F \cap C \neq \emptyset,\ F \subseteq \overline{C}\}, \quad k = 0, \ldots, n$. Then (9) can be rewritten as $\tilde{\chi}^n(\mathbb{F}(\mathcal{G}) \sqcap X) = \sum_{x \in \mathbb{L}^n} \sum_{k=0}^{n} (-1)^k \#(\mathcal{F}_0^k \sqcap X_{-x})$.

For $n = 3$ let $x_0, \ldots, x_7$ denote the vertices of the unit cell C (defined as in the examples of Sect. 3.3). Following the technique described in [17], we define a function

$$g(x) = \sum_{i=0}^{7} 2^i \mathbf{1}_X(x + x_i), \quad x \in \mathbb{L}^3,$$

with integer values $0, \ldots, 255$. This means that depending on whether the vertex x_i hits the object or the background it gets the weight 2^i or 0, respectively, see also Fig. 5a). Furthermore, let $W \subset \mathbb{R}^3$ be a compact window with $X \oplus \overline{C} \subseteq W$. We introduce a vector h with coefficients

$$h_i = \sum_{x \in \mathbb{L}^3 \cap W} \delta_i(g(x)), \quad i = 0, \ldots, 255,$$

where δ denotes Kronecker's delta, $\delta_i(j) = 1$ for $i = j$ and $\delta_i(j) = 0$ otherwise. Then (10) can be rewritten as

$$\tilde{\chi}^3(\mathbb{F}(\mathcal{G}) \sqcap X) = \frac{1}{8} hw, \tag{11}$$

where w is given in Tables 1 and 2 for the adjacencies $\mathbb{F}_6, \mathbb{F}_{14.1}, \mathbb{F}_{14.2}$, and $\mathbb{F}_{26}$.

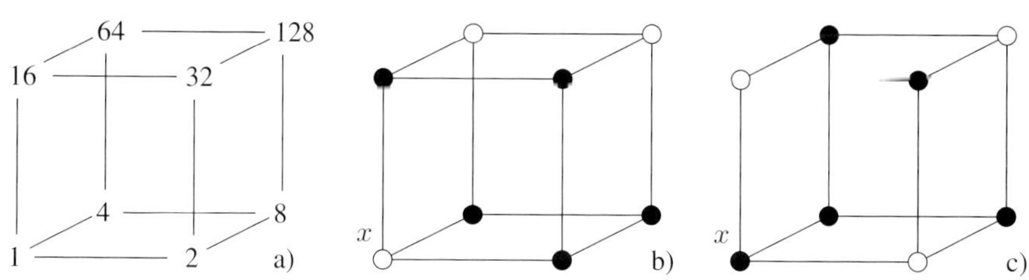

Fig. 5. (a) Weights used for the coding of the pixel configurations of a 3d image , (b) and (c) pixel configurations coded as $g(x) = 62$ and $g(x) = 109$, respectively. In these examples the corresponding component w_i of the vector w in (11) depends considerably on the chosen adjacency. We get the values $-3,\ -3,\ -7$, and 1 for the configuration (b) and $-1,\ -5,\ -1$, and 3 for (c), respectively, see also Table 1.

3.5 Pairs of Complementary Adjacencies

It is a usual requirement in image analysis that any adjacency for the 'object' X has to be in coherence and consistency with an appropriate 'complementary adjacency' used for the 'background' X^c. Usually, the criterion is some discrete Jordan surface theorem. Our motivation is consistency (6) of the approximations of the Euler number.

Definition 5. *The pair* $(\mathbb{F}(\mathcal{G}), \mathbb{F}(\mathcal{G}_c))$ *of adjacency systems is* complementary *if for all bounded sets* $X \subset \mathbb{R}^n$

(i) $F \cap F_c = \emptyset$ *for all* $F \in \mathbb{F}(\mathcal{G}) \sqcap X$ *and* $F_c \in \mathbb{F}(\mathcal{G}_c) \sqcap X^c$ *and*
(ii) $\tilde{\chi}^n(\mathbb{F}(\mathcal{G}) \sqcap X) = (-1)^{n+1}\tilde{\chi}^n(\mathbb{F}(\mathcal{G}_c) \sqcap X^c)$.

Notice that (i) is a nontrivial requirement since it is not even allowed that the discretizations of X and of X^c – as topologically closed polyhedra – touch in any point. Also (ii) is not an obvious consequence of (1) since the discretizations of X and X^c do not fill the space.

Remark 4. Herman introduced in [6] the concept of *Jordan pair*. Two adjacencies are a Jordan pair if endowing object and background with these adjacencies yields a digital Jordan surface theorem. (Very roughly, a closed surface divides the space properly into two components. Paradoxes as in [6, Proposition 3.1] are prevented.) In [6, 11, 13, 27] Jordan surface theorems are proved for different concepts of digital surface and the pairs $(6, 26)$, $(26, 6)$, $(6, 18)$, and $(18, 6)$, in [13] also for $(14, 6)$, $(6, 14)$. The surface concept in [6, 13, 27] is motivated by surface rendering algorithms and not suitable in our context. However, provided the right definition of a surface, we assume that a Jordan surface theorem holds for our $(14.1, 14.1)$ and $(14.2, 14.2)$, too. However, this is beyond the scope of this paper and will be investigated in a future paper.

An important consequence of (i) is that an edge of $\mathbb{F}$ and an edge of $\mathbb{F}_c$ may either coincide or meet in a lattice point or be disjoint, but never may intersect in a single point not being a lattice point. This limits the choice of diagonals on the facets of the lattice cells as well as of spatial diagonals through the cells. (In the 2d case the 8 neighborhood can only be accompanied by the 4 neighborhood, since in the 8 neighborhood already both diagonals are edges in the corresponding $\mathbb{F}$.)

On the other hand, (ii) demands some kind of maximization of the sets of edges and of the refinement of the tessellation of one of the adjacencies if the complementary one is given. E.g. it has to be guaranteed that two points of the discretized set $\mathbb{F} \sqcap X$ are either connected within this set or they are separated by the discretization $\mathbb{F}_c \sqcap X^c$ of the complement. (In the 2d case the 4 neighborhood has to be combined with the 8 neighborhood for the complement in order to fulfill (ii).)

It is not yet clear whether there exists a complementary adjacency for every adjacency and whether this would be unique. However, it is easy to see that if $\mathbb{F}(\mathcal{G}_c)$ is a complementary adjacency of $\mathbb{F}(\mathcal{G})$ then also $\mathbb{F}(\mathcal{G})$ is complementary to $\mathbb{F}(\mathcal{G}_c)$. An adjacency system $\mathbb{F}(\mathcal{G})$ is said to be *self-complementary* if $\tilde{\chi}^n(\mathbb{F}(\mathcal{G}) \sqcap X^c) = (-1)^{n+1}\tilde{\chi}^n(\mathbb{F}(\mathcal{G}) \sqcap X)$ for all X.

Theorem 3. *The pair* $(\mathbb{F}_6, \mathbb{F}_{26})$ *of adjacency systems is a pair complementary adjacency systems, and* $(\mathbb{F}_{26}, \mathbb{F}_6)$ *is complementary, too.*

Proof: In order to show (i) of Definition 5 note that $(X \cap \mathbb{L}^3) \cap (X^c \cap \mathbb{L}^3) = \emptyset$, i.e. the sets of vertices of the discretizations of X and of X^c are disjoint. Since the edges of $\mathbb{F}_6$

either coincide or meet in an edge or are disjoint to all edges of $\mathbb{F}_{26}$ we can conclude that the union of all edges of the discretization of X is disjoint to that of X^c. From the fact that the tessellations of $\mathbb{F}_{26}$ are refinements of the tessellation $\mathcal{G}_0$ one can now derive (i). Namely, since each 2d face belonging to $\mathbb{F}_6$ is the union of 2d faces in $\mathbb{F}_{26}$, it is clear that if a 2d face belongs to $\mathbb{F}_6 \sqcap X^c$ then no subset of it (of any dimension) may belong to $\mathbb{F}_{26} \sqcap X^c$. Analogous arguments can be used for the 3d cells.

Due to the additivity of the Euler number it is sufficient to show that the consistency condition (ii) in Definition 5 is fulfilled in each cell of the lattice. This can be checked considering all 256 possible point configurations on the vertices of a cell. Of course, it has to be taken into account that each 2d facet of a cell belongs to two cells, each edge belongs to four cells and each vertex belongs to 8 cells. We use the same edge correction as in (10).

It is sufficient for the validity of the consistency relation (ii) that $w_i^{(a)} = w_{255-i}^{(d)}$ for $i = 1, \ldots, 128$, see Tables 1 and 2. □

Theorem 4. *The pairs* $(\mathbb{F}_{14.1}, \mathbb{F}_{14.1})$ *and* $(\mathbb{F}_{14.2}, \mathbb{F}_{14.2})$ *are complementary pairs of adjacency systems.*

Proof: Condition (i) of Definition 5 can be proved using the same arguments as for the $(6, 26)$ neighborhood. To prove the consistency (ii), we consider the possible point configurations in single cells as in the proof for the $(6, 26)$ adjacencies. The evaluations in columns (b) and (c) of Tables 1 and 2 show that $w_i^{(b)} = w_{255-i}^{(b)}$ for $\mathbb{F}_{14.1}$ and $w_i^{(c)} = w_{255-i}^{(c)}$ for $\mathbb{F}_{14.2}$. □

Intuitively, the reason for the self-complementarity of $\mathbb{F}_{14.1}$ as well as of $\mathbb{F}_{14.2}$ is the fact that a tetrahedral tessellation cannot be further refined and that also the induced tessellations on the 2d facets of the cells are translation invariant, i.e. the tessellations on the contact faces of adjacent cells coincide. This is not true for all tetrahedral tessellations. A counterexample is, for instance, any tessellation containing the tetrahedron with the vertices x_0, x_3, x_5, x_6.

3.6 Sufficient Conditions for Correct Approximation

Consider a pair $(\mathbb{F}(\mathcal{G}), \mathbb{F}(\mathcal{G}_c))$ of adjacencies with respect to the superpositions of admissible tessellations $\mathcal{G}$ and $\mathcal{G}_c$, respectively. Let $\mathcal{S}(\mathcal{G})$ denote the set of all segments $[x, y]$ for which the edge $\langle x, y \rangle$ belongs to the neighborhood graph Γ of $\mathbb{F}(\mathcal{G})$.

Theorem 5. *Let* $(\mathbb{F}(\mathcal{G}), \mathbb{F}(\mathcal{G}_c))$ *be one of the four pairs of complementary adjacencies* $(\mathbb{F}_6, \mathbb{F}_{26})$, $(\mathbb{F}_{26}, \mathbb{F}_6)$, $(\mathbb{F}_{14.1}, \mathbb{F}_{14.1})$ *and* $(\mathbb{F}_{14.2}, \mathbb{F}_{14.2})$. *If* X *is morphologically closed with respect to all segments of* $\mathcal{S}(\mathcal{G})$, *and its complement* X^c *is morphologically closed with respect to all segments of* $\mathcal{S}(\mathcal{G}_c)$ *then the approximation* $\tilde{\chi}^3(\mathbb{F}(\mathcal{G}) \sqcap X)$ *is correct for* $\chi^3(X)$, *i.e.* $\tilde{\chi}^3(\mathbb{F}(\mathcal{G}) \sqcap X) = \chi^3(X)$.

Table 1. The components $w_i,\ i = 0, \ldots, 127$, of the vector w: column (a) for the adjacency $\mathbb{F}_6$, (b) for $\mathbb{F}_{14.1}$, (c) for $\mathbb{F}_{14.2}$, and (d) for $\mathbb{F}_{26}$.

i	(a)	(b)	(c)	(d)	i	(a)	(b)	(c)	(d)	i	(a)	(b)	(c)	(d)	i	(a)	(b)	(c)	(d)
0	0	0	0	0	32	1	1	1	1	64	1	1	1	1	96	2	2	2	-2
1	1	1	1	1	33	2	-2	-2	-2	65	2	-2	2	-2	97	3	-5	-1	-1
2	1	1	1	1	34	0	0	0	0	66	2	2	2	-6	98	1	1	1	-3
3	0	0	0	0	35	-1	-1	-1	-1	67	1	-3	1	-3	99	0	-4	0	0
4	1	1	1	1	36	2	2	2	-6	68	0	0	0	0	100	1	1	1	-3
5	0	0	0	0	37	1	-3	-3	-3	69	-1	-1	-1	-1	101	0	-4	-4	0
6	2	2	2	-2	38	1	1	1	-3	70	1	1	1	-3	102	0	0	0	0
7	-1	-1	-1	-1	39	-2	-2	-2	-2	71	-2	-2	-2	-2	103	-3	-3	-3	1
8	1	1	1	1	40	2	2	-2	-2	72	2	2	2	-2	104	3	3	-1	-1
9	2	-2	-2	-2	41	3	-5	-1	-1	73	3	-5	-1	-1	105	4	-8	0	4
10	0	0	0	0	42	-1	-1	-1	-1	74	1	1	1	-3	106	0	0	0	0
11	-1	-1	-1	-1	43	-2	-2	-2	-2	75	0	-4	0	0	107	-1	-5	-1	3
12	0	0	0	0	44	1	1	-3	-3	76	-1	-1	-1	-1	108	0	0	-4	0
13	-1	-1	-1	-1	45	0	-4	0	0	77	-2	-2	-2	-2	109	-1	-5	-1	3
14	-1	-1	-1	-1	46	-2	-2	-2	-2	78	-2	-2	-2	-2	110	-3	-3	-3	1
15	0	0	0	0	47	-1	-1	-1	-1	79	-1	-1	-1	-1	111	-2	-2	-2	2
16	1	1	1	1	48	0	0	0	0	80	0	0	0	0	112	-1	-1	-1	-1
17	0	0	0	0	49	-1	-1	-1	-1	81	-1	-1	-1	-1	113	-2	-2	-2	-2
18	2	2	2	-2	50	-1	-1	-1	-1	82	1	1	1	-3	114	-2	-2	-2	-2
19	-1	-1	-1	-1	51	0	0	0	0	83	-2	-2	-2	-2	115	-1	-1	-1	-1
20	2	2	-2	-2	52	1	1	-3	-3	84	-1	-1	-1	-1	116	-2	-2	-2	-2
21	-1	-1	-1	-1	53	-2	-2	-2	-2	85	0	0	0	0	117	-1	-1	-1	-1
22	3	3	-1	-1	54	0	0	-4	0	86	0	0	0	0	118	-3	-3	-3	1
23	-2	-2	-2	-2	55	-1	-1	-1	-1	87	-1	-1	-1	-1	119	0	0	0	0
24	2	2	2	-6	56	1	1	-3	-3	88	1	1	1	-3	120	0	0	-4	0
25	1	-3	-3	-3	57	0	-4	0	0	89	0	-4	-4	0	121	-1	-5	-1	3
26	1	1	1	-3	58	-2	-2	-2	-2	90	0	0	0	0	122	-3	-3	-3	1
27	-2	-2	-2	-2	59	-1	-1	-1	-1	91	-3	-3	-3	1	123	-2	-2	-2	2
28	1	1	-3	-3	60	0	0	-8	0	92	-2	-2	-2	-2	124	-3	-3	-7	1
29	-2	-2	-2	-2	61	-3	-3	1	1	93	-1	-1	-1	-1	125	-2	-2	2	2
30	0	0	-4	0	62	-3	-3	-7	1	94	-3	-3	-3	1	126	-6	-6	-6	2
31	-1	-1	-1	-1	63	0	0	0	0	95	0	0	0	0	127	1	1	1	1

The proof will be given for $\mathbb{F}(\mathcal{G}) = \mathbb{F}_{14.1}$, $\mathbb{F}_{14.2}$ and $\mathbb{F}_{26}$ respectively. The case $\mathbb{F}(\mathcal{G}) = \mathbb{F}_6$ can be reduced to $\mathbb{F}(\mathcal{G}) = \mathbb{F}_{26}$ for the complement with the help of the consistency relation (6).
We write $\cup\mathbb{F}(\mathcal{G}) \sqcap X = \bigcup\{F \in \mathbb{F}(\mathcal{G}) \sqcap X\}$ for short. Due to the additivity of the Euler number, it is sufficient to prove $\tilde{\chi}^k(X \cap P) = \tilde{\chi}^k(\cup\mathbb{F}(\mathcal{G}) \sqcap X \cap P),\ k = 0, \ldots 3$, where P is an arbitrary tetrahedron of the tessellation according to the 14.1 or 14.2 adjacency, or a lattice cell $\overline{C} + x,\ x \in \mathbb{L}^3$, if $\mathbb{F} = \mathbb{F}_{26}$ is regarded, or a k-face of the mentioned polytope, respectively. In the following let P be one of these particular polytopes of dimension k.

As a preliminary step we prove the following assertion.

Table 2. The components w_i, $i = 128, \ldots, 255$, of the vector w: column (a) for the adjacency $\mathbb{F}_6$, (b) for $\mathbb{F}_{14.1}$, (c) for $\mathbb{F}_{14.2}$, and (d) for $\mathbb{F}_{26}$.

i	(a)	(b)	(c)	(d)	i	(a)	(b)	(c)	(d)	i	(a)	(b)	(c)	(d)	i	(a)	(b)	(c)	(d)
128	1	1	1	1	160	0	0	0	0	192	0	0	0	0	224	-1	-1	-1	-1
129	2	-6	-6	-6	161	1	-3	-3	-3	193	1	-3	-7	-3	225	0	0	-4	0
130	2	-2	2	-2	162	-1	-1	-1	-1	194	1	-3	1	-3	226	-2	-2	-2	-2
131	1	-3	-7	-3	163	-2	-2	-2	-2	195	0	0	-8	0	227	-3	1	-3	1
132	2	-2	-2	-2	164	1	-3	-3	-3	196	-1	-1	-1	-1	228	-2	-2	-2	-2
133	1	-3	-3	-3	165	0	0	0	0	197	-2	-2	-2	-2	229	-3	1	1	1
134	3	-5	-1	-1	166	0	-4	-4	0	198	0	-4	0	0	230	-3	-3	-3	1
135	0	0	-4	0	167	-3	1	1	1	199	-3	1	-3	1	231	-6	2	2	2
136	0	0	0	0	168	-1	-1	-1	-1	200	-1	-1	-1	-1	232	-2	-2	-2	-2
137	1	-3	-3	-3	169	0	0	0	0	201	0	0	-4	0	233	-1	3	-1	3
138	-1	-1	-1	-1	170	0	0	0	0	202	-2	-2	-2	-2	234	-1	-1	-1	-1
139	-2	-2	-2	-2	171	-1	-1	-1	-1	203	-3	1	-3	1	235	-2	2	-2	2
140	-1	-1	-1	-1	172	-2	-2	-2	-2	204	0	0	0	0	236	-1	-1	-1	-1
141	-2	-2	-2	-2	173	-3	1	1	1	205	-1	-1	-1	-1	237	-2	2	2	2
142	-2	-2	-2	-2	174	-1	-1	-1	-1	206	-1	-1	-1	-1	238	0	0	0	0
143	-1	-1	-1	-1	175	0	0	0	0	207	0	0	0	0	239	1	1	1	1
144	2	-2	-2	-2	176	-1	-1	-1	-1	208	-1	-1	-1	-1	240	0	0	0	0
145	1	-3	-3	-3	177	-2	-2	-2	-2	209	-2	-2	-2	-2	241	-1	-1	-1	-1
146	3	-5	-1	-1	178	-2	-2	-2	-2	210	0	-4	0	0	242	-1	-1	-1	-1
147	0	0	-4	0	179	-1	-1	-1	-1	211	-3	1	-3	1	243	0	0	0	0
148	3	-5	-1	-1	180	0	-4	0	0	212	-2	-2	-2	-2	244	-1	-1	-1	-1
149	0	0	0	0	181	-3	1	1	1	213	-1	-1	-1	-1	245	0	0	0	0
150	4	-8	0	4	182	-1	-5	-1	3	214	-1	-5	-1	3	246	-2	-2	-2	2
151	-1	3	-1	3	183	-2	2	2	2	215	-2	2	-2	2	247	1	1	1	1
152	1	-3	-3	-3	184	-2	-2	-2	-2	216	-2	-2	-2	-2	248	-1	-1	-1	-1
153	0	0	0	0	185	-3	1	1	1	217	-3	1	1	1	249	-2	2	2	2
154	0	-4	-4	0	186	-1	-1	-1	-1	218	-3	-3	-3	1	250	0	0	0	0
155	-3	1	1	1	187	0	0	0	0	219	-6	2	2	2	251	1	1	1	1
156	0	-4	0	0	188	-3	-3	1	1	220	-1	-1	-1	-1	252	0	0	0	0
157	-3	1	1	1	189	-6	2	2	2	221	0	0	0	0	253	1	1	1	1
158	-1	-5	-1	3	190	-2	-2	2	2	222	-2	-2	-2	2	254	1	1	1	1
159	-2	2	2	2	191	1	1	1	1	223	1	1	1	1	255	0	0	0	0

Lemma 1. *Under the assumptions of Theorem 5 we have*

$$X \cap P = \emptyset \iff X \cap \mathcal{F}^0(P) = \emptyset \tag{12}$$

and

$$X \cap \mathcal{F}^0(P) \neq \emptyset \iff \tilde{\chi}^k(\cup \mathbb{F}(\mathcal{G}) \sqcap X \cap P) = 1. \tag{13}$$

Proof:
The nontrivial part of (12) can be shown by induction over the dimension j of the faces of P. If $X \cap \mathcal{F}^0(P) = \emptyset$, i.e. $\mathcal{F}^0(P) \subseteq X^c$, then the morphological closure with the edges of P yields $\mathcal{F}^1(P) \subseteq X^c$. For those particular polytopes P which are considered, this procedure of morphological closing can be iterated to show $P \subseteq X^c$.

The equivalence (13) is a consequence of $\cup\mathbb{F}(\mathcal{G}) \sqcap X \cap P = \mathrm{conv}(\mathcal{F}^0(P) \cap X)$, the convex hull of the vertices belonging to X. (Notice that this is true for $\mathbb{F}_{14.1}$, $\mathbb{F}_{14.2}$, $\mathbb{F}_{26}$ but not for $\mathbb{F}_6$.) □

Proof of Theorem 5:
Regarding the lemma above, it is sufficient to show that

$$\tilde{\chi}^k(X \cap P) = 1 \Longleftrightarrow X \cap P \neq \emptyset. \tag{14}$$

This can be done with Hadwiger's recursive formula (3). The crucial point is that for any j-dimensional plane L^j parallel to a j-dimensional face of P and any $(j-1)$-dimensional plane $L \subset L^j$ (L can be regarded as a linear subspace in L^j) parallel to a $(j-1)$-dimensional face of P

$$\begin{aligned} &\tilde{\chi}^j(X \cap P \cap L^j) = \\ &\sum_{x \in L^\perp \cap L^j} \lim_{\varepsilon \to 0} \left[\tilde{\chi}^{j-1}(X \cap P \cap L_x) - \tilde{\chi}^{j-1}(X \cap P \cap L_{x+\varepsilon u})\right] = 1 \end{aligned} \tag{15}$$

if $X \cap P \cap L^j \neq \emptyset$, with $u \in L^\perp \cap L^j$ and $j = 1, \ldots, k$.
This formula will now be shown in the following parts (a) to (d).

(a) The assumption that X^c is morphologically closed with respect to all segments in $\mathcal{S}(\mathcal{G}_c)$ implies:

$$\begin{aligned} &\text{For all } y \in X \text{ and all } F \in \mathcal{S}(\mathcal{G}_c) \text{ there exists a } t \in \mathbb{R}^3 \text{ with} \\ &\qquad y \in (F+t) \subseteq X, \end{aligned} \tag{16}$$

and analogously, if X is morphologically closed with respect to all segments in $\mathcal{S}(\mathcal{G})$, then

$$\begin{aligned} &\text{for all } y \in X^c \text{ and all } F \in \mathcal{S}(\mathcal{G}) \text{ there exists a } t \in \mathbb{R}^3 \text{ with} \\ &\qquad y \in (F+t) \subseteq X^c, \end{aligned} \tag{17}$$

see [31], II.D., relations between morphologically closed and open sets and the characterization of morphologically open sets.

(b) Let L^j and L be the same as used in (15) and $x_1, x_2, x \in L^\perp \cap L^j$. If $P \cap L_{x_1} \neq \emptyset$ and $P \cap L_{x_2} \neq \emptyset$ then

$$\begin{aligned} &X \cap P \cap L_{x_1} = \emptyset \wedge X \cap P \cap L_{x_2} = \emptyset \Longrightarrow \\ &X \cap P \cap L_x = \emptyset \text{ for all } L_x \text{ between } L_{x_1}, L_{x_2}. \end{aligned} \tag{18}$$

Assume $X \cap P \cap L_x \neq \emptyset$, i.e. there is a $y \in X \cap P \cap L_x$. Then (16) provides that there is a segment $y \in (F+t) \subseteq X$. This segment can be chosen as a translate of an edge of P which hits L_{x_1} and L_{x_2}. Hence $(F+t)$ hits at least one of the sets $L_{x_1} \cap P$ or $L_{x_2} \cap P$. This contradicts the assumption, and thus (18) is shown.

(c) With the same notation as in (b) and analogously to (18)

$$\begin{aligned} &X \cap P \cap L_{x_1} \neq \emptyset \wedge X \cap P \cap L_{x_2} \neq \emptyset \Longrightarrow \\ &X \cap P \cap L_x \neq \emptyset \text{ for all } L_x \text{ between } L_{x_1}, L_{x_2}. \end{aligned} \tag{19}$$

Let $y_1 \in X \cap P \cap L_{x_1}$, $y_2 \in X \cap P \cap L_{x_2}$.
Firstly consider the case that $L_{x_1}, L_{x_2} \subset L^2$ are lines. Let $S \subseteq P$ be the smallest triangle in the plane L^2 with y_2 as one vertex and y_1 an element in the opposite edge, say F_S which is parallel to an edge $F \in \mathcal{S}(\mathcal{G}_c)$, and all edges of S parallel to one of the segments in $\mathcal{S}(\mathcal{G})$. (Notice that it is not a misprint that once the index c occurs and once not. This distinction is important for the pair of adjacencies $(26, 6)$.) Then, according to (16) there is a translated segment $F + t \subseteq X$ through y_1. Since F_S is not longer than F, the translate $F + t$ hits at least one of the endpoints of F_S, say y_S. Since X is assumed to be morphologically closed with respect to all segments in $\mathcal{S}(\mathcal{G})$, also the segment $[y_2, y_S] \subseteq X$ since this is not longer than the parallel segment from $\mathcal{S}(\mathcal{G})$. Thus it is shown that the points y_1 and y_2 are connected by a chain of segments $[y_1, y_S]$ and $[y_S, y_2]$ which belong to $X \cap P$. This yields (19).
If L_{x_1}, L_{x_2} are planes, then a minimal tetrahedron S exists with y_2 as one vertex and y_1 in the opposite facet of S, and all edges parallel to appropriate segments in $\mathcal{S}(\mathcal{G})$. With an iterated application of the same ideas and arguments as they were used for the triangle above, it is possible to show the existence of a sequence of segments inside $X \cap P$ which connects y_1 and y_2.

(d) Summarizing (b) and (c) implies that $\{x \in L^{\perp} \cap L^j : X \cap P \cap L_x \neq \emptyset\}$ is a closed convex set, i.e. a segment on the line $L^{\perp} \cap L^j$. Hence, if $X \cap P \cap L^j \neq \emptyset$, the sum in (15) has exactly one item different from 0, and its value is 1. Induction over $j = 1, \ldots, k$ yields (15) and the proof of (14) is complete.

□

If nothing can be assumed about the regularity of X or X^c, both $X \cap \mathbb{L}^n$ and $X^c \cap \mathbb{L}^n$ should be treated equally. In particular, self-complementary adjacencies can be a good choice. On the other hand, our simulation studies in Sect. 4 illustrate that the self-complementary 14.1 and 14.2 adjacencies can induce a considerably large approximation error.

Assume now for example that X is a finite union of compact and convex sets X_i, $i = 1, \ldots, m$, having smooth surfaces. If the X_i are morphologically open with respect to the segments of $\mathcal{S}(\mathcal{G}_c)$ then their union $X = \cup_{i=1}^m X_i$ is morphologically open with respect to these segments, too. But X^c is not necessarily open with respect to the elements of $\mathcal{S}(\mathcal{G}_c)$. As a consequence, the pair $(\mathbb{F}_6, \mathbb{F}_{26})$ of adjacency systems is the best choice for the measurement of $\chi^n(X)$; i.e. $\tilde{\chi}^n(\mathbb{F}_6 \sqcap X)$ is 'better' than the approximation of $\chi^n(X)$ with respect to a self-complementary adjacency system.

The following lemma states that, if both X and X^c are 'regular enough', their Euler number is independent of the choice of the adjacency system.

Lemma 2. *Let X be a compact set which is morphologically open and morphologically closed with respect to all segments $[x, y]$ with $x, y \in \mathcal{F}^0(\overline{C})$, $x \neq y$, and $(x, y) \subset C$. Then $\tilde{\chi}^n(\mathbb{F}(\mathcal{G}) \sqcap X) = \chi^n(X)$ for all superpositions of admissible tessellations $\mathcal{G}$.*

Proof: Notice that the set of segments mentioned in this lemma is equivalent to $\mathcal{S}(\mathcal{G}_{max})$. The proof follows from Theorem 5, since $\mathcal{S}(\mathcal{G}) \subset \mathcal{S}(\mathcal{G}_{max})$. □.

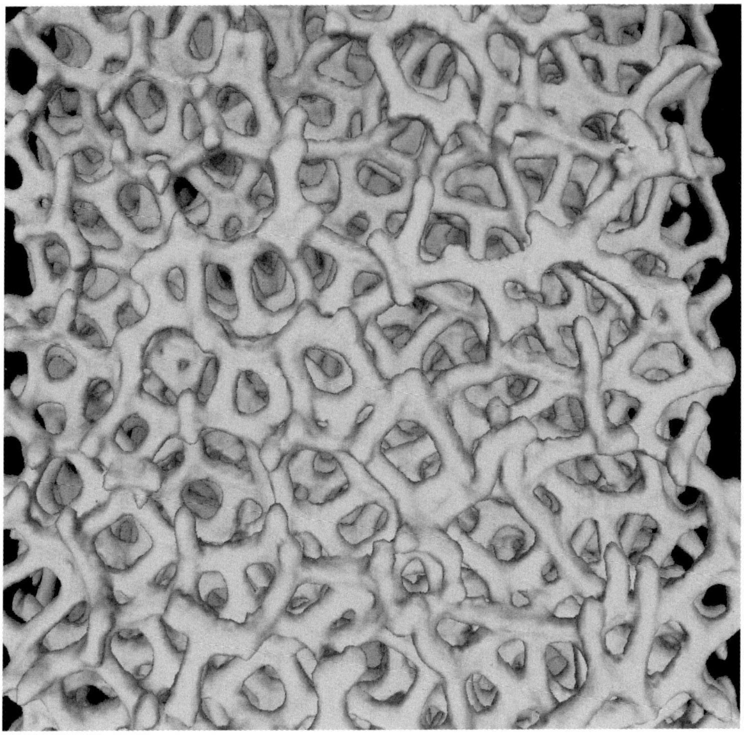

Fig. 6. The microstructure of an open nickel foam. The XCT scan consists of $341 \times 320 \times 511$ pixels (simple cubic lattice, $|u| = 10\ \mu$m).

4 Applications and a Simulation Study

As applications we consider the microstructures of three porous materials: a Fontainebleau sandstone, an open nickel foam, and an autoclaved aerated concrete (AAC). Samples of these microstructures are given as 3d images. For visualizations of the data see Figs. 1, 6, and 7, respectively. The Fontainebleau sandstone is a standard example in papers on simulation of fluid flow through porous media. (Notice that different data sets are used.) The three microstructures are given as 3d images obtained by X-ray microtomography.

Each component of the considered microstructures can be modeled as a macroscopically homogeneous random set Ξ as described in [24]. More precisely, we assume that with probability one the random set $\Xi \cap W$ belongs to the convex ring (for every compact set W with nonempty interior), and the distribution law of Ξ is invariant with respect to translations $x \in \mathbb{R}^n$. Then the density of the Euler number (which is also called the connectivity density) is given by

$$\chi_V(\Xi) = \lim_{\varepsilon \to \infty} \frac{\tilde{\chi}^n \left(\mathbb{F} \sqcap (\Xi \cap \varepsilon W) \right)}{|\varepsilon W|}. \tag{20}$$

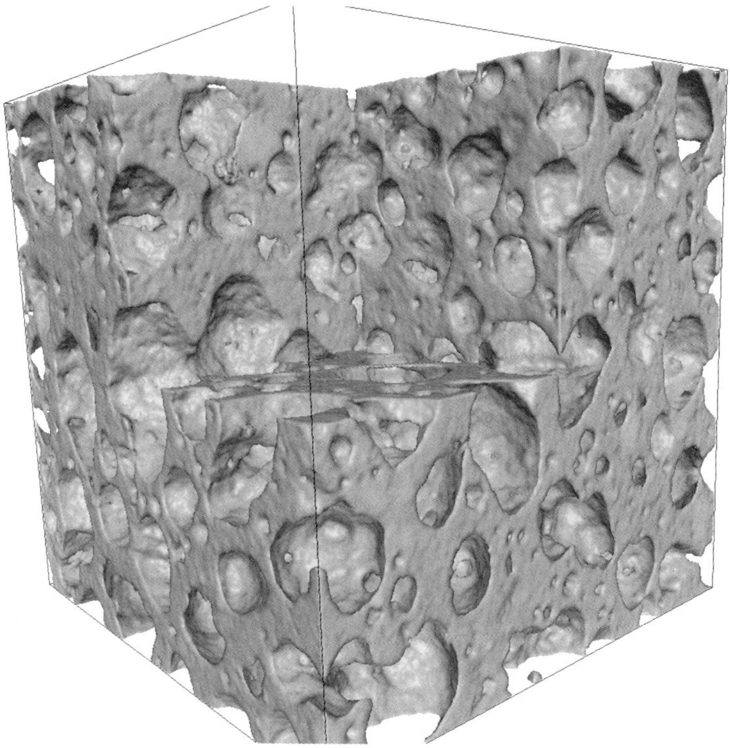

Fig. 7. The microstructure of an autoclaved aerated concrete (AAC). The XCT scan consists of $320 \times 330 \times 330$ pixels (simple cubic lattice, $|u| = 31.3\ \mu\text{m}$).

Here $|\cdot|$ denotes n-dimensional volume. The density of the Euler number can be estimated from the set Ξ observed in a compact window W using

$$\widehat{\chi_V}(\Xi) = \frac{1}{\#(\mathbb{L}^n \cap W_0)\,|C|} \sum_{x \in \mathbb{L}^n \cap W_0} \sum_{k=0}^{n} (-1)^k \sum_{\ell=k}^{n} 2^{\ell-n} \#(\mathcal{F}^{k\ell} \sqcap \Xi_{-x}), \qquad (21)$$

where W_0 is the reduced window, $W_0 = W \ominus \check{C}$. Notice that $\#(\mathbb{L}^n \cap W_0)\,|C| \approx |W_0|$. Furthermore, we remark that the estimator $\widehat{\chi_V}(\Xi)$ may be biased but it is free of edge effects.

Table 3 shows that experimental values for $\chi_V(\Xi)$ can depend highly on the chosen adjacency. Differences in the measurement values are a consequence of the irregularity of the microstructures. It is an obvious consequence of Theorem 5 and Lemma 2 that the constituents of the microstructures are not morphologically regular with respect to the edges, face diagonals, and space diagonals of C.

In the first example we consider the Fontainebleau sandstone shown in Fig. 1. The 3d data are equivalent to that considered in [7]. In particular, for these data our algorithm leads to large differences between the estimates and, hence, they are only of low value for practical use. Obviously, the lateral resolution in the image is not sufficient for an accurate analysis of this very 'tiny' microstructure, i.e. the regularity conditions formulated

Table 3. Experimental data for the density χ_V of the Euler number with respect to the adjacencies $\mathbb{F}_6$, $\mathbb{F}_{14.1}$, $\mathbb{F}_{14.2}$, and $\mathbb{F}_{26}$ for some microstructures and realizations of Boolean models. The data are given in mm^{-3}. For the microstructures the true values are unknown. For the Boolean models the true values (expectations) were determined from the model parameters using Miles' formulae.

microstructure/ geometric model	$\mathbb{F}_6$	$\mathbb{F}_{14.1}$	$\mathbb{F}_{14.2}$	$\mathbb{F}_{26}$	true value
Fontainebleau sandstone, Fig. 1	-198	-273	-366	-369	unknown
open nickel foam, Fig. 6	-105	-105	-106	-113	unknown
AAC, Fig. 7	4.65	5.46	5.09	6.21	unknown
Boolean model, $\lambda = 500\,\mathrm{mm}^{-3}$	50	-46	-108	-76	107
Boolean model, $\lambda = 1\,000\,\mathrm{mm}^{-3}$	-308	-515	-679	-592	-188
Boolean model, $\lambda = 2\,000\,\mathrm{mm}^{-3}$	-861	-1079	-1420	-979	-970
Boolean model, $\lambda = 4\,000\,\mathrm{mm}^{-3}$	-365	-261	-579	240	-603

in Theorem 5 are not fulfilled. In this case the image acquisition (i.e. the discretization) induced a certain loss of information with considerable consequences for the measurement of the Euler density. For comparison see the contribution by Knackstedt in this volume, where the Euler number computed from another image of the Fontainebleau sandstone.

For the open foam shown in Fig. 6 the differences between the estimates are small. Since these differences are neglectable, one can believe that the foam is morphologically regular with respect to segments of lengths close to the lattice spacings. Thus we can also expect low bias of the estimates, i.e. the measurement values are assumed to be close to the true Euler density $\chi_V(\Xi)$.

Additionally to the analysis of these sample data we performed a simulation study. To compare theoretical (true) values of the density of the Euler number with estimates, we consider particular Boolean models in the Euclidean space for which the theoretical values can be computed from the model parameters using Miles' formulae, see [22],[31, p. 495], or [29, p. 217]. The Boolean models were simulated with balls of constant diameter $d = 100\,\mu\mathrm{m}$ and the densities $\lambda = 500\,\mathrm{mm}^{-3}$, $1\,000\,\mathrm{mm}^{-3}$, $2\,000\,\mathrm{mm}^{-3}$, and $4\,000\,\mathrm{mm}^{-3}$ of the corresponding Poisson fields of germs. The corresponding realizations are observed on a simple cubic lattice with uniform spacings $|u| = 10\,\mu\mathrm{m}$. The experimental data were computed form 32 images (i.e. discretizations of simulated Boolean models) where each image was of size $512 \times 512 \times 512$ pixels (i.e. each sampling window was a cube of volume $134.2\,\mathrm{mm}^3$). The experimental values given in Table 3 are mean values taken over 32 images (realizations). The standard deviation of the means was less than $1.6\,\mathrm{mm}^{-3}$. This means that the distances of each experimental value from the corresponding true value can be considered as the bias of estimation. Notice that for large sampling windows, the estimation variance of the Euler-Poincaré characteristic of Boolean models can also be computed by formulae given in [5].

This simulation study shows that there can be a considerable bias in the estimates caused by discretization. It is perhaps surprising that a large bias can even occur in cases when the primary grains are much larger than the lattice cells. Notice that in our example the balls' diameter is $d = 10|u|$. Thus the bias is mainly due to tiny structure in the complement X^c when balls are very close together without overlapping.

Depending on the point density λ, the minimal bias is obtained for different adjacencies: $\mathbb{F}_6$ for $\lambda = 500\,\mathrm{mm}^{-3}$, $1000\,\mathrm{mm}^{-3}$, $\mathbb{F}_{26}$ for $\lambda = 2000\,\mathrm{mm}^{-3}$, and $\mathbb{F}_{14.2}$ for $\lambda = 4000\,\mathrm{mm}^{-3}$. That means, there is no clear advantage of one adjacency over the others and without further knowledge about the geometry of Ξ, an appropriate choice of $\mathbb{F}$ seems to be impossible.

In practical applications, the Euler number should be measured with respect to several adjacencies. Then the differences between the results provides an impression of the bias of the measurements. Of course, other adjacencies could be useful besides the four examples suggested in Sect. 3.3. Admissible tessellations and their superpositions yield a wide variety of adjacencies. A constructive method for finding complementary adjacencies has yet to be found.

Acknowledgements

The authors thank the editors for valuable hints on a first draft of this paper. We are indebted to E. Schlegel for supplying the AAC specimen shown in Fig. 7, S. Gondrom for scanning the nickel foam and the AAC, R. Hilfer for providing the data of the Fontainebleau sandstone, and H. Seibert for some ideas and helpful comments. The authors J.O. and K.S. are grateful to the *Stiftung Rheinland-Pfalz für Innovation* for financial support.

References

1. Arns, C. H., A. Knackstedt, W. V. Pinczewski, K. Mecke (2001): 'Euler-Poincaré characteristics of classes of disordered media'. *Phys. Rev.* **E 63**, pp. ???
2. Bieri, H., W. Nef (1984): 'Alogorithms for the Euler Characteristic and Related Additive Functionals of Digital Objects'. *Comput. Vision Graphics Image Process.* **28**, pp. 166–175
3. Gruber, P. M. (1993): 'Geometry of numbers'. In: *Handbook of convex geometry*, edited by P. M. Gruber, J. M. Wills (North Holland, Amsterdam), pp. 739–763
4. Hadwiger, H. (1957): *Vorlesungen über Inhalt, Oberfläche und Isoperimetrie* (Springer, Heidelberg)
5. Heinrich, L. (2002) 'Asymptodic normality of the Euler-Poincaré characteristic of Boolean models in large sampling windows.' *Report No.* **449**, Universität Augsburg, Institut für Mathematik.
6. Herman, G. T. (1992): 'Discrete Multidimensional Jordan Surfaces'. *Graph. Mod. Image Proc.* **54**, pp. 507–515
7. Hilfer, R. (2000): 'Local porosity theory and stochastic reconstruction'. In: *Statistical Physics and Spatial Statistics*, edited by K. R. Mecke, D. Stoyan (Springer, Heidelberg), pp. 203–241
8. Imiya, A., U. Eckhardt (1997): 'The Euler Characteristic of Discrete Object'. *Discrete Geometry for Computer Imagery* , pp. 163–174
9. Jernot, J. P., P. Jouannot-Chesney, C. Lantuéjoul (2001): 'Determination of the connectivity number of a set using a tessellation'. Manuscript.
10. Khalimsky, E., R. Kopperman, P. Meyer (1990): 'Computer Graphics and connected topologies on finite ordered sets'. In: *Topology and its Application*, vol. 36 (Elsevier Science Publisher B.V), pp. 1–17
11. Kong, T. Y., A. W. Roscoe (1985): 'Continuous analogs of axiomatized digital surfaces'. *Comput. Vision Graphics Image Process.* **29**, pp. 60–86

12. Kong, T. Y., A. Rosenfeld (1989): 'Digital Topology: Introduction and Survey'. *Comput. Vision Graphics Image Process.* **48**, pp. 357–393
13. Kong, T. Y., J. K. Udupa (1992): 'A justification of a fast surface tracking algorithm'. *CVGIP: Graph. Models Image Process.* **54**, pp. 162–170
14. Lang, C., J. Ohser, R. Hilfer (2001): 'On the analysis of spatial binary images'. *J. Microsc.* **202**, pp. 1–12
15. Lee, C., T. Poston, A. Rosenfeld (1991): 'Winding and Euler Numbers for 2D and 3D Digital Images'. *CVGIP: Graph. Models Image Process.* **53**(6), pp. 522–537
16. Lee, C., T. Poston, A. Rosenfeld (1993): 'Holes and Genus of 2D and 3D Digital Images'. *CVGIP: Graph. Models Image Process.* **55**(1), pp. 20–47
17. Lohmann, G. (1998): *Volumetric Image Analysis* (Wiley-Teubner, Chichester, New York)
18. Matheron, G. (1975): *Random Sets and Integral Geometry* (Wiley, New York)
19. Mecke, K. R. (1993): *Integralgeometrie in der Statistischen Physik* (Harri Deutsch, Frankfurt a. M.)
20. Mecke, K. R. (2000): 'Additivity, convexity, and beyond: applications of Minkowski functionals in statistical physics'. In: *Statistical Physics and Spatial Statistics*, edited by K. R. Mecke, D. Stoyan (Springer, Heidelberg), pp. 111–184
21. Mecke, J., R. G. Schneider, D. Stoyan, W. R. R. Weil (1990): *Stochastische Geometrie* (Birkhäuser, Basel, Boston, Berlin)
22. Miles, R. (1976): 'Estimating aggregate and overall characterictics from thick sections by transmission microscopy'. *J. Microscopy* **107**, pp. 227–233
23. Nagel, W., J. Ohser, K. Pischang (2000): 'An integral-geometric approach for the Euler-Poincaré characteristic of spatial images'. *J. Microsc.* **189**, pp. 54–62
24. Ohser, J., F. Mücklich (2000): *Statistical Analysis of Materials Structures* (J Wiley & Sons, Chichester, New York)
25. Ohser, J., W. Nagel (1996): 'The estimation of the Euler-Poincaré characteristic from observations on parallel sections'. *J. Microsc.* **184**, pp. 117–126
26. Rosenfeld, A., A. Kak (1979): *Digital Picture Processing* (Academic Press, New York)
27. Rosenfeld, A., T. Y. Kong, A. Y. Wu (1991): 'Digital surfaces'. *CVGIP: Graph. Models Image Process.* **53**, pp. 305–312
28. Schneider, R. (1993): *Convex Bodies: The Brunn-Minkowski Theory*, vol. 44 of *Encyclopedia of Mathematics and Its Application* (Cambridge University Press, Cambridge)
29. Schneider, R., W. Weil (2000): *Stochastische Geometrie* (Teubner, Stuttgart)
30. Serra, J. (1969): *Introduction á la Morphologie Mathématique*, (Cahiers du Centre de Morphologie Mathématique, Booklet No. 3, E.N.S.M.P., Paris)
31. Serra, J. (1982): *Image Analysis and Mathematical Morphology*, vol. 1 (Academic Press, London)
32. Voss, K. (1993): *Discrete Images, Objects, and Functions in* Z^n (Springer-Verlag, New York)
33. Webster, R. J. (1994): *Convexity* (Oxford University Press, New York)
34. Weil, W. (2000): 'Mixed measures and homogeneous Boolean models'. In: *Statistical Physics and Spatial Statistics*, edited by K. R. Mecke, D. Stoyan (Springer, Heidelberg), pp. 95–110

Shape Statistics for Random Domains and Particles

Dietrich Stoyan, Ashot Davtyan, and Daulet Turetayev

Institut für Stochastik, TU Bergakademie Freiberg
D 09596 Freiberg

Abstract. This paper surveys ideas of statistical analysis of planar images of particles such as powder particles or sand grains, domains such as monolayer domains on water or water droplets on planar surfaces and biological cells or vesicles. For a simple and fast discrimination between collectives of particles, shape ratios or indices such as 'area:perimeter' ratio are powerful tools. A more detailed description is possible by means of various functions such as radius–vector function, tangent–angle function and erosion function. A deeper understanding of particle shape and size is possible by studying the relevant physical processes which generate them, such as fracture, abrasion and growth by aggregation.

The second part of the paper discusses a particular stochastic model, called Gibbs pixel-particle. It produces two-dimensional connected lattice figures, often called lattice animals, the distribution of which depends on an energy function which is controled by particle area and boundary length and roughness. These pixel-particles vary in a broad spectrum of possible shapes and sizes.

1 Introduction

An important but still little developed branch of geometrical statistics is shape statistics of geometrical objects which can be called particles or domains. Practical examples are powder particles, sand grains, pebbles, worn stones or rock fragments and other objects forming granular matter. It is clear that their shape has some influence on physicomechanical properties of collectives of such objects as agglomeration, packing, permeability, and suspension stability [19]. Remarkable fluctuations of shape are also observable in the study of galaxies (see the paper by Beisbart et al. in this volume) or of biological cells, cell nuclei or vesicles (see the paper by Schwarz and Gompper in this volume). Planar objects of variable size and shape are spots, droplets (e.g. of water on glass surfaces as in the paper by Döge and Stoyan in this volume) or phase domains (e.g. monolayers on water as in the paper by Lösche and Krüger in this volume).

In statistical analysis these objects are given usually as images, as *planar data*, perhaps resulting from planar sections or projections of three-dimensional bodies. They are typically characterized by their boundaries or outlines. In the case discussed here, the objects are featureless, the outlines do not contain particular points. By the way, such points, often called 'landmarks', play an important role in biological applications. (Look at your left hand and observe the lot of landmarks on its outline which you could use to describe its shape.) Today there exists a broad literature on landmark statistics, which is, by the way, rather complicated in its algebra and geometry, see the books [4],

[8], [24], [31], and [44]. But this article discusses methods which do not use landmarks, and the planar objects studied are called *particles*, regardless of their origin.

Mathematically, particles could be described as compact sets or, in another approach, as closed curves. However, typically position and orientation of a particle are uninteresting. Therefore, particles will be considered here as equivalence classes of congruent sets or curves. (In practical statistics, however, it is convenient to introduce a location, may be by using the centre of gravity as origin, and an orientation, perhaps by defining two principal axes.)

The statistics of random particles is still in the beginning. Some book references are [2], [3], [19], [48] and [51]. It is typical that many results of particle statistics are scattered irregularly in the literature and still there is no standard reference.

2 A Survey of Methods in Particle Statistics

2.1 Shape Ratios

The problems in particle statistics are quite different. Sometimes it is sufficient to discriminate between samples of particles, while in other situations a detailed description is needed, which helps to understand the process of particle formation or to predict the particle behaviour in physical processes.

For the purpose of discrimination it is very often sufficient to use *shape ratios* or indices, see [48] for a collection of a lot of them. A well-known example is the area-perimeter ratio

$$f_{AU}(X) = \frac{4\pi A(X)}{U(X)^2} \tag{1}$$

where $A(X)$ and $U(X)$ are area and perimeter of the particle X. For any disk it is $f_{AU}(X) = 1$, while for all other particles $f_{AU}(X) < 1$, and the smaller $f_{AU}(X)$ is, the greater is the deviation from circular shape. A three-dimensional analogue of f_{AU} is the 'reduced volume' $v = V(X)/[(4\pi/3)(S(X)/4\pi)^{3/2}]$, where $S(X)$ is the surface area of X and $V(X)$ the volume.

A refined ratio is

$$f_L(X) = \frac{4A(X)}{\pi L(X)^2}$$

introduced in [29] for the case of non-convex particles. Here $L(X)$ is the geodetic diameter, the maximum of the lengths of all shortest curves (geodetics) completely contained in X and connecting two points of X. Other shape ratios are based on lengths of suitable principal axes.

In [48] a series of such shape ratios is collected and discussed. They serve for characterization of various geometrical properties of particles, e.g. deviations from convexity, circularity and symmetry. Also the papers by Beisbart et al., Arns et al., Ohser et al., Robins, and Vogel in this volume use shape indices.

Note the following fact: A particle model may be 'isotropc', i. e. the particles are generated by a physical process which respects isotropy. Nevertheless, the shape ratios may not reflect the isotropy and may yield values as for anisotropic objects like ellipses.

2.2 Shape Functions

More detailed but also more expensive forms of particle description use functions. Then to each particle of a sample a function is assigned and the statistician is confronted with functional data which can be analysed by statistical methods for stochastic processes or by special methods for functional data [43].

The perhaps most natural function is the radius-vector function. Here inside the particle the origin o of a Cartesian coordinate system is chosen, usually the centre of gravity. Then $r(\varphi)$ is the distance from o to the outline of X in direction φ, where φ is the angle between the x-axis and a ray starting in the origin. (Clearly, the choice of the x-axis plays some role.) This $r(\varphi)$ is uniquely defined if the particle is 'starshaped' with respect to o, i.e. if every ray intersects the outline in one and only one point. The function $r(\varphi)$ is periodic with period 2π and for its characterization parameters are of particular interest which are invariant if $r(\varphi)$ is replaced by $r(\varphi + \varphi_o)$ for any φ_o.

The radius-vector approach is used in many papers; basic ideas are systematically collected in [20] and [48]. It is very popular to describe $r(\varphi)$ by a Fourier series and to use Fourier coefficients as shape parameters. There are also stochastic models and corresponding statistical methods for random radius-vector functions, see [20] and [48]. The first paper studies in detail a model which produces particles being the result of random deformations of templates, which can be also non-circular.

A weak point of the radius-vector approach is that the origin o plays such a prominent role, which is only introduced by the observer and has nothing to do with the particle's morphology. (An exception may be a domain which results from diffusion from a given and known centre or a cell with a clear nucleus centre.) The question which interesting functionals of $r(\varphi)$ are independent of the choice of o has been not yet systematically studied. Clearly, area and perimeter can be expressed by $r(\varphi)$.

There are three other functional approaches which are free of this disadvantage, but which are more sophisticated and do not enable an easy particle reconstruction.

The first is the tangent-angle function $t(s)$, [48], p. 69. It is assumed that the particle has a smooth outline $(x(s), y(s))$, where s is the boundary arclength measured in anti-clockwise direction beginning in some starting point, $0 \leq s \leq L(X)$. The corresponding tangent-angle is then given by

$$t(s) = \arctan \frac{\dot{y}(s)}{\dot{x}(s)} \quad \text{for} \quad 0 \leq s \leq L(X).$$

Its derivative yields the curvature, and therefore $t(s)$ was used in [34] for describing the curvature radius of discrete surfaces. It may be convenient to reparametrize so that $t(s)$ is a function on $[0, 2\pi]$. Cleary, $t(s)$ yields only shape information.

The second approach uses the erosion function $E_s(r)$, [48], p. 123. Here a reconstruction of the particle X from the function is difficult or impossible, but the numerical determination of the function is easy since well-known methods of mathematical morphology [45] can be used. The function is defined as

$$E_s(r) = A(X \ominus b(o, r)) \quad \text{for} \quad r \geq 0, \tag{2}$$

i.e., it is the area of any set congruent to X eroded by the disk of radius r centred at the origin o. The definition can be extended to negative values of r by

$$E_s(r) = A(X \oplus b(o, -r)) \quad \text{for} \quad r < 0.$$

It is possible to replace the Minkowski subtraction sign $\ominus$ by the opening sign $\circ$ to obtain an analogous function, which could be called 'opening function'. Following the spirit of K. Mecke's Minkowski functional approach [35], also functions where A (= area) is replaced by L = length of particle boundary or by χ = Euler–Poincaré characteristic can make sense.

Finally, chord length distribution functions $L(r)$ are useful [48], p. 117. The random chord length of a set is the length of the chord obtained by intersecting the set with a random line. (In the non-convex case a chord can consist of a series of segments.) Usually such statistical data result from X-ray investigations [13] or stereological work [50]. There typically the observed chord length distribution is a mixture of such distribution functions, since samples of different particles are analysed simultaneously. In a series of papers, Gille has derived the formulas for chord length distribution functions for many convex bodies, see [13].

Stochastic geometers have tried to define and compute *means* for samples of particles, see [47] and [48]. These means are 'set-valued', i. e. they are particles as the elements of the sample. In the case of an underlying isotropic model, one of these means, the Aumann mean, is a disc, while the means discussed in [47] are anisotropic objects.

2.3 Modeling Growth, Fracture and Abrasion

The functions described above are merely formal statistical descriptors, they are in many cases not related to the processes which lead to the particles in the given shape and size. The aim of this subsection is to sketch briefly some ways of constructing suitable stochastic models for random particles.

Many particles are the results of *growth* processes such as biological growth, crystal growth or coagulation (snow-balling). One possible model in this context is the shot noise process. It was originally developed for a physical process on the time axis with random events ('shots') and corresponding impulses, which are superposed [7]. In the particle context the 'shots' are associated with small subparticles which are added to the growing particle. The corresponding radius vector function has the form

$$r(\varphi) = \sum\nolimits_{x \in \Pi}^{w} a_x v(\varphi - x) \quad \text{for} \quad 0 \leq \varphi \leq 2\pi\,.$$

The points x belong to a Poisson print process Π on $[0, 2\pi]$ and correspond to the centres of the subparticles, which have the random masses a_x; $v(z)$ is a probability density function on $[0, 2\pi]$ which describes the shape of the subparticles. The 'w' at the sum sign means that the addition is made in the sense of a wrapping around the unit circle, i.e. $a_x v(\varphi - x)$ means $a_x[v(\varphi - x) + v(\varphi - x + 2\pi) + v(\varphi - x - 2\pi)]$. For this model under independence and stationarity assumptions formulas for mean and covariance function of $r(\varphi)$ can be derived. A more realistic model would consider the dynamic effect that two identical subparticles added to the same particle in different stages of the growth process cause different changes of shape because of growing perimeters.

Many particles have their origin in a *fracture* process, which appears in nature or in machines like mills. The particles resulting from such a process can perhaps be modeled as polygons, perhaps even as cells of tessellations of the plane. An elegant model of this kind is the Poisson line process, for the cells of which many formulas are known [48], [50]. Figure 1 shows some such cells obtained by simulation and particles which result from them by opening operations with disks of growing radii.

In many situations, given particles result from *abrasion* processes, for example sand grains, worn stones or particles in tumbling barrels. The paper [27] reports on an analysis of a big data set of limestone particles processed in a tumbling barrel. There the operations of mathematical morphology are used to describe size reduction and rounding. Particle size could be well described by erosion, i.e. using the model

$$X(t) = X \ominus b(o, r(t)) \quad \text{for} \quad t \geq 0\,,$$

where $X(t)$ is the particle after t time units in the tumbling barrel ($X = X(0)$) and $r(t)$ an erosion radius, for which empirically a plausible form was found. Thus the erosion function was useful. Even consequences of the equation

$$(X \ominus b(o, r_1)) \ominus b(o, r_2) = X \ominus b(o, r_1 + r_2)$$

turned out to be satisfied. Of course, the shape change by rounding cannot be modeled by the operation erosion and also opening did not help. A satisfactory approximation was obtained by the formal model

$$X(t) = (X \ominus b(o, r_1(t)) \oplus b(o, r_2(t)) \quad \text{for} \quad t \geq 0$$

where the radius difference $r_1(t) - r_2(t)$ describes the speed of abrasion.

This model can be considered as a simplified version of *curvature flow* in $\mathbb{R}^d$. This is a mathematically more sophisticated model of abrasion and rounding processes. It can be formulated as follows. Given a closed hypersurface $\overrightarrow{B}(s)$ in $\mathbb{R}^d,\ s \in S^{d-1}$, the solution of the initial value problem

$$\frac{\partial \overrightarrow{B}(s,t)}{\partial t} = -k(s,t)\overrightarrow{N}(s,t), \tag{3}$$

$$\overrightarrow{B}(s,0) = \overrightarrow{B}(s)$$

is studied, where k is the (mean or Gaussian) curvature and $\overrightarrow{N}$ is the outer unit normal at $\overrightarrow{B}(s,t)$. [6], [9], [21] and [52] show that under convexity and smoothness conditions (3) has a unique solution $\overrightarrow{B}(s,t)$ for $0 \leq t < T$ with $T = V/db_d$, where V is the volume enclosed by $\overrightarrow{B}(s)$ and by b_d the volume of unit sphere in $\mathbb{R}^d$. As t tends to T, the hypersurfaces approach the shape of a sphere and finally shrink to a point.

3 Gibbs Pixel-Particles

3.1 Energies and Shapes

This section considers a special class of particle models, which are based on a combination of geometrical and physical principles. The idea follows the general approach

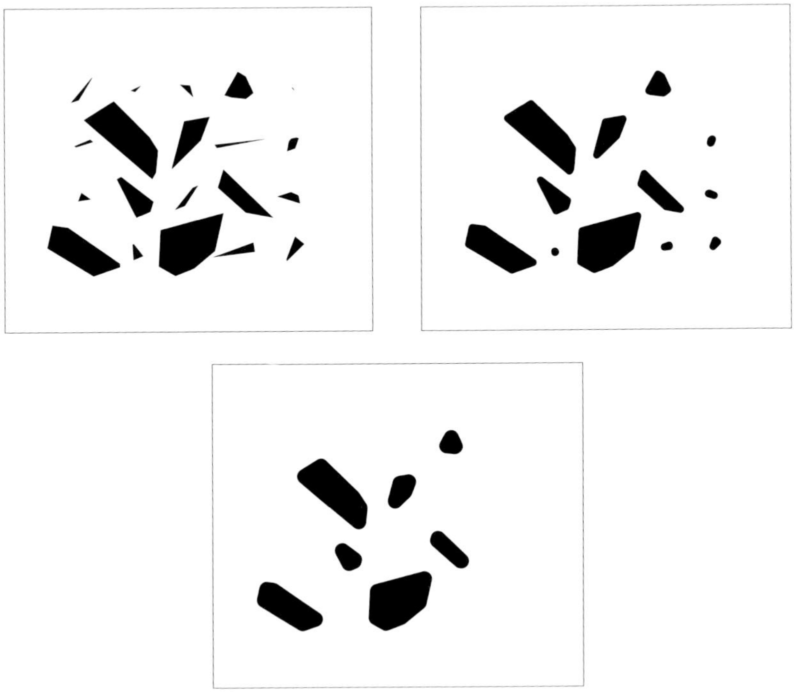

Fig. 1. Some simulated Poison polygons and corresponding smoothed objects obtained by opening with disks of various radii (Courtesy of C. Lantuejoul).

to use *geometrical energies*. There some energy is assigned to a particle ([28], [33]) or another geometrical object like a surface or membrane ([15], [16], [17], [18], [25], [32], [41], [42]) which is in close relation to its geometry. It may contain terms depending on curvature, surface tension, bending and differences to templates, mean or equilibrium shapes. This approach is attractive since it starts from first principles and is not a priori based on measurement or description methods. The papers by Beisbart et al., Lösche and Krüger, Schwarz and Gompper and Graner in this volume are impressive examples for the application of this idea. Note also the use of quermass or Minkowski functionals in [35].

There are two approaches. The first yields objects which represent minima of energy. These minima are determined by various numerical methods, often based on discretizations of the objects [22], [32], [42]. A possibility of modeling random objects is simulated annealing. Here an inhomogeneous Markov chain is considered which moves from some initial state to an absorbing state, which is a (local) minimum of the energy function. Usually different minima are attainable in dependence on the random minimization process and so one can speak about random objects and particles. Another method uses a random start configuration and a deterministic minimization procedure [28], [33].

The alternative approach considers distributions of the form exp(– energy). They correspond to equilibrium states or to random fluctuations around equilibrium states. The simulation of the corresponding geometrical objects uses the Monte Carlo method, i.e. an ergodic homogeneous Markov chain is simulated the unique stationary distribution of which is the target distribution. For example, [20] and [25] study random polygons the energy of which characterizes deviations from a given template. The papers [15] and [16] are excellent surveys on results for random surfaces of fixed area content, see also [17]. There the energy depends on surface curvature as well as on material parameters like bending rigidity.

Also the present paper considers a model of the second type, which appears with other energies already in [1]. The particles are discrete planar sets, formed by pixels, and there is no origin or another reference point. The energy is extremely simple, it only depends on geometrical functionals which are motion invariant, such as area (= pixel number), boundary length or some roughness parameter. The aim is to show the great flexibility of the corresponding model and thus its applicability in various situations. It produces shapes similar to those of branched polymers as well as compact bodies with rough boundaries.

Statistical problems, in particular parameter estimation, have been not yet attacked; probably Monte Carlo likelihood [11] is the appropriate method.

3.2 Pixel-Particles and their Parameters

The sets studied in this section are subsets of the lattice $\mathbb{Z}^2$, the set of all pixels (i, j) with i and j being integers. Each pixel is associated with its Wigner-Seitz cell, the square in $\mathbb{R}^2$ of unit area and sides parallel to the lines $x = const$ and $y = const$, with centre in the pixel.

A set X of pixels is called *connected*, if the union set (which is also denoted by X) of all its associated squares has the property that the set $X \ominus b(o, \epsilon)$ is connected for every ϵ with $0 < \epsilon < 1/2$. If additionally the set X does not have holes, X is called *pixel-particle*. (In the physical literature also pixel-particles with holes are considered [54]). In the literature of discrete geometry, such sets are called *n-minoes*, if they consist of n pixels, see [14], [26], [38] and [39], while in percolation theory the name *lattice animal* is used, see [46]. The boundary of a pixel particle is called 'lattice polygon' or 'self-avoiding ring'. A similar model is the bead-and-tether model created by Leibler, Singh and Fisher [30].

Pixel-particles can be characterized by geometrical parameters. The *area* of X is the number of its pixels or the area of X. The boundary of X is described by various numbers. The first is the *number of vertices* $V(X)$. A vertex of X is simply a vertex of X, as explained in Fig. 2.

It makes sense to distinguish between outer (convex) and inner (concave) vertices. An outer vertex is the vertex of a single square while an inner vertex belongs to three squares.

A simple length parameter is the *real perimeter* $C(X)$, the perimeter length of S. Clearly, $C(X)$ is a natural number, the number of outer square sides of S. Unfortunately, it is not a good idea to use $C(X)$ as the perimeter of X; in many cases it would be unreasonably long, and for square and included discrete discs the same value of $C(X)$ is obtained.

There are various possibilities to define a 'perimeter' which has better properties than $C(X)$. The starting point for the perimeter length definition used in the present

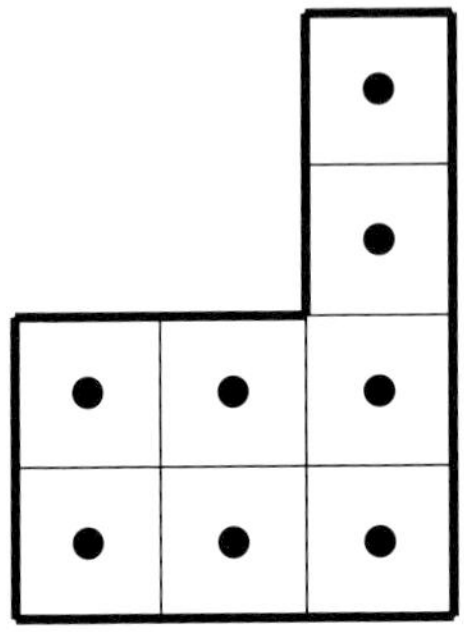

Fig. 2. A pixel-particle consisting of 5 pixels. It has 6 vertices, 5 outer and 1 inner.

paper is the well-known integral-geometric formula for the length $L(X)$ of a curve X

$$L(X) = \frac{1}{2}\int_0^{2\pi} P_\varphi(X)\mathrm{d}\varphi, \tag{4}$$

where $P_\varphi(X)$ is the total projection length of X in direction φ, see [40]. Ohser and Mücklich [37] had the idea to use (4) in image analysis and discretized the integral in formula (4) by considering eight directions, as shown in Fig. 3, and introduced the *discrete perimeter* $U(X)$,

$$U(X) = \frac{\pi}{8}\sum_{i=0}^{7} P_i(X),$$

where $P_i(X)$ is the total projection length in the ith direction; the integral is replaced by a sum corresponding to the 8 directions, where each has weight $\frac{\pi}{4}$.

The term 'total projection length in direction i' needs explanation. The total projection length of an outer pixel side is 1 or 0 in the four horizontal and vertical directions, while for all four diagonal directions it is $1/\sqrt{2}$. The total projection *set* of a pixel-particle in a given direction is equal to the set of exits from all boundary pixels to the exterior of the pixel-particle. The total projection length is then the number of elements of the total projection set in the case of four horizontal and vertical directions and the same number multiplied by $1/\sqrt{2}$ in the case of diagonal directions.

Proposition. *The discrete perimeter $U(X)$ can be obtained by*

$$U(X) = \frac{\pi}{8(\sqrt{2}-1)}\left[C(X) - V(X)(1-\frac{1}{\sqrt{2}})\right]$$

where $C(X)$ is the real perimeter X and $V(X)$ the number of vertices of X.

Proof. Clearly, for the four horizontal and vertical directions the sum of total projection lengths of X is $C(X)$. In the case of the four diagonal directions, it is necessary to determine the total projection sets and then the corresponding numbers of exits from inside of X to outside in these directions, as shown for an example in Fig. 4. The number of exits can be expressed through $C(X)$ and $V(X)$. Every square side on the boundary

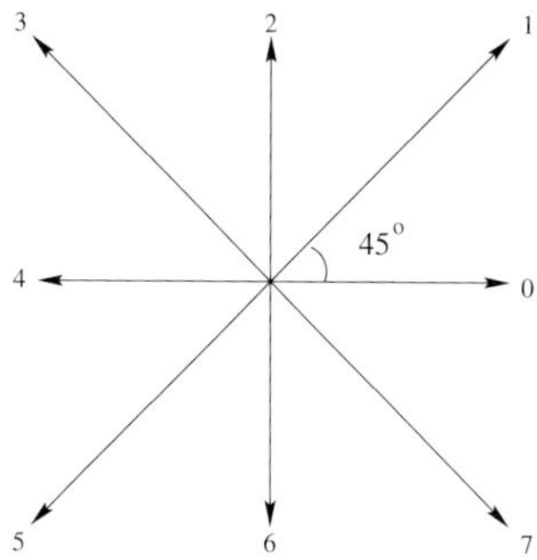

Fig. 3. Eight directions for obtaining total projections of pixel-particles.

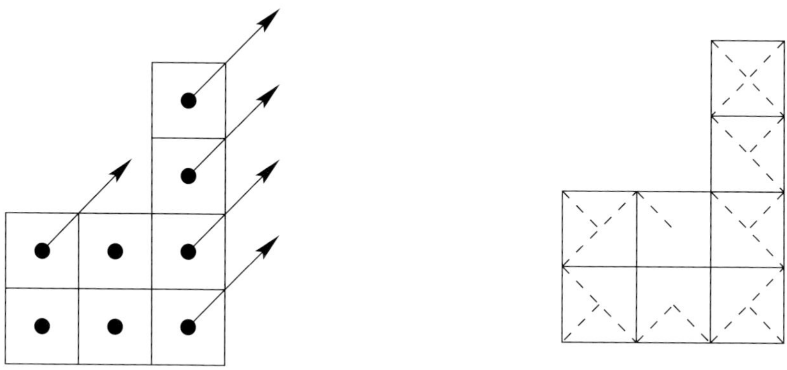

Fig. 4. Left: The total projection set of a pixel-particle for direction 1. The corresponding total projection length is $5/\sqrt{2}$. Right: The same pixel-particle with exits from inside to outside in the four diagonal directions

corresponds to two exits, if the end points of the side are not vertices. In the case of outer vertices one exit must be taken away, that going to the vertex, because it corresponds simultaneously to two adjacent sides. In the case of inner vertices there are two adjacent sides which as an one of their end points have a common inner vertex, each with only one exit. Therefore, it is necessary to add one exit coming from the square with vertex adjacent at the inner vertex. Consequently, the number of exits from inside of X to outside in the four diagonal directions is $2C(X) - V(X)$.

Therefore the sum of total projection lengths for the four diagonal directions is

$$\frac{2C(X) - V(X)}{\sqrt{2}}.$$

This yields

$$U(X) = \frac{\pi}{8}\left(C(X) + \frac{2C(X) - V(X)}{\sqrt{2}}\right) = \frac{\pi}{8(\sqrt{2}-1)}\left[C(X) - V(X)(1 - \frac{1}{\sqrt{2}})\right].$$

This $U(X)$ is by no means ideal. It is of course not alsways $U(\lambda X) = \lambda U(X)$ and for a square not the value $C(X)$ is obtained. Nevertheless, the discrete perimeter $U(X)$

is closer to the idea of isotropy and leads to nearly circular objects as shown in Figs. 5 and 8.

As a roughness measure for the boundary of pixel-particles this paper uses the *number $T(X)$ of towers and hollows*. This idea is inspired by [53] where, however, the case of rough fracture surfaces was considered, i.e., the aim was to describe deviations from ideal planar surfaces. A 'tower' is defined as follows. Let the sides of the squares forming S be characterized by one of the four numbers: 1, 2, 3 and 4, denoting bottom, left, top and right, respectively. Then combinations of sides (in anti-clockwise order) on the boundary of S such as 432 or 44322 are called 'towers'. Correspondingly, combinations such as 412 or 44122 are called 'hollows'.

3.3 Energy Functions for Gibbs Pixel-Particles

The random particles considered in the following have a Gibbs distribution in a suitable configuration space. The elements of this space are not pixel-particles introduced in Sect. 3.2 above but instead classes of congruent pixel-particles. That means, congruent pixel-particles are identified. In the following these classes are denoted by X and called 'pixel-particles' again.

The probability distribution P of a random pixel-particle is given by

$$P(X) = \exp(-E(X))/Z, \tag{5}$$

where Z is the partition function and $E(X)$ is the energy function. For the energy function two very simple variants are considered:

$$E(X) = aA(X)\,,$$

where $A(X)$ is the area of X and a is a model parameter, and

$$E(X) = \frac{(A(X)-\mu_A)^2}{2\sigma_A^2} + \frac{(U(X)-\mu_U)^2}{2\sigma_U^2} + \frac{(T(X)-\mu_T)^2}{2\sigma_T^2}. \tag{6}$$

Here $U(X)$ and $T(X)$ are perimeter and number of 'towers' and 'hollows' of the set X. Furthermore, μ_A, μ_U and μ_T are fixed values of the geometrical characteristics and together with σ_A, σ_U and σ_T the model parameters.

In [1] the energy

$$E(X) = uU(X) + s\sqrt{A(X)}$$

was considered.

The parameters μ_A, μ_U and μ_T can be considered as prefered area, perimeter and tower number; deviations from these values cost energy [42]. They could be called 'spontaneous' area etc.

More interesting energies, which may lead to more interesting particle shapes, are considered in [28], [33]. But the figures there do not correspond to a Gibbs distribution but instead result from an energy minization.

3.4 The Number of Polyonominoes

The distribution of a particular Gibbs pixel-particle model is closely related to a problem sometimes studied in discrete geometry, namely counting polyminoes, see [10], [14], [23], [38], and [39]. In terms of the present paper this problem is as follows: How many distinct pixel-particles consisting of n pixels are there? Pixel-particles are considered distinct if they have different shapes; their orientation and location in the plane is of no importance.

Denote by A_n this number. With big numerical expense, [39] gives the values up to $n = 24$. For the growth of A_n with n there are bounds, namely

$$Ce^{1.36n} \leq A_n \leq Ce^{1.54n}, \tag{7}$$

see [26].

Consider now a Gibbs pixel-particle with energy $E(X) = \alpha A(X) = \alpha n$, where α is a positive model parameter. Its pixel-number distribution is given by

$$P_n = P(A(X) = n) = A_n e^{-\alpha n}/Z_\alpha \quad \text{for} \quad n = 0, 1, \ldots . \tag{8}$$

If A_n would have the form $A_n = Ce^{\beta n}$, then the P_n would have the form $Ce^{(\beta-\alpha)n}/Z_\alpha$. Consequently, the distribution P_n would exist only if $\alpha > \beta$, otherwise the Gibbs pixel-particle would become infinitely large.

Own simulations of such Gibbs pixel-particles using the method described in Sect. 3.5 yielded the critical value $\beta = 1.3745$. That means that for $\alpha > 1.3745$ the Gibbs pixel-particle distribution does not degenerate. The critical value is between the bounds of (7), close to the lower value. This result supports the assumption that there are no existence problems for Gibbs pixel-particles with an energy containing a term $(A(X) - \mu_A)^2$. For an energy such as

$$E(X) = a\sqrt{A(X)}$$

one cannot expect existence of the distribution. If there is not a misunderstanding, the results in [1] are doubtful.

Further theoretical results for pixel-particles can be found in [10]. The number of pixel particles with m boundary edges grows asymptotically as $e^{\kappa m}$ with $\kappa = 0.97$.

3.5 A MCMC Scheme for Simulating Gibbs Pixel-Particles

The probability distribution $\mathcal{P}$ of Gibbs pixel-particles seems to be too complicated for analytical investigations. Therefore it is necessary to generate samples from the distribution $\mathcal{P}$. This can be done by Markov chain Monte Carlo as described in [12]. The idea is to construct an ergodic Markov chain the stationary distribution of which is just $\mathcal{P}$. The states of this chain are pixel-particles.

In the case considered here, the Metropolis-Hastings algorithm as described in [12] is used. The transition from a pixel-particle X_t to a pixel-particle X_{t+1} is performed as follows. First a candidate pixel-particle Y from a proposal distribution $q(\cdot \mid X_t)$ is sampled as a proposal for the new state and then by means of the Hastings ratio r it is decided whether or not the proposal is accepted. If it is accepted then the new state of the chain is $X_{t+1} = Y$, otherwise $X_{t+1} = X_t$.

The proposal pixel-particle is constructed as follows. (Note that this construction is one of many possibilities; it is the form of the Hastings ratio which ensures that the stationary distribution of the chain is P.) Generate a uniform on $[0, 1]$ random variable ω_1. If $\omega_1 \leq 1/2$ then add a pixel to X_t, otherwise subtract a pixel.

In the case of pixel addition the location of the new pixel is obtained by adding it with uniform distribution to the outer side of any outer pixel. (A pixel is an outer pixel if one of its sides belongs to the perimeter.)

In the case of pixel subtraction the pixel to be deleted is one of the outer pixels chosen with uniform distribution, but under the condition that the received figure is a pixel-particle.

The corresponding transition probability is $q(Y \mid X)$ with $Y = X \cup \{z\}$. The assumption that the new pixel z is located with uniform probability at the outer square sides of X implies that

$$q(Y \mid X) = \frac{C(z, X)}{C(X)},$$

where $C(z, X)$ is the number of joint sides of z and X. Obviously, $C(z, X)$ can take the values 1, 2 or 3, and it is easy to see that

$$C(z, X) = \begin{cases} 1 \text{ if } C(X) - C(Y) = -2, \\ 2 \text{ if } C(X) - C(Y) = 0, \\ 3 \text{ if } C(X) - C(Y) = 2. \end{cases}$$

Since joint sides of z and X are not included in the perimeter of Y and any pixel has four sides, it is

$$C(z, Y) = 4 - C(z, X).$$

Consequently, $q(X \mid Y)$ satisfies

$$q(X \mid Y) = \frac{4 - C(z, X)}{C(Y)}.$$

The case with subtraction of a pixel can be considered analogously.

The Hastings ratio r is given by

$$r = \min \left\{ \exp(E(X) - E(Y)) \frac{q(X \mid Y)}{q(Y \mid X)}, 1 \right\}.$$

3.6 A Gallery of Simulated Gibbs Pixel-Particles

The following figures show series of simulated Gibbs pixel-particles with energy (6) and various parameters. The aim of the first series (Fig. 5) was to come close to the shapes of sand grains (which, however, do probably not follow a Gibbs process model but instead an abrasion process model). The second in Fig. 6 presents particles of longer perimeter, which have more variable shapes than the objects of Fig. 5. Figure 7 tries to show the influence of the parameter μ_U, which controls the perimeter and thus particle shape

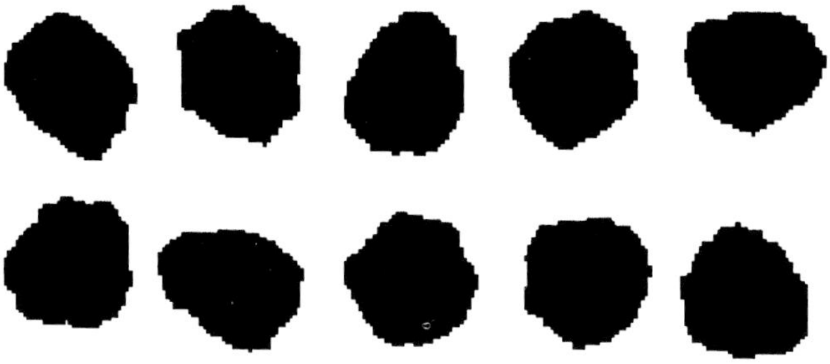

Fig. 5. Ten simulated Gibbs pixel-particles with the same parameters: $\sigma_U = 2.24, \mu_U = 100, \sigma_A = 7.07, \mu_A = 900, \sigma_T = 0.79, \mu_T = 0$. There is obviously some variability of shape and very little of area and perimeter. The mean area and perimeter are Ā = 890 and Ū = 117, the corresponding standard deviations are s_A = 7.1 and s_U = 1.1. The smallness of s_A and s_U is a result of the smallness of the variability parameters σ_A and σ_U.

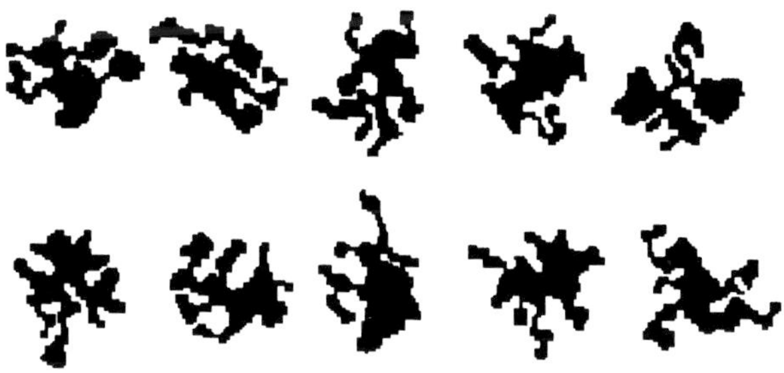

Fig. 6. Ten simulated Gibbs pixel-particles with the same parameters as in Fig. 5 but with larger variability parameter $\sigma_U = 20$. The mean area and perimeter are Ā = 900 and Ū = 344, the corresponding standard deviations are s_A = 5.1 and s_U = 20.5.

Fig. 7. Ten simulated Gibbs pixel-particles with the same parameters: $\sigma_U = 2.24, \mu_U = 120, \sigma_A = 7.07, \mu_A = 900, \sigma_T = 0.79, \mu_T = 0$. The μ_U, which is larger than that in the case of Fig. 6, makes that the mean perimeter is longer and the shapes are more variable. The mean area and perimeter is longer and the shapes are more variable. The mean area and perimeter are Ā = 896 and Ū = 128, the corresponding standard deviations are s_A = 6.8 and s_U = 1.5.

variability. Figure 8 shows particles with rougher boundaries, resulting by the parameter choice $\sigma_T = \infty$.

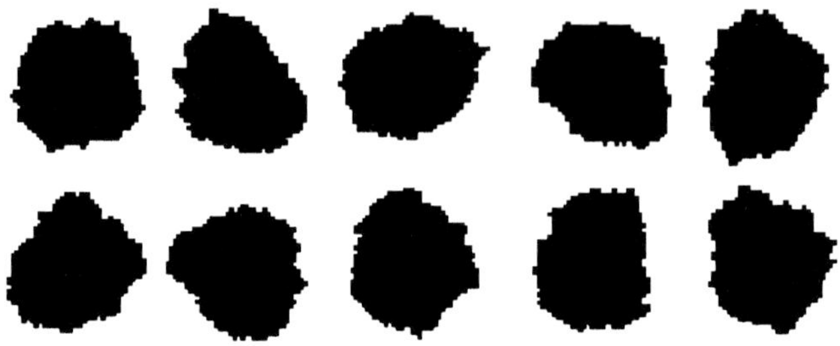

Fig. 8. Ten simulated Gibbs pixel-particles with the same parameters: $\sigma_U = 2.24, \mu_U = 120, \sigma_A = 7.07, \mu_A = 900, \sigma_T = \infty, \mu_T = 0$. The choice $\sigma_T = \infty$ makes that now more towers and hollows appear than in Fig.6 and 7. The mean area and perimeter are $\bar{A} = 893$ and $\bar{U} = 130$, the corresponding standard deviations are $s_A = 6.6$ and $s_U = 1.7$.

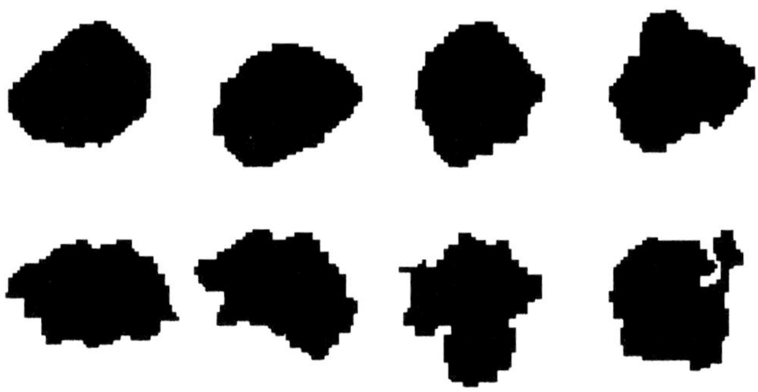

Fig. 9. A series of simulated Gibbs pixel-particles with parameters: $\sigma_U = 2.24, \sigma_A = 7.07, \mu_A = 900, \sigma_T = \infty, \mu_T = 0$ and $\mu_U = 90, 100, ..., 160$.

Fig. 10. Ten simulated Gibbs pixel-particles with the same parameters: $\sigma_U = 2.24, \mu_U = 100, \sigma_A = 7.07, \mu_A = 100, \sigma_T = \infty, \mu_T = 0$. The small parameter μ_A makes that the particles have now an irregular shape with several thin branches. The mean area and perimeter are $\bar{A} = 109$ and $\bar{U} = 102$.

In the case of the energy $E(X) = (A(X) - \mu_A)^2/2\sigma_A^2$, which may be associated with infinite 'variances' σ_U^2 and σ_T^2, the formulas of Sect. 3.4 enable the analytical determination of the distribution of $A(X)$. However this case without limitation on perimeter and roughness leads to a chaotic pixel-particle, which follows a uniform distribution of shapes given the area distribution.

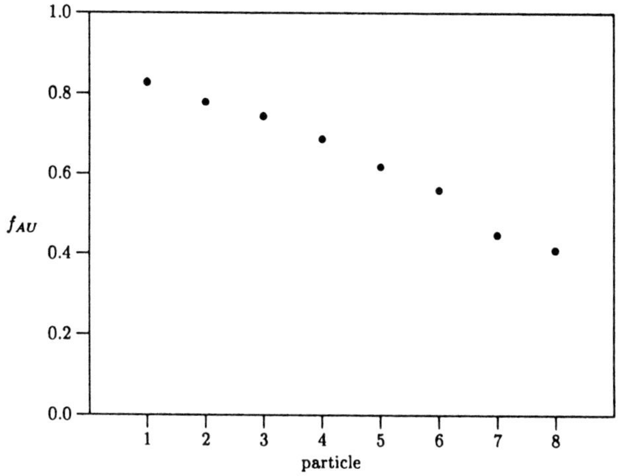

Fig. 11. Area-perimeter ratios $f_{AU}(X)$ defined by (1), for the particles of Fig.9

Figure 11 shows the area-perimeter ratios of the eight particles of Fig. 9. Since the perimeter length is increasing, $f_{AU}(X)$ decreases.

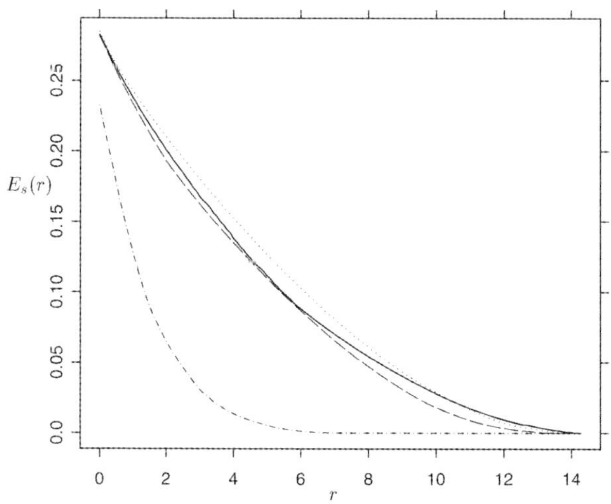

Fig. 12. Erosion functions $E_s(r)$ for the first particles of Figs. 5 - 8.
—— = 5, — · — = 6, – – = 7, · · · = 8.

Figure 12 shows the erosion functions $E_s(r)$ defined by (2) for the first particles of Figs. 5 - 8. It is clear that $E_s(r)$ decreases rather fast for the particle of Fig. 6, since the thin tentacles are eroded away already for small values of r.

In the physical literature many other energies have been studied. Close to the present paper is [5], where pixel-particles on a honey-comb lattice were considered. The energy was $E(X) = aA(X)$, where a was interpreted as pressure difference. For the case of fixed $C(X)$ the mean area $A(X)$ was determined. In [54] in the same situation the influence of the parameter a was discussed. For $a > 0$ (inflation) circular objects (as on Fig. 5) appear, while for $a < 0$ (deflation) the objects resemble branched polymers (as the objects on Fig. 10).

References

1. Anastassiou, D., D. J. Sakrison (1981): 'A probability model for simple closed random curves', *IEEE Transact. Inform. Theor.* **27**, pp. 376–381
2. Beddow, J.K. (1980): *Particle Science and Technology* (Chemical Publishers, New York)
3. Beddow, J.K. (1984): *Particle Characterization in Technology* (CRC Press, Boca Raton)
4. Bookstein, F.L. (1991): *Morphometric Tools for Landmark Data: Geometry and Biology* (Cambridge University Press, Cambridge)
5. Cardy, J. (1994): 'Mean area of self-avoiding loops', *Phys. Rev. Lett.* **72**, pp. 1580–1583
6. Chow, B. (1985): 'Deforming convex hypersurfaces by the nth root of the Gaussian curvature', *Differential Geometry* **22**, pp. 117–138
7. Cox, D.R., V. Isham (1980): *Point Processes* (Chapman & Hall, London–New York)
8. Dryden, I.L., K.V. Mardia (1998): *Statistical Shape Analysis* (Wiley, Chichester)
9. Firey, W.J. (1974): 'Shapes of worn stones', *Mathematika* **21**, pp.1–11
10. Fisher, M.E., A.J. Guttmann, S.G. Whittington (1991): 'Two-dimensional lattice vesicles and polygons', *J. Phys. A: Math. Gen.* **24**, pp. 3095–3105
11. Geyer, C. (1999): 'Likelihood inference for spatial point processes'. In: *Stochastic Geometry. Likelihood and Computation*, ed by O.E. Barndorff-Nielsen, W.S. Kendall, M.N.M. van Lieshout (Chapman & Hall/CRC, Boca Raton/ London), pp.79–140
12. Gilks, W.R., S. Richardson, D.G. Spiegelhalter (1996): *Introducing Markov Chain Monte Carlo* (Chapman& Hall, London)
13. Gille, W. (2000): 'Chord length distributions and small-angle scattering', *Europ. Phys. J. B* **17**, pp. 371-383
14. Golomb, S. (1994): *Polyominoes* (Princeton Univ. Press, Princeton)
15. Gompper, G., D.M. Kroll (1997a): 'Network models of fluid, hexatic and polymerized membranes', *J. Phys.: Condens. Matter* **9**, pp. 8795–8834
16. Gompper, G., D.M. Kroll (1997b): 'Fluctuations of polymerized, fluid and hexatic membranes: continuum models and simulations', *Current Opinion in Colloid & Interface Science* **2**, pp. 373–381
17. Gompper, G., D.M. Kroll (1998): 'Membranes with fluctuating topology: Monte Carlo simulations', *Phys. Rev. Lett.* **81**, pp. 2284–2287
18. Grenander, U., M.I. Miller (1994): 'Representations of knowledge in complex systems (with discussion)', *J. R. Statist. Soc. B* **56**, pp. 549–603
19. Hawkins, A.E. (1993): *The Shape of Powder-Particle Outlines* (Wiley, New York)
20. Hobolth, A., J. Pedersen, E.B. Vedel-Jensen (2000): 'A continuous parametric shape model.' Research Report 13, Laboratory of Computational Stochastics, Department of Mathematical Sciences, University of Aarhus.
21. Huisken, G. (1984): Flow by mean curvature of convex surfaces into spheres. *J. Differential Geometry* **20**, pp. 237–266

22. Jaric, M., U. Seifert, W. Wintz, M. Wortis (1995): 'Vesicular instabilities: The prolate-to-oblate transition and other shape instabilities of fluid bilayer membranes', *Phys. Rev. E.* **52**, pp. 6623–6634
23. Jensen, I., A. J. Guttmann (2000): 'Statistics of lattice animals (polyominoes) and polygons', *J. Phys. A: Math. Gen.* **33**, pp. L257–L263
24. Kendall, D.G., D. Barden, T.K. Carne, H. Le (1999): *Shape and Shape Theory* (Wiley, Chichester)
25. Kent, J.T., I.L. Dryden, C.R. Anderson (2000): 'Using circulant symmetry to model featureless objects', *Biometrika* **87**, pp. 527–544
26. Klarner, D.A. (1997): 'Polyominoes'. In: *Handbook of Discrete and Computational Geometry*, ed by J.E. Goodman, J.O. Rourke (CRC Press, Boca Raton), pp.225–240
27. Körner, R., D. Stoyan (1999): 'Modeling abrasion processes by operations of mathematical morphology', *Particle & Particle Systems Characterization* **14**, pp.66–70
28. Krüger, P., M. Lösche (2000): 'Molecular chirality and the domain shapes in lipid monolayers on aqueous surfaces', *Phys. Rev. E.* **62**, pp. 7031–7043
29. Lantuejoul, C., F. Maisonneuve (1984): 'Geodesic methods in quantative image analysis' *Pattern Recognition* **17**, pp. 177–187
30. Leibler, S., R.R.P. Singh, M.E. Fisher (1987): 'Thermodynamic behavior of two-diemnsional vesicles', *Phys. Rev. Lett.* **59**, pp. 1989–1992
31. Lele, S.R., J.T. Richtsmeier (2000): *An Invariant Approach to Statistical Analysis of Shapes* (Chapman & Hall/CRC, Boca Raton/ London)
32. Lipowsky, R., E. Sackmann (eds.) (1995): *Structure and Dynamics of Membranes* (Elsevier Science Publ., Amsterdam)
33. Lösche, M., P. Krüger (2002): 'Quantitative evaluation of Langmuir monolayer phase morphologies using a Minkowski function analysis', **this volume**
34. Mecke, K.R. (1995): 'Bending rigidity of fluctuating membranes' *Z. Phys. B* **97**, pp. 379-387
35. Mecke, K. (2000): 'Additivity, convexity and beyond: Applications of Minkowski functionals in statistical physics', in: [36], pp. 111–184
36. Mecke, K. , D. Stoyan (2000): *Statistical Physics and Spatial Statistics* Lecture Notes in Physics 554 (Springer-Verlag)
37. Ohser, J., F. Mücklich (2000): *Statistical Analysis of Microstructures in Materials Science* (Wiley, Chichester)
38. Quaisser, E. (1994): *Diskrete Geometrie* (Spektrum Akademislar Verlag, Heidelberg)
39. Redelmeier, D.H. (1981): 'Counting polyominoes: yet another attack' *Discrete Math.* **36**, pp. 191–203
40. Santaló, L.A. (1976): *Geometry and Geometric Probability* (Addison-Wesley Publishing Company)
41. Schwarz, U.S., G. Gompper (2002): 'Bicontinuous surfaces in self-assembling amphiphilic systems', **this volume**
42. Seifert, U. (2000): 'Giant vesicles: A theoretical perspective'. In: *Giant Vesicles*, ed. by P.L. Liusi, P. Walde (J. Wiley New York), pp. 71–91
43. Silverman, B.W., J.O. Ramsey (1997): *Functional Data Analysis* (Springer–Verlag, New York)
44. Small, C. (1996): *The Statistical Theory of Shape* (Springer–Verlag, New York)
45. Soille, P. (1999): *Morphological Image Analysis* (Springer-Verlag, Heidelberg)
46. Stauffer D., A. Aharony (1994): *Introduction to Percolation Theory. Revised 2nd edition* (Taylor & Francis, London)
47. Stoyan, D., I. Molchanov (1997): 'Set-valued means of random particles' *J. Math. Imaging and Vision* **7**, pp. 111–121
48. Stoyan, D., H. Stoyan (1994): *Fractals, Random Shapes and Point Fields* (Wiley, Chichester)

49. Stoyan, D. (2000): 'Basic ideas of spatial statistics'. In: [36], pp. 3–21
50. Stoyan, D., W.S. Kendall, J. Mecke (1995): *Stochastic Geometry and its Applications* (Wiley, Chichester)
51. Toussaint, G.T. (ed.) (1988): *Computational Morphology. A Computational Geometric Approach to the Analysis of Form* (North Holland, Amsterdam)
52. Tso, K. (1985): 'Deforming a hypersurface by its Gauss-Kronecker curvature', *Comm. Pure Appl. Math.* **38**, pp. 867–882 **21**, pp. 1–11
53. Underwood, E. (1987): 'The analysis of nonplanar surfaces using stereological and other methods' *Acta Stereol* **6/III**, pp. 855-876
54. Vanderzande, C. (1993): 'Vesicles, the tricritical-0-state Potts model, and the collapse of branched polymers', *Phys. Rev. Lett.* **70**, pp. 3595–3598

A Survey on Contact Distributions

Daniel Hug[1], Günter Last[2], and Wolfgang Weil[3]

[1] Mathematisches Institut, Universität Freiburg, Eckerstr. 1, D-79104 Freiburg, Germany,
e-mail: daniel.hug@math.uni-freiburg.de
[2] Institut für Mathematische Stochastik, Universität Karlsruhe (TH), Englerstr. 2, D-76128 Karlsruhe,
e-mail: g.last@math.uni-karlsruhe.de
[3] Mathematisches Institut II, Universität Karlsruhe (TH), Englerstr. 2, D-76128 Karlsruhe,
e-mail: weil@math.uni-karlsruhe.de

Abstract. This survey describes contact distributions of random structures from a geometric point of view. Various extensions of the classical contact distribution functions are considered. As a rule, we explain all concepts first in a simple situation, under the assumption of stationarity, and for Poisson point processes. In a second step we proceed to more complicated models. Particular emphasis is given to random patterns involving clustering.

1 Introduction

Contact distributions are tools to describe distributional properties of random spatial structures from outside the structure. For simplicity and in order to give a unified presentation, we consider random structures in a space $\mathbb{R}^d$ of general dimension $d \geq 2$. (Most results will remain true for $d = 1$ if interpreted properly). The two types of random structures which we discuss are simple point processes Φ and random closed sets Z, where the second concept can be considered as a generalization of the first.

Suppose that $B \subset \mathbb{R}^d$ is a compact set which contains the origin 0 (we call B a structuring element) and put $B^* := -B$. For a reference point $x \in \mathbb{R}^d$ and a random set Z, we consider the (random) number $\rho \geq 0$ for which the first contact of the "growing" set $\rho B + x$ with Z occurs. We can also adopt a slightly different point of view and define ρ to be the random time it takes the growing set $Z + tB^*$ to reach x. The distribution of ρ is a typical example of a contact distribution which is based on (generalized) distance measurements. If Z is stationary, the contact distribution is independent of x, we then can choose $x = 0$. If Z has interior points, then typically (for example in the stationary case) $x \in Z$ with positive probability. In this situation, we will mostly consider the conditional distribution of ρ, given that $x \notin Z$. If B is the unit ball, ρ gives the Euclidean distance of x to Z. For general structuring elements B, we therefore interpret ρ as a B-*distance*. Besides the distribution of the B-distance, various other contact information can be taken into account, e.g. the direction in which the distance is attained or local geometric properties of Z at the point of contact.

We will study these different types of contact distributions for random sets Z and point processes Φ. For many applications it is justified to assume that these random structures are stationary/spatially homogeneous (i.e. their distribution is translation invariant) or even stationary and isotropic, at least when considered on a suitable scale, e.g. as in cosmological models (see the contribution by C. Beisbart, M. Kerscher and K. Mecke, in Chap. 4 of this volume), or in appropriate subregions where the influence of the boundary can be neglected. For this reason, and since stationary models are easier to treat, we start with and for large parts concentrate on stationary structures. The formulae which we present also simplify for special classes of random sets resp. point processes. We therefore begin with Poisson processes in $\mathbb{R}^d$, then study particle processes and Boolean models, continue with cluster models and then treat the case of more general random sets. Some applications of contact distributions in the porous media literature are mentioned in [11, 26] (see also Chap. 1 'Shape of solids: porous and composite materials', in this volume), classical applications to forestry are reported in [43].

Extensions of results for contact distributions to not necessarily stationary models are more difficult to handle, since they usually involve mixed functionals [50] (compare also [26]). Therefore these generalizations will be discussed in a second step after the stationary case has been described carefully. It is evident that such extensions are required for the statistical analysis and modelling of inhomogeneous media [33]. Such inhomogeneous structures arise for instance if effects of external magnetic or gravitational fields cannot be neglected (compare [26]) or in materials formed by layers with a structural gradient near to their boundaries (gradient structures; see [5]).

The selection of the material for this tutorial paper was strongly influenced by our attempt to explain ideas related to our recent research and by trying to avoid too much overlap with existing introductions to the subject.

2 B-Distances and Related Notions

2.1 B-Distances

Let $\mathcal{C}^d$ denote the class of all compact sets in $\mathbb{R}^d$. We fix a *structuring element* $B \in \mathcal{C}^d$ with $0 \in B$ and define the B-distance (the distance relative to B) of a point $x \in \mathbb{R}^d$ to the set $A \subset \mathbb{R}^d$ by

$$d_B(x, A) := \inf\{r \geq 0 : (x + rB) \cap A \neq \emptyset\}.$$

The set on the right side may be empty (e.g. if 0 is a boundary point of B or $A = \emptyset$), in which case $d_B(x, A) = \infty$. The distance $d_B(x, A)$ is translation invariant in the sense that $d_B(x + z, A + z) = d_B(x, A)$ for all $z \in \mathbb{R}^d$. Clearly, $d_B(x, A) \leq r$ if and only if x is contained in the *generalized outer parallel set* $A + rB^*$ of A. Here, $C + D := \{x + y : x \in C, y \in D\}$ is the (Minkowski) sum of two sets $C, D \subset \mathbb{R}^d$. For convenience we abbreviate $d_B(x, y) := d_B(x, \{y\})$, then

$$d_B(x, y) = d_{B^*}(y, x), \quad x, y \in \mathbb{R}^d.$$

We usually assume in the following that B is a compact convex set with $0 \in B$. In this case we refer to B as a *gauge body*. If B is, in addition, full dimensional and centrally symmetric (w.r.t. 0), then $(x, y) \mapsto d_B(x, y)$ is a metric on $\mathbb{R}^d$ induced by the norm $x \mapsto d_B(x, 0)$, and the pair $(\mathbb{R}^d, d_B)$ is called a Minkowski space. The choice $B = B^d$ (the Euclidean unit ball) yields the usual Euclidean metric which is based on the scalar product $\langle \cdot \, , \cdot \rangle$; we then write $d(x, A) := d_{B^d}(x, A)$.

2.2 Contact Vectors

If the B-distance $d_B(x, A)$ of a point $x \notin A$ is attained in a unique point y in the boundary ∂A of A (that means, if $(x + d_B(x, A)B) \cap A = \{y\}$), then we define the *contact direction vector* $u_B(x, A)$ as the element of ∂B given by

$$u_B(x, A) := \frac{y - x}{d_B(x, A)}.$$

The points $x \in \mathbb{R}^d \setminus A$ for which the distance $d_B(x, A)$ is attained in more than one point of A (and for which $u_B(x, A)$ is therefore not defined) form the *exoskeleton* $\mathrm{exo}_B(A)$ of A (see [15] and also [40, 41]). In the Euclidean case, and if A is a finite or locally finite set, $\mathrm{exo}_{B^d}(A)$ is (the boundary of) the *Voronoi tessellation* generated by A. Subsequently, we write $V_d(A)$ for the Lebesgue measure of a Borel set $A \subset \mathbb{R}^d$. Then, if A is a closed subset of $\mathbb{R}^d$ and $B \subset \mathbb{R}^d$ is compact, strictly convex and has 0 as an interior point, we have $V_d(\mathrm{exo}_B(A)) = 0$ (see [12, 14]).

3 Spatial Random Structures

In this section, we give a short introduction to random sets and (different types of) point processes.

3.1 Point Processes

Point processes are not only fundamental objects of modern probability theory, but also provide basic models in applied probability; see [4] for a detailed exposition of the interesting and rich theory of point processes. Loosely speaking, a (simple) *point process* Φ on $\mathbb{R}^d$ is a finite or countable collection $\xi_1, \xi_2, \ldots$ of (mutually different) random points scattered in the d-dimensional Euclidean space $\mathbb{R}^d$. Each random variable ξ_n is called a *point* of Φ. We assume that the points of Φ do not accumulate in bounded sets, i.e. each bounded subset of $\mathbb{R}^d$ contains only a finite number of the points of Φ. The points then form a *locally finite* closed subset $\{\xi_1, \xi_2, \ldots\}$ of $\mathbb{R}^d$.

To give a more formal definition of a point process, we introduce the space $\mathbf{N}$ of all locally finite sets $\varphi = \{x_1, x_2, \ldots\} \subset \mathbb{R}^d$. It is convenient to interpret a set $\varphi \in \mathbf{N}$ also as a counting measure, namely as $\varphi = \sum_x \delta_x$, where δ_x denotes the Dirac measure located at $x \in \mathbb{R}^d$ and where the sum extends over all x which belong to φ. For simplicity, we write $\varphi = \sum_{n=1}^{\infty} \delta_{x_n}$, in the following, even if the sum is a finite one. Such a measure φ is *simple*, that is each point has at most mass 1, and *locally finite*, that is the number $\varphi(A)$ of points of φ in $A \subset \mathbb{R}^d$ is finite for all bounded Borel sets A. A point process Φ is now a random element in $\mathbf{N}$ governed by an underlying probability measure $\mathbb{P}$. (To be mathematically more precise, one assumes some abstract probability space $(\Omega, \mathcal{A}, \mathbb{P})$ in the background and supplies $\mathbf{N}$ with the σ-field $\mathcal{N}$ generated by the mappings $\varphi \mapsto \varphi(A)$, where A varies through all Borel sets in $\mathbb{R}^d$. Then, Φ is a measurable mapping from Ω to $\mathbf{N}$ and we assume hereafter that all random variables are defined on this probability space.) According to our conventions we regard Φ as a locally finite random closed set and, at the same time, as a simple random counting measure. Then $\Phi(A)$ is equal to the number of points of Φ that are in the set A, and $\mathbb{P}(\Phi(A) = n)$ is the probability that there are exactly n points of Φ in A.

The *intensity measure* Λ of Φ is the measure on $\mathbb{R}^d$ defined by

$$\Lambda(A) := \mathbb{E}\,\Phi(A) = \sum_{n=1}^{\infty} n\mathbb{P}(\Phi(A) = n),$$

where $A \subset \mathbb{R}^d$ is a Borel set and the symbol $\mathbb{E}$ denotes expectation with respect to the underlying probability measure $\mathbb{P}$, i.e. $\mathbb{E}\,\Phi(A) = \langle \Phi(A) \rangle$ in the notation used by physicists. The point process Φ is said to be *stationary*, if its distribution is invariant under translations, which means that $\mathbb{P}(\Phi + x \in C) = \mathbb{P}(\Phi \in C)$, for all $x \in \mathbb{R}^d$ and all events $C \in \mathcal{N}$. In this case, $\mathbb{P}(x \in \Phi)$ does not depend on $x \in \mathbb{R}^d$, so that Fubini's theorem implies

$$\mathbb{P}(0 \in \Phi) = \int_{[0,1]^d} \mathbb{P}(x \in \Phi) dx = \mathbb{E}\,V_d(\Phi \cap [0,1]^d) = 0,$$

where the dx always refers to integration with respect to the Lebesgue measure on the underlying space (in this case $\mathbb{R}^d$). The intensity measure Λ of a stationary point process Φ fulfills

$$\Lambda(A) = \gamma V_d(A),$$

for all Borel sets A, where the *intensity* γ of Φ is defined as the mean number of points falling in a set of volume 1, i.e.

$$\gamma := \mathbb{E}\,\Phi([0,1]^d).$$

We always assume $\gamma < \infty$.

The contributions by C. Beisbart et al. and by V. Robins in Chap. 3, and by G. Döge and D. Stoyan in Chap. 4 of this volume contain applications of point processes in the natural sciences.

3.2 Particle Processes

Point processes can be considered on quite general metric spaces. In particular, we can define a point process X on $\mathcal{C}^d$, the class of compact subsets of $\mathbb{R}^d$, to be a locally finite random subset of $\mathcal{C}^d$ (equivalently, a locally finite, random counting measure on $\mathcal{C}^d$). Such a point process X is called a *particle process*. In the present context, X is said to be locally finite if

$$X(\{C \in \mathcal{C}^d : C \cap D \neq \emptyset\}) < \infty, \quad D \in \mathcal{C}^d. \tag{1}$$

Condition (1) is stronger than $X(A) < \infty$, for each compact subset $A \subset \mathcal{C}^d$ (here compactness refers to the Hausdorff metric on $\mathcal{C}^d$).

More general processes are possible where the 'particles' are closed sets, for example affine subspaces (flats) of a certain dimension $k \in \{0, \ldots, d-1\}$ (the case $k = 0$ brings us back to the ordinary point processes on $\mathbb{R}^d$). Although contact distributions can also be considered for such *k-flat processes*, we will concentrate on compact particles, in the following.

Since the translation group operates on $\mathcal{C}^d$ in the obvious way, we can define invariance properties of particle processes X. In particular, we call X *stationary*, if the distribution of X is translation invariant.

3.3 Particle Processes as Marked Point Processes

In general, a *marked point process* Ψ on $\mathbb{R}^d$ (with a suitable metric space $\mathbf{K}$ as mark space) is a point process on $\mathbb{R}^d \times \mathbf{K}$ such that $\Psi(C \times \mathbf{K}) < \infty$ almost surely for any compact set $C \subset \mathbb{R}^d$. The points of Ψ can be enumerated in the form $(\xi_1, \kappa_1), (\xi_2, \kappa_2), \ldots$, where the ξ_n are random elements of $\mathbb{R}^d$ and the κ_n are random elements of $\mathbf{K}$. Here we will only consider marked point processes for which $\xi_n \neq \xi_m$ for $n \neq m$, that is, for which the unmarked point process $\Phi := \Psi(\cdot \times \mathbf{K})$ is simple. The intensity measure of a marked point process is a measure on $\mathbb{R}^d \times \mathbf{K}$. Its value on a Borel set $A \subset \mathbb{R}^d \times \mathbf{K}$ is again the expected number $\mathbb{E}\,\Psi(A)$ of points of Ψ falling into A. In case $A = B \times L$, this is the mean number of marked points $(x, K) \in \Psi$ with $x \in A$ and mark $K \in L$. A marked point process is said to be *stationary* if the distribution of $\{(\xi_n + x, \kappa_n) : n = 1, 2, \ldots\}$ is independent of $x \in \mathbb{R}^d$. In that case we define the intensity γ of Ψ as the intensity of the unmarked process Φ. The intensity measure of a stationary marked point process Ψ with finite intensity γ is of the form $\mathbb{E}\,\Psi(\cdot) = \gamma V_d \otimes \mathbb{Q}$, where $\mathbb{Q}$ is a probability measure on $\mathbf{K}$. If the marks are independent and identically distributed as well as independent of the point process Φ, we say that Ψ is an *independent marking* of Φ or (somewhat sloppy) that Ψ has *independent marks*. In this case we have the decomposition $\mathbb{E}\,\Psi(\cdot) = \Lambda \otimes \mathbb{Q}$, where Λ is the intensity measure of Φ and $\mathbb{Q}$ is the distribution of the marks. Notice that, in general, stationarity does not imply independent marks; an important example where this implication holds is the Poisson process discussed in later sections. In the cases where Ψ is stationary or obtained by independent marking, we call $\mathbb{Q}$ the *mark distribution* of Ψ.

Physical applications of marked point processes can be found in the contribution by C. Beisbart, M. Kerscher and K. Mecke in Chap. 4 of this volume, as well as in the article by F. Schmid and N.H. Phuong in Chap. 2.

There is an important connection between particle processes and marked point processes with mark space $\mathcal{C}^d$ (instead of marks we speak of *grains* here). Each particle process X can be transformed into a marked point process Ψ on $\mathbb{R}^d \times \mathcal{C}^d$, if compact sets C are represented as pairs (x, D) (where we think of x as the 'location' of C and where $D := C - x$ is considered to be the 'shape' of C). Such a representation is always possible, if we have chosen a *center map* $c : \mathcal{C}^d \to \mathbb{R}^d$, that is a measurable mapping which is translation covariant in the sense that $c(C + x) = c(C) + x$ is satisfied, for $x \in \mathbb{R}^d$ and $C \in \mathcal{C}^d \setminus \{\emptyset\}$, and $c(\emptyset) = 0$. For a given center map c, the mapping $C \mapsto (c(C), C - c(C))$ transforms X into a marked point process Ψ. Here we assume that different particles have different centers. (The finiteness condition for marked point processes is another extra assumption; if $c(C) \in C$ for all $C \in X$, then it is induced by (1).) The marked point process Ψ is stationary if and only if X is stationary. Further, in the stationary case, the grain distribution $\mathbb{Q}$ is concentrated on $\mathcal{C}^d_0 := \{C \in \mathcal{C}^d : c(C) = 0\}$.

Also then the intensity of Ψ can be interpreted as the mean number of particles of X per unit volume.

The representation of a particle process as a marked point process is frequently suggested by the given physical situation, and then a center map is given in a natural way. For example, we may choose $c(C)$ as the midpoint of the circumsphere of C in which case the particles $C + x$, $(x, C) \in \Psi$, can be interpreted as grains grown around the germs x. A further center map of physical nature is the center of mass. In statistical applications often other center maps are natural, for example the lower/upper left tangent point. In the stationary case, different representations of a particle process X as marked point processes Ψ_1, Ψ_2 (according to different center maps c_1, c_2) lead to the same intensities, but to different (though closely related) grain distributions $\mathbb{Q}_1, \mathbb{Q}_2$. In the following it is convenient to represent a particle process X as

$$X = \{C + x : (x, C) \in \Psi\} = \{Z_n + \xi_n : n \in \mathbb{N}\}, \tag{2}$$

where $\Psi = \{(\xi_n, Z_n) : n \in \mathbb{N}\}$ is a marked point process which is locally finite, i.e. satisfies the condition

$$\int_{\mathbb{R}^d \times \mathcal{C}^d} \mathbf{1}\{(C + x) \cap D \neq \emptyset\} \Psi(d(x, C)) < \infty, \quad D \in \mathcal{C}^d, \tag{3}$$

equivalent to (1).

3.4 Random Sets and Grain Models

Point processes in $\mathbb{R}^d$ can be interpreted as locally finite random sets. Now we consider more general random (closed) sets Z in $\mathbb{R}^d$. Formally, a *random closed set* Z can be defined as a random element of the space $\mathcal{F}^d$ of all closed subsets of $\mathbb{R}^d$ (again, a mathematical description requires a suitable σ-field on $\mathcal{F}^d$ such that Z becomes a measurable mapping $Z : \Omega \to \mathcal{F}^d$). For the general theory, see [24, 39, 42]. We say that Z is *stationary*, if Z and $Z + x$ have the same distribution, for all $x \in \mathbb{R}^d$. A general example of a random set Z is the union set of a particle process X or, equivalently,

$$Z = \bigcup_{n \in \mathbb{N}} (Z_n + \xi_n), \tag{4}$$

where $\Psi = \{(\xi_n, Z_n) : n \in \mathbb{N}\}$ is the marked point process representing X as in (2). Condition (3) (resp. (1)) guarantees that Z is closed. A random set Z represented as in (4) is also referred to as a *grain model*. The reader should be warned that grain models have been defined in different ways in the literature (see e.g. [29, 39, 42]).

Since a non-empty stationary random closed set Z is almost surely either empty or its convex hull coincides with the whole space $\mathbb{R}^d$, a stationary grain model is almost surely either empty or the union of infinitely many grains.

Grain models are still of quite general nature. In fact, every random closed set Z in $\mathbb{R}^d$ allows a representation (4). Moreover, if Z is stationary, one can assume that the marked point process Ψ is also stationary. In order to get a smaller (and more interesting) class, we may assume that Ψ has independent marks, that means, the grains $Z_1, Z_2, \ldots$

are independent and identically distributed as well as independent of the germ process Φ. In this case, or if Ψ is stationary, we call the mark distribution $\mathbb{Q}$ the *distribution of the typical grain* and fix a random closed set Z_0 (the *typical grain*) with this distribution. The most prominent example of a grain model with independent grains is the *Boolean model*, where the underlying germ process Φ is a *Poisson process*. In the next section, we study this class of random sets (and point processes) in more detail under the assumption of stationarity; more general point processes and random sets are discussed subsequently.

3.5 Grain Models with Convex Grains

Of special interest are random sets Z with realizations in the *extended convex ring*

$$\mathcal{S}^d := \{A \in \mathcal{F}^d \ : \ A \cap K \in \mathcal{R}^d \text{ for all } K \in \mathcal{K}^d\}.$$

Here, $\mathcal{K}^d$ denotes the class of all compact convex subsets of $\mathbb{R}^d$ (*convex bodies*), and the *convex ring* $\mathcal{R}^d$ is the class of finite unions of convex bodies (polyconvex bodies). Random $\mathcal{S}^d$-sets provide a sufficiently general framework to cover most situations which arise in practical applications (see [32, 42]). If a grain model Z is given for which all the grains are convex, then Z is a random $\mathcal{S}^d$-set. This model is of particular importance, since every random $\mathcal{S}^d$-set Z can be represented as a grain model (4) with convex grains Z_i (and with stationary germ process Φ, if Z is stationary) (see [39] and [52]). Whereas the representation of Z as a grain model with compact grains is rather simple, it is more difficult to construct a decomposition into random convex grains which have the same invariance properties as Z.

3.6 Integrability Conditions

For a grain model Z, the assumption (3) on the underlying marked point process Ψ (or the corresponding particle process (2)) is implied by

$$\int_{\mathbb{R}^d \times \mathcal{C}^d} \mathbf{1}\{(C + x) \cap D \neq \emptyset\} \tilde{\Lambda}(d(x, C)) < \infty, \quad D \in \mathcal{C}^d, \tag{5}$$

where $\tilde{\Lambda}$ is the intensity measure of Ψ. By *Campbell's theorem*, the above integral is just the expectation of (3). In the case of stationarity, condition (5), for all $D \in \mathcal{C}^d$, amounts to the same as requiring that the typical grain Z_0 fulfills

$$\mathbb{E}\, V_d(Z_0 + D) < \infty, \quad D \in \mathcal{C}^d, \tag{6}$$

(see also [8] or [39]).

4 Contact Distribution Functions

4.1 The Capacity Functional

By a simple measure theoretic argument, one sees that the distribution of a random set Z is completely determined by the *capacity functional* T_Z of Z. It is defined on $\mathcal{C}^d$ by

$$T_Z(C) := \mathbb{P}(Z \cap C \neq \emptyset) = 1 - \mathbb{P}(Z \cap C = \emptyset), \quad C \in \mathcal{C}^d.$$

In the case of a point process Φ on $\mathbb{R}^d$ we have $T_\Phi(C) = 1 - \mathbb{P}(\Phi(C) = 0)$ for $C \in \mathcal{C}^d$. The distribution of a random closed set Z need not be determined by the restriction of T_Z to $\mathcal{K}^d$ (this fact applies already to point processes). However, the values $T_Z(K)$, $K \in \mathcal{K}^d$, still contain interesting information and are easier to treat from both an analytical and a statistical point of view. They are closely connected to the contact distribution functions of Z which we introduce next.

4.2 Contact Distributions

Let $B \in \mathcal{K}^d$ with $0 \in B$ be fixed. For a stationary random set Z the formal definition of the contact distribution function with structuring element B is

$$H_B(r) := \mathbb{P}(rB \cap Z \neq \emptyset \,|\, 0 \notin Z), \quad r \geq 0, \tag{7}$$

or equivalently,

$$H_B(r) = \mathbb{P}(d_B(0, Z) \leq r \,|\, 0 \notin Z), \quad r \geq 0, \tag{8}$$

(we always assume $\mathbb{P}(0 \notin Z) > 0$, i.e. we exclude the uninteresting case $Z = \mathbb{R}^d$ $\mathbb{P}$-a.s., where the contact distribution is not well defined). H_B is a distribution function in the extended sense, it may have an atom at infinity.

A closely related concept is the *empty space function*

$$F_B(r) := \mathbb{P}(d_B(0, Z) \leq r) = T_Z(rB), \quad r \geq 0,$$

of Z. It has an atom at 0 of size

$$p := \mathbb{P}(0 \in Z).$$

Due to stationarity and Fubini's theorem we have

$$p = \int_{[0,1]^d} \mathbb{P}(x \in Z) dx = \mathbb{E}\, V_d([0,1]^d \cap Z),$$

which explains why p is called the *volume fraction* of Z (the mean volume of Z per unit volume of $\mathbb{R}^d$). Since $0 \in B$ we have

$$(1 - p)(1 - H_B(r)) = \mathbb{P}(rB \cap Z = \emptyset) = 1 - F_B(r).$$

If $p = 0$ (this occurs for lower dimensional sets Z), H_B and F_B coincide.

For a given realization of Z, H_B can be estimated from volume measurements. Namely, we have

$$\begin{aligned} H_B(r) &= \mathbb{P}(d_B(0, Z) \leq r \,|\, 0 \notin Z) = \mathbb{P}(rB \cap Z \neq \emptyset \,|\, 0 \notin Z) \\ &= 1 - \frac{\mathbb{P}(0 \notin Z + rB^*)}{\mathbb{P}(0 \notin Z)} = \frac{\mathbb{P}(0 \in Z + rB^*) - p}{1 - p} \\ &= \frac{\mathbb{E}\,[V_d((Z + rB^*) \cap [0,1]^d) - p]}{1 - p}, \quad r \geq 0. \end{aligned}$$

A straightforward generalization is

$$H_B(r) = \frac{\mathbb{E}\left[V_d((Z + rB^*) \cap W) - V_d(Z \cap W)\right]}{V_d(W) - \mathbb{E}\, V_d(Z \cap W)}, \tag{9}$$

for an arbitrary Borel set $W \subset \mathbb{R}^d$ of positive volume.

If $Z = \Phi$ is a point process, we have $p = 0$, hence the condition $0 \notin \Phi$ (required in the definition of a contact distribution function) is fullfilled almost surely. Also, in this case, $H_B \equiv 0$ if B is lower dimensional. If B has a non-empty interior, then

$$(1 - p)(1 - H_B(\infty)) = \mathbb{P}(Z = \emptyset),$$

where $H_B(\infty) := \lim_{t\to\infty} H_B(t)$. Hence $H_B(\infty) < 1$ in case $\mathbb{P}(Z = \emptyset) > 0$. For point processes there exist notions related to the contact distribution function such as the nearest-neighbour distribution function or the J-function introduced in [47] and extended in [48]; see also [42]. For the latter, an extension to certain grain models is suggested in [20]. Other extensions concern the n-neighbour distance distribution, see [16] and [44] for further details.

The values of the contact distribution functions H_B essentially depend on the choice of the gauge body B. Therefore, we may choose different gauge bodies to gain additional information about a given random structure. Two contact distribution functions, which have received particular attention in the literature, are the *spherical contact distribution function* $H_s := H_{B^d}$ and the *linear contact distribution function* $H_{[0,u]}$ in direction $u \in S^{d-1}$, where $S^{d-1} := \partial B^d$ is the Euclidean unit sphere (in the literature often H_l is used for the linear contact distribution; however, this does not reflect the dependence on the direction u which is important if Z is not isotropic). The connection of the linear contact distribution to chord length distributions is explained in [42]; see also the references given in Sect. 12. For an application of this connection to the study of contact and chord length distributions of stationary Voronoi tessellations see [10] and [31]. The following sections contain a more detailed discussion of these special contact distributions.

4.3 Estimators of Contact Distribution Functions

Formula (9) suggests

$$\frac{V_d((Z + rB^*) \cap W) - V_d(Z \cap W)}{V_d(W) - V_d(Z \cap W)}$$

as a simple estimator of $H_B(r)$ (plus sampling). It requires, however, knowledge from outside the window W. A traditional approach to cope with this problem is to consider only those points in W having a certain minimal distance from the boundary of W (minus sampling). Further information on this and more sophisticated methods can be found in [1, 42, 45]. In [3] an instructive comparison is provided between Hanisch type (Horvitz-Thompson style) and Kaplan-Meier type estimators for the empty space function of a stationary random closed set.

The estimators of H_B discussed in the literature are based on a single observation of Z and can be computed (or rather approximated) with the help of digital image analysers. They serve as convenient and useful summary statistics in the exploratory analysis of spatial patterns (see e.g. [2, 3, 6, 7, 9]). For instance one might plot these estimators against theoretically known contact distribution functions, such as those of the Boolean model with convex grains (see Sect. 5). If Z is *ergodic*, then H_B can in principle be determined from a single realization of Z by extending the observation window (see [9]).

4.4 Regularity Properties

The contact distribution functions of stationary random closed sets with respect to general gauge bodies enjoy an important regularity property. In [7] it was shown that the contact distribution function H_B of a stationary random closed set Z is absolutely continuous if 0 is an interior point of the gauge body B. Hence it makes sense to define and estimate the density (and the hazard rate). More specifically, in the spherical case the density H'_s of H_s satisfies

$$(1-p)H'_s(r) = \mathbb{E}\left[\mathcal{H}^{d-1}(\partial(Z + rB^d) \cap [0,1]^d)\right], \quad r > 0, \tag{10}$$

where $\mathcal{H}^{d-1}(C)$ is the $(d-1)$-dimensional Hausdorff measure of C. (If C is a smooth $(d-1)$-dimensional surface, then $\mathcal{H}^{d-1}(C)$ is just the surface area of C.) Independently, and by a completely different approach, in [21] not only the absolute continuity of H_s for a random $\mathcal{S}^d$-set was proved, but also a formula expressing the density in terms of certain geometric mean values. We will resume this issue later in this paper.

4.5 Generalizations of Contact Distributions

We have already mentioned that the contact distribution functions do not determine the distribution of Z. Hence, it is natural to extend their definition by taking into account more information than just the distance $d_B(0, Z)$. Using the *contact direction vector* $u_B(0, Z)$, we can define a generalized *contact distribution function* of Z,

$$H_B(r, C) := \mathbb{P}(d_B(0, Z) \le r, u_B(0, Z) \in C \,|\, 0 \notin Z), \tag{11}$$

where $r \ge 0$ and $C \subset \partial B$ is a Borel set. This definition is subject to the assumption that the vector $u_B(0, Z)$ is $\mathbb{P}$-a.s. well-defined given that $0 \notin Z$. If Z is a random closed set and B is strictly convex, containing 0 in its interior, then this is indeed the case. This follows from $V_d(\text{exo}_B(Z)) = 0$ (compare the remark at the end of Sect. 2) and the stationarity of Z. The function $H_B(\cdot, \cdot)$ determines the conditional distribution of the pair $(d_B(0, Z), u_B(0, Z))$ and hence that of the *contact vector* $d_B(0, Z)u_B(0, Z)$. For each fixed r the function $H_B(r, \cdot)$ is a measure on ∂B. Note that $H_B(\cdot, \partial B)$ is the contact distribution function defined by (7).

In Sect. 11 further extensions of (generalized) contact distributions are discussed; there, in particular, local (second order) information about Z at the point of contact is taken into account.

5 Poisson Processes and Boolean Models

In this section we provide explicit formulas for contact distributions of stationary Poisson processes and Boolean models with respect to various structuring elements B. Applications of these models in physics may be found in [25, 26] or in the contributions by C. Arns, M. Knackstedt and K. Mecke (Chap. 1) and by V. Robins (Chap. 3) in this volume.

5.1 The Stationary Poisson Process

The most important instance of a stationary point process on $\mathbb{R}^d$ is a stationary *Poisson process*. It is the mathematical model for a collection of points that are distributed "completely random" in space. For a Poisson process Φ on $\mathbb{R}^d$, $\Phi(A_1), \ldots, \Phi(A_k)$ are stochastically independent random variables, whenever the measurable sets $A_1, \ldots, A_k \subset \mathbb{R}^d$ are pairwise disjoint. Moreover,

$$\mathbb{P}(\Phi(A) = m) = \frac{(\gamma V_d(A))^m}{m!} \exp[-\gamma V_d(A)], \quad m = 0, 1, \ldots,$$

for all Borel sets $A \subset \mathbb{R}^d$, i.e. $\Phi(A)$ is Poisson distributed with parameter $\gamma V_d(A)$. It follows that γ is the intensity of Φ and that the distribution of a stationary Poisson process Φ is completely determined by its intensity.

From the uniqueness result mentioned in the previous section we obtain that a point process Φ on $\mathbb{R}^d$ is a stationary Poisson process of intensity γ if and only if

$$\mathbb{P}(\Phi(A) = 0) = \exp[-\gamma V_d(A)],$$

for all Borel sets $A \subset \mathbb{R}^d$. In this case we have

$$H_B(r) = 1 - \mathbb{P}(\Phi(rB) = 0) = 1 - \exp[-\gamma V_d(B) r^d], \quad r \geq 0. \tag{12}$$

5.2 The Stationary Boolean Model

A stationary *Boolean model* Z is a grain model

$$Z = \bigcup_{n \in \mathbb{N}} (Z_n + \xi_n),$$

with independent, identically distributed grains Z_n, where the underlying point process $\Phi = \{\xi_n : n = 1, 2, \ldots\}$ is a stationary Poisson process that is independent of $\{Z_n : n = 1, 2 \ldots\}$. In this case the finiteness condition (3) is even equivalent to (6). An alternative definition of the Boolean model in terms of a stationary Poisson particle process X will be given later in Sect. 10.3.

The distribution of a stationary Boolean model Z is determined by the pair $(\gamma, \mathbb{Q})$, where γ is the intensity of Φ and $\mathbb{Q}$ is the common distribution of the grains Z_i. We again consider a random set Z_0 with distribution $\mathbb{Q}$ and call this the *typical grain* of Z. The capacity functional T_Z of Z has the form

$$\begin{aligned} T_Z(C) &= 1 - \mathbb{P}(Z \cap C = \emptyset) \\ &= 1 - \exp\left\{-\gamma \int_{\mathcal{C}^d} V_d(K + C^*)\mathbb{Q}(dK)\right\} \\ &= 1 - \exp\{-\gamma \mathbb{E}\, V_d(Z_0 + C^*)\}, \end{aligned} \tag{13}$$

and hence we have

$$p = \mathbb{P}(0 \in Z) = 1 - \exp\{-\gamma \mathbb{E}\, V_d(Z_0)\}.$$

Reversely the distribution of Z determines the intensity γ and the distribution $\mathbb{Q}$ (up to the centering of the particles, see [39]). The stationary Boolean model will be generalized later in this paper. If not stated otherwise a Boolean model will always be assumed to be stationary.

5.3 The Spherical Contact Distribution of the Boolean Model

By (13) the spherical contact distribution H_s of the stationary Boolean model Z is given by

$$H_s(r) = 1 - \exp\left\{-\gamma \int_{\mathcal{C}^d} [V_d(K + rB^d) - V_d(K)]\mathbb{Q}(dK)\right\}. \tag{14}$$

At this point, we can proceed further if we assume convexity of the grains, since this allows us to use the *Steiner formula* for convex bodies,

$$V_d(K + rB^d) = \sum_{j=0}^{d} r^j b_j V_{d-j}(K). \tag{15}$$

Here, b_j is the (j-dimensional) volume of the Euclidean unit ball in $\mathbb{R}^j$ and the coefficients $V_0(K), \dots, V_d(K)$ are the *intrinsic volumes* or *Minkowski functionals* of the convex body K. Note that these functionals occur in the literature with different normalization, different enumeration and also different names. We have chosen here the notation which is standard in convex geometry (see Schneider's book [37]). In particular, $V_d(K)$ is the volume of K, $V_{d-1}(K)$ is half the surface area, $V_{d-2}(K)$ is proportional to the integral mean curvature, ..., $V_1(K)$ is proportional to the mean width of K, and $V_0(K)$ is the Euler characteristic (which, for convex K, is 1 if K is non-empty and 0 if K is the empty set). Combining (14) with (15), we arrive at

$$H_s(r) = 1 - \exp\left\{-\gamma \sum_{j=1}^{d} r^j b_j \mathbb{E}\, V_{d-j}(Z_0)\right\}. \tag{16}$$

The usefulness of (16) for the statistical analysis of Z is obvious. Fitting a polynomial of order d (and without constant term) to the (empirical) function corresponding to $-\ln(1 - H_s(r))$, yields estimators for $\gamma \mathbb{E}\, V_0(Z_0), \dots, \gamma \mathbb{E}\, V_{d-1}(Z_0)$ and can also be used to check the Boolean hypothesis (see [40] or [42]). Note that here $\gamma \mathbb{E}\, V_0(Z_0) = \gamma$, hence we obtain in particular an estimator of the intensity γ of Φ. Note also that the remaining mean value $\gamma \mathbb{E}\, V_d(Z_0)$ can be estimated directly from the volume fraction p of Z.

5.4 General Structuring Elements

If we aim to estimate other functionals of the stationary Boolean model Z, in particular, functionals which reflect the anisotropy of Z, then the spherical contact distribution H_s is no longer sufficient and we are forced to use more general structuring elements B (e.g. linear contact distributions H_B where $B = [0, u]$ is a segment). The above formula for the capacity functional of a Boolean model (with convex grains) immediately implies

$$H_B(r) = 1 - \exp\left\{-\gamma \int_{\mathcal{K}^d} [V_d(K + rB^*) - V_d(K)]\mathbb{Q}(dK)\right\}. \tag{17}$$

For a further investigation of the right side, we again assume convexity, not only of the grains K, but also of B. The reason for this is that if B is a convex body, then a generalization of the Steiner formula can be used. More explicitly, given convex bodies K, B, the generalized Steiner formula reads

$$V_d(K + B) = \sum_{k=0}^{d} \binom{d}{k} V(K\,[k], B\,[d-k])$$

with mixed volumes

$$V(K\,[k], B\,[d-k]) := V(\underbrace{K, \ldots, K}_{k}, \underbrace{B, \ldots, B}_{d-k})$$

(sometimes we write $V(K, B\,[d-1])$ instead of $V(K\,[1], B\,[d-1])$). Hence, in this case, we obtain

$$H_B(r) = 1 - \exp\left\{-\gamma \sum_{k=1}^{d} \binom{d}{k} r^k \int_{\mathcal{K}^d} V(K\,[d-k], B^*\,[k])\mathbb{Q}(dK)\right\}. \tag{18}$$

Relation (16) is a special case of (18) as can be seen from

$$V_j(K) = \frac{\binom{d}{j}}{b_{d-j}} V(K\,[j], B^d\,[d-j]), \quad j = 0, \ldots, d.$$

Therefore measurements of the contact distribution function $H_B(r)$, $r \geq 0$, for convex bodies $B \in \mathcal{K}^d$ as structuring elements lead to estimators for

$$\gamma \int_{\mathcal{K}^d} V(K\,[d-k], B^*\,[k])\mathbb{Q}(dK), \quad k = 1, \ldots, d.$$

These formulas can be specified further by special choices of B and under additional assumptions on the grains. For instance, one can choose B to be a cube and the typical grain to be $\mathbb{P}$-a.s. a rectangular parallelepiped; compare [45].

5.5 The Linear Contact Distribution of the Boolean Model

The linear contact distribution functions are defined with a gauge body B that is a segment $[0,u]$, $u \in S^{d-1}$. For $B = [0,u]$, the summands in (18) corresponding to $k \geq 2$ all vanish, and hence

$$\begin{aligned} H_{[0,u]}(r) &= 1 - \exp\left\{-\gamma r d \int_{\mathcal{K}^d} V([0,u], K[d-1])\mathbb{Q}(dK)\right\} \\ &= 1 - \exp\left\{-\gamma r \int_{\mathcal{K}^d} V_{d-1}(K \mid u^\perp)\mathbb{Q}(dK)\right\} \\ &= 1 - \exp\left\{-\gamma \frac{r}{2} \int_{\mathcal{K}^d} \int_{S^{d-1}} |\langle u, v\rangle| S_{d-1}(K; dv)\mathbb{Q}(dK)\right\}, \end{aligned}$$

due to classical formulas for the mixed volume $V([0,u], K[d-1])$ (compare §5.3 in [37]). Here, $V_{d-1}(K \mid u^\perp)$ is the $(d-1)$-volume of the orthogonal projection of K onto the hyperplane $u^\perp$, $\langle u, v\rangle$ is the scalar product of the (unit) vectors u, v, and $S_{d-1}(K; \cdot)$ denotes the surface area measure of K (of order $d-1$), which is a finite measure on S^{d-1} (see Sect. 7). If we define the mean surface area measure of the typical grain by

$$\bar{S}_{d-1} := \int_{\mathcal{K}^d} S_{d-1}(K; \cdot)\mathbb{Q}(dK) \tag{19}$$

as a measure over S^{d-1}, we finally obtain

$$H_{[0,u]}(r) = 1 - \exp\left\{-\gamma \frac{r}{2} \int_{S^{d-1}} |\langle u, v\rangle| \bar{S}_{d-1}(dv)\right\}.$$

If Z (and hence Z_0) is isotropic, $\mathbb{Q}$ is rotation invariant, and therefore $H_l := H_{[0,u]}$ is independent of u. We obtain

$$H_l(r) = 1 - \exp\left\{-\gamma r \frac{2b_{d-1}}{d b_d} \mathbb{E}\, V_{d-1}(Z_0)\right\}.$$

5.6 The Mean Normal Distribution

The measure $\mu := \gamma \bar{S}_{d-1}$ is also called the *oriented mean normal measure* of the underlying Poisson particle process X, $X := \{Z_1 + \xi_1, Z_2 + \xi_2, \dots\}$ (with $\Phi = \{\xi_1, \xi_2, \dots\}$ and with convex grains Z_i). The oriented mean normal measure of X is the surface area measure of a convex body (unique up to translations, if μ is not degenerated), which is called the *Blaschke body* $B(X)$ of X. The normalized measure

$$\mathbb{M} := \frac{\mu}{\mu(S^{d-1})} = \frac{\mathbb{E}\, S_{d-1}(Z_0; \cdot)}{2\mathbb{E}\, V_{d-1}(Z_0)}$$

is called *oriented mean normal distribution*, it describes the distribution of the outer normal vector in a typical boundary point of the grains (more precisely, a typical point of $\bigcup_{n=1}^{\infty}(\partial Z_n + \xi_n)$). We emphasize that $\mathbb{M}$ contains oriented directional information about the grains, it distinguishes between outer and inner normals in the boundary points. The corresponding unoriented *mean normal distribution* is given by

$$\widetilde{\mathbb{M}}(A) := \frac{1}{2}(\mathbb{M}(A) + \mathbb{M}(A^*)), \quad A \subset S^{d-1}.$$

If the grains are almost surely $(d-1)$-dimensional, then the oriented and the unoriented mean normal distributions coincide. They both vanish if the grains have dimension $\leq d-2$. However, for Boolean models Z with full dimensional grains (or surfaces as grains), the measures $\mathbb{M}$ and $\widetilde{\mathbb{M}}$ contain relevant directional information about the grains. It is therefore important to know, how far the measures $\widetilde{\mathbb{M}}$ (resp. $\mathbb{M}$) are determined by linear contact distribution functions $H_{[0,u]}$, $u \in S^{d-1}$. Since

$$H_{[0,u]}(r) = 1 - \exp\left\{-\gamma r \mathbb{E}\, V_{d-1}(Z_0) \int_{S^{d-1}} |\langle u, v\rangle| \widetilde{\mathbb{M}}(dv)\right\}, \tag{20}$$

we find that the collection of all linear contact distribution functions $\{H_{[0,u]}, u \in S^{d-1}\}$ determines the unoriented mean normal distribution $\widetilde{\mathbb{M}}$ uniquely (but not the oriented mean normal distribution $\mathbb{M}$). The reason for this result is that the spherical integral transform (the so-called *Cosine transform*)

$$\rho \mapsto T\rho, \qquad T\rho(u) := \int_{S^{d-1}} |\langle u, v\rangle| \rho(dv),$$

which maps finite measures ρ on S^{d-1} to continuous functions, is injective on *even* measures (but is not injective in general, since $T\rho = T\rho^*$, where ρ^* is the reflection of ρ). If we restrict T to the even C^∞ functions, then the inverse transform T^{-1} can be obtained using expansions into spherical harmonics. However, the inverse transform is not stable, which means that it is quite sensitive to small perturbations of the given data. For example, if the linear contact distributions $H_{[0,u]}(r)$ of Z are determined for a fixed r and finitely many directions $u = u_1, \ldots, u_k$, and if the resulting values $H_{[0,u_1]}(r), \ldots, H_{[0,u_k]}(r)$ are interpolated to obtain a smooth function f on S^{d-1}, the resulting estimator

$$-\frac{1}{\gamma r \mathbb{E}\, V_{d-1}(Z_0)} T^{-1} \ln(1 - f(\cdot))$$

for $\widetilde{\mathbb{M}}$ will typically be a signed measure, which may be far from the true one. This problem can be overcome by means of more sophisticated estimators (based on linear programming methods or the EM algorithm), it is discussed in a quite similar situation (the estimation of fibre processes from hyperplane sections) in [17].

6 Poisson Cluster Processes

Poisson processes can be used to build more general point process models. An important example are *Poisson cluster processes* whose spherical contact distribution functions are studied in the present section. This subject is continued in Sect. 8 with the investigation of Poisson cluster models.

6.1 Definition of a Poisson Cluster Process

We start with the construction of Poisson cluster processes. There is a stationary Poisson process Φ_p (of *parents*) with finite intensity γ_p and a family $\{N_x : x \in \Phi_p\}$ of finite point processes on $\mathbb{R}^d$ such that $\{(x, N_x) : x \in \Phi_p\}$ is an independent marking of Φ_p. The conditional distribution of N_x given Φ_p is the same for all $x \in \Phi_p$ and we let N_0 denote a *typical cluster*, i.e. a point process with this distribution. We also assume that the mean number γ_c of cluster points,

$$\gamma_c := \mathbb{E}\, N_0(\mathbb{R}^d),$$

is finite. Then

$$\Phi := \bigcup_{x \in \Phi_p} (N_x + x) \tag{21}$$

is a stationary point process with intensity

$$\gamma = \gamma_p \gamma_c.$$

It is called Poisson cluster process based on Φ_p and the typical cluster N_0 (we remark that the points of different clusters are almost surely different, i.e. we do not have multiple points).

6.2 Contact Distributions of a Poisson Cluster Process

A Poisson cluster process Φ can be interpreted as a Boolean model with typical (compact but in general not convex) grain N_0. Using (13), we obtain that

$$T_\Phi(C) = 1 - \exp\{-\gamma_p \mathbb{E}\, V_d(N_0 + C^*)\} \tag{22}$$

for $C \in \mathcal{C}^d$. This shows that the distribution of Φ is determined by γ_p and by the distribution of N_0. In particular, the contact distribution of Φ with respect to a fixed structuring element B is given by

$$H_B(r) = 1 - \exp\{-\gamma_p \mathbb{E}\, V_d(N_0 + rB^*)\}\,. \tag{23}$$

To exploit this we need a formula for the generalized parallel volume $V_d(N_0 + rB^*)$. We start with the Euclidean case $B = B^d$, the general case will be treated later. For $\varphi \in \mathbf{N}$ we define

$$\varphi^{(0)} := \varphi \setminus \{0\}\,.$$

A straightforward argument yields for all $\varphi \in \mathbf{N}$ and all $r > 0$ the formula

$$\begin{aligned} V_d(\varphi + rB^d) = \int_0^r \int_{S^{d-1}} \int_{\mathbb{R}^d} & \mathbf{1}\{B^d(tu, t) \cap (\varphi - x)^{(0)} = \emptyset\} \\ & \times t^{d-1} \varphi(dx) \mathcal{H}^{d-1}(du)dt, \end{aligned} \tag{24}$$

where S^{d-1} is the unit sphere in $\mathbb{R}^d$ and $B^d(x,r)$ is the Euclidean ball with radius r and center at x. Formally, this is a very special case of the generalized *Steiner formula* in [21], Theorem 3.1. Inserting (24) into (23), we obtain that

$$1 - H_s(r) = \exp\left\{-\int_0^r \lambda_s(t)dt\right\}, \tag{25}$$

where

$$\lambda_s(t) := \gamma_p t^{d-1} \int_{S^{d-1}} \mathbb{E}\left[\int_{\mathbb{R}^d} \mathbf{1}\{(N_0 - x)^{(0)} \cap B^d(tu,t) = \emptyset\} N_0(dx)\right] \mathcal{H}^{d-1}(du). \tag{26}$$

In case $N_0 = \{0\}$ we have $\Phi = \Phi_p$ and

$$\lambda_s(t) = \gamma d b_d t^{d-1}, \tag{27}$$

in accordance with (12). (Note that here $\gamma_c = 1$.) Adapting the terminology used in reliability theory to our situation, we call λ_s the *empty space hazard* of Φ (see [7]). Loosely speaking, we have

$$\mathbb{P}(d(0,\Phi) \in (t, t+h] \mid d(0,\Phi) \geq t) \approx \lambda_s(t)h$$

for small $h > 0$. Formula (25) has been proved in [20]. Additional formulae for the spherical contact distribution functions of some special Poisson cluster processes can be found in [35] and [44].

6.3 Gauss–Poisson Processes

In this subsection we assume that the cluster N_0 has at most 2 points, hence $\eta := N_0(\mathbb{R}^d) \leq 2$. Then we can write

$$N_0 = \begin{cases} \emptyset & \text{if } \eta = 0, \\ \{Y_0\} & \text{if } \eta = 1, \\ \{Y_1, Y_2\} & \text{if } \eta = 2, \end{cases} \tag{28}$$

where Y_0, Y_1, Y_2 are random vectors in $\mathbb{R}^d$ with $Y_1 \neq Y_2$. If $Y_0 = Y_1 = 0$ and η is independent of Y_2, then Φ is called *Gauss-Poisson* process. We have

$$\begin{aligned} \mathbb{E}\left[\int_{\mathbb{R}^d} \mathbf{1}\{(N_0 - x)^{(0)} \cap B^d(tu,t) = \emptyset\} N_0(dx)\right] &= \mathbb{P}(\eta = 1) \\ + \mathbb{P}(\eta = 2)\mathbb{P}(Y_1 - Y_2 \notin B^d(tu,t) \mid \eta = 2) \\ + \mathbb{P}(\eta = 2)\mathbb{P}(Y_2 - Y_1 \notin B^d(tu,t) \mid \eta = 2) \end{aligned}$$

(if $\mathbb{P}(\eta = 2) > 0$), and therefore (26) simplifies to

$$\begin{aligned} t^{1-d}\lambda_s(t) &= \gamma_p d b_d \mathbb{P}(\eta = 1) \\ &\quad + 2\mathbb{P}(\eta = 2)\gamma_p \int_{S^{d-1}} \mathbb{P}(Y_1 - Y_2 \notin B^d(tu,t) \mid \eta = 2)\mathcal{H}^{d-1}(du) \end{aligned}$$

(see also Example 5.2 in [20]).

6.4 Neyman–Scott Processes

We now consider the case that the typical cluster is given by

$$N_0 = \begin{cases} \emptyset & \text{if } \eta = 0, \\ \{Y_{i,n} : i = 1, \ldots, n\} & \text{if } \eta = n \geq 1, \end{cases} \tag{29}$$

where the random cluster size η is a random element of $\mathbb{N}_0$, and the $Y_{i,n}$, $n \in \mathbb{N}$, $i = 1, \ldots, n$, are independent and identically distributed random elements of $\mathbb{R}^d$ that are independent of η. In this case $\gamma_c = \mathbb{E}\,\eta$ which we have assumed to be finite. The Poisson cluster process Φ is then called a *Neyman–Scott process* (see also §16.2 in [44]). Let $\mathbb{V}$ denote the distribution of $Y_{1,1}$. We have

$$\begin{aligned}
&\mathbb{E}\left[\int_{\mathbb{R}^d} \mathbf{1}\{(N_0 - x)^{(0)} \cap B^d(tu,t) = \emptyset\} N_0(dx)\right] \\
&= \mathbb{P}(\eta = 1) + \sum_{n=2}^{\infty} n\mathbb{P}(\eta = n)\mathbb{P}(Y_{i,n} - Y_{1,n} \notin B^d(tu,t), i = 2, \ldots, n) \\
&= \mathbb{P}(\eta = 1) + \sum_{n=2}^{\infty} n\mathbb{P}(\eta = n) \int_{\mathbb{R}^d} \mathbb{P}(Y_{1,1} - x \notin B^d(tu,t))^{n-1}\mathbb{V}(dx) \\
&= \int_{\mathbb{R}^d} g'(\mathbb{P}(Y_{1,1} - x \notin B^d(tu,t)))\mathbb{V}(dx),
\end{aligned}$$

where g' is the derivative of the probability generating function g of η. Hence we conclude from (26) that

$$\lambda_s(t) = \gamma_p t^{d-1} \int_{S^{d-1}} \int_{\mathbb{R}^d} g'(\mathbb{P}(Y_{1,1} - x \notin B^d(tu,t)))\mathbb{V}(dx)\mathcal{H}^{d-1}(du), \tag{30}$$

correcting a statement in Example 5.1 of [20].

6.5 Asymptotic Behaviour of the Empty Space Hazard

Let Φ be a general Poisson cluster process as in Sects. 6.1 and 6.2. Formula (26) gives the empty space hazard λ_s of Φ in terms of the parent intensity γ_p and the distribution of the typical cluster N_0. Although informative, this formula is still rather complicated. It is therefore instructive to compare λ_s as given by (26) with the right side of (27). From monotone convergence

$$\lim_{t \to 0} t^{1-d} \lambda_s(t) = \gamma d b_d, \tag{31}$$

i.e. for small values of t the empty space hazard of a Poisson cluster process behaves approximately like the empty space hazard of a Poisson process with the same intensity. As $t \to \infty$ we have $B^d(tu,t) \uparrow \{0\} \cup \{x \in \mathbb{R}^d : \langle x, u \rangle > 0\}$ for each $u \in S^{d-1}$ and it easily follows that

$$\lim_{t\to\infty} t^{1-d}\lambda_s(t) = \gamma_p \int_{S^{d-1}} \mathbb{E}\left[\int_{\mathbb{R}^d} \mathbf{1}\{(N_0 - x)^{(0)} \cap H_u^+ = \emptyset\} N_0(dx)\right] \mathcal{H}^{d-1}(du), \tag{32}$$

where $H_u^+ := \{x \in \mathbb{R}^d : \langle x, u\rangle \geq 0\}$. It is an interesting fact that the convergence in (31) and (32) is monotone, i.e. that $t^{1-d}\lambda_s(t)$ is monotone decreasing in t. The right-hand side of (32) can be considerably simplified. Splitting the expectation according to the events $N_0(\mathbb{R}^d) = n$, $n \in \mathbb{N}$, we have to determine the integrals

$$\int_{S^{d-1}} \left[\sum_{k=1}^{n} \mathbb{P}(\langle Y_j - Y_k, u\rangle \leq 0 \text{ for all } j \neq k, N_0(\mathbb{R}^d) = n)\right] \mathcal{H}^{d-1}(du),$$

for all $n \geq 2$, where $Y_1, \dots, Y_n$ are pairwise distinct (random) vectors in $\mathbb{R}^d$. Fixing $u \in S^{d-1}$ and assuming without loss of generality that $\langle Y_j - Y_k, u\rangle \neq 0$ for $j \neq k$, it can be shown by induction that the above sum in brackets is just $\mathbb{P}(N_0(\mathbb{R}^d) = n)$. Hence we obtain from (32) that

$$\lim_{t\to\infty} t^{1-d}\lambda_s(t) = \mathbb{P}(N_0 \neq \emptyset)\gamma_p d b_d, \tag{33}$$

irrespective of any specific assumptions on the typical cluster N_0. This is the same asymptotics as that of the Poisson process $\{x \in \Phi_p : N_x \neq \emptyset\}$. The latter has intensity $\mathbb{P}(N_0 \neq \emptyset)\gamma_p$. This means in a sense, that different cluster points cannot be distinguished from a very far distance.

7 Local Geometric Concepts

7.1 Support Measures

The Minkowski functionals are important characteristics of convex bodies and of sets in the convex ring (*polyconvex bodies*). For convex bodies, they can be obtained as coefficients of a Steiner formula, for polyconvex bodies they are determined by their additivity properties. We now describe how these functionals can be localized to measure-valued functionals which appear as coefficient measures of a local Steiner formula. We have already seen in the second part of Sect. 5 that it is important to gain local directional information about random $\mathcal{S}^d$-sets. The (Euclidean) *support measures* (or *generalized curvature measures*) $\Theta_i(K;\cdot)$, $i = 0, \dots, d-1$, of a convex body $K \in \mathcal{K}^d$, which we introduce now, are essential geometric tools useful to quantify this local information. The support measures are finite measures on $\mathbb{R}^d \times S^{d-1}$ which can be obtained as coefficients of a local Steiner formula, and whose definition is based on the underlying Euclidean structure and therefore on the geometric properties of the Euclidean unit ball. These measures, and in particular the *surface area measures* $S_i(K;\cdot) := \Theta_i(K; \mathbb{R}^d \times \cdot)$, $i = 0, \dots, d-1$, have been used in the literature since quite a while. (Recall that we already introduced the surface area measure $S_{d-1}(K;\cdot)$ in Sect. 5.) Of more recent vintage are (relative) support measures which are defined with respect to a more general convex gauge body B (with $0 \in B$) which replaces the Euclidean unit ball.

In order to be consistent with the literature, we first set $d_B(K,x) := d_{B^*}(x,K)$ and $u_B(K,x) := -u_{B^*}(x,K)$ for $x \in \mathbb{R}^d \setminus K$. Furthermore, we assume that K and B are in general relative position, which means that K and B do not contain parallel segments in parallel and equally oriented support (hyper)planes. A sufficient condition is that K or B is strictly convex, but the condition is also satisfied for an arbitrary convex body K and almost all rotational images of an arbitrary convex body B. For convex bodies K and B in general relative position (with $0 \in B$), there are finite measures $\Theta_0(K;B;\cdot),\dots,\Theta_{d-1}(K;B;\cdot)$ on $\mathbb{R}^d \times \partial B$ which satisfy the local Steiner formula

$$\begin{aligned} V_d(\{x \in (K+rB)\setminus K : (x - d_B(K,x)u_B(K,x), u_B(K,x)) \in \cdot\}) \quad (34) \\ = \frac{1}{d}\sum_{j=0}^{d-1} r^{d-j}\binom{d}{j}\Theta_j(K;B;\cdot), \end{aligned}$$

for all $r \geq 0$. The crucial point here is that the local parallel volume on the left side is a polynomial in the distance parameter r, and the coefficients of this polynomial are measures. These measures are called *relative support measures* of K with respect to B; for further details see [14, 15, 18, 38]. Note that $\Theta_j(K;B;\cdot)$ is concentrated on $\partial K \times \partial B$ and that

$$\Theta_j(K;\cdot) := \Theta_j(K;B^d;\cdot)$$

just defines the Euclidean support measures. The *relative surface area measures* $S_j(K;B;\cdot) := \Theta_j(K;B;\mathbb{R}^d \times \cdot)$ and the ordinary surface area measures $S_j(K;\cdot)$ arise as image measures of (relative) support measures. Especially, we get

$$\Theta_j(K;B;\partial K \times \partial B) = dV(K\,[j], B\,[d-j])$$

and, in particular,

$$\Theta_j(K;\mathbb{R}^d \times S^{d-1}) = \frac{d b_{d-j}}{\binom{d}{j}} V_j(K),$$

for $j = 0,\dots d-1$.

7.2 Smooth Convex Bodies

For sufficiently smooth convex bodies, the Euclidean surface area measures can be described in terms of basic curvature functions. Let $h_K(u) := \max\{\langle x,u\rangle : x \in K\}$, $u \in \mathbb{R}^d$, denote the *support function* of $K \in \mathcal{K}^d$ at u. We say that K is of class C^2_+ if ∂K is a hypersurface of class C^2 with everywhere positive Gauss curvature. Now let $K \in \mathcal{K}^d$ be of class C^2_+. This assumption implies that h_K is of class C^2 and K is strictly convex. The gradient $\nabla h_K(u)$ is the unique boundary point $\tau(K,u)$ of K with exterior unit normal vector u. The *principal radii of curvature* of K in (direction) u are the reciprocals of the principal curvatures of K at $\tau(K,u)$. Writing $s_j(K,u)$, $j \in \{1,\dots,d-1\}$, for the j-th normalized elementary symmetric function of the principal radii of curvature of K in u and setting $s_0(K,u) := 1$, for all $u \in S^{d-1}$, we get for any $K \in \mathcal{K}^d$ of class C^2_+ that

$$S_j(K;\cdot) = \int_{S^{d-1}} \mathbf{1}\{u \in \cdot\} s_j(K,u)\mathcal{H}^{d-1}(du).$$

This equation provides a relationship between the coefficients of the local Steiner formula for the convex body K and second order (local) geometric information on the bounding hypersurface ∂K. For more details we refer to [37].

7.3 Additive Extensions of Support Measures

An important property which the relative support measures share with the Minkowski functionals is additivity. As a consequence these measures can be extended as additive, signed, measure-valued functionals to the (extended) convex ring, and we use the same notation for these extensions. Later we will make use of $\Theta_{d-1}(K;B;\cdot)$, $K \in \mathcal{S}^d$. Now we provide some information on these measures, but we have to refer to [14] for the details. We assume that the gauge body B is strictly convex and contains the origin in its interior. Then the measure $\Theta_{d-1}(K;B;\cdot)$ is non-negative and concentrated on the relative (generalized) normal bundle $\mathcal{N}_B(K)$ of K which is defined by

$$\mathcal{N}_B(K) := \{(x - d_B(K,x)u_B(K,x), u_B(K,x)) : x \notin K \cup \mathrm{exo}_{B^*}(K)\}.$$

Furthermore, it can be expressed in terms of the Euclidean support measure $\Theta_{d-1}(K;\cdot)$ and the support function h_B of B:

$$\Theta_{d-1}(K;B;\cdot) = \int_{\mathcal{N}_{B^d}(K)} \mathbf{1}\{(x, \nabla h_B(u)) \in \cdot\} h_B(u) \Theta_{d-1}(K; d(x,u)). \tag{35}$$

Assume that K is *regular*, i.e. coincides with the closure of its interior. Then $\Theta_{d-1}(K;\cdot)$ is just an integral over the boundary of K:

$$\Theta_{d-1}(K;\cdot) = \int_{\partial K} \mathbf{1}\{(x, \nu(K,x)) \in \cdot\} \mathcal{H}^{d-1}(dx).$$

Here $\nu(K,x)$ denotes the outer normal of K at the boundary point x, which is uniquely defined for $\mathcal{H}^{d-1}$-a.e. $x \in \partial K$.

8 Poisson Cluster Models

In this section we consider generalized contact distribution functions of grain models which are constructed from stationary Poisson cluster processes with independent marking. Especially, we describe formulas for the local contact distributions of Poisson cluster processes and Boolean models.

8.1 Definition of a Poisson Cluster Model

We continue the discussion of Poisson cluster processes by considering the grain model Z (with convex grains) defined by (4), where Φ is a stationary Poisson cluster process (compare Sect. 6) and $Z_1, Z_2, \ldots$ is a sequence of independent and identically distributed random convex bodies which is independent of $\Phi = \{\xi_i : i = 1, 2, \ldots\}$. Again, we write $\mathbb{Q}$ for the common distribution of the grains and Z_0 for the typical grain, i.e. a random convex body with distribution $\mathbb{Q}$. From Sect. 3 we recall the standard assumption (6) which ensures that $\{Z_i + \xi_i : i = 1, 2, \ldots\}$ is a locally finite particle process. It can be easily shown (compare [8]) that the capacity functional of Z is given by

$$T_Z(C) = 1 - \exp\left\{ - \gamma_p \int \left(1 - \mathbb{E}\left[\prod_{x \in N_0 + y} \mathbb{P}(Z_0 \cap (C - x) = \emptyset)\right]\right) dy \right\} \tag{36}$$

for all $C \in \mathcal{C}^d$. On the other hand, the random closed set Z thus obtained can also be interpreted as a stationary Boolean model with typical (compact, but not necessarily convex) grain

$$\bigcup_{x \in N_0} (Z(x) + x), \tag{37}$$

where $\{Z(x) : x \in N_0\}$ is a family of random convex bodies that are conditionally independent given N_0 and where the conditional distribution of $Z(x)$ given N_0 is $\mathbb{Q}$. In order to check this, one shows that the capacity functional of such a Boolean model is equal to the functional (36). The capacity functional of Z can be used to derive formulas for the contact distribution functions of Z. However, in order to include contact directions, we have to pursue a different approach.

The *reduced second moment measure* of N_0 is the measure on $\mathbb{R}^d$ defined by

$$\alpha_0 := \mathbb{E}\left[\sum_{\substack{x,y \in N_0 \\ x \neq y}} \mathbf{1}\{x - y \in \cdot\} \right]. \tag{38}$$

Throughout the section we will assume that this measure is absolutely continuous.

8.2 The Spherical Contact Distribution of a Poisson Cluster Model

In this subsection we study the generalized spherical contact distribution function defined by (11) with $B = B^d$, that is

$$H_s(r, C) := \mathbb{P}(d(0, Z) \leq r, u(0, Z) \in C \mid 0 \notin Z),$$

where $r \geq 0$ and $C \subset S^{d-1}$ is a Borel set. First we define the means of the support measures with respect to the distribution $\mathbb{Q}$ of the typical grain (reflected at the origin),

$$\bar{\Theta}_j^* := \int \Theta_j(K^*; \cdot)\mathbb{Q}(dK).$$

By our integrability assumption on $\mathbb{Q}$, we thus obtain finite measures on $\mathbb{R}^d \times S^{d-1}$. For given $\psi = \sum_{n=1}^m \delta_{x_n}$, we let $\Gamma(\psi, \cdot)$ denote the distribution of the random closed set $\bigcup_{n=1}^m (Z_n + x_n)$, where $Z_1, \ldots, Z_m$ are independent with distribution $\mathbb{Q}$. We also set $\Gamma(\emptyset, \cdot) := \delta_\emptyset$. Finally, we define

$$\mu_0 := \mathbb{E}\left[\int_{\mathbb{R}^d} \int_{\mathcal{F}^d} \mathbf{1}\{A \in \cdot\} \Gamma((N_0 - y)^{(0)}, dA) N_0(dy) \right],$$

which is a finite measure on $\mathcal{F}^d$. Then

$$H_s(r,C) = \gamma_p \sum_{j=0}^{d-1} \binom{d-1}{j} \int_0^r \int_{\mathbb{R}^d \times C} \int_{\mathcal{F}^d} (1 - H_s(t)) t^{d-1-j} \times \mathbf{1}\{d(-x - tu, A) > t\} \mu_0(dA) \bar{\Theta}_j^*(d(x,u)) dt. \quad (39)$$

This equation has been proved in [20] in the special case $C = \mathbb{R}^d$. For a general measurable set $C \subset \mathbb{R}^d$, (39) is a special case of Proposition 4.27 and Example 4.28 in [14]. Defining

$$\lambda_s(t,C) := \gamma_p \sum_{j=0}^{d-1} \binom{d-1}{j} t^{d-1-j} \int_{\mathbb{R}^d \times C} \int_{\mathcal{F}^d} \times \mathbf{1}\{d(-x - tu, A) > t\} \mu_0(dA) \bar{\Theta}_j^*(d(x,u)), \quad (40)$$

we see that $H_s(\cdot, C)$ has the density $\lambda_s(t,C)(1 - H_s(t))$. The function $\lambda_s(\cdot, C)$ can be interpreted as a direction dependent empty space hazard, because

$$\mathbb{P}(d(0,Z) \in (t, t+h], u(0,Z) \in C \,|\, d(0,Z) \geq t) \approx \lambda_s(t,C)h$$

for small $h > 0$. The choice $C = S^{d-1}$ leads to

$$H_s(r) = 1 - \exp\left\{ - \int_0^r \lambda_s(t) dt \right\}, \quad (41)$$

where $\lambda_s(t) := \lambda_s(t, S^{d-1})$. If Φ is Poisson, then Z is a Boolean model (with typical grain Z_0) and (41) simplifies to (16). Moreover, we have in this case

$$\lambda_s(t,C) = \gamma \sum_{j=0}^{d-1} t^{d-j-1} \binom{d-1}{j} \bar{S}_j^*(C), \quad (42)$$

where

$$\bar{S}_j^* := \int \Theta_j(K^*; \mathbb{R}^d \times \cdot) \mathbb{Q}(dK) = \bar{\Theta}_j^*(\mathbb{R}^d \times \cdot), \quad (43)$$

is the mean of the j-th surface area measure $S_j(K^*; \cdot)$ with respect to $\mathbb{Q}$ and $\gamma = \gamma_p$ since $\gamma_c = 1$.

As $t \to 0$, the inner integral of (40) tends to

$$\int_{\mathcal{F}^d} \mathbf{1}\{x \notin A^*\} \mu_0(dA).$$

Furthermore,

$$\lim_{t \to \infty} t^{1-d} \lambda_s(t,C) = \gamma_p \int_{\mathbb{R}^d \times C} \int_{\mathcal{F}^d} \mathbf{1}\{(H_{u,x} \cup \{x\}) \cap A^* = \emptyset\} \mu_0(dA) \bar{\Theta}_0^*(d(x,u)),$$

where $H_{u,x} := \{x + y : y \in \mathbb{R}^d, \langle y, u \rangle > 0\}$.

8.3 Examples

We now discuss cluster models based on the two special cluster processes introduced in Sects. 6.3 and 6.4. First we consider the typical cluster given by (28). In order to guarantee the absolute continuity of the measure (38) we assume that the conditional distribution of $Y_1 - Y_2$ given that $\eta = 2$ is absolutely continuous. Equation (40) takes the form

$$\begin{aligned}\lambda_s(t,C) &= \gamma_p \mathbb{P}(\eta = 1) \sum_{j=0}^{d-1} \binom{d-1}{j} t^{d-1-j} \bar{S}_j^*(C) \\ &+ \gamma_p \mathbb{P}(\eta = 2) \sum_{j=0}^{d-1} \binom{d-1}{j} t^{d-1-j} \\ &\times \int_{\mathbb{R}^d \times C} \{\mathbb{P}((Z_1 + Y_2 - Y_1) \cap B^d(-x - tu, t) = \emptyset \mid \eta = 2) \\ &+ \mathbb{P}((Z_2 + Y_1 - Y_2) \cap B^d(-x - tu, t) = \emptyset \mid \eta = 2)\} \bar{\Theta}_j^*(d(x,u)),\end{aligned} \tag{44}$$

where Z_1, Z_2 have distribution $\mathbb{Q}$ and $Y_1 - Y_2, Z_i$ are independent for $i = 1, 2$.

Next we consider a Neyman–Scott process Φ as defined at (29) and assume that the distribution of $Y_{1,2} - Y_{2,2}$ is absolutely continuous. The measure (38) is then absolutely continuous and we obtain similarly as above

$$\begin{aligned}\lambda_s(t,C) &= \gamma_p \sum_{j=0}^{d-1} \binom{d-1}{j} t^{d-j-1} \int_{\mathbb{R}^d \times C} \int_{\mathbb{R}^d} \\ &\times g'(\mathbb{P}((Z_0 + Y_0 - y) \cap B^d(-x - tu, t) = \emptyset)) \mathbb{V}(dy) \bar{\Theta}_j^*(d(x,u)),\end{aligned} \tag{45}$$

where Y_0 has the distribution $\mathbb{V}$ of the $Y_{i,n}$ and is independent of the typical grain Z_0. This corrects a statement in Example 5.3 of [20].

Formulas (44) and (45) can be used to compute $\lambda_s(t) = \lambda_s(t, S^{d-1})$ via Monte-Carlo integration. Clearly this requires more specific assumptions on the typical grain. A convenient and popular choice is a spherical grain $Z_0 = \tau B^d$, where τ is a positive random variable. In [20], it has been shown (for $d = 3$) that the empty space hazard might be an appropriate tool to reveal clustering phenomena. However, if the typical cluster N_0 contains at most two points (as in the Gauss-Poisson case, for instance) then the empty space hazard can very much resemble that of a Boolean model. If, on the other hand, N_0 contains three or even more points, then the empty space between the grains of the cluster cannot be neglected anymore and a quadratic function cannot be fitted to the rate in a satisfactory manner.

8.4 General Gauge Bodies

Now we turn to the generalized contact distribution of Z with respect to a general gauge body B. Since for our analysis we have to use relative support measures with respect to B, we now make the further asumption that B^* and the typical grain Z_0 are almost surely in general relative position. If B is strictly convex, or if $\mathbb{Q}$ is concentrated on the set of strictly convex bodies, then this assumption is automatically satisfied. Extending a previous definition, we set

$$\bar{\Theta}_j^*(B;\cdot) := \int \Theta_j(K^*;B;\cdot)\mathbb{Q}(dK), \quad j=0,\ldots,d-1,$$

and obtain finite measures on $\mathbb{R}^d \times \partial B$. As a generalization of (39), we have

$$\begin{aligned} H_B(r,C) &= \gamma_p \sum_{j=0}^{d-1} \binom{d-1}{j} \int_0^r \int_{\mathbb{R}^d \times C} \int_{\mathcal{F}^d} (1-H_B(t))t^{d-1-j} \\ &\quad \times \mathbf{1}\{d_B(-x-tu,A)>t\}\mu_0(dA)\bar{\Theta}_j^*(B;d(x,u))dt, \end{aligned} \tag{46}$$

for all $r \geq 0$ and Borel sets $C \subset \partial B$. For smooth B this has been proved in [14]. The present more general result can be obtained by combining a special case of Theorem 5.1 in [15] with Example 4.28 in [14]. In Sect. 10, we will describe a further extension to instationary models.

The special case $Z_0 = \{0\}$ of (39) deserves some discussion. In that case, $Z = \Phi$ and

$$\begin{aligned} H_B(r,C) = \gamma_p &\int_0^r \int_C (1-H_B(t))t^{d-1} \\ &\times \mathbb{E}\left[\int_{\mathbb{R}^d} \mathbf{1}\{d_B(-tu,(N_0-y)^{(0)})>t\}N_0(dy)\right] \Theta_B(du)dt, \end{aligned} \tag{47}$$

where

$$\Theta_B := d\int_B \mathbf{1}\{x/d_B(0,x) \in \cdot\}dx. \tag{48}$$

This is just the measure $\Theta_0(\{0\};B;\{0\}\times\cdot)$ with total mass $d \cdot V_d(B)$. If B is lower dimensional, then Θ_B is the zero measure and the right side of (47) vanishes identically, in accordance with $\mathbb{P}(d_B(0,\Phi)<\infty)=0$. In case $B=B^d$ we may use polar coordinates to see that Θ_{B^d} is the Hausdorff measure $\mathcal{H}^{d-1}$ on S^{d-1}. More generally, for any $B \in \mathcal{K}^d$ with 0 as an interior point one has

$$\Theta_B = \int_{\partial B} \mathbf{1}\{u \in \cdot\}h_B(\nu(B,u))\mathcal{H}^{d-1}(du).$$

Suppose that the expectation

$$\mathbb{E}\left[\int_{\mathbb{R}^d} \mathbf{1}\{d_B(-tu,(N_0-y)^{(0)})>t\}N_0(dy)\right]$$

does not depend on $u \in \partial B$ for all $t>0$. This is the case if the cluster distribution is *isotropic*, i.e. invariant under rotations around the origin. Equation (47) then implies the independence of $d_B(0,\Phi)$ and $u_B(0,\Phi)$. In the Poisson case (where isotropy is fulfilled automatically),

$$H_B(r,C) = \gamma \Theta_B(C) \int_0^r (1 - H_B(t)) t^{d-1} dt, \tag{49}$$

or

$$H_B(r,C) = (dV_d(B))^{-1} \Theta_B(C)(1 - \exp[-\gamma r^d V_d(B)]).$$

Other special cases of (46) and (47) can be discussed as before.

9 General Stationary Random Sets

In this section, we consider a general stationary random $\mathcal{S}^d$-set Z and discuss the information contained in the contact distribution function H_B of Z, where B is a general gauge body. Even under these weak assumptions, we can deduce some regularity properties of H_B.

9.1 The General Form of Direction Dependent Contact Distributions

As explained in Sect. 3 we may represent Z as a stationary grain model $Z = \bigcup_{n\in\mathbb{N}}(Z_n + \xi_n)$ with underlying stationary marked point process $\Psi = \{(\xi_n, Z_n) : n \in \mathbb{N}\}$ on $\mathbb{R}^d \times \mathcal{K}^d$. Since Ψ is stationary, the intensity measure $\tilde{\Lambda}$ of Ψ can be decomposed in the form $\tilde{\Lambda} = \gamma V_d \otimes \mathbb{Q}$, where γ is the intensity and $\mathbb{Q}$ is the grain distribution of Ψ (compare Sect. 3.3). We assume that the structuring element B and K^* are in general relative position for $\mathbb{Q}$-almost all K and that Ψ satisfies a weak regularity condition (for details, see [14, 15]). Then we obtain

$$\begin{aligned}(1-p)H_B(r,C) = \gamma \sum_{j=0}^{d-1} \binom{d-1}{j} \int_0^r \int_{\mathcal{K}^d} \int_{\mathbb{R}^d \times C} t^{d-1-j} \\ \times \mathbb{P}^{(0,K)}(\{\psi : d_B(-y - tu, Z(\psi, K)) > t\}) \Theta_j(K^*; B; d(y,b)) \mathbb{Q}(dK) dt,\end{aligned} \tag{50}$$

where

$$Z(\psi, K) := \bigcup_{(x,L)\in\psi\setminus\{(0,K)\}} (L + x),$$

and where the probability measures $\mathbb{P}^{(0,K)}$, $K \in \mathcal{K}^d$, are the Palm distributions of Ψ. The latter can be interpreted as conditional distributions of Ψ given that $(0, K) \in \Psi$; see §3.4 and §4.3 in [39]. In [15], a more general result was deduced without the assumption of stationarity. The statement of such a result requires the general Palm distributions $\mathbb{P}^{(x,K)}$, $(x, K) \in \mathbb{R}^d \times \mathcal{K}^d$, of Ψ. Relation (50) can be deduced either directly or from the more general result by means of

$$\mathbb{P}^{(x,K)} = \int \mathbf{1}\{\psi + x \in \cdot\} \mathbb{P}^{(0,K)}(d\psi)$$

which holds for $\tilde{\Lambda}$-a.e. $(x, K) \in \mathbb{R}^d \times \mathcal{K}^d$, where $\psi + x := \{(y + x, K) : (y, K) \in \psi\}$. From equation (50) we can deduce several further consequences. A first observation is that (50) implies that $H_B(\cdot, C)$ is absolutely continuous and a version of the density is given explicitly. In particular, we see that the (right) derivative of $H_B(\cdot, C)$ at 0 exists and can be expressed in the form

$$(1-p)H'_B(0,C) = \gamma \int_{\mathcal{K}^d} \int_{\mathbb{R}^d \times C} \mathbb{P}^{(0,K)}(\{\psi : -y \notin Z(\psi, K)\}) \\ \times \Theta_{d-1}(K^*; B; d(y,b))\mathbb{Q}(dK)dt.$$

A related version of such a result is stated as Proposition 5.4 in [14], the additional assumptions on B (such as smoothness), which are adopted in [14], can be removed. The absolute continuity of H_B was already shown in [2] and [7] for rather general random closed sets, though in a less explicit and in a nonlocal form.

9.2 First Derivatives and Surface Intensities

In the remaining part of this section we assume that B is strictly convex and 0 is in the interior of B. Using the mean of the support measure $\Theta_{d-1}(Z; B^*; \cdot)$ (see Sect. 7.3) the derivatives $H'_B(0, C)$, $C \subset \partial B$, can be expressed as

$$(1-p)H'_B(0,C) = \mathbb{E}\,\Theta_{d-1}(Z; B^*; [0,1]^d \times C^*). \tag{51}$$

This formula holds without any reference to a marked point process Ψ from which Z may be derived. The right-hand side of this formula can be interpreted as a direction dependent (relative) *surface density* of Z. Since $\mathbb{E}\,\Theta_{d-1}(Z; B^*; \cdot \times \partial B^*)$ is a locally finite (by assumption), translation invariant, non-negative measure, it follows that

$$\mathbb{E}\,\Theta_{d-1}(Z; B^*; \cdot \times \partial B^*) = \lambda_{d-1;B} V_d,$$

where the constant $\lambda_{d-1;B} := \mathbb{E}\,\Theta_{d-1}(Z; B^*; [0,1]^d \times \partial B^*)$ is the (relative) *surface density* of Z. (By assumption this number is finite.) Hence

$$H'_B(0) = \lambda_{d-1;B}/(1-p). \tag{52}$$

In case $B = B^d$, the number $\lambda_{d-1;B}$ is just the classical surface density of Z.

Define $C^d := [0,1]^d$ and let $\partial^+ C^d$ denote the upper right boundary of C^d. Then, under a suitable integrability assumption,

$$\mathbb{E}\,\Theta_{d-1}(Z; B^*; [0,1]^d \times \cdot) = \lim_{r\to\infty} \frac{\mathbb{E}\,S_{d-1}(Z \cap rW; B^*; \cdot)}{V_d(rW)} \tag{53}$$
$$= \mathbb{E}\,S_{d-1}(Z \cap C^d; B^*; \cdot) - \mathbb{E}\,S_{d-1}(Z \cap \partial^+ C^d; B^*; \cdot)$$

for any $W \in \mathcal{K}^d$ with $V_d(W) > 0$; moreover, if Z is ergodic, then the expectation on the right side of (53) can be omitted and we obtain $\mathbb{P}$-a.s. equality. Furthermore, we have

$$(1-p)H'_B(0) = \int h(B^*, u)\mathbb{E}\,\Theta_{d-1}(Z; [0,1]^d \times du), \tag{54}$$

which is a consequence of (35) and (51) that will be used below. All these assertions follow from results which were obtained in a general non-stationary setting in [14],

Sects. 4 and 5, and from [13], Sect. 3. Finally, we remark that if Z is assumed to be ergodic, then by a modification of the argument for Theorem 3.4 in [13] we get $\mathbb{P}$-a.s.

$$(1-p)H'_B(0) = \lim_{r\to\infty} \limsup_{\epsilon\downarrow 0} \frac{V_d([(Z+\epsilon B^*)\setminus Z]\cap rW)}{\epsilon V_d(rW)}. \tag{55}$$

Relation (55) remains true if '$\limsup$' is replaced by '$\liminf$'.

9.3 Mean Normal Measure and Dilation Volumes of Stationary Boolean Models

We now assume that Z is a stationary Boolean model as defined in Sect. 5.2. In this case relation (50) and Slivnyak's theorem for marked point processes (see Theorem 3.4.9 in [39]) directly yield that

$$H_B(r,C) = \gamma \sum_{j=0}^{d-1} \binom{d-1}{j} \bar{S}_j^*(B;C) \int_0^r t^{d-1-j}(1-H_B(t))dt, \tag{56}$$

where

$$\bar{S}_j^*(B;\cdot) := \bar{\Theta}_j^*(B;\mathbb{R}^d\times\cdot) = \int \Theta_j(K^*;B;\mathbb{R}^d\times\cdot)\mathbb{Q}(dK)$$

are the relative versions of the measures defined by (43). Relation (56) can also be obtained as a special case of equation (46). A more general result for a stationary Boolean model Z with grains in $\mathcal{R}^d$ is contained in [13], Theorem 2.4. In particular, (56) implies that

$$H'_B(0) = \gamma \bar{S}_{d-1}^*(B;\partial B) = \gamma \int h(B,u)\bar{S}_{d-1}^*(du), \tag{57}$$

which can also be deduced from equation (18). Moreover, since $Z+tB$ is a stationary Boolean model, for all $t\geq 0$, and therefore ergodic, we get $\mathbb{P}$-a.s.

$$H_B(t) = \lim_{r\to\infty} \frac{V_d([(Z+tB^*)\setminus Z]\cap rW)}{V_d(rW\setminus Z)}, \tag{58}$$

for any $t\geq 0$ and an arbitrary $W\in\mathcal{K}^d$ with $V_d(W)>0$. As in (19) we let $\bar{S}_{d-1}$ denote the mean normal measure of the typical grain of a stationary marked Poisson process Ψ from which Z is derived. Then, combining (57) and (58), we obtain for an arbitrary gauge body B with $0\in B$ that $\mathbb{P}$-a.s.

$$\gamma \int h(B,u)\bar{S}_{d-1}(du) = \lim_{\epsilon\downarrow 0}\lim_{r\to\infty} \frac{V_d([(Z+\epsilon B)\setminus Z]\cap rW)}{\epsilon V_d(rW\setminus Z)}. \tag{59}$$

On the other hand, using (54), (55) and Theorem 4.1 in [49], we also get that $\mathbb{P}$-a.s.

$$\gamma \int h(B,u)\bar{S}_{d-1}(du) = \lim_{r\to\infty}\limsup_{\epsilon\downarrow 0} \frac{V_d([(Z+\epsilon B)\setminus Z]\cap rW)}{\epsilon V_d(rW\setminus Z)}, \tag{60}$$

for any $W\in\mathcal{K}^d$ with $V_d(W)>0$. Equation (60) holds for all convex bodies B which contain the origin as a relative interior point; moreover, '$\limsup$' can be replaced by 'lim inf' in (60). Note that the limits on the right sides of (59) and (60) are taken in reversed order. Since the dilation volumes on the right sides of these equations can be determined in principle from observations of the Boolean model for all admissible convex bodies B, the oriented mean normal measure is determined. It is obvious now that these relations can be used to estimate the (unoriented) mean normal distribution of the underlying stationary Poisson particle process X, but the problems with the inversion of the Cosine transform which we mentioned in Sect. 5 occur here as well.

9.4 Second Derivatives

Assume that Z is a locally finite union of segments in $\mathbb{R}^d$, $d \geq 3$. (Such random sets Z are also called segment processes, see e.g. [42].) Then the first derivative $H_s'(0, C)$ vanishes, as follows from (52). However, for $d = 3$, Corollary 5.2 in [14] implies that

$$H_s''(0, C) = 2\pi L_V \mathcal{R}_1(C).$$

Here, L_V is the length density of Z and $\mathcal{R}_1(\cdot)$ is the (normalized) rose of directions of Z. Although not yet available in the literature, this result remains true for the union sets of stationary fibre processes in $\mathbb{R}^3$ (i.e. processes of curves with suitable regularity properties). Analogous results should hold for k-dimensional random sets Z, $k \in \{2, \ldots, d-2\}$.

For full-dimensional grains in $\mathbb{R}^d$ the situation becomes more complicated. In [23], a further step was made by considering the second derivative $H_s''(0)$ of the spherical contact distribution at 0. As the authors show, for stationary grain models with smooth convex grains, and under some mild conditions on the size and relative position of the grains, the derivative $H_s''(0)$ exists (in the sense of Radon–Nikodym), but does not coincide with one of the standard functionals of Z. More precisely,

$$H_s''(0) = \frac{2\pi}{1-p}(\lambda_{d-2}^+ - \mu_{d-2}),$$

where λ_{d-2}^+ is the intensity of the non-negative $(d-2)$-nd curvature measure of Z (see [14, 21]) and μ_{d-2} is a corresponding intensity of a measure which integrates the tangens of the 'outer' angle over all singular boundary points of Z (see [23]). A heuristic discussion of the second derivative of the parallel volume $V_2(K + rB^2)$, $K \in \mathcal{R}^2$, is given in [25].

10 The Instationary Case

10.1 Introduction

So far we mainly considered stationary point processes. If Φ is an arbitrary point process, then we can still define direction dependent contact distribution functions with respect to a structuring element B by

$$H_B(x, r, C) := \mathbb{P}(d_B(x, \Phi) \leq r, u_B(x, \Phi) \in C \mid x \notin \Phi), \tag{61}$$

where $r \geq 0$ and $C \subset \partial B$ is a Borel set, provided that $\mathbb{P}(x \notin \Phi) > 0$ and $u_B(x, \Phi)$ is well-defined $\mathbb{P}$-a.s. Notice that we now take into account the location $x \in \mathbb{R}^d$, i.e. we consider $x + tB$, $t > 0$, as the growing set. In a similar way the contact distribution of an arbitrary random closed set is defined. In contrast to the stationary case, these functions depend on the choice of a reference point $x \in \mathbb{R}^d$. They can be studied by means of the family of Palm distributions of Φ (see [14]).

10.2 The Instationary Poisson Process

A general Poisson process Φ on $\mathbb{R}^d$ is a point process which has the same independence properties as a stationary Poisson process (i.e. for pairwise disjoint Borel sets $A_1, \dots, A_k \subset \mathbb{R}^d$, the random variables $\Phi(A_1), \dots, \Phi(A_k)$ are independent) and which satisfies

$$\mathbb{P}(\Phi(A) = m) = \frac{\Lambda(A)^m}{m!} \exp[-\Lambda(A)], \quad m = 0, 1, \dots,$$

for all Borel sets $A \subset \mathbb{R}^d$, where $\Lambda(A) = \mathbb{E}\,\Phi(A)$ is the mean number of points of Φ in the set A. Λ is a measure on $\mathbb{R}^d$, which is called the *intensity measure* of Φ. Since we have assumed from the very beginning that $\Phi(\{x\}) \leq 1$ for all $x \in \mathbb{R}^d$, it follows that the intensity measure Λ is diffuse, i.e. that $\Lambda(\{x\}) = 0$ for all $x \in \mathbb{R}^d$. If the intensity measure Λ of Φ is a multiple of Lebesgue measure, then Φ is a stationary Poisson process.

Assume now that Φ is a Poisson process with an absolutely continuous intensity measure Λ. Hence, for all Borel sets $A \subset \mathbb{R}^d$,

$$\Lambda(A) = \int_A \gamma(x)dx, \tag{62}$$

where $\gamma : \mathbb{R}^d \to [0, \infty)$ is a measurable function, which may be called the *intensity function* of Φ. This assumption ensures that $\mathbb{P}(x \in \Phi) = 0$ and that $u_B(x, \Phi)$ is almost surely uniquely determined; moreover,

$$H_B(x, r, C) = \int_0^r \int_C (1 - H_B(x, t)) t^{d-1} \gamma(x + tu) \Theta_B(du) dt, \tag{63}$$

where $H_B(x, t) := H_B(x, t, \partial B)$ and the measure Θ_B has been defined by (48).

10.3 General Poisson Processes and Boolean Models

A Poisson process can be defined on a general metric space $\mathbf{X}$. Given a diffuse and σ-finite measure Λ on $\mathbf{X}$, the definition in the preceding subsection applies verbatim.

Consider a Poisson process Φ on $\mathbb{R}^d$ with intensity measure Λ and let $\Psi = \{(\xi_n, Z_n) : n \in \mathbb{N}\}$ be an independent marking of Φ with mark distribution $\mathbb{Q}$ on $\mathcal{C}^d$. It is one of the nice and fundamental properties of the Poisson process (see e.g. [19, 39]) that Ψ is a Poisson process on $\mathbb{R}^d \times \mathcal{C}^d$ with intensity measure $\tilde{\Lambda} = \Lambda \otimes \mathbb{Q}$. From the mapping theorem for Poisson processes (see [19]) we obtain that the particle process $X := \{Z_n + \xi_n : n \in \mathbb{N}\}$ is a Poisson process on $\mathcal{C}^d$ (provided X has a locally finite intensity measure). It is natural to generalize the definition in Sect. 5.2 and to call

$$Z := \bigcup_{n\in\mathbb{N}} (Z_n + \xi_n) \tag{64}$$

again a (inhomogeneous) Boolean model.

Motivated by the preceding observations we can define a general Boolean model as $Z = \cup_{K\in X} K$, where X is a *Poisson particle process* , i.e. a Poisson process on $\mathcal{C}^d$ with a locally finite intensity measure. Alternatively (and almost equivalently) we can start with a Poisson process $\Psi = \{(\xi_n, Z_n) : n \in \mathbb{N}\}$ on $\mathbb{R}^d \times \mathcal{C}^d$, whose intensity measure $\tilde{\Lambda}$ satisfies (5), and define a Boolean model by (64). We take the second approach. The capacity functional of a general Boolean model Z is given by the formula

$$T_Z(C) = 1 - \exp\left\{-\int_{\mathbb{R}^d\times\mathcal{C}^d} \mathbf{1}\{(K+y)\cap C \neq \emptyset\}\tilde{\Lambda}(d(y,K))\right\}. \tag{65}$$

Therefore the distribution of Z determines the intensity measure (restricted to $\mathcal{C}^d \setminus \{\emptyset\}$) and hence the distribution of the associated Poisson particle process $X = \{Z_n+\xi_n : n \in \mathbb{N}\}$ (assuming that X has a diffuse intensity measure). However it does not determine the intensity measure of Ψ. For an instationary random closed set Z, the volume fraction is defined as a function of $x \in \mathbb{R}^d$ by

$$p(x) := \mathbb{P}(x \in Z) = T_Z(\{x\}).$$

Spezializing (65), we thus obtain for a Boolean model Z that

$$p(x) = 1 - \exp\left\{-\tilde{\Lambda}(\{(y,K) : x \in K+y\})\right\}, \quad x \in \mathbb{R}^d.$$

Let us fix a Boolean model as defined by (64) in terms of a Poisson process $\Psi = \{(\xi_n, Z_n) : n \in \mathbb{N}\}$ having intensity measure $\tilde{\Lambda}$. We assume that the measure $\Lambda := \tilde{\Lambda}(\cdot \times \mathcal{C}^d)$ is locally finite and diffuse. Then $\Phi := \{\xi_n : n \in \mathbb{N}\}$ is a Poisson process with intensity measure Λ. Given Φ, the grains $Z_1, Z_2, \ldots$ are conditionally independent, the conditional distribution being described by a *stochastic kernel* κ from $\mathbb{R}^d$ to $\mathcal{C}^d$. This kernel is defined by

$$\tilde{\Lambda} = \int_{\mathbb{R}^d}\int_{\mathcal{C}^d} \mathbf{1}\{(x,K) \in \cdot\}\kappa(x,dK)\Lambda(dx).$$

Given that x is a point of Φ, the probability measure $\kappa(x,\cdot)$ is the conditional distribution of the grain associated with x. The marked point process is an independent marking of Φ if and only if there is some probability measure $\mathbb{Q}$ (the mark distribution) such that $\kappa(x,\cdot) = \mathbb{Q}$ for Λ-a.e. $x \in \mathbb{R}^d$. In this case we call Z a (inhomogeneous) Boolean model with *independent grains*. This definition depends not only on the distribution of Z, but also on the distribution of Ψ. It is possible that Z can be represented in terms of two Poisson processes, where the first has independent marks and the second not. (In a sense this is the rule rather than the exception.)

Later (in Sect. 10.5) we will assume that

$$\tilde{\Lambda} = \int_{\mathbb{R}^d}\int_{\mathcal{K}^d} \mathbf{1}\{(x,K) \in \cdot\}\gamma(x,K)dx\mathbb{Q}(dK), \tag{66}$$

for some (measurable) function γ on $\mathbb{R}^d \times \mathcal{K}^d$ and a probability measure $\mathbb{Q}$ on $\mathcal{K}^d$. This happens if and only if $\kappa(x, \cdot)$ is for Λ-a.e. $x \in \mathbb{R}^d$ absolutely continuous with respect to $\mathbb{Q}$. Then Ψ has independent marks if and only if $\gamma(x, K)$ is independent of K for $\mathbb{Q}$-a.e. K and Λ-a.e. $x \in \mathbb{R}^d$. In this case γ is the intensity function of the Poisson process Φ.

10.4 Instationary Cluster Models

In this subsection we describe the direction dependent contact distribution function of an instationary Poisson cluster model. Let Φ_p be a Poisson process in $\mathbb{R}^d$ with absolutely continuous intensity measure as in (62) and intensity function γ_p. Let $\{N_x : x \in \Phi_p\}$ be a family of point processes that are conditionally independent given Φ_p and such that the conditional distribution of N_x is the same for all $x \in \Phi_p$. As in the stationary case we may then introduce a *typical cluster* N_0 which satisfies $\mathbb{E}\, N_0(\mathbb{R}^d) < \infty$ and for which the reduced second moment measure of N_0 is absolutely continuous (compare Sect. 8.1). Then we define the cluster process Φ as in (21) and impose the assumption that $\mathbb{E}\,\Phi(C) < \infty$ for all $C \in \mathcal{C}^d$. It is easy to see that

$$\mathbb{E}\,\Phi(A) = \int_A \int_{\mathbb{R}^d} \gamma_p(x-y)\mathbb{E}\, N_0(dy)dx,$$

for Borel sets $A \subset \mathbb{R}^d$, hence the intensity measure of Φ is absolutely continuous. The Poisson cluster model Z is now obtained from Φ by independent marking as in the beginning of Sect. 8. Here we impose the integrability assumption

$$\int_{\mathbb{R}^d} \mathbb{P}((Z_0 + x) \cap C \neq \emptyset)\mathbb{E}\,\Phi(dx) < \infty$$

for all $C \in \mathcal{C}^d$, which is just assumption (5). Similarly to the stationary situation, Z can be interpreted as an inhomogeneous Boolean model with independent grains and a typical grain as in (37). (Again this interpretation can be justified by a comparison of capacity functionals.) From this we can deduce a formula for $H_B(x, r)$. In order to obtain a local result, we have to proceed in a different way and combine results from [14] and [15]. Using the probability measures $\Gamma(\psi, \cdot)$ introduced in Sect. 8.2, we define

$$\mu_0'(\cdot) := \mathbb{E}\left[\int_{\mathbb{R}^d} \int_{\mathbb{R}^d \times \mathcal{F}^d} \mathbf{1}\{(y, A) \in \cdot\}\Gamma((N_0 - y)^{(0)}, dA)N_0(dy)\right].$$

This is a finite measure on $\mathbb{R}^d \times \mathcal{F}^d$. Assuming (as in Sect. 8.4) that B^* and the typical grain Z_0 are almost surely in general relative position, we obtain

$$\begin{aligned} H_B(x, r, C) = \sum_{j=0}^{d-1} \binom{d-1}{j} \int_0^r \int_{\mathbb{R}^d \times C} \int_{\mathbb{R}^d \times \mathcal{F}^d} &(1 - H_B(x, t))t^{d-1-j} \\ &\times \mathbf{1}\{d_B(-z - tu, A) > t\}\gamma_p(x + z + tu - y) \\ &\times \mu_0'(d(y, A))\bar{\Theta}_j^*(B; d(z, u))dt. \end{aligned} \tag{67}$$

This result encompasses all results concerning Poisson cluster models or processes which we mentioned so far. It is a special case of Proposition 4.27 in [14] except that we do not have to assume that B is smooth (see Theorem 5.1 in [15]).

10.5 Direction Dependent Contact Distributions of the General Boolean Model with Convex Grains

We consider a general Boolean model Z as defined in Sect. 10.3. We assume that the underlying Poisson process Ψ is concentrated on $\mathbb{R}^d \times \mathcal{K}^d$ and that it has an intensity measure $\tilde{\Lambda}$ of the form (66). Finally, we fix a gauge body B and assume (as previously) that B^* and the typical grain Z_0 are almost surely in general relative position. Then

$$H_B(x,r,C) = \sum_{j=0}^{d-1} \binom{d-1}{j} \int_0^r \int_{\mathcal{K}^d} \int_{\mathbb{R}^d \times C} (1 - H_B(x,t)) t^{d-1-j} \\ \times \gamma(x+z+tu, K) \Theta_j(K^*; B; d(z,u)) \mathbb{Q}(dK) dt, \tag{68}$$

a result which has been proved in [15]. In the case of a Boolean model with independent grains the function $\gamma(x,K) \equiv \gamma(x)$ does not depend on K and we obtain

$$H_B(x,r,C) = \sum_{j=0}^{d-1} \binom{d-1}{j} \int_0^r \int_{\mathbb{R}^d \times C} (1 - H_B(x,t)) t^{d-1-j} \\ \times \gamma(x+z+tu) \bar{\Theta}_j^*(B; d(z,u)) dt. \tag{69}$$

Clearly, this formula is also a special case of (67), obtained by putting $N_0 = \{0\}$. Equation (63) is recovered if $Z_i = \{0\}$ and $N_0 = \{0\}$ $\mathbb{P}$-a.s.

For a Boolean model with independent grains we now explore the kind of information about $(\gamma, \mathbb{Q})$ which we can retrieve from knowledge of the contact distribution functions of Z. Under suitable continuity and integrability assumptions we obtain from (69) that

$$\frac{H'_B(x,t,C)}{1 - H_B(x,t)} = \sum_{j=0}^{d-1} \binom{d-1}{j} t^{d-1-j} \int_{\mathbb{R}^d \times C} \gamma(x+z+tu) \bar{\Theta}_j^*(B; d(z,u)), \tag{70}$$

where the derivative of $H_B(x,t,C)$ is taken with respect to t. This already indicates that in its present form, the contact distribution functions may determine certain mean values for any choice of a measurable set $C \subset \mathbb{R}^d$ and a strictly convex body $B \in \mathcal{K}^d$. However, in order to make significant further progress and to determine spatial densities or distributions, it seems to be necessary to extend the notion of a contact distribution function by including additional second order information. This subject will be pursued more thoroughly in the next section.

10.6 A Boolean Model with Spherical Grains

The following special but important example shows the sort of information we can hope to extract from the contact distribution in its present form. Let Z be an inhomogeneous Boolean model with independent grains. We assume that the typical grain is of the (randomly scaled) form τB^d, where τ has finite moments up to order $d-1$. We also assume that the intensity measure of the germ process has a continuous and bounded density $\gamma(x)$. In this special situation, we can deduce from (70) that the local contact distribution functions $H_{B^d}(x,t,C)$ (actually the local hazard rates), for $x \in \mathbb{R}^d$, $t \geq 0$ and measurable sets $C \subset S^{d-1}$, determine

$$\sum_{j=0}^{d-1} \binom{d-1}{j} t^{d-1-j} \mathbb{E}\left[\tau^j \gamma(x - \tau u - tu)\right] \tag{71}$$

for all $x \in \mathbb{R}^d$, $t \geq 0$ and $u \in S^{d-1}$. Setting $t = 0$, we find that $\mathbb{E}\left[\tau^{d-1}\gamma(x - \tau u)\right]$ is determined for all $x \in \mathbb{R}^d$ and $u \in S^{d-1}$. Replacing in this expression x by $x - tu$, we find that $\mathbb{E}\left[\tau^{d-1}\gamma(x - \tau u - tu)\right]$ is determined as well. Hence we can conclude that

$$\sum_{j=0}^{d-2} \binom{d-1}{j} t^{d-2-j} \mathbb{E}\left[\tau^j \gamma(x - \tau u - tu)\right]$$

is also determined for all $x \in \mathbb{R}^d$, $t \geq 0$ and $u \in S^{d-1}$. This argument can be repeated until we finally get that $\mathbb{E}\left[\gamma(x - \tau u)\right]$ is determined for all $x \in \mathbb{R}^d$ and $u \in S^{d-1}$. Thus we obtain at least some information on γ of convolution type. Assuming additionally that γ is constant along all lines of a known direction $u_0 \in S^{d-1}$, we can infer that γ and the moments $\mathbb{E}\,\tau^j$, $j = 1, \ldots, d-1$, are determined. Another simple case where γ is determined arises if τ is deterministic and known. We remark that nonspherical convex randomly scaled grains can be treated similarly, if an appropriate gauge body B is chosen instead of the unit ball B^d. In general, however, we cannot expect to determine γ or the distribution of τ, on the basis of the information which is encoded in the kind of contact distribution functions we have considered up to now.

11 Smooth Boolean Models and Second Order Information

11.1 Smooth Boolean Models

The example at the end of the preceding section shows that even for a very special Boolean model Z, the contact distribution functions studied so far do not contain enough information to determine all characteristic quantities of Z. Therefore we extend the notion of a contact distribution again by including additional, second order information about Z in its definition. At the same time, we will restrict our attention to certain classes of Boolean models which are still sufficiently general to describe many situations which are considered in practice.

For the following considerations, we concentrate on (non-stationary) Boolean models Z with independent convex grains which fulfill certain smoothness conditions. We assume that the intensity measure $\tilde{\Lambda}$ of the underlying Poisson process Ψ has the form

$$\tilde{\Lambda} = \int_{\mathcal{K}^d} \int_{\mathbb{R}^d} \mathbf{1}\{(x,K) \in \cdot\}\gamma(x)dx\,\mathbb{Q}(dK), \tag{72}$$

with the spatial intensity function γ and the distribution $\mathbb{Q}$ of the typical grain. In addition, we assume that

(i) γ is bounded and continuous,
(ii) K is of class C^2_+ for $\mathbb{Q}$-a.e. $K \in \mathcal{K}^d$.

We then call Z a *smooth Boolean model.*

11.2 Contact Distributions with Local Information

Let Z be a smooth Boolean model and $x \in \mathbb{R}^d$. For almost all realizations of Z such that $x \notin Z$ there is a unique boundary point (*contact point*) $y \in Z$ such that $d(x, Z) = d(x, y)$ and y lies in the boundary of a unique particle $K(x, Z) = L + z$, $(z, L) \in \Psi$. Note that $L + z$ has outer normal $-u(x, Z)$ at y. Then we define

$$\tilde{s}(x, Z) := (\tilde{s}_1(x, Z), \ldots, \tilde{s}_{d-1}(x, Z)) \in \mathbb{R}^{d-1},$$

where $\tilde{s}_j(x, Z) := s_j(K(x, Z), -u(x, Z))$ is the j-th normalized elementary symmetric function of the principal radii of curvature of the convex body $K(x, Z)$ in direction $-u(x, Z)$. The (extended) generalized contact distribution function of Z is now given as

$$H(x, r, C \times D) := \mathbb{P}(d(x, Z) \leq r, u(x, Z) \in C, \tilde{s}(x, Z) \in D \mid x \notin Z), \qquad (73)$$

for all $x \in \mathbb{R}^d$, $r \geq 0$ and measurable sets $C \subset S^{d-1}$, $D \subset \mathbb{R}^{d-1}$. Extended generalized contact distributions of a similar kind have been introduced and studied for much more general random closed sets Z; see [15].

Knowing all the conditional probabilities in (73) is equivalent to knowing the conditional expectations

$$\mathbb{E}\left[\mathbf{1}\{d(x, Z) \leq r\} h(u(x, Z), \tilde{s}(x, Z)) \mid x \notin Z\right]$$

for all $x \in \mathbb{R}^d$ and all measurable functions $h : S^{d-1} \times \mathbb{R}^{d-1} \to [0, \infty)$. In addition to distributional information about the position of the contact point in Z realizing the distance from x to Z, these conditional probabilities and expectations also contain second order information about Z in a (arbitrarily small) neighbourhood of this contact point (local information). In the remaining part of this section, we describe the extent to which γ and $\mathbb{Q}$ are determined by the contact distribution functions of the form (73), i.e. we study the sort of information about the characteristics of X which is implicitly included in (73).

The crucial relationship which we use for our analysis is

$$\begin{aligned} &\mathbb{E}\left[\mathbf{1}\{d(x, Z) \leq r\} h(u(x, Z), \tilde{s}(x, Z)) \mid x \notin Z\right] \\ &\quad = \sum_{j=0}^{d-1} \binom{d-1}{j} \int_0^r t^{d-1-j}(1 - H(x, t)) \int_{\mathcal{K}^d} \int_{S^{d-1}} h(-u, s(K, u)) \qquad (74) \\ &\qquad \times \gamma(x - tu - \tau(K, u)) s_j(K, u) \mathcal{H}^{d-1}(du) \mathbb{Q}(dK) dt, \end{aligned}$$

where $s(K, u) := (s_1(K, u), \ldots, s_{d-1}(K, u)) \in \mathbb{R}^{d-1}$. More general formulas, relating conditional expectations of locally defined functionals of Z to mean values of γ (of convolution type) with respect to $\mathbb{Q}$, have been established in [15]. For instance, as in the preceding sections one can consider other gauge bodies than the unit ball, drop the smoothness assumptions on $(\gamma, \mathbb{Q})$, use local functionals of Z different from $\tilde{s}(x, Z)$, and one can even avoid the Poisson assumption on Ψ. Relation (74) implies that the contact distribution functions (73) determine the mean values

$$\int_{\mathcal{K}^d} \gamma(x - \tau(K,u)) g(s(K,u)) \mathbb{Q}(dK), \tag{75}$$

for all $x \in \mathbb{R}^d$, $u \in S^{d-1}$ and all measurable functions $g : \mathbb{R}^{d-1} \to [0,\infty)$. Subsequently, we consider the difficult inverse problem of retrieving information about $(\gamma, \mathbb{Q})$ from these mean values.

11.3 Independent Marking with Respect to Tangent Points

A situation of a very special nature is given if there is a direction $u_0 \in S^{d-1}$ such that the mark distribution $\mathbb{Q}$ of Ψ is concentrated on the set $\{K \in \mathcal{K}^d : \tau(K, u_0) = 0\}$. This situation corresponds to the choice of the center map $c(K) = \tau(K, u_0)$, $K \in \mathcal{K}^d$, (compare Sect. 3.3); hence, $c(K)$ is the unique 'tangent point' of K with exterior unit normal u_0. Note that $c(K)$ is uniquely defined for $\mathbb{Q}$-a.e. $K \in \mathcal{K}^d$. Then, for the choice $u = u_0$, the expression in (75) simplifies to

$$\gamma(x) \int_{\mathcal{K}^d} g(s(K, u_0)) \mathbb{Q}(dK),$$

for all $x \in \mathbb{R}^d$ and measurable functions $g : \mathbb{R}^{d-1} \to [0,\infty)$, since $\tau(K, u_0) = 0$ for $\mathbb{Q}$-almost all $K \in \mathcal{K}^d$. Therefore the spatial density γ and the distribution $\mathbb{Q}(\{K : s(K, u_0) \in \cdot\})$ are uniquely determined by (73). A particular case arises if Z is stationary, since then we may choose $c(K) = \tau(K, u)$, for any $u \in S^{d-1}$. In this case, we thus find that the intensity γ of X and the distribution of the function $s(Z_0, u)$ of the typical grain Z_0 of Ψ are determined by (73). This result then holds for any other center map as well since the intensity γ remains the same and the grain distribution given for the new center map is a well-defined image of $\mathbb{Q}$. (Note that $s_j(K, u) = s_j(K + x, u)$ for all $K \in \mathcal{K}^d$ of class C^2_+, $x \in \mathbb{R}^d$ and $u \in S^{d-1}$.)

Note that for a stationary Boolean model, it is known that the point process of tangent points determines all parameters of the Boolean model (see [28, 30]). However, this point process is not used directly in the preceding investigation.

11.4 The Case of Integrable Intensity Functions

We return to a general smooth Boolean model, but assume now that γ is integrable. Then, integrating with respect to $x \in \mathbb{R}^d$, using Fubini's theorem and exploiting the translation invariance of Lebesgue measure, we see that (75) implies that the distribution

$$\mathbb{Q}(\{K : s(K,u) \in \cdot\}) \tag{76}$$

is determined for all $u \in S^{d-1}$. As a simple consequence we obtain that the mean surface measures $\bar{S}_j$ of the particles are determined by (73), since

$$\int_{\mathcal{K}^d} S_j(K, C) \mathbb{Q}(dK) = \int_C \int_{\mathcal{K}^d} s_j(K,u) \mathbb{Q}(dK) \mathcal{H}^{d-1}(du),$$

for all measurable sets $C \subset S^{d-1}$. The above argument can be extended to the class of intensity functions which are integrable on a subspace and enjoy a partial invariance property with respect to the orthogonal subspace; see [15] for a precise statement.

11.5 Homothetic Particles

Another special case where we obtain further results is that of a smooth Boolean model (with independent grains) where the particles are multiples of certain fixed shapes and the spatial intensity function γ is integrable.

In a first example of that kind, we consider a typical grain Z_0 of the form

$$Z_0 = \tau\, Z_0', \tag{77}$$

where $\tau > 0$ is a random variable and Z_0' is a random convex body (of class C_+^2) with $Z_0' \in \{K_1, \ldots, K_m\}$. In addition, we assume that there is some $i \in \{1, \ldots, m\}$ such that

$$\mathbb{P}(Z_0' = K_i) > 0 \quad \text{and} \quad \mathbb{P}(\tau \le \epsilon \mid Z_0' = K_i) > 0, \tag{78}$$

for all $\epsilon > 0$. This condition means that arbitrarily small scaled copies of a particular shape can be observed with positive probability. Under these assumptions, not only the distribution in (76), but also the spatial density γ is uniquely determined by (73).

A second example is given by (77) with $Z_0' \in \{K_1, K_2\}$, where K_1, K_2 are two convex bodies that cannot be transformed into each other by translations and dilatations. Set $\alpha_i := \mathbb{P}(Z_0' = K_i)$, $i \in \{1, 2\}$, and assume that

$$\limsup_{n \to \infty} \mathbb{E}\left[\tau^n \mid Z_0' = K_i\right]^{\frac{1}{n}} / n < \infty \tag{79}$$

whenever $\alpha_i > 0$. Then, (73) determines α_1, α_2 and $\mathbb{P}(\tau \in \cdot \mid Z_0' = K_i)$ whenever $\alpha_i > 0$. Moreover, if the bodies K_1, K_2 are known, then $\mathbb{Q} = \mathbb{P}(Z_0 \in \cdot)$ is determined as well. Under the additional assumption (78), γ is also determined (compare the first example). Here, for $\alpha_i > 0$, the weak growth condition (79) ensures that the conditional distribution $\mathbb{P}(\tau \in \cdot \mid Z_0' = K_i)$ is determined by its moments $\mathbb{E}\left[\tau^n \mid Z_0' = K_i\right]$, $n \in \mathbb{N}$.

In our final example, we consider a single shape, i.e. we assume that $Z_0 = \tau K_0$, where K_0 is a fixed (but in general unknown) convex body and τ is normalized in the sense that $\mathbb{E}\,\tau = 1$. Then γ, $\mathbb{Q}$ and K_0 are determined by the generalized contact distribution functions of Z of the form (73). In order to illustrate the sort of information which is required on the right side of (73), let us consider the special case where we already know that $K_0 = B^d$ is the Euclidean ball (compare the example at the end of Sect. 10). Hence Z is a Boolean model with spherical grains. Then, for a point $x \notin Z$ we have to determine the direction $u(x, Z)$ of the unit vector pointing from x to the nearest boundary point in Z, which is uniquely determined $\mathbb{P}$-a.s. Further, in this nearest point we have to measure the reciprocals of the principal curvatures of ∂Z in order to get the value of $\tilde{s}_1(x, Z)$ (say); in the case of spherical grains, this value is equal to the radius of the $\mathbb{P}$-a.s. unique spherical particle of Z nearest to x. Assuming that we can perform all these measurements for almost all points $x \notin Z$, we find that the spatial density γ and the radial distribution of the spherical grains are determined by these measurements.

We remark that the above examples are special cases of more general results in [15].

12 Final Remarks

In this section we discuss some further extensions (and modifications) of contact distributions and indicate potential topics of future research. Throughout we fix a stationary random closed set $Z \subset \mathbb{R}^d$.

12.1 General Sampling Schemes

Let $B \in \mathcal{K}^d$, $0 \in B$, be a gauge body. The contact distribution function H_B can be interpreted as the distribution function of the relative distance of a "randomly chosen" point in the complement $\mathbb{R}^d \setminus Z$ of Z from Z. It was proposed in [34] (see also [36]) to generalize this sampling scheme by choosing the reference point as the "typical point" of a random measure ζ on $\mathbb{R}^d$. To give a rigorous definition we assume that ζ has a finite and positive intensity $\mathbb{E}\,\zeta([0,1]^d)$ and that ζ is stationary *jointly* with Z. (The joint distribution of Z and ζ is invariant under translations.) Let $\mathbb{P}^0_\zeta$ denote the *Palm probability measure* of ζ defined on the underlying sample space $(\Omega, \mathcal{A})$. The probability $\mathbb{P}^0_\zeta(A)$ of an event $A \in \mathcal{A}$ can be interpreted as the conditional probability of A given that 0 is a typical point of ζ (see [22] for more formal definitions and further references). We then define

$$H_{B,\zeta}(r) := \mathbb{P}^0_\zeta(d_B(0, Z \setminus \{0\}) \leq r), \quad r \geq 0. \tag{80}$$

Choosing ζ as the Lebesgue measure restricted to the complement of Z, we obtain the classical contact distribution function H_B discussed in this paper. But there are many other interesting special cases. If, for instance, $Z = \zeta$ is a point process, then $H_{B^d,\zeta}$ is the nearest-neighbour distance distribution function of Z. As another example we might choose ζ as the $(d-1)$-dimensional Hausdorff-measure restricted to the exoskeleton $\mathrm{exo}_B(Z)$ of Z. (This requires suitable regularity properties of $\mathrm{exo}_B(Z)$.)

Definition (80) can be further extended by allowing random gauge bodies. A convenient way to accomplish this extension is to use a *random field* $Y = \{Y(x) : x \in \mathbb{R}^d\}$ taking values in the set of all gauge bodies and being stationary jointly with Z and ζ. We may then define

$$H_{Y,\zeta}(r) := \mathbb{P}^0_\zeta(d_{Y(0)}(0, Z \setminus \{0\}) \leq r), \quad r \geq 0. \tag{81}$$

For instance we might assume that the random gauge body $Y(0)$ is under $\mathbb{P}^0_\zeta$ independent of Z. Then $H_{Y,\zeta}$ is just a mixture of the distribution functions $H_{B,\zeta}$. A natural choice in such a setting (discussed in [34]) is $Y(0) = [0, U]$ (or $Y(0) = [-U, U]$), where U is a random element of S^{d-1}. Another special case is to take ζ as the $(d-1)$-dimensional Hausdorff-measure restricted to the boundary of Z and to define $Y(x) = [0, \nu_Z(x)]$, where $\nu_Z(x)$ is the outer normal of Z at the boundary point x. Then $H_{Y,\zeta}$ is the distribution of the distance of a typical boundary point x of Z from $Z \setminus \{x\}$ in outer normal direction. In this case $Y(0)$ and Z are *not* independent under $\mathbb{P}^0_\zeta$.

12.2 Intrinsic Volumes of Parallel Sets

Up to affine transformation (see (9)) the contact distribution function $H_B(r)$ with respect to a gauge body B is the volume fraction of the generalized parallel set

$$Z_{\oplus r} := Z + rB^* .$$

Instead one can also study the means of other geometric functionals applied to (the suitably restricted) random set $Z_{\oplus r}$. An appropriate mathematical concept are the *densities* of additive functionals. Assume that Z is a random $\mathcal{S}^d$-set, let B be a strictly convex gauge body B containing 0 in its interior, and let $W \in \mathcal{K}^d$ with $V_d(W) > 0$. Under suitable integrability conditions the limit

$$H_{B,j}(r, C) = \lim_{t\to\infty} \frac{\mathbb{E}\left[\Theta_j(Z_{\oplus r} \cap tW; B^*; \mathbb{R}^d \times C)\right]}{V_d(tW)} \tag{82}$$

exists (and is finite) for any $j \in \{0, \ldots, d-1\}$, $r \geq 0$, and $C \subset \partial B^*$ (see [13] and [39]). Moreover, $H_{B,j}(r, \cdot)$ is a signed measure on ∂B^*. For $j = d-1$ this measure has been discussed in Sect. 9.2.

The idea to study the behaviour of intrinsic volumes as a function of a threshold-parameter was first formulated in [27] in the context of continuous percolation. Further very fruitful applications of such and related ideas to the analysis of complex spatial structures occuring in statistical physics can be found in [25] and [26] (see also the contributions by H-J. Vogel (Chap. 1), by M. Lösche and P. Krüger (Chap. 2), by C. Beisbart et al. (Chap. 3), and by C. Beisbart, M. Kerscher and K. Mecke (Chap. 4), in this volume). In [16], the functions $H_{s,j}(r) := H_{B^d,j}(r, S^{d-1})$ are used as summary statistics for the analysis of point processes Z composed by the centers of galaxy clusters. Moreover, in [46] it is suggested to analyse means of various geometric functionals of eroded or dilated porous media in order to gain insight into their physical properties or for testing goodness-of-fit of potential models for such structures.

If Z is a (stationary) Boolean model with typical convex grain Z_0, then $Z_{\oplus r}$ is a Boolean model with typical grain $Z_0 + rB^*$. If Z_0 has an isotropic distribution it is possible to use general results for Boolean models (see e.g. [25, 39, 50]) to express the functions $H_{s,j}$ in terms of the intensity of the underlying Poisson process and the mean volumes $\mathbb{E}\, V_j(Z_0)$. In the non-isotropic case the formulas for $H_{s,1}, \ldots, H_{s,d-2}$ become more complicated (see [50, 51]). We are not aware of any other (more or less) explicit analytic results.

References

1. Baddeley, A.J. (1999): 'Spatial sampling and censoring'. In: *Stochastic Geometry: Likelihood and Computation*, ed. by O. Barndorff-Nielsen, W. Kendall, M.N.M. van Lieshout (Chapman & Hall/CRC, Boca Raton), pp. 37–78.
2. Baddeley, A.J., Gill, R.D. (1997): 'Kaplan Meier estimators of distance distributions for spatial point processes'. *Ann. Statist.* **25**, pp. 263–292.
3. Chiu, S.N., Stoyan, D. (1998): 'Estimators of distance distributions for spatial patterns'. *Stat. Neerl.* **52**, pp. 239–246.
4. Daley, D.J., Vere-Jones, D. (1988): *An Introduction to the Theory of Point Processes.* (Springer, New York)
5. Hahn, U., Micheletti, A., Pohlink, R., Stoyan, D., Wendrock, H. (1999): 'Stereological analysis and modelling of gradient structures'. *Journal of Microscopy* **195**, pp. 113–124.

6. Hansen, M.B., Baddeley, A.J., Gill, R.D. (1996): 'Kaplan–Meier type estimators for linear contact distributions'. *Scand. J. Statist.* **23**, pp. 129–155.
7. Hansen, M.B., Baddeley, A.J., Gill, R.D. (1999): 'First contact distributions for spatial patterns: regularity and estimation'. *Adv. in Appl. Probab.* **31**, pp. 15–33.
8. Heinrich, L. (1992): 'On existence and mixing properties of germ-grain models'. *Statistics* **23**, pp. 271–286.
9. Heinrich, L. (1993): 'Asymptotic properties of minimum contrast estimators for parameters of Boolean models'. *Metrika* **40**, pp. 67–94.
10. Heinrich, L. (1998): 'Contact and chord length distribution of a stationary Voronoi tessellation'. *Adv. in Appl. Probab.* **30**, pp. 603–618.
11. Hilfer, R. (2000): 'Local porosity theory and stochastic reconstruction for porous media'. In: *Statistical Physics and Spatial Statistics*, Lecture Notes in Physics **554**, ed. by K. Mecke, D. Stoyan (Springer, Berlin), pp. 203–241.
12. Hug, D. (1999): *Measures, curvatures and currents in convex geometry*, Habilitationsschrift. (Albert-Ludwigs-Universität, Freiburg)
13. Hug, D. (2000): 'Contact distributions of Boolean models'. *Rend. Circ. Mat. Palermo (2) Suppl.* **65**, pp. 137–181.
14. Hug, D., Last, G. (2000): 'On support measures in Minkowski spaces and contact distributions in stochastic geometry'. *Ann. Probab.* **28**, pp. 796–850.
15. Hug, D., Last, G., Weil, W. (2002): 'Generalized contact distributions of inhomogeneous Boolean models'. *Adv. in Appl. Probab.* **34**, pp. 21–47.
16. Kerscher, M. (2000): 'Statistical analysis of large-scale structure in the universe'. In: *Statistical Physics and Spatial Statistics*, Lecture Notes in Physics, ed. by K. Mecke, D. Stoyan (Springer, Berlin), pp. **36–71**.
17. Kiderlen, M. (2001): 'Non-parametric estimation of the directional distribution of stationary line and fibre processes'. *Adv. in Appl. Probab.* **33**, pp. 6–24.
18. Kiderlen, M., Weil, W. (1999): 'Measure-valued valuations and mixed curvature measures of convex bodies'. *Geom. Dedicata* **76**, pp. 291–329.
19. Kingman, J. F. (1995): *Poisson Processes* (Oxford, Clarendon)
20. Last, G., Holtmann, M. (1999): 'On the empty space function of some germ-grain models'. *Pattern Recognition* **32**, pp. 1587–1600.
21. Last, G., Schassberger, R. (1998): 'On the distribution of the spherical contact vector of stationary germ-grain models'. *Adv. in Appl. Probab.* **30**, pp. 36–52.
22. Last, G., Schassberger, R. (2000): 'On stationary stochastic flows and Palm probabilities of surface processes'. *Ann. Appl. Probab.* **10**, pp. 463–492.
23. Last, G., Schassberger, R. (2001): 'On the second derivative of the spherical contact distribution function of smooth grain models'. *Probab. Theory and Related Fields* **121**, pp. 49–72.
24. Matheron, G. (1975): *Random Sets and Integral Geometry*. (Wiley, New York)
25. Mecke, K. (1994): *Integralgeometrie in der Statistischen Physik.* (Verlag Harri Deutsch, Frankfurt am Main)
26. Mecke, K. (2000): 'Additivity, convexity and beyond: applications of Minkowski functionals in statistical physics'. In: *Statistical Physics and Spatial Statistics*, Lecture Notes in Physics **554**, ed. by K. Mecke, D. Stoyan (Springer, Berlin), pp. 112–184
27. Mecke, K., Wagner, H. (1991): 'Euler characteristics and related measures for random geometric sets'. *J. Stat. Phys.* **64**, pp. 843–850.
28. Molchanov, I. (1995): 'Statistics of the Boolean model: from the estimation of means to the estimation of distributions'. *Adv. in Appl. Probab.* **27**, pp. 63–86.
29. Molchanov, I. (1997): *Statistics of the Boolean model for Practitioners and Mathematicians.* (Wiley, Chichester)

30. Molchanov, I., Stoyan, D. (1994): ‘Asymptotic properties of estimators for parameters of the Boolean model’. *Adv. in Appl. Probab.* **26**, pp. 301–323.
31. Muche, L., Stoyan, D. (1992): ‘Contact and chord length distribution of the Poisson Voronoi tessellation’. *J. Appl. Probab.* **29**, pp. 467–471.
32. Ohser, J., Mücklich, F. (2000): *Statistical Analysis of Microstructures in Materials Science*. (Wiley, Chichester)
33. Quintanilla, J., Torquato, S. (1997): ‘Microstructure functionals for a model of statistically inhomogeneous random media’. *Physical Review E* **55**, pp. 1558–1565.
34. Rataj, J., (1993): ‘Random distances and edge correction’. *Statistics* **24**, pp. 377–385.
35. Rataj, J., Saxl, I. (1995): ‘Boolean cluster models: mean cluster dilations and spherical distances’. *Mathematica Bohemica* **122**, pp. 21–36.
36. Saxl, I. (1993): ‘Contact distances and random free paths’. *Journal of Microscopy* **170**, pp. 53–64.
37. Schneider, R. (1993): *Convex Bodies: the Brunn-Minkowski Theory*, Encyclopedia of Mathematics and its Applications 44. (Cambridge University Press, Cambridge)
38. Schneider, R. (1994): ‘An extension of the principal kinematic formula of integral geometry’. *Rend. Circ. Mat. Palermo*, II. Ser., **35**, pp. 275–290.
39. Schneider, R., Weil, W. (2000): *Stochastische Geometrie*. (Teubner, Stuttgart)
40. Serra, J. (1982): *Image Analysis and Mathematical Morphology*. (Academic Press, London)
41. Soille, P. (1999): *Morphological Image Analysis: Principles and Applications*. (Springer, Berlin)
42. Stoyan, D., Kendall, W.S., Mecke, J. (1995): *Stochastic Geometry and its Applications*, Second Edition. (Wiley, Chichester)
43. Stoyan, D., Penttinen, A. (2000): ‘Recent applications of point process methods in forestry statistics’. *Statistical Science* **15**, pp. 61–78.
44. Stoyan, D., Stoyan, H. (1994): *Fractals, Random Shapes and Point Fields: Methods of Geometrical Statistics*. (Wiley, Chichester)
45. Stoyan, D., Stoyan, H., Tscheschel, A., Mattfeld, T. (2001): ‘On the estimation of distance distribution functions for point processes and random sets’. *Image Anal. Stereol.* **20**, pp. 65–69.
46. Tscheschel, A., Stoyan, D., Hilfer, R. (2000): ‘Erosion-dilation analysis for experimental and synthetic microstructures of sedimentary rock’. *Physica A* **284**, pp. 46–58.
47. Van Lieshout, M.N.M., Baddeley, A.J. (1995): ‘A non-parametric measure of spatial interaction in point-patterns’. *Statist. Neerland.* **49**, pp. 344–361.
48. Van Lieshout, M.N.M., Baddeley, A.J. (1999): ‘Indices of dependence between types in multivariate point patterns’. *Scand J. Statist.* **26**, pp. 511–532.
49. Weil, W. (1997): ‘Mean bodies associated with random closed sets’ *Rend. Circ. Mat. Palermo (2) Suppl.* **50**, pp. 387–412.
50. Weil, W. (2000): ‘Mixed measures and inhomogeneous Boolean models’. In: *Statistical Physics and Spatial Statistics*, Lecture Notes in Physics **554**, ed. by K. Mecke, D. Stoyan (Springer, Berlin), pp. 95–110
51. Weil, W. (2001): ‘Densities of mixed volumes for Boolean models’ *Adv. in Appl. Probab.* **33**, pp. 39–60.
52. Weil, W., Wieacker, J.A. (1988): ‘A representation theorem for random sets’. *Probab. Math. Statist.* **9**, pp. 147–151.

Mark Correlations: Relating Physical Properties to Spatial Distributions

Claus Beisbart[1], Martin Kerscher[2], and Klaus Mecke[3,4]

[1] University of Oxford, Nuclear & Astrophysics Laboratory, Keble Road, Oxford OX1 3RH, Great Britain
[2] Ludwig-Maximilians-Universität, Sektion Physik, Theresienstraße 37, D-80333 München, Germany
[3] Max-Planck-Institut für Metallforschung, Heisenbergstr. 1, D-70569 Stuttgart, Germany
[4] Institut für Theoretische und Angewandte Physik, Fakultät für Physik, Universität Stuttgart, Pfaffenwaldring 57, D-70569 Stuttgart, Germany

Abstract. Mark correlations provide a systematic approach to look at objects both distributed in space and bearing intrinsic information, for instance on physical properties. The interplay of the objects' properties (marks) with the spatial clustering is of vivid interest for many applications; are, e.g., galaxies with high luminosities more strongly clustered than dim ones? Do neighbored pores in a sandstone have similar sizes? How does the shape of impact craters on a planet depend on the geological surface properties? In this article, we give an introduction into the appropriate mathematical framework to deal with such questions, i.e. the theory of marked point processes. After having clarified the notion of segregation effects, we define universal test quantities applicable to realizations of a marked point processes. We show their power using concrete data sets in analyzing the luminosity-dependence of the galaxy clustering, the alignment of dark matter halos in gravitational N-body simulations, the morphology- and diameter-dependence of the Martian crater distribution and the size correlations of pores in sandstone. In order to understand our data in more detail, we discuss the Boolean depletion model, the random field model and the Cox random field model. The first model describes depletion effects in the distribution of Martian craters and pores in sandstone, whereas the last one accounts at least qualitatively for the observed luminosity-dependence of the galaxy clustering.

1 Marked Point Sets

Observations of spatial patterns at various length scales frequently are the only point where the physical world meets theoretical models. In many cases these patterns consist of a number of comparable objects distributed in space such as pores in a sandstone, or craters on the surface of a planet. Another example is given in Fig. 1, where we display the galaxy distribution as traced by a recent galaxy catalogue. The galaxies are represented as circles centered at their positions, whereas the size of the circles mirrors the luminosity of a galaxy. In order to test to which extent theoretical predictions fit the empirically found structures of that type, one has to rely on quantitative measures describing the physical information. Since theoretical models mostly do not try to explain the structures individually, but rather predict some of their generic properties, one has to adopt a *statistical point of view* and to interpret the data as a realization of a random process. In a first step one often confines oneself to the spatial distribution of the objects constituting the patterns and investigates their clustering thereby thinking of it as a realization of a *point process*. Assuming that perspective, however, one neglects a possible linkage between the spatial clustering and the intrinsic properties of the objects. For instance, there are strong indications that the clustering of galaxies depends on their luminosity as well as on their morphological type. Considering Fig. 1, one might infer that luminous galaxies are more strongly correlated than dim ones. Effects like that are referred to as *mark segregation* and provide insight into the generation and interactions of, e.g., galaxies or other objects under consideration. The appropriate statistical framework to describe the relation between the spatial distribution of physical objects and their inner

Fig. 1. The galaxy distribution as traced by the Southern Sky Redshift Survey 2 (SSRS 2). We show a part of the sample investigated, projected down into two dimensions. Each circle represents a galaxy, its radius is proportional to the galaxy's luminosity. For further details see Sect. 2.1.

properties are *marked point processes*, where discrete, scalar-, or vector-valued marks are attached to the random points.

In this contribution we outline how to describe marked point processes; along that line we discuss two notions of independence (Sect. 1) and define corresponding statistics that allow us to quantify possible dependencies. After having shown that some empirical data sets show significant signals of mark segregation (Sect.2), we turn to analytical models, both motivated by mathematical and physical considerations (Sect. 3).

Contact distribution functions as presented in the contribution by D. Hug et al. in this volume are an alternative technique to measure and statistically quantify distances which finally can be used to relate physical properties to spatial structures. Mark correlation functions are useful to quantify molecular orientations in liquid crystals (see the contribution by F. Schmid and N. H. Phuong in this volume) or in self-assembling amphiphilic systems (see the contribution by U. S. Schwarz and G. Gompper in this volume). But also to study anisotropies in composite or porous materials, which are essential for elastic and transport properties (see the contributions by D. Jeulin, C. Arns et al. and H.-J. Vogel in this volume), mark correlations may be relevant.

1.1 The Framework

The empirical data – the positions $\mathbf{x}_i$ of some objects together with their intrinsic properties m_i – are interpreted as a realization of a marked point process $\{(\mathbf{x}_i, m_i)\}_{i=1}^N$ (Stoyan, Kendall and Mecke, 1995). For simplicity we restrict ourselves to homogeneous and isotropic processes.

The hierarchy of joint probability densities provides a suitable tool to describe the stochastic properties of a marked point process. Thus, let $\varrho_1^{\mathcal{SM}}((\mathbf{x}, m))$ denote the probability density of finding a point at $\mathbf{x}$ with a mark m. For a homogeneous process this splits into $\varrho_1^{\mathcal{SM}}((\mathbf{x}, m)) = \varrho\mathcal{M}_1(m)$ where ϱ denotes the mean number density of points in space and $\mathcal{M}_1(m)$ is the probability density of finding the mark m on an arbitrary point. Later on we need moments of this mark distribution; for real-valued marks the kth-moment of the mark-distribution is defined as

$$\overline{m^k} = \int \mathrm{d}m\, \mathcal{M}_1(m) m^k; \tag{1}$$

the mark variance is $\sigma_M^2 = \overline{m^2} - \overline{m}^2$.

Accordingly, $\varrho_2^{\mathcal{SM}}((\mathbf{x}_1, m_1), (\mathbf{x}_2, m_2))$ quantifies the probability density to find two points at $\mathbf{x}_1$ and $\mathbf{x}_2$ with marks m_1 and m_2, respectively (for second-order theory of marked point processes see [58, 60]). It effectively depends only on m_1, m_2, and the pair separation $r = |\mathbf{x}_2 - \mathbf{x}_1|$ for a homogeneous and isotropic process. Two-point properties certainly are the simplest non-trivial quantities for homogeneous random processes, but it may be necessary to move on to higher correlations in order to discriminate between certain models.

1.2 Two Notions of Independence

In the following we will discuss two notions of independence, which may arise for marked point patterns. For this, consider two Renaissance families, call them the Sforza and the Gonzaga. They used to build castles spread out more or less homogeneously over Italy. In order to describe this example in terms of a marked point process, we consider the locations of the castles as points on a map of Italy, and treat a castle's owner as a discrete mark, S and G, respectively. There are many ways how the castles can be built and related to each other.

Independent sub-point processes: For example, the Sforza may build their castles regardless of the Gonzaga castles. In that case the probability of finding a Sforza castle at $\mathbf{x}_1$ and a Gonzaga castle at $\mathbf{x}_2$ factorizes into two one-point probabilities and we can think of the Sforza and the Gonzaga castles as uncorrelated sub-point processes. In the language of marked point processes this means, e.g., that

$$\begin{aligned}\varrho_2^{\mathcal{S},\mathcal{M}}\left((\mathbf{x}_1,m_1),(\mathbf{x}_2,m_2)\right) &= \varrho_1^{\mathcal{S}\mathcal{M}}\left((\mathbf{x}_1,m_1)\right)\varrho_1^{\mathcal{S}\mathcal{M}}\left((\mathbf{x}_2,m_2)\right)\\ &= \varrho^2\mathcal{M}_1(m_1)\mathcal{M}_1(m_2),\end{aligned} \tag{2}$$

for any $m_1 \neq m_2$. If all the joint n-point densities factorize into a product of n'-point densities of one type each, then we speak of *independent sub-point processes*. Dependent sub-point processes indicate *interactions* between points of different marks; for instance, the Gonzaga may build their castles close to the Sforza ones in order to avoid that a region becomes dominated by the other family's castles.

Mark-independent clustering: A second type of independence refers to the question whether the different families have different styles to plan their castles. For instance, the Gonzaga may distribute their castles in a grid-like manner over Italy, whereas the Sforza may incline to build a second castle close to each castle they own. Rather than asking whether two sub-point processes (namely the Gonzaga and the Sforza castles, respectively) are independent ("independent sub-point processes"), we are now discussing whether they are *different* as regards their statistical clustering properties. Any such difference means that the clustering *depends* on the intrinsic mark of a point.

Whenever the two-point probability density of finding two objects at $\mathbf{x}_1$ and $\mathbf{x}_2$ depends on the objects' intrinsic properties we speak of *mark-dependent clustering*. It is useful to rephrase this statement by using Bayes' theorem and the conditional mark probability density

$$\mathcal{M}_2(m_1,m_2|\mathbf{x}_1,\mathbf{x}_2) = \frac{\varrho_2^{\mathcal{S},\mathcal{M}}\left((\mathbf{x}_1,m_1),(\mathbf{x}_2,m_2)\right)}{\varrho_2^{\mathcal{S}}\left(\mathbf{x}_1,\mathbf{x}_2\right)}, \tag{3}$$

in case the spatial product density $\varrho_2^{\mathcal{S}}(\cdot)$ does not vanish. $\mathcal{M}_2(m_1,m_2|\mathbf{x}_1,\mathbf{x}_2)$ is the probability density of finding the marks m_1 and m_2 on objects located at $\mathbf{x}_1$ and $\mathbf{x}_2$, given that there are objects at these points. Clearly, $\mathcal{M}_2(m_1,m_2|\mathbf{x}_1,\mathbf{x}_2)$ depends only on the pair separation $r = |\mathbf{x}_1 - \mathbf{x}_2|$ for homogeneous and isotropic point processes. We speak of *mark-independent* clustering, if $\mathcal{M}_2(m_1,m_2|r)$ factorizes

$$\mathcal{M}_2(m_1,m_2|r) = \mathcal{M}_1(m_1)\mathcal{M}_1(m_2) \tag{4}$$

and thus does not depend on the pair separation. That means that regarding their marks, pairs with a separation r are not different from any other pairs. On the contrary, mark-dependent clustering or *mark segregation* implies that the marks on certain pairs show deviations from the global mark distribution.

In order to distinguish between both sorts of independencies, let us consider the case where we are given a map of Italy only showing the Gonzaga castles. If the distribution of castles in Italy can be understood as consisting of independent sub-point processes, we cannot infer anything about the Sforza castles from the Gonzaga ones. However, if $\varrho_2^{\mathcal{S},\mathcal{M}}\left((\mathbf{x}_1,S),(\mathbf{x}_2,G)\right) > \varrho^2\mathcal{M}_1(S)\mathcal{M}_1(G)$, Sforza castles are likely to be found close to Gonzaga ones. Here, $\mathcal{M}_1(S)$ and $\mathcal{M}_1(G)$ are the probabilities that a castle belongs to the Sforza or Gonzaga family. If, on the other hand, mark-independent clustering applies, typical clustering properties such as the spatial clustering strength are equal for both castle distributions, and the Gonzaga castles are in the statistical sense already representative of the whole castle distribution in Italy. That means in particular that, if the Gonzaga castles are clustered, so are the Sforza ones.

Before we turn to applications, we have to develop practical test quantities in order to test for segregation effects in real data and to describe them in more detail.

1.3 Investigating the Independence of Sub-point Processes

To investigate correlations between sub-point processes, suitably extended nearest neighbor distribution functions or K-functions have been employed [16, 20]. Also the (conditional) cross-correlation functions can be used (see (8)), for a further test see [60], p. 302. Here we consider a multivariate extension of the J-function [68], as suggested by [69].

For this, consider the nearest neighbor's distance distribution from an object with mark m_i to other objects with mark m_j, $G_{ij}(r)$ ("i to j", for details see [69]). Let $G_{i\circ}(r)$ denote the distribution of the nearest neighbor's distance from an object of type i to any other object (denoted by $\circ$). Finally, $G_{\circ\circ}(r)$ is the nearest neighbor distribution of all points. Similar extensions of the empty space function are possible, too. Let $F_i(r)$ denote the distribution of the nearest i-object's distance from an arbitrary position, whereas $F_\circ(r)$ is the nearest object's distance distribution from a random point in space to any object in the sample. We consider the following quantities:

$$J_{ij}(r) = \frac{1 - G_{ij}(r)}{1 - F_j(r)}, \; J_{i\circ}(r) = \frac{1 - G_{i\circ}(r)}{1 - F_\circ(r)}, \; J(r) = \frac{1 - G_{\circ\circ}(r)}{1 - F_\circ(r)}, \tag{5}$$

They are defined whenever $F_j(r), F_\circ(r) < 1$. If two sub-point processes, defined by marks $i \neq j$, are independent then one gets [69]

$$J_{ij}(r) = 1. \tag{6}$$

Note, that the J_{ij} depend on higher-order correlations functions, similar to the J-function [35]. Suitable estimators for these J-functions are derived from estimators of the F and G-functions [58, 4].

1.4 Investigating Mark Segregation

In order to quantify the mark-dependent clustering or to look for the mark segregation, it proves useful to integrate the conditional probability density $\mathcal{M}_2(m_1, m_2|r)$ over the marks weighting with a test function $f(m_1, m_2)$ [55, 58]. This procedure reduces the number of variables and leaves us with the weighted pair average:

$$\langle f \rangle_{\rm P} = \int \mathrm{d}m_1 \int \mathrm{d}m_2 \; f(m_1, m_2) \mathcal{M}_2(m_1, m_2|r). \tag{7}$$

The choice of an appropriate weight-function depends on whether the marks are non-quantitative labels or continuous physical quantities.

1. For labels only combinations of indicator-functions are possible, the integral degenerates into a sum over the labels. Supposed the marks of our objects belong to classes labelled with $i, j, \ldots$, the conditional cross-correlation functions are given by

$$C_{ij}(r) \equiv \langle \delta_{m_1 i}\delta_{m_2 j} + (1-\delta_{ij})\delta_{m_2 i}\delta_{m_1 j}\rangle_{\mathrm{P}}(r), \tag{8}$$

   with the Kronecker $\delta_{m_1 i} = 1$ for $m_1 = i$ and zero otherwise. Mark segregation is indicated by $C_{ij} \neq 2\varrho_i\varrho_j/\varrho^2$ for $i \neq j$ and $C_{ii} \neq \varrho_i^2/\varrho^2$, where ϱ_i denotes the number density of points with label i. The C_{ij} are cross-correlation functions under the *condition* that two points are separated by a distance of r (compare [60], p. 264, for applications see the Martian crater distribution studied in Sect. 2.3 and Fig. 7 in particular).
2. For positive real-valued marks m, the following pair averages prove to be powerful and distinctive [51, 7]:
   a) One of the most simplest weights to be used is the mean mark:

$$k_m(r) \equiv \frac{\langle m_1 + m_2\rangle_{\mathrm{P}}(r)}{2\,\overline{m}}. \tag{9}$$

   quantifies the deviation of the mean mark on pairs with separation r from the overall mean mark $\overline{m}$. A $k_m > 1$ indicates mark segregation for point pairs with a separation r, specifically their mean mark is then larger than the overall mark average.
   Closely related is Stoyan's k_{mm} function using the squared geometric mean of the marks as a weight [55, 60]

$$k_{mm}(r) \equiv \frac{\langle m_1 m_2\rangle_{\mathrm{P}}(r)}{\overline{m}^2}. \tag{10}$$

   b) Accordingly, higher moments of the marks may be used to quantify mark segregation, like the mark fluctuations

$$\mathit{var}(r) \equiv \left\langle \left(m_1 - \langle m_1\rangle_{\mathrm{P}}(r)\right)^2\right\rangle_{\mathrm{P}}(r), \tag{11}$$

   or the mark-variogram [70, 61]:

$$\gamma(r) \equiv \left\langle \tfrac{1}{2}\left(m_1 - m_2\right)^2\right\rangle_{\mathrm{P}}(r), \tag{12}$$

   c) The mark covariance [17] is

$$\mathit{cov}(r) \equiv \langle m_1 m_2\rangle_{\mathrm{P}}(r) - \langle m_1\rangle_{\mathrm{P}}(r)\,\langle m_2\rangle_{\mathrm{P}}(r). \tag{13}$$

   Mark segregation can be detected by looking whether $\mathit{cov}(r)$ differs from zero. A $\mathit{cov}(r)$ larger than zero, e.g., indicates that points with separation r tend to have similar marks. Sometimes the mark covariance is normalized by the fluctuations [33]: $\mathit{cov}(r)/\mathit{var}(r)$.

These conditional mark correlation functions can be calculated from only three independent pair averages [51]: $\langle m\rangle_{\rm P}(r)$, $\langle m_1 m_2\rangle_{\rm P}(r)$, and $\langle m^2\rangle_{\rm P}(r)$. Thus the above mentioned characteristics are not independent, e.g. $var(r) = \gamma(r) + cov(r)$. We apply these mark correlation functions to the galaxy distribution in Sect. 2.1 (Fig. 3), to Martian craters in Sect. 2.3 (Fig. 7) and to pores in sandstones considered in Sect. 2.4.

3. Also vector-valued information $\mathbf{l}_i$, describing, e.g., the orientation of an anisotropic object at position $\mathbf{x}_i$ may be available. It is therefore interesting to consider vector marks such as done by [45, 49, 60] who use a mark correlation function to quantify the alignment of vector marks. Here we suggest three mark correlation functions quantifying geometrically different possibilities of an alignment. In order to ensure coordinate-independence of our descriptors, we focus on scalar combinations of the vector marks in using the scalar product $\cdot$ and the cross product $\times$. Different from the case of scalar marks, it is a non-trivial task to find a set of vector-mark correlation functions which contain all possible information (at least up to a fixed order in mark space). We provide a systematic account of how to construct suitable vector-mark correlation functions in a complete and unique way for general dimensions in the Appendix.
Here we only cite the most important results. For that we need the distance vector between two points, $\mathbf{r} \equiv \mathbf{x}_1 - \mathbf{x}_2$, the normalized distance vector, $\hat{\mathbf{r}} \equiv \mathbf{r}/r$, and the normalized vector mark: $\hat{\mathbf{l}}_i \equiv \mathbf{l}_i/l_i$ with $l_i = |\mathbf{l}_i|$. The following conditional mark correlation functions will be used to quantify alignment effects:
   a) $\mathcal{A}(r)$ quantifies the $\mathcal{A}$lignment of the two vector marks $\mathbf{l}_1$ and $\mathbf{l}_2$:
$$\mathcal{A}(r) = \frac{1}{\bar{l}^2}\left\langle \mathbf{l}_1 \cdot \mathbf{l}_2 \right\rangle_{\rm P}(r) \quad . \tag{14}$$
It is proportional to the cosine of the angle between $\mathbf{l}_1$ and $\mathbf{l}_2$. We normalize with the mean $\bar{l}$. For purely independent vector marks $\mathcal{A}(r)$ is zero, whereas $\mathcal{A}(r) > 0$ means that the marks of pairs separated by r tend to align parallel to each other. – In some applications, e.g. for the orientations of ellipsoidal objects, the vector mark is only defined up to a sign, i.e. $\mathbf{l}$ and $-\mathbf{l}$ mean actually the same. In this case the absolute value of the scalar product is useful:
$$\mathcal{A}'(r) \equiv \frac{1}{\bar{l}^2}\left\langle |\mathbf{l}_1 \cdot \mathbf{l}_2| \right\rangle_{\rm P}(r) \quad . \tag{15}$$
For uncorrelated random vectors we get $\mathcal{A}'(r) = 1/2$. $\mathcal{A}$ and $\mathcal{A}'$ can readily be generalized to any dimension d, where we expect $\mathcal{A}' = \pi^{-\frac{1}{2}}\frac{\Gamma(\frac{d}{2})}{\Gamma(\frac{d+1}{2})}$ for uncorrelated random orientations. In two dimensions $\mathcal{A}'$ is proportional to k_d as defined by [60].
   b) $\mathcal{F}(r)$ quantifies the $\mathcal{F}$ilamentary alignment of the vectors $\mathbf{l}_1$ and $\mathbf{l}_2$ with respect to the line connecting both halo positions:
$$\mathcal{F}(r) \equiv \frac{1}{2\,\bar{l}}\left\langle |\mathbf{l}_1 \cdot \hat{\mathbf{r}}| + |\mathbf{l}_2 \cdot \hat{\mathbf{r}}| \right\rangle_{\rm P}(r), \tag{16}$$

$\mathcal{F}(r)$ is proportional to the cosine of the angle between $\mathbf{l}_1$ and the distance vector $\hat{\mathbf{r}}$ connecting the points. For uncorrelated random vector marks, we expect again $\mathcal{F}(r) = 1/2$; $\mathcal{F}(r)$ becomes larger than that, whenever the vector marks of the objects tend to point to objects separated by r – an example is provided by rod-like metallic grains in an electric field: they concentrate along the field lines and orient themselves parallel to the field lines.

c) $\mathcal{P}(r)$ quantifies the $\mathcal{P}$lanar alignment of the vectors and the distance vector. $\mathcal{P}(r)$ is proportional to the volume of the rhomb defined by $\mathbf{l}_1$, $\mathbf{l}_2$ and $\hat{\mathbf{r}}$:

$$\mathcal{P}(r) = \frac{1}{2\bar{l}^2} \left\langle \left| \mathbf{l}_1 \cdot \frac{\mathbf{l}_2 \times \hat{\mathbf{r}}}{|\hat{\mathbf{l}}_2 \times \hat{\mathbf{r}}|} \right| + \left| \mathbf{l}_2 \cdot \frac{\mathbf{l}_1 \times \hat{\mathbf{r}}}{|\hat{\mathbf{l}}_1 \times \hat{\mathbf{r}}|} \right| \right\rangle_{\mathrm{P}} (r), \tag{17}$$

Quite obviously, this quantity can not be generalized to arbitrary dimensions; the deeper reason for that will become clear in the Appendix. – We get $\mathcal{P}(r) = 1/2$ for randomly oriented vectors, whereas it is becoming larger for the case that $\mathbf{l}_2$ is perpendicular to $\mathbf{l}_1$ as well as to $\hat{\mathbf{r}}$.

Applications of vector marks can be found in Sect. 2.2 (Fig. 4) where we consider the orientation of dark matter halos in cosmological simulations. But one can think of other applications: mark correlation functions may serve as orientational order parameters in liquid crystals in order to discriminate between nemetic and smectic phases (see the contribution by F. Schmid and N. H. Phuong in this volume). They can also quantify the local orientation and order in liquids such as the recently measured five-fold local symmetry found in liquid lead [50]. As a further application one could try to measure the signature of hexatic phases in two-dimensional colloidal dispersions and in 2D melting scenarios occurring in experiments and simulations of hard-disk systems (for a review on hard sphere models see [39]. Finally, the orientations of anisotropic channels in sandstone (see the contribution by C. Arns et al. in this volume) are relevant for macroscopic transport properties, therefore their quantitative characterization in terms of mark correlation functions might be interesting.

Before we move on to applications a few general remarks are in order: First, the definition of these mark characteristics based on the conditional density $\mathcal{M}_2(\cdot)$ leads to ambiguities at r equal zero as discussed by [51], but there is no problem for $r > 0$. – Furthermore, suitable estimators for our test quantities are based on estimators for the usual two-point correlation function [60, 13, 7].

Mark-dependent clustering can also be defined at any n-point level. Mark-independent clustering at every order is called the random labelling property [16]. Mark correlation functions based on the n-point densities may be used. For discrete marks the multivariate J-functions (see ((5))) are an interesting alternative, sensitive to higher-order correlations. The random labelling property then leads to the relation

$$J_{i\circ}(r) = J, \tag{18}$$

which may be used as a test [69].

2 Describing Empirical Data: Some Applications

In many cases already the question whether one or the other type of dependence as outlined above applies to certain data sets is a controversial issue. In the following we will apply our test quantities to a couple of data sets in order to probe whether there is an interplay between some objects' marks and their positions in space. Other applications to biological, ecological, mineralogical, geological data can be found in [57, 60, 43, 20].

2.1 Segregation Effects in the Distribution of Galaxies

The distribution of galaxies in space shows a couple of interesting features and challenges theoretical models trying to understand cosmological structure formation (see e.g. [34]). There has been a long debate, whether and how strongly the clustering of galaxies depends on their luminosity and their morphological type (see, e.g. [28, 30, 27]). The methods which have been used so far to establish such claims were based on the spatial two-point correlation function; it was estimated from different subsamples that were drawn from a catalogue and defined by morphology or luminosity. However, some authors claimed that the signal of luminosity segregation observed by others was a spurious effect, caused by inhomogeneities in the sample and an inadequate choice of the statistics [64]. [7] could show that methods based on the mark-correlation functions, as discussed in Sect. 1.4, are not impaired by inhomogeneities, and found a clear signal of luminosity and morphology segregation.

In order to quantify segregation effects in the galaxy distribution we consider the Southern Sky Redshift Survey 2 (SSRS 2, [18]), which maps a significant fraction of the sky and provides us with the angular sky positions, the distances (determined via the redshifts), and some intrinsic properties of the galaxies such as their flux and their morphological type. As marks we consider either a galaxy's luminosity estimated from its distance and flux, or its morphological type. In the latter case we effectively divide our sample into early-type galaxies (mainly elliptical galaxies) and late-type galaxies (mainly spirals). In order to analyze homogeneous samples, we focus on a volume-limited sample of $100h^{-1}$Mpc depth[5] [7].

In a first step we ask whether the early- and the late-type galaxies form independent sub-processes. In Fig. 2 we show J_{el} as function of the distance r being far away from the value of one. Recalling ((6)), we conclude that the morphological types of galaxies are not distributed independently on the sky. Not surprisingly, the inequality $J_{el} < 1$ indicates positive interactions between the galaxies of both morphological types; indeed galaxies attract each other through gravity irrespective of their morphological types.

After having confirmed the presence of interactions between the different types of galaxies, we tackle the issue whether the clustering of galaxies is different for different galaxies. We consider the luminosities as marks (see Fig. 1). In Fig. 3 we show some of the mark-weighted conditional correlation functions. Already at first glance, they show

[5] One Mpc equals roughly 3.26 million light years. The number h accounts for the uncertainty in the measured Hubble constant and is about $h \approx 0.65$. Volume-limited samples are defined by a limiting depth and a limiting luminosity. One considers only those galaxies which could have been observed if they were located at the limiting depth of the sample.

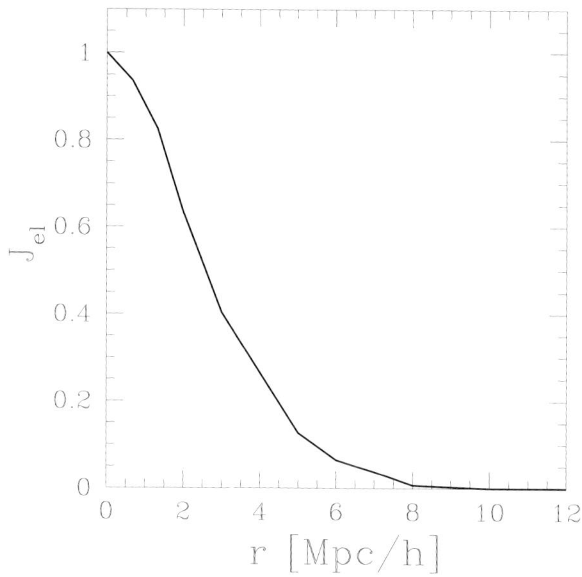

Fig. 2. The J_{el} function of early-type (e) and late-type (l) galaxies vs. the galaxy separation r in a volume-limited sample of $100h^{-1}$Mpc depth from the SSRS 2 catalogue.

evidence for luminosity segregation, relevant on scales up to $15h^{-1}$Mpc. To strengthen our claims, we redistribute the luminosities of the galaxies within our sample randomly, holding the galaxy positions fixed. In that way we mimic a marked point process with the same spatial clustering and the same one-point distribution of the luminosities, but without luminosity segregation. Comparing with the fluctuations around this null hypothesis, we see that the signal within the SSRS 2 is significant.

The details of the mark correlation functions provide some further insight into the segregation effects. The mean mark $k_m(r) > 1$ indicates that the luminous galaxies are more strongly clustered than the dim ones. Our signal is scale-dependent and decreasing for higher pair separations. The stronger clustering of luminous galaxies is in agreement with earlier claims comparing the correlation amplitude of several volume-limited samples [73].

The $var(r)$ being larger than the mark variance of the whole sample, σ_M^2, shows that on galaxy pairs with separations smaller than $15h^{-1}$Mpc the luminosity fluctuations are enhanced. The fact that the mark segregation effect extends to scales of up to $15h^{-1}$Mpc is interesting on its own. In particular, it indicates that galaxy clusters are not the only source of luminosity segregation, since typically galaxy clusters are of the size of $3h^{-1}$Mpc.

The signal for the covariance $cov(r)$, however, could be due to galaxy pairs inside clusters. It is relevant mainly on scales up to $4h^{-1}$Mpc indicating that the luminosities on galaxy pairs with small separations tend to assume similar values. – Our results in part confirm claims by [9], who compared the correlation functions ξ_2 for different

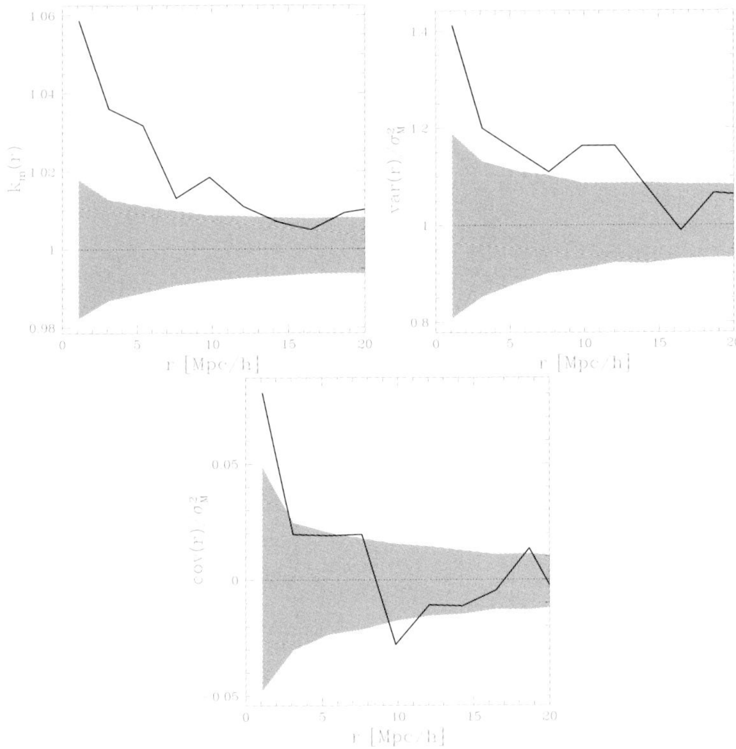

Fig. 3. The luminosity-weighted correlation functions for a volume-limited subsample of the SSRS 2 with a depth of $100h^{-1}$Mpc. The shaded areas denote the range of one-σ fluctuations for randomized marks around the case of no mark segregation. The fluctuations were estimated from 1000 reshufflings of the luminosities.

volume-limited subsamples and different luminosity classes of the SSRS 2 catalog (see also [8]).

2.2 Orientations of Dark Matter Halos

Many structures found in the Universe such as galaxies and galaxy clusters show anisotropic features. Therefore one can assign orientations to them and ask whether these orientations are correlated and form coherent patterns. Here we discuss a similar question on the base of numerical simulations of large scale structure (e.g., [10, 36]).

In such simulations the trajectories of massive particles are numerically integrated. These particles represent the dominant mass component in the Universe, the dark matter. Through gravitational instability high density peaks ("halos") form in the distribution of the particles; these halos are likely to be the places where galaxies originate. In the following we will report on alignment correlations between such halos [22], for a further application of mark correlation functions in this field see [25].

The halos used by [22] stem from a N-body simulation in a periodic box with a side length of $500h^{-1}$Mpc. The initial and boundary conditions were fixed according to a

ΛCDM cosmology (for a discussion of cosmological models see [48, 15]). Halos were identified using a friend-of-friends algorithm in the dark matter distribution. Not all of the halos found were taken into account; rather the mass range and the spatial number density of the selected halos were chosen to resemble the properties of observed galaxy clusters in the REFLEX catalogue [12]. Typically our halos show a prolate distribution of their dark matter particles.

For each halo the direction of the elongation is determined from the major axis of the mass-ellipsoid. This leads to a marked point set where the orientation $\mathbf{l}_i$ is attached to each halo position $\mathbf{x}_i$ as a vector mark with $|\mathbf{l}_i| = 1$. Details can be founds in [22].

In Fig. 4 the vector-mark correlation functions as defined in (14), (16), and (17) are shown. Since only the orientation of the mass ellipsoids can be determined, we use $\mathcal{A}'(r)$ (15) instead of $\mathcal{A}(r)$. The signal in $\mathcal{A}'(r)$ indicates that pairs of halos with a distance smaller than $30h^{-1}$Mpc show a tendency of parallel alignment of their orientations $\mathbf{l}_1, \mathbf{l}_2$. The deviation from a pure random alignment is in the percent range but clearly outside the random fluctuations. The alignment of the halos' orientations $\mathbf{l}_1, \mathbf{l}_2$ with the connecting vector $\hat{\mathbf{r}}$ quantified by $\mathcal{F}(r)$ is significantly stronger; it is particularly interesting that this alignment effect extends to scales of about $100h^{-1}$Mpc.

In a qualitative picture this may be explained by halos aligned along the filaments of the large scale structure. Indeed such filaments are prominent features found in the galaxy distribution [32] and in N-body simulations [41], often with a length of up to $100h^{-1}$Mpc. The lowered $\mathcal{P}(r)$ indicates that the volume of the rhomboid given by $\mathbf{l}_1, \mathbf{l}_2$ and $\hat{\mathbf{r}}$ is reduced for halo pairs with a separation below $80h^{-1}$Mpc. Already a preferred alignment of $\mathbf{l}_1, \mathbf{l}_2$ along $\hat{\mathbf{r}}$ leads to such a reduction, similar to a plane-like arrangement of $\mathbf{l}_1, \mathbf{l}_2, \hat{\mathbf{r}}$. For the halo distribution the signal in $\mathcal{P}(r)$ seems to be dominated by the filamentary alignment.

The question whether there are non-trivial orientation patterns for galaxies or galaxy clusters has been discussed for a long time. [11] reported a significant alignment of the observed galaxy clusters out to $100h^{-1}$Mpc. [62, 63], however claimed that this effect is small and likely to be caused by systematics; [67] find no indication for alignment effects at all. Subsequently several authors purported to have found signs of alignments in the galaxy and galaxy cluster distribution (see e.g. [21, 37, 24, 29]). Our Fig. 4 shows that from simulations significant large-scale correlations are to be expected in the orientations of galaxy clusters, in agreement with the results by [11]. These results are also supported by a simulation study carried out by [46].

2.3 Martian Craters

Let us now turn to another, still astrophysical, but significantly closer object: the Mars (see Fig. 5). Many planets' surfaces display impact craters with diameters up to ~ 260 km and a broad range of inner morphologies. These craters are surrounded by ejecta forming different types of patterns. The craters and their ejecta are likely to be caused by asteroids and periodic comets crossing the planets' orbits, falling down onto the planet's surface, and spreading some of the underlying surfaces material around the original impact crater. A variety of different crater morphologies and a wide range of ejecta patterns can be found. In principle, either the different impact objects (especially their energies) or the various surface types of the planet may explain the repertory of patterns observed.

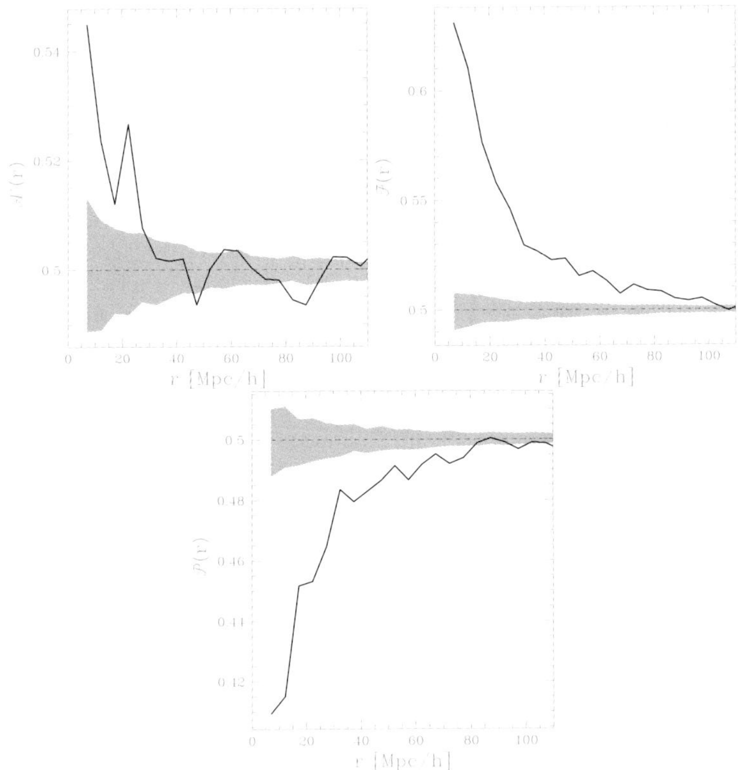

Fig. 4. The correlations of halo orientations in numerical simulations. The orientation of each dark matter halo, specified by the direction of the major axis 1 of the mass ellipsoid, is used as a vector mark. The dashed area is obtained by randomizing the orientations among the halos.

Whereas the energy variations of impact objects do not cause any peculiarities in the spatial distribution of the craters (apart from a possible latitude dependence), geographic inhomogeneities are expected to originate inhomogeneities in the craters' morphological properties.

We try to answer the question for the ejecta patterns' origin using data collected by [6] who already found correlations between crater characteristics and the local surface type employing geologic maps of the Mars. Complementary to their approach, we investigate two-point properties without any reference to geologic Mars maps. We restrict ourselves only to craters which have a diameter larger than 8 km and whose ejecta pattern could be classified, ending up with 3527 craters spread out all over the Martian surface. We use spherical distances for our analysis of pairs.

In a first step we divide the ejecta patterns into two broad classes consisting of either the simple patterns (single and double lobe morphology, i.e. SL and DL in terms of the classification by [6]; we speak of "simple craters") or the remaining, more complex configurations ("complex craters"). Using our conditional cross correlation functions C_{ij} as defined in (8), we see a highly significant signal for mark correlations (Fig. 6).

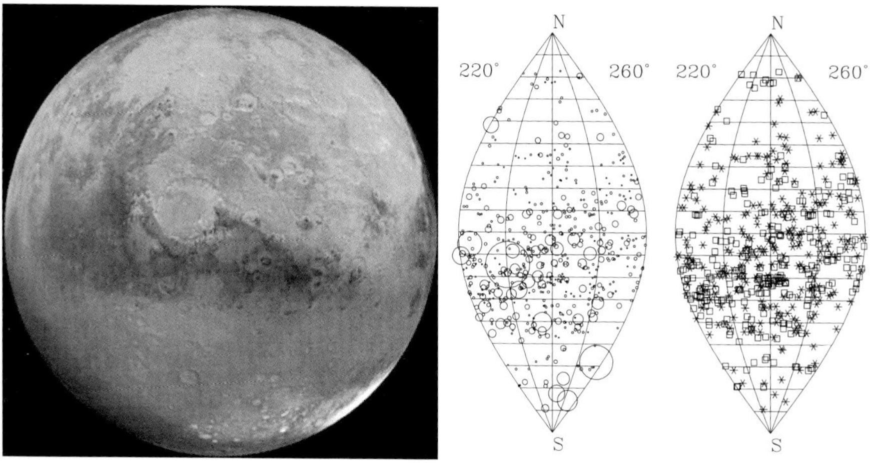

Fig. 5. The Martian surface with its craters. Whereas the left panel (from http://pds.jpl.nasa.gov/planets/captions/mars/schiap.htm) illustrates the various geological settings to be found on the planet's surface, the other panels focus on a small patch and show the craters together with their radii (middle panel, the size of the symbols are proportional to the radii of the craters) and together with the craters' types (right panel, simple morphology as quadrangles and more complex craters as stars). The latter viewgraphs rely on the data by [6].

At small separations, crater pairs are disproportionally built up of simple craters at the expense of cross correlations. This can be explained assuming that crater formation depends on the local surface type: if the simple craters are more frequent in certain geological environments than in others, then there are also more pairs of them to be found as far as one focuses on distances smaller than the typical scale of one geological surface type. Cross pairs are suppressed, since typical pairs with small separations belong to one geological setting where the simple craters either dominate or do not. Only a small, positive segregation signal occurs for the complex craters. Hence our analysis indicates that the broad class of complex craters is distributed quite homogeneously over all of the geologies. On top of this there are probably simple craters, their frequency significantly depending on the surface type.

If the ejecta patterns were independent of the surface, no mark segregation could be observed (other sources of mark segregation are unlikely, since the Martian craters are a result of a long bombardment history diluting any eventual peculiar crater correlations). In this sense, the signal observed indicates a surface-dependence of crater formation. This result is remarkable, given that we did not use any geological information on the Mars at all. The picture emerging could be described using the random field model, where a field (here the surface type) determines the mark of the points (see below).

In a second step, we analyze the interplay between the craters' diameters and their spatial clustering. Now the diameter serves as a continuous mark. The results in Fig. 7 show a clear signal for mark segregation in k_m and cov at small scales. The latter signals that pairs with separations in a broad range up to 1700 km tend to have similar diameters; this is in agreement with the earlier picture: as [6] showed, the simple craters

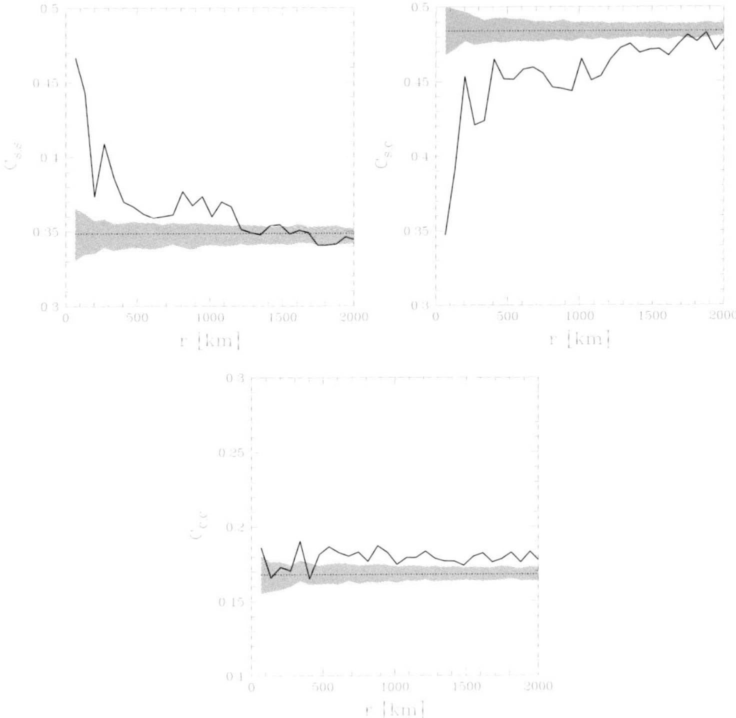

Fig. 6. The conditional cross-correlation functions for Martian craters. We split the sample of craters into two broad classes according to their ejecta types: simple morphologies (S) consisting of SL and DL types, and complex morphologies (C) with all other types (see [6] for details). The results indicate, that at scales up to about 1500 km the clustering of the simple craters is enhanced at expense of cross correlations. The shaded areas denote the one-σ fluctuations for randomized marks estimated from 100 realizations of the mark reshuffling.

are mostly small-sized. Pairs with relatively small separations thus often stem from the same geological setting and therefore have similar diameters and similar morphological type.

Also the signal of k_m seems to support this picture: since the simple craters are more strongly clustered than the other ones and since they have smaller diameters, one could expect $k_m < 1$. As we shall see in Sect. 3, however, a $k_m \neq 1$ contradicts the random field model; therefore, the mark-dependence on the underlying surface type (thought of as a random field) cannot account for the signal observed. Thus, we have to look for an alternative explanation: it seems reasonable, that, whenever a crater is found somewhere, no other crater can be observed close nearby (because an impact close to an existing crater will either destroy the old one or cover it with ejecta such that it is not likely to be observed as a crater). This results in a sort of effective hard-core repulsion. This repulsion should be larger for larger craters. Thus, pairs with very small separations can only be formed by small craters, therefore $k_m < 1$ for tiny r. The scale beyond which $k_m(r) \sim 1$ should somehow be hidden within the crater diameter distribution. Indeed,

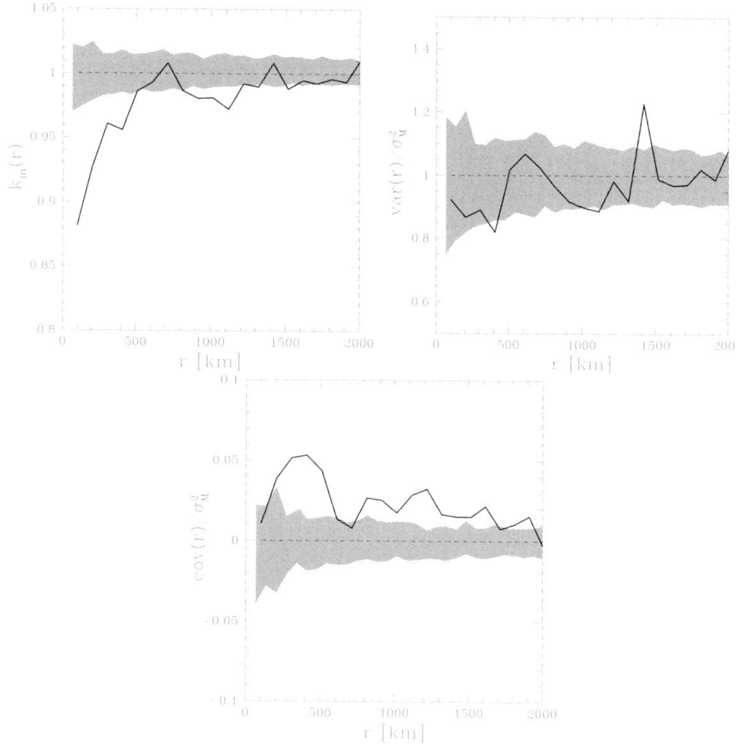

Fig. 7. The radius-weighted correlation functions for craters on Mars. The radius of each crater serves as a mark. r is the spherical distance. The shaded areas denote the one-σ fluctuations for randomized marks estimated from 100 realizations of the mark reshuffling.

at about 500 km the segregation vanishes, which is about twice the largest diameter in our sample. Taking into account that the ejecta patterns extend beyond the crater, this seems to be a reasonable agreement. As shown in Sect. 3.1 a model based on these consideration is able to produce such a depletion in the $k_m(r)$. This effect could also in turn explain part of the cross correlations observed earlier in Fig. 6. A similar effect is to be expected for the mark variance. Close pairs are only accessible to craters with a smaller range of diameters; therefore, their variance is diminished in comparison to the whole sample. However, an effect like this is barely visible in the data.

Altogether, the crater distribution is dominated by two effects: the type of the ejecta pattern and the crater diameter depend on the surface, in addition, there is a sort of repulsion effect on small scales.

2.4 Pores in Sandstone

Now we turn to systems on smaller scales. Sandstone is an example of a porous medium and has extensively been investigated, mainly because oil was found in the pore network of similar stones. In order to extract the oil from the stone one can try to wash it out using

Fig. 8. The pores within a Fontainbleau sandstone sample. Note, that this is a negative image, where the pores are displayed in grey. The geometrical features of the pore network are important for macroscopic properties of the stone. In this sample the pores occupy 13% of the volume. The size of the whole sample shown is about 1.5 mm^3 (Courtesy M. Knackstedt).

a second liquid, e.g. water. Therefore, one tries to understand from a theoretical point of view, how the microscopic geometry of the pore network determines the macroscopic properties of such a multi-phase flow. Especially the topology and connectivity of the microcaves and tunnels prove to be crucial for the flow properties at macroscopic scales. Details are given, for instance, in the contributions by C. Arns et al., H.-J. Vogel et al. and J. Ohser in this volume. A sensible physical model, therefore, in the first place has to rely on a thorough description of the pore pattern.

One way to understand the pore network is to think of it as a union of simple geometrical bodies. Following [53], one can identify distinct pores together with their position and their pore radius or extension. This allows us to understand the pore structure in terms of a marked point process, where the marks are the pore radii.

In the following, we consider three-dimensional data taken from one of the Fontainbleau sandstone samples through synchrotron X-ray tomography. These data trace a 4.52 mm diameter cylindrical core extracted from a block with bulk porosity $\phi = 13\%$,, where the bulk porosity is the volume fraction occupied by the pores. A piece with 2.91mm length (resulting in a 46.7 mm^3 volume) of the core was imaged and tomographically reconstructed [23, 54, 3, 2]. Further details of this sample are presented in the contribution by C. Arns et al. in this volume. Based on the reconstructed images the positions of pores and their radii were identified as described in [53].

In our results for the mark correlation functions a strong depletion of $k_m(r)$ and $var(r)$ is visible for $r < 200\mu$m in Fig. 10. This small-scale effect may be explained similarly to the Martian craters: large pores are never found close to each others, since

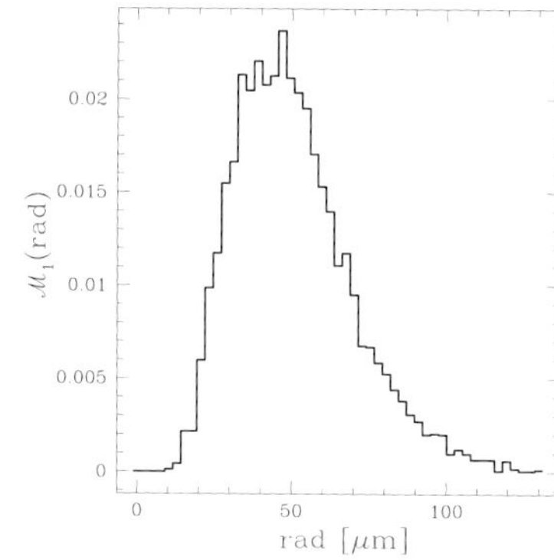

Fig. 9. The empirical one-point distribution $\mathcal{M}_1$ of the pore sizes.

they have to be separated by at least the sum of their radii. The histogram of the pore radii in Fig. 9 shows that most of the pores have radii smaller than 100μm, and consequently this effect is confined to $r < 200\mu$m. In Sect. 3.1 we discuss the Boolean depletion model which is based on this geometric constraints and is able to produce such a reduction in the $k_m(r)$. This purely geometric constraint also explains the reduced $var(r)$ and increased covariance $cov(r)$. For separations larger than 200μm there is no signal from the covariance, but both $k_m(r)$ and $var(r)$ show a small increase out to $\sim 1000\mu$m. This indicate that pairs of pores out to these separations tend to be larger in size and show slightly increased fluctuations. However, this effect is small (of the order of 1%) and may be explained by the definition of the holes, which may lead to "artificial small pores" as "bridges" between larger ones. This hypothesis has to be tested using different hole definitions. In any case the main conclusion seems to be that apart from the depletion effect at small scales there are no other mark correlations.

3 Models for Marked Point Processes

Given the significant mark correlations found in various applications, one may ask how these signals can be understood in terms of stochastic models. A thorough understanding of course requires a physical modeling of the individual situation. There are, however, some generic models, which we will focus on in the following: in Sect. 3.1 we introduce the Boolean depletion model, which is able to explain some of the features observed in the distribution of craters and pores in sandstone. Another generic model is the *random field model* where the marks of the points stem from an independent random field (Sect. 3.2). In Sect. 3.3 we generalize the idea behind the random field model further in order to get the *Cox random field model*, which allows for correlations between the point set and the random field. Other model classes and their applications are discussed by e.g. [20, 44, 17, 60, 71].

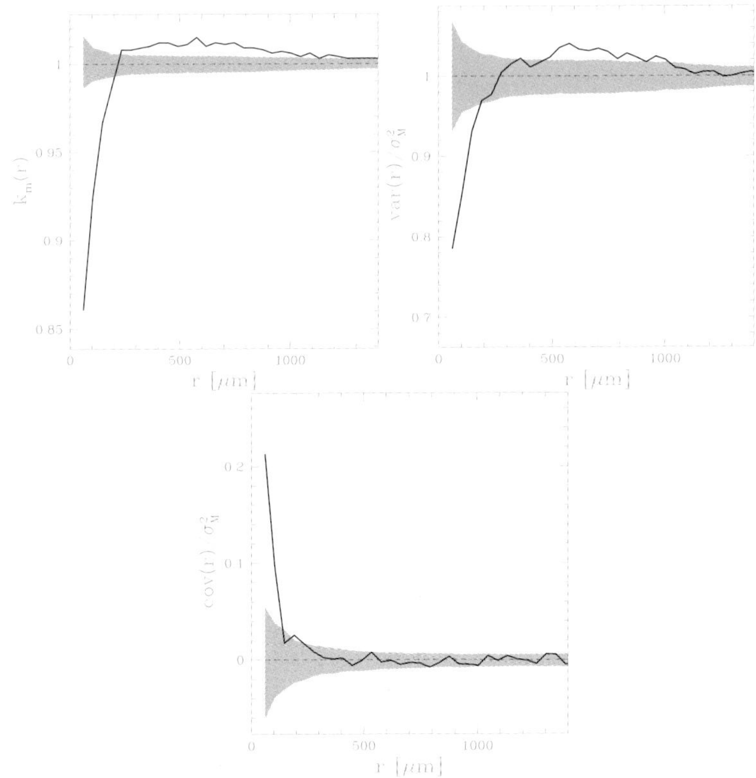

Fig. 10. The mark-weighted correlation functions from the holes in the Fontainbleau sandstone. The pores' radii serve as marks. The k_m being smaller than one indicates a depletion effect. The shaded areas again denote the range of one-σ fluctuations for randomized marks around the case of no mark segregation. The fluctuations were estimated from 200 reshufflings of the radii.

3.1 The Boolean Depletion Model

In our analysis of the Martian craters and the holes in sandstone, we found that for small separations only small craters, or small holes in the sandstone, could be found. We interpreted this as a pure geometric selection effect. The Boolean depletion model is able to quantify this effect, but also shows further interesting features.

The starting point is the Boolean model of overlapping spheres $B_R(\mathbf{x})$ (see also the contributions by C. Arns et al. and D. Hug in this volume as well as [56]). For that, the spheres' centers $\mathbf{x}_i$ are generated randomly and independently, i.e. according to a Poisson process of number density ϱ_0. The radii R of the spheres are then chosen independently according to a distribution function $F_0(R)$, i.e. with probability density $f_0(R) = \frac{\partial F_0(R)}{\partial R}$. The main idea behind the depletion is to delete spheres which are covered by other spheres. To make this procedure unique we remove only those spheres which are completely covered by a (notably larger) sphere[6]. The positions and radii of the remaining spheres define a marked point process. Note, that this depletion mechanism is minimal in the sense that a lot of overlapping spheres may remain. This Boolean depletion model may be considered as the low-density limit of the well-known Widom-Rowlinson model, or (more generally) of non-additive hard sphere mixtures (see [72, 40, 39]).

The probability that a sphere of radius R is not removed is then given by

$$\begin{aligned} f_{\rm nr}(R) &= \lim_{N,\Omega\to\infty} \prod_{i=1}^{N} \int_0^\infty \mathrm{d}R_i \; f_0(R_i) \left(1 - \frac{4\pi}{3} \frac{(R_i - R)^3}{|\Omega|} \Theta(R_i - R)\right) \\ &= \exp\left(-\varrho_0 \omega_d \int_0^\infty \mathrm{d}x \; f_0(R+x) x^d\right), \end{aligned} \tag{19}$$

with the step function $\Theta(x) = 0$ for $x < 1$ and $\Theta(x) = 1$ otherwise, and the volume of the d-dimensional unit ball ω_d ($\omega_1 = 2$, $\omega_2 = \pi$, $\omega_3 = 4\pi/3$). The limit in ((19)) is performed by keeping $\varrho_0 = N/|\Omega|$ constant, with N the initial number of spheres and $|\Omega|$ the volume of the domain.

The number density of the remaining spheres reads

$$\varrho = \varrho_0 \int_0^\infty \mathrm{d}R \; f_0(R) f_{\rm nr}(R) \quad , \tag{20}$$

where the one-point probability density $\mathcal{M}_1(R)$ that a sphere has radius R is given by

$$\mathcal{M}_1(R) = f_0(R) f_{\rm nr}(R) \frac{\varrho_0}{\varrho}. \tag{21}$$

The probability that one or both of the spheres $B_{R_1}(\mathbf{x}_1)$ and $B_{R_2}(\mathbf{x}_2)$ are not removed is given by

$$f_{\rm nr}(\mathbf{x}_1, R_1; \mathbf{x}_2, R_2) = \begin{cases} 0 & \text{if } r < |R_2 - R_1|, \\ \exp\left(-\varrho_0 g_{\rm nr}(\mathbf{x}_1, R_1; \mathbf{x}_2, R_2)\right) & \text{otherwise} , \end{cases} \tag{22}$$

with $B_{R<0}(\mathbf{x}) = \emptyset$ and

$$g_{\rm nr}(\mathbf{x}_1, R_1; \mathbf{x}_2, R_2) = \int_0^\infty \mathrm{d}x \; V(B_{x-R_1}(\mathbf{x}_1) \cup B_{x-R_2}(\mathbf{x}_2)) \; f_0(x). \tag{23}$$

At this point we have to consider the set union of two spheres with radii $b_1 \equiv x - R_1$ and $b_2 \equiv x - R_2$, respectively; the volume of this geometrical configuration can be calculated; in three dimensions , e.g., we have:

$$\begin{aligned} V(B_{b_1}(\boldsymbol{x}_1) \cup B_{b_2}(\boldsymbol{x}_2)) &= \frac{2\pi}{3}(b_1^3 + b_2^3) \\ &\quad - \frac{2\pi}{3}\left(\frac{r^3}{8} - \frac{3}{4} r (b_1^2 + b_2^2) - \frac{3}{8r}(b_2^2 - b_2^2)^2\right) \end{aligned} \tag{24}$$

[6] This process can be thought of as a dilution of the original Poisson process, for some general remarks on diluting Poisson processes see [58], p. 163. A comparable model was considered by [59].

for $|b_2 - b_1| \leq r = |\mathbf{x}_2 - \mathbf{x}_2| \leq b_1 + b_2$. Otherwise this volume reduces either to the volume of the larger sphere ($r < |b_2 - b_1|$) or to the sum of both spherical volumes ($r > b_1 + b_2$).
Similarly as in ((20)) the spatial two-point density turns out to be

$$\varrho_2^S(\mathbf{x}_1, \mathbf{x}_2) = \varrho_0^2 \int_0^\infty \mathrm{d}R_1 \int_0^\infty \mathrm{d}R_2 f_0(R_1) f_0(R_2) f_{\mathrm{nr}}(\mathbf{x}_1, R_1; \mathbf{x}_2, R_2) \quad , \tag{25}$$

such that the conditional two-point mark density simply reads

$$\mathcal{M}_2(R_1, R_2|\mathbf{x}_1, \mathbf{x}_2) = f_0(R_1) f_0(R_2) f_{\mathrm{nr}}(\mathbf{x}_1, R_1; \mathbf{x}_2, R_2) \frac{\varrho_0^2}{\varrho_2^S(\mathbf{x}_1, \mathbf{x}_2)} . \tag{26}$$

From this we can derive all of the mark correlation functions from Sect. 1.4.

A bimodal distribution: In order to get an analytically tractable model we adopt a bimodal radius distribution in the original Boolean model and start therefore with

$$f_0(R) = \alpha_0 \delta(R - R_1) + (1 - \alpha_0) \delta(R - R_2) \quad , \tag{27}$$

where we assume that $R_1 < R_2$. Due to the depletion the number density ϱ of the spheres as well as the probability α to find the smaller radius R_2 at a given point are then lowered; we get

$$\alpha = \alpha_0 \frac{\mathrm{e}^{-n}}{1 - \alpha_0 + \alpha_0 \mathrm{e}^{-n}} \leq \alpha_0, \tag{28}$$

$$\varrho = \varrho_0 \left(1 - \alpha_0 + \alpha_0 \mathrm{e}^{-n}\right) = \varrho_0 \frac{1 - \alpha_0}{1 - \alpha} \tag{29}$$

with $n = \varrho(1 - \alpha) \frac{4\pi}{3} (R_2 - R_1)^3$. Altogether, the bimodal model can be parameterized in terms of the radii R_1 R_2, the ratio $\alpha_0 \in [0, 1]$ and the density $\varrho_0 \in \mathbf{R}^+$. The latter two quantities, however are not observable from the final point process, therefore we convert them into the parameters $\alpha \in [0, 1]$ and $\varrho \in \mathbf{R}^+$, so that all other quantities can be expressed in terms of these, for instance, $\alpha_0 = \frac{\alpha}{\alpha + (1-\alpha)\mathrm{e}^{-n}} \geq \alpha$, and $\varrho_0 = \varrho \alpha \mathrm{e}^n + \varrho(1 - \alpha)$,
From ((21)) we determine the mean mark, i.e. the mean radius of the spheres

$$\overline{m} = \overline{R} = \alpha R_1 + (1 - \alpha) R_2, \tag{30}$$

and from ((25)) the spatial product density

$$\varrho_2^S(r) = \varrho^2 \begin{cases} (1 - \alpha)^2 R_2 + \alpha^2 R_1 \exp\left(n I(x)\right) & 0 \leq x < 1, \\ 1 + \alpha^2 \left[\exp\left(n I(x)\right) - 1\right] & 1 \leq x < 2, \\ 1 & 2 \leq x, \end{cases} \tag{31}$$

with the normalized inter-sectional volume $I(x) = 1 - \frac{3}{4}x + \frac{1}{16}x^3$ of two spheres and $x = \frac{r}{|R_2 - R_1|}$. Finally, using ((26)) one can calculate the mark correlation functions, e.g.

$$k_m(r) = \begin{cases} 1 - \alpha^2(1-\alpha)\frac{R_2-R_1}{\overline{R}}\frac{\exp(nI(x))-\alpha^{-1}+1}{(1-\alpha)^2+\alpha^2\exp(nI(x))} & 0 \le x < 1, \\ 1 - \alpha^2(1-\alpha)\frac{R_2-R_1}{\overline{R}}\frac{\exp(nI(x))-1}{1+\alpha^2[\exp(nI(x))-1]} & 1 \le x < 2, \\ 1 & 2 \le x. \end{cases} \tag{32}$$

In Fig. 11 the $k_m(r)$ function from the Boolean depletion model is shown. The model with the solid line illustrates that a reduced $k_m(r)$ for small radii can be obtained by simply removing smaller spheres. At least qualitatively this model is able to explain the depletion effects we have seen both in the distribution of Martian craters (Fig. 7) and in the distribution of pores in sandstone (Fig. 10). The jump at $r = R_2 - R_1$ is a relict of the strictly bimodal distribution with only two radii. Figure 11 also shows that the Boolean depletion model is quite flexible, allowing for a $k_m(r) < 1$, but also $k_m(r) > 1$ is possible.

Without ignoring the considerable difference of this Boolean depletion model to the pore size distribution in real sandstones (see Figs. 8-10) one may still recognize some interesting similarities: This simple model explains naturally a decrease of $k_m(r)$ if the distribution of the radii is symmetric ($\alpha = 1/2$). As visible in Fig. 9 this is approximately the case for the pore radii. Moreover, note that even quantitative features are captured correctly indicating that the decrease of $k_m(r)$ visible in Fig. 10 is indeed due to a depletion effect. For instance, the decrease starts at $r \approx R_M$ where $R_M \approx 100\mu m$ is the largest occurring radius (see the histogram in Fig. 9) and the value of $k_m(0) \approx 0.8$ at $r = 0$ is in accordance with (32) assuming that $R_2 - R_1 \approx \overline{R}$ and the normalized density of pores $n \approx 1$ necessary for a connected network. Of course a more detailed analysis is necessary based on (21) and (26) and the histogram shown in Fig. 9.

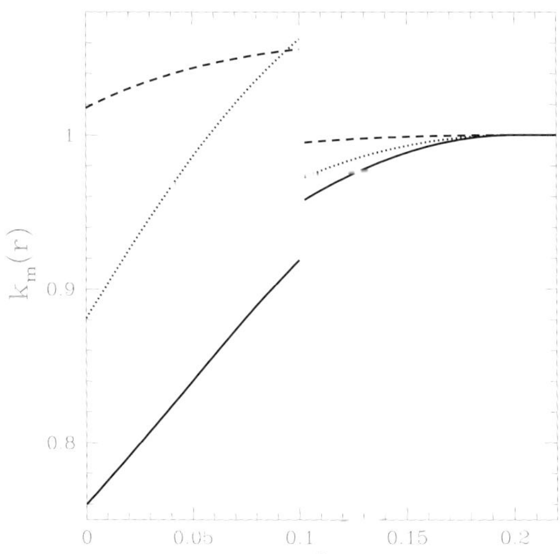

Fig. 11. The $k_m(r)$ function for the Boolean depletion model with parameters $R_1 = 0.05$, $R_2 = 0.15$, $\varrho = 500$, and $\alpha = 0.5$ (solid line), $\alpha = 0.3$ (dotted line), $\alpha = 0.1$ (dashed line).

3.2 The Random Field Model

The "random-field model" covers a class of models motivated from fields such as geology (see, e.g., [70]). The level of the ground water, for instance, is thought of as a realization of a random field which may be directly sampled at points (hopefully) independent from the value of the field or which may influence the size of a tree in a forest.

In general, a realization of the random field model is constructed from a realization of a point process and a realization of a random field $u(\mathbf{x})$. The mark of each object located at $\mathbf{x}_i$ traces the accompanying random field via $m_i = u(\mathbf{x}_i)$. The crucial assumption is that the point process is stochastically independent from the random field.

We denote the mean value of the homogeneous random field by $\overline{u} = \mathbb{E}[u(\mathbf{x})] = \overline{u^1}$ and the moments by $\overline{u^k} = \int \mathrm{d}u\, w(u) u^k$, with the one-point probability density w of the random field and $\mathbb{E}$ the expectation over realizations of the random field. The product density of the random field is $\rho_2^u(r) = \mathbb{E}\big[u(\mathbf{x}_1)u(\mathbf{x}_2)\big]$ with $r = |\mathbf{x}_1 - \mathbf{x}_2|$. For a general discussion of random field models, see [1].

In this model the one-point density of the marks is $\mathcal{M}_1(m) = w(m)$, and $\overline{m^k} = \overline{u^k}$ etc. The conditional mark density is given by

$$\mathcal{M}_2(m_1, m_2|\mathbf{x}_1, \mathbf{x}_2) = \mathbb{E}\big[\delta(m_1 - u(\mathbf{x}_1))\delta(m_2 - u(\mathbf{x}_2))\big], \tag{33}$$

where δ is the Dirac delta distribution. Clearly, this expression is only well-defined under a suitable integral over the marks. With ((7)) one obtains

$$\langle m_1 \rangle_{\mathrm{P}}(r) = \overline{u}, \quad \left\langle m_1^2 \right\rangle_{\mathrm{P}}(r) = \overline{u^2}, \quad \langle m_1 m_2 \rangle_{\mathrm{P}}(r) = \rho_2^u(r), \tag{34}$$

and the mark-correlation functions defined in Sect. 1.4 read

$$\begin{aligned} k_m(r) = 1,\ k_{mm}(r) = \rho_2^u(r)/\overline{u}^2,\ \gamma(r) = \overline{u^2} - \rho_2^u(r), \\ cov(r) = \rho_2^u(r) - \overline{u}^2,\ var(r) = \overline{u^2} - \overline{u}^2 = \sigma_M^2. \end{aligned} \tag{35}$$

Therefore, there are some explicit predictions for the random field model: an empirically determined k_m significantly differing from one not only indicates mark segregation, but also that the data is incompatible with the random field model. Looking at Fig. 3 we see immediately that the galaxy data are not consistent with the random field model. Similar tests based on the relation between k_{mm} and the mark-variogram γ were investigated by [70] and [52]. The failure of the random field model to describe the luminosity segregation in the galaxy distribution allows the following plausible physical interpretation: the galaxies do not merely trace an independent luminosity field; rather the luminosities of galaxies depend on the clustering of the galaxies. We shall try to account for this with a better model in the following section.

3.3 The Cox Random Field Model

In the random field model, the field was only used to generate the points' marks. In the Cox random field model, on the contrary, the random field determines the spatial distribution of the points as well. As before, consider a homogeneous and isotropic random field $u(\mathbf{x}) \geq 0$. The point process is constructed as a Cox-process (see e.g. [58]). The mean number of points in a set B is given by the intensity measure

$$\Lambda(B) = \int_B \mathrm{d}\mathbf{x}\; a\; u(\mathbf{x}), \tag{36}$$

where a is a proportionality factor fixing the mean number density $\varrho = a\overline{u}$. The (spatial) product density of the point distribution is

$$\varrho_2^S(\mathbf{x}_1, \mathbf{x}_2) = a^2\; \rho_2^u(r) = a^2\; \overline{u}^2(1 + \xi_2^u(r)), \tag{37}$$

where again $\rho_2^u(r)$ denotes the product density of the random field. ξ_2^u is the normalized two-point cumulant of the random field (see below). We will also need the n-point densities of the random field:

$$\rho_n^u(\mathbf{x}_1, \ldots, \mathbf{x}_n) = \mathbb{E}\big[u(\mathbf{x}_1)\cdots u(\mathbf{x}_1)\big]. \tag{38}$$

Like in the random field model, the marks trace the field, but this time rather in a probabilistic way than in a deterministic one: the mark m_i on a galaxy located at $\mathbf{x}_i$ is a random variable with the probability density $p(m_i|u(\mathbf{x}_i))$ depending on the value of the field $u(\mathbf{x}_i)$ at $\mathbf{x}_i$. This can be used as a stochastic model for the genesis of galaxies depending on the local matter density.

In order to calculate the conditional mark correlation functions we define the conditional moments of the mark distribution given the value u of the random field:

$$\overline{m^k}(u) = \int \mathrm{d}m\; p(m|u)\; m^k. \tag{39}$$

The spatial mark product-density is

$$\varrho_2^{SM}((\mathbf{x}_1, m_1), (\mathbf{x}_2, m_2)) = a^2\mathbb{E}\big[p(m_1|u(\mathbf{x}_1))p(m_2|u(\mathbf{x}_2))\; u(\mathbf{x}_1)u(\mathbf{x}_2)\big]. \tag{40}$$

and with ((3))

$$\mathcal{M}_2(m_1, m_2|\mathbf{x}_1, \mathbf{x}_2) = \frac{1}{\rho_2^u(r)}\mathbb{E}\big[p(m_1|u(\mathbf{x}_1))p(m_2|u(\mathbf{x}_2))\; u(\mathbf{x}_1)u(\mathbf{x}_2)\big], \tag{41}$$

for $\rho_2^u(r) \neq 0$ and zero otherwise. The mark correlation functions can therefore be expressed in terms of weighted correlations of the random field:

$$\begin{aligned}
\langle m\rangle_{\mathrm{P}}(r) &= \frac{1}{\rho_2^u(r)}\mathbb{E}\big[\overline{m}(u(\mathbf{x}_1))\; u(\mathbf{x}_1)u(\mathbf{x}_2)\big], \\
\langle m^2\rangle_{\mathrm{P}}(r) &= \frac{1}{\rho_2^u(r)}\mathbb{E}\Big[\overline{m^2}(u(\mathbf{x}_1))\; u(\mathbf{x}_1)u(\mathbf{x}_2)\Big], \\
\langle m_1 m_2\rangle_{\mathrm{P}}(r) &= \frac{1}{\rho_2^u(r)}\mathbb{E}\big[\overline{m}(u(\mathbf{x}_1))\overline{m}(u(\mathbf{x}_2))\; u(\mathbf{x}_1)u(\mathbf{x}_2)\big].
\end{aligned} \tag{42}$$

A special choice for $p(m|u)$: To proceed further, we have to specify $p(m|u)$. As a simple example we choose m_i equal to the value of the field $u(\mathbf{x}_i)$ at the point $\mathbf{x}_i$, such as in the random field model. Thinking of the random field as a mass density field and the mark of a galaxy luminosity, that means that the galaxies trace the density field and that their luminosities are directly proportional to the value of the field. With $p(m|u) = \delta(m-u)$ the conditional mark moments become $\overline{m^k}(u) = u^k$. The moments of the unconstrained mark distribution read $\overline{m^k} = \overline{u^{k+1}}/\overline{u}$, and the three basic pair averages are

$$\langle m_1 \rangle_{\mathrm{P}}(r) = \frac{\rho_3^u(\mathbf{x}_1,\mathbf{x}_1,\mathbf{x}_2)}{\rho_2^u(r)}, \; \langle m_1^2 \rangle_{\mathrm{P}}(r) = \frac{\rho_4^u(\mathbf{x}_1,\mathbf{x}_1,\mathbf{x}_1,\mathbf{x}_2)}{\rho_2^u(r)}$$

$$\langle m_1 m_2 \rangle_{\mathrm{P}}(r) = \frac{\rho_4^u(\mathbf{x}_1,\mathbf{x}_1,\mathbf{x}_2,\mathbf{x}_2)}{\rho_2^u(r)}. \tag{43}$$

Hence, the mark correlation functions defined in Sect. 1.4 are determined by the higher-order correlations of the random field. With the Cox random field model we go beyond the random field model, e.g.

$$k_m(r) = \frac{\langle m \rangle_{\mathrm{P}}(r)}{\overline{m}} = \frac{\overline{u}\,\rho_3^u(\mathbf{x}_1,\mathbf{x}_1,\mathbf{x}_2)}{\overline{u^2}\,\rho_2^u(r)} \tag{44}$$

is not equal to one any more.

Hierarchical field correlations: At this point, we have to specify the correlations of the random field $u(\mathbf{x})$. The simplest choice, a Gaussian random field, is not feasible here, since a number density (cp. (36)) has to be strictly positive, whereas the Gaussian model allows for negative values. Instead, we will use the hierarchical ansatz: we first express the two- and three-point correlations in terms of normalized cumulants ξ_2 and ξ_3 (see, e.g., [19, 5, 35]),

$$\rho_2^u(\mathbf{x}_1,\mathbf{x}_2) = \overline{u}^2\Big(1+\xi_2^u(\mathbf{x}_1,\mathbf{x}_2)\Big),$$

$$\rho_3^u(\mathbf{x}_1,\mathbf{x}_2,\mathbf{x}_3) = \overline{u}^3\Big(1+\xi_2^u(\mathbf{x}_1,\mathbf{x}_2)+\xi_2^u(\mathbf{x}_2,\mathbf{x}_3)+\xi_2^u(\mathbf{x}_1,\mathbf{x}_3)+\xi_3^u(\mathbf{x}_1,\mathbf{x}_2,\mathbf{x}_3)\Big). \tag{45}$$

In order to eliminate ξ_3^u we use the hierarchical ansatz (see e.g. [47]):

$$\xi_3^u(\mathbf{x}_1,\mathbf{x}_2,\mathbf{x}_3) = Q\Big(\xi_2^u(\mathbf{x}_1,\mathbf{x}_2)\xi_2^u(\mathbf{x}_2,\mathbf{x}_3)+\xi_2^u(\mathbf{x}_2,\mathbf{x}_3)\xi_2^u(\mathbf{x}_1,\mathbf{x}_3) + \xi_2^u(\mathbf{x}_1,\mathbf{x}_2)\xi_2^u(\mathbf{x}_1,\mathbf{x}_3)\Big). \tag{46}$$

This ansatz is in reasonable agreement with data from the galaxy distribution, provided Q is of the order of unity ([65]). Several choices for $\xi_2(r)$ and Q lead to well-defined Cox point process models based on the random field $u(\mathbf{x})$ [5, 66]. Now we can express $k_m(r)$ from ((44)) entirely in terms of the two-point correlation function $\xi_2^u(r)$ of the random field:

$$k_m(r) = \frac{1 + 2\xi_2^u(r) + \xi_2^u(0) + Q\Big(\xi_2^u(r)^2 + 2\xi_2^u(r)\xi_2^u(0)\Big)}{\Big(1 + \xi_2^u(r)\Big)\Big(1 + \xi_2^u(0)\Big)}, \tag{47}$$

where we made use of the fact that $\sigma_u^2 = \overline{u^2} - \overline{u}^2 = \overline{u}^2\xi_2^u(0)$. Inserting typical parameters found from the spatial clustering of the galaxy distribution we see from Fig. 12 that the Cox random field model allows us to qualitatively describe the observed luminosity segregation in Fig. 3. But the amplitude of k_m predicted by this model is too high. The Cox random field model, however, is quite flexible in allowing for different choices for $p(m|u)$; also different models for the higher-order correlations of the random field may be used, e.g. a log-normal random field [14, 42]. Clearly more work is needed to turn this into viable model for the galaxy distribution.

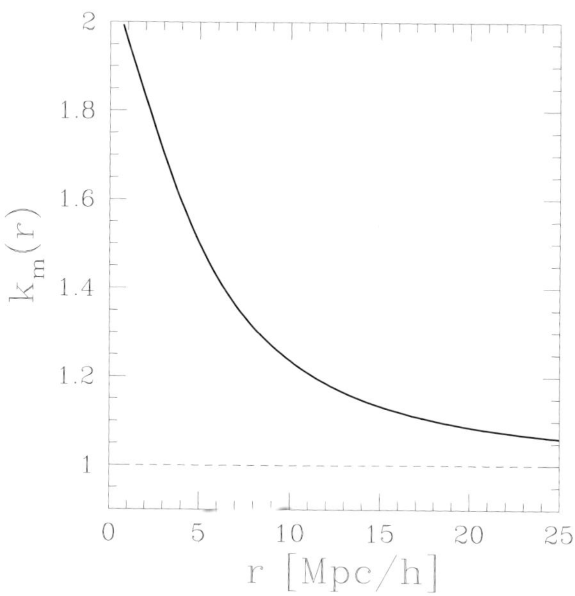

Fig. 12. The $k_m(r)$ function for the Cox random field model according to ((47)). We use $Q = 1$ and $\xi_2^u(r) = (5h^{-1}\mathrm{Mpc}/r)^{1.7}$ truncated on small scales at $\xi_2^u(r < 0.1h^{-1}\mathrm{Mpc}) = \sigma_u^2/\overline{u}^2 = \xi_2^u(0) \sim 750$.

4 Conclusions

Whenever objects are sampled together with their spatial positions and some of their intrinsic properties, marked point processes are the stochastic models for those data sets. Combining the spatial information and the objects' inner properties one can constrain their generation mechanism and their interactions.

Developing the framework of marked point processes further and outlining some of their general notions is thus of interest for physical applications. Let us therefore look at mark correlations again from both a statistical and a physical perspective. We focused on two kinds of dependencies.

On the one hand, one can always ask, whether objects of different types "know" from each other. From a statistical point of view, this is the question whether the marked point process consists of two completely independent sub-point processes. Physically, this concerns the question whether the objects have been generated together and whether they interact with each other.

On the other hand, it is often interesting to know whether the spatial distribution of the objects changes with their inner properties. For the statistician, this translates into the question whether mark segregation or mark-independent clustering is present. For the physicist such a dependency is interesting since one can learn from them whether and how the interactions distinguish between different object classes or whether the formation of the objects' mark depends on the environment.

We discussed statistics capable of probing to which extent mark correlations are present in a given data set, and showed how to assess the statistical significance. Applying our statistics to real data, we could demonstrate, that the clustering of galaxies depends on their luminosities. Large scale correlations of the orientations of dark matter halos were found. Using the Mars data we could validate a picture of crater generation on the Martian surface: mainly, the local geological setting determines the crater type. We also could show that the sizes of pores in sandstone are correlated.

In order to understand empirical data sets in detail, we need models to compare to. As generic models the Boolean depletion model, the random field model and its extension, the Cox random field models are of interest.

Further application of the mark correlations properties may inspire the development of further models. It seems therefore that marked point processes could spark interesting interactions between physicists and mathematicians. Certainly, the distributions of physicists and mathematicians in coffee breaks at the Wuppertal conference were clustered, each. But could one observe positive cross-correlations? Using mark correlations we argue, that, even more, there is lots of space for positive interactions.. . . .

Acknowledgments

We would like to thank Andreas Faltenbacher, Stefan Gottlöber and Volker Müller for allowing us to present some results from the orientation analysis of the dark matter halos (Sect. 2.2). For providing the sandstone data (Sect. 2.4) and discussion we thank Mark Knackstedt. Herbert Wagner provided constant support and encouragement, especially we would like to thank him for introducing us to the concepts of geometric algebra as used in the Appendix.

Appendix: Completeness of Mark Correlation Functions

In order to form versatile test functions for describing mark segregation effects, we integrated the conditional mark probability density $\mathcal{M}_2(m_1, m_2|r)$ twice in mark space thereby weighting with a function of the marks $f(m_1, m_2)$ (see (7)). Such a pair-averaging reduces the full information present in $\mathcal{M}_2(m_1, m_2|r)$. So one may ask, whether or in which sense the mark correlation functions give a complete picture of the present two-point mark correlations.

For scalar marks m_i this task is trivial. With a polynomial weighting function $f(m_1, m_2) \sim m_1^{n_1} m_2^{n_2}$ $(n_1, n_2 = 0, 1, ..)$ we consider moments of $\mathcal{M}_2(m_1, m_2|r)$, hence, we can be complete only up to a given polynomial order in the marks m_1 and m_2. At first order there is only the mean $\langle m \rangle_{\rm P}(r)$. At second order we have $\langle m^2 \rangle_{\rm P}(r)$ and $\langle m_1 m_2 \rangle_{\rm P}(r)$. All the mark correlation functions discussed in Sect. 1.4 can be constructed from these three pair averages[7]. Higher-order moments of the marks involve more and more cross-terms.

For vector-valued marks, however, it is not obvious that the test quantities proposed in Sect. 1.4 trace all possible correlations between the vectors up to third order. To settle this case we have to consider the framework of geometric algebra, also called Clifford algebra. A detailed introduction to geometric algebra is given in [31], shorter introductions are [26, 38]. In geometric algebra one assigns a unique meaning to the geometric product (or Clifford product) of quantities like vectors, directed areas, directed volumes, etc. The geometric product $\mathbf{ab}$ of two vectors $\mathbf{a}$ and $\mathbf{b}$ splits into its symmetric and antisymmetric part

$$\mathbf{ab} = \mathbf{a} \cdot \mathbf{b} + \mathbf{a} \wedge \mathbf{b}. \tag{48}$$

Here $\mathbf{a} \cdot \mathbf{b}$ denotes the usual scalar product; in three dimensions, the wedge product $\mathbf{a} \wedge \mathbf{b}$ is closely related to the cross product between these two vectors. However, $\mathbf{a} \wedge \mathbf{b}$ is not a vector like $\mathbf{a} \times \mathbf{b}$, but a bivector – a directed area. Higher products of vectors can be simplified according to the rules of geometric algebra (for details see [31]).

Let us consider the situation where objects situated at $\mathbf{x}_1$ and $\mathbf{x}_2$ bear vector marks $\mathbf{l}_1$ and $\mathbf{l}_2$, respectively, and let the normalized distance vector be $\hat{\mathbf{r}} = (\mathbf{x}_1 - \mathbf{x}_2)/r$. Note, that $\hat{\mathbf{r}}$ is not a mark at all, rather it can be thought of as another vector which may be useful for constructing mark correlation functions.

For many applications it is reasonable to assume isotropy in mark space, i.e. all of the mark correlation functions are invariant under common rotations of the marks. For galaxies, e.g., there does not seem to be an a priori preferred direction for their orientation. In more detail we have then

$$\mathcal{M}_1(\mathbf{l}) = \mathcal{M}_1(R\mathbf{l}) = \mathcal{M}_1(|\mathbf{l}|) \quad ,$$
$$\mathcal{M}_2(\mathbf{l}_1, \mathbf{l}_2|r) = \mathcal{M}_2(R\mathbf{l}_1, R\mathbf{l}_2|r) \quad ,$$

and so on, where R is an arbitrary rotation in mark space. This means that the mark correlation functions depend only on rotationally invariant combinations of the vector marks. Therefore, only rotationally invariant combinations of vectors are sensible building blocks for weighting functions. We thus can restrict ourselves to scalar weighting functions, which result in coordinate-independent vector-mark correlation functions.

Again we proceed by considering mixed moments as basic combinations. We restrict ourselves to scalar quantities being polynomial in the vector components. One may also

[7] This completeness of $\langle m^2 \rangle_{\rm P}(r)$ and $\langle m_1 m_2 \rangle_{\rm P}(r)$ at the two-point level, however, does not imply that one should not consider linear combinations of them. For instance, it may well be the case, that only certain linear combinations yield significant results.

discuss moments in a broader sense allowing for vector moduli. In this wider sense, for example, $|\mathbf{l}_1|$ or $|\mathbf{l}_1 \times (\mathbf{l}_1 \times \hat{\mathbf{r}})|$ would be allowed. We do not consider such quantities here, because they are not polynomial in the vector components. Their squares anyway appear at higher orders. Furthermore, it turns out that the characterization we will provide depends on the embedding dimension. The first- and second-order moments are identical in two and three dimensions, but at the third order they start to differ.

1. In the strict sense of scalar quantities being linear in the vector components there are no first-order moments for vectors.
2. At second order we encounter the following products: $\mathbf{l}_1\mathbf{l}_1$, $\hat{\mathbf{r}}\hat{\mathbf{r}}$, $\mathbf{l}_1\mathbf{l}_2$, $\hat{\mathbf{r}}\mathbf{l}_1$. Note, that, e.g., $\mathbf{l}_1\hat{\mathbf{r}}$ and $\mathbf{l}_2\hat{\mathbf{r}}$ do not make any difference as regards the mark correlation functions, since the pair averages implicitly render the indices symmetric; moreover, although the geometrical product is non-commutative, $\mathbf{l}_1 \wedge \hat{\mathbf{r}}$ and $\hat{\mathbf{r}} \wedge \mathbf{l}_1$ do not lead to different mark correlation functions. Furthermore, $\hat{\mathbf{r}}\hat{\mathbf{r}} = 1$. $\mathbf{l}_1\mathbf{l}_1 = \mathbf{l}_1 \cdot \mathbf{l}_1 = l_1^2$ provides us with higher moments of the modulus of the vectors. To investigate these kinds of correlations already scalar marks would be sufficient. New information is encoded in the other products.
   Consider $\mathbf{l}_1\mathbf{l}_2 = \mathbf{l}_1 \cdot \mathbf{l}_2 + \mathbf{l}_1 \wedge \mathbf{l}_2$. The symmetric part $\mathbf{l}_1 \cdot \mathbf{l}_2$ is clearly a scalar and defines the alignment $\mathcal{A}(r)$ (14). The antisymmetric part $\mathbf{l}_1 \wedge \mathbf{l}_2$ is a bivector. Its – unique – modulus (see again [31]), $|\mathbf{l}_1 \wedge \mathbf{l}_2| = \sqrt{l_1^2 l_2^2 - (\mathbf{l}_1 \cdot \mathbf{l}_2)^2}$, may be useful, but is no longer a polynomial in the vector components. $|\mathbf{l}_1 \wedge \mathbf{l}_2|^2$ appears at the fourth order. In a completely analogous way we can treat $\mathbf{l}_1\hat{\mathbf{r}} = \mathbf{l}_1 \cdot \hat{\mathbf{r}} + \mathbf{l}_1 \wedge \hat{\mathbf{r}}$. The symmetric part $\mathbf{l}_1 \cdot \hat{\mathbf{r}}$ defines $\mathcal{F}(r)$. Hence at second order, the only possible vector-mark correlation functions are $\mathcal{A}(r)$ and $\mathcal{F}(r)$.
3. At third order we have to consider products of three vectors. In general the product of three vectors $\mathbf{a}, \mathbf{b}, \mathbf{c}$ splits into

$$\mathbf{abc} = \mathbf{a}(\mathbf{b} \cdot \mathbf{c}) + (\mathbf{a} \cdot \mathbf{b})\mathbf{c} - (\mathbf{a} \cdot \mathbf{c})\mathbf{b} + \mathbf{a} \wedge (\mathbf{b} \wedge \mathbf{c}). \tag{49}$$

   i.e., a vector (consisting of the three first terms), and a pseudo-scalar, a directed volume. In two dimensions the pseudo-scalar $\mathbf{a} \wedge (\mathbf{b} \wedge \mathbf{c})$ vanishes.
   Now we have to form all possible products of the three vectors $\mathbf{l}_1, \mathbf{l}_2, \hat{\mathbf{r}}$ and to derive scalars. In three dimensions the only new combination is the pseudo-scalar $\mathbf{l}_1 \wedge (\mathbf{l}_2 \wedge \hat{\mathbf{r}})$ giving the oriented volume $\mathbf{l}_1 \cdot (\mathbf{l}_2 \times \hat{\mathbf{r}})$. Unfortunately, this oriented volume averages out to zero. Thus, in a strict sense, there are no interesting third-order quantities. Closely related, however, is the modulus of the pseudoscalar $|\mathbf{l}_1 \cdot (\mathbf{l}_2 \times \hat{\mathbf{r}})|$ proportional to our $\mathcal{P}(r)$. This expression is invariant under permutations of the vectors.
4. At third order and in two dimensions all of the relevant combinations are products of first- and second-order combinations; no specifically new combination appears. This is different from the case of three dimensions, where at third order an entirely new geometric object, the pseudo-scalar $\mathbf{l}_1 \wedge (\mathbf{l}_2 \wedge \hat{\mathbf{r}})$ can be constructed. There is a general scheme behind this argument: since in d dimensions any geometrical product of more than d vectors vanishes, all relevant combinations of vectors at orders higher than d are essentially products of combinations of lower-order factors.

References

1. Adler, R. J. (1981): *The Geometry of Random Fields* (John Wiley & Sons, Chichester)
2. Arns, C., M. Knackstedt, W. Pinczewski, K. Mecke (2001): 'Characterisation of irregular spatial structures by prallel sets', In press
3. Arns, C., M. Knackstedt, W. Pinczewski, K. Mecke (2001): 'Euler-poincaré characteristics of classes of disordered media', *Phys. Rev. E* **63**, p. 31112
4. Baddeley, A. J. (1999): 'Sampling and censoring'. In: *Stochastic Geometry, Likelihood and Computation*, ed. by O. Barndorff-Nielsen, W. Kendall, M. van Lieshout, volume 80 of *Monographs on Statistics and Applied Probability*, chapter 2 (Chapman and Hall, London)
5. Balian, R., R. Schaeffer (1989): 'Scale–invariant matter distribution in the Universe I. counts in cells', *Astronomy & Astrophysics* **220**, pp. 1–29
6. Barlow, N. G., T. L. Bradley (1990): 'Martian impact craters: Correlations of ejecta and interior morphologies with diameter, latitude, and terrain', *Icarus* **87**, pp. 156–179
7. Beisbart, C., M. Kerscher (2000): 'Luminosity– and morphology–dependent clustering of galaxies', *Astrophysical Journal* **545**, pp. 6–25
8. Benoist, C., A. Cappi, L. Da Costa, S. Maurogordato, F. Bouchet, R. Schaeffer (April 1999): 'Biasing and high-order statistics from the southern-sky redshift survey', *Astrophysical Journal* **514**, pp. 563–578
9. Benoist, C., S. Maurogordato, L. Da Costa, A. Cappi, R. Schaeffer (December 1996): 'Biasing in the galaxy distribution', *Astrophysical Journal* **472**, p. 452
10. Bertschinger, E. (1998): 'Simulations of structure formation in the universe', *Ann. Rev. Astron. Astrophys.* **36**, pp. 599–654
11. Binggeli, B. (1982): 'The shape and orientation of clusters of galaxies', *Astronomy & Astrophysics* **107**, pp. 338–349
12. Böhringer, H., P. Schuecker, L. Guzzo, C. Collins, W. Voges, S. Schindler, D. Neumann, G. Chincharini, R. Cruddace, A. Edge, H. MacGillivray, P. Shaver (2001): 'The ROSTA-ESO flux limited X-ray (REFLEX) galaxy cluster survey I: The construction of the cluster sample', *Astronomy & Astrophysics* , p. 826
13. Capobianco, R., E. Renshaw (1998): 'The autocovariance function of marked point processes: A comparison between two different approaches', *Biom. J.* **40**, pp. 431–446
14. Coles, P., B. Jones (January 1991): 'A lognormal model for the cosmological mass distribution', *MNRAS* **248**, pp. 1–13
15. Coles, P., F. Lucchin (1994): *Cosmology: The Origin and Evolution of Cosmic Structure* (John Wiley & Sons, Chichester)
16. Cox, D., V. Isham (1980): *Point Processes* (Chapman and Hall, London)
17. Cressie, N. (1991): *Statistics for Spatial Data* (John Wiley & Sons, Chichester)
18. da Costa, L. N., C. N. A. Willmer, P. Pellegrini, O. L. Chaves, C. Rite, M. A. G. Maia, M. J. Geller, D. W. Latham, M. J. Kurtz, J. P. Huchra, M. Ramella, A. P. Fairall, C. Smith, S. Lipari (1998): 'The Southern Sky Redshift Survey', *AJ* **116**, pp. 1–7
19. Daley, D. J., D. Vere-Jones (1988): *An Introduction to the Theory of Point Processes* (Springer, Berlin)
20. Diggle, P. J. (1983): *Statistical Analysis of Spatial Point Patterns* (Academic Press, New York and London)
21. Djorgovski, S. (1987): 'Coherent orientation effects of galaxies and clusters'. In: *Nearly Normal Galaxies. From the Planck Time to the Present*, ed. by S. M. Faber (Springer, New York), pp. 227–233
22. Faltenbacher, A., S. Gottlöber, M. Kerscher, V. Müller (2002): 'Correlations in the orientation of galaxy clusters', submitted to Astronomy & Astrophysics

23. Flannery, B. P., H. W. Deckman, W. G. Roberge, K. L. D'amico (1987): 'Three–dimensional X–ray microtomography', *Science* **237**, pp. 1439–1444
24. Fuller, T. M., M. J. West, T. J. Bridges (1999): 'Alignments of the dominant galaxies in poor clusters', *Astrophysical Journal* **519**, pp. 22–26
25. Gottlöber, S., M. Kerscher, A.V. Kravtsov, A. Faltenbacher, A. Klypin, V. Müller (2002): 'Spatial distribution of galactic halos and their merger histories', *Astronomy & Astrophysics* **387**, pp. 778
26. Gull, S., A. Lasenby, C. Doran (1993): 'Imaginary numbers are nor real. – the geometric algebra of spacetime', *Found. Phys.* **23(9)**, p. 1175
27. Guzzo, L., J. Bartlett, A. Cappi, S. Maurogordato, E. Zucca, G. Zamorani, C. Balkowski, A. Blanchard, V. Cayatte, G. Chincarini, C. Collins, D. Maccagni, H. MacGillivray, R. Merighi, M. Mignoli, D. Proust, M. Ramella, R. Scaramella, G. Stirpe, G. Vettolani (2000): 'The ESO Slice Project (ESP) galaxy redshift survey. VII. the redshift and real-space correlation functions', *Astronomy & Astrophysics* **355**, pp. 1–16
28. Hamilton, A. J. S. (August 1988): 'Evidence for biasing in the cfa survey', *Astrophysical Journal* **331**, pp. L59–L62
29. Heavens, A. F., A. Refregier, C. Heymans (2000): 'Intrinsic correlation of galaxy shapes: implications for weak lensing measurements', *MNRAS* **319**, pp. 649–656
30. Hermit, S., B. X. Santiago, O. Lahav, M. A. Strauss, M. Davis, A. Dressler, J. P. Huchra (1996): 'The two–point correlation function and the morphological segregation in the optical redshift survey', *MNRAS* **283**, p. 709
31. Hestens, D. (1986): *New Foundations for Classical Mechanics* (D. Reidel Publishing Company, Dordrecht, Holland)
32. Huchra, J. P., M. J. Geller, V. De Lapparent, H. G. Corwin Jr. (1990): 'The CfA redshift survey – data for the NGP + 30 zone', *Astrophysical Journal Supplement* **72**, pp. 433–470
33. Isham, V. (1985): 'Marked point processes and their correlations'. In: *Spatial Processes and Spatial Time Series Analysis*, ed. by F. Droesbeke (Publications des Facultés universitaires Sain-Louis, Bruxelles)
34. Kerscher, M. (2000): 'Statistical analysis of large–scale structure in the Universe'. In: *Statistical Physics and Spatial Statistics: The Art of Analyzing and Modeling Spatial Structures and Pattern Formation*, ed. by K. R. Mecke, D. Stoyan, Number 554 in Lecture Notes in Physics (Springer, Berlin), astro-ph/9912329
35. Kerscher, M. (2001): 'Constructing, characterizing and simulating Gaussian and high–order point processes', *Phys. Rev. E* **64(5)**, p. 056109, astro-ph/0102153
36. Klypin, A. A. (2000): 'Numerical simulations in cosmology i: Methods', in 'Lecture at the Summer School "Relativistic Cosmology: Theory and Observations"', Astro-ph/0005502
37. Lambas, D. G., E. J. Groth, P. Peebles (1988): 'Statistics of galaxy orientations: Morphology and large–scale structure', *Astronomical Journal* **95**, pp. 975–984
38. Lasenby, J., A. N. Lasenby, C. J. Doran (2000): 'A unified mathematical language for physics and engineering in the 21st century', *Phil. Trans. R. Soc. London A* **358**, pp. 21–39
39. Löwen, H. (2000): 'Fun with hard spheres'. In: *Statistical Physics and Spatial Statistics: The Art of Analyzing and Modeling Spatial Structures and Pattern Formation*, ed. by K. R. Mecke, D. Stoyan, Number 554 in Lecture Notes in Physics (Springer, Berlin)
40. Mecke, K. (2000): 'Additivity, convexity, and beyond: Application of minkowski functionals in statistical physics'. In: *Statistical Physics and Spatial Statistics: The Art of Analyzing and Modeling Spatial Structures and Pattern Formation*, ed. by K. R. Mecke, D. Stoyan, Number 554 in Lecture Notes in Physics (Springer, Berlin)
41. Melott, A. L., S. F. Shandarin (1990): 'Generation of large-scale cosmological structures by gravitational clustering', *Nature* **346**, pp. 633–635.
42. Møller, J., A. R. Syversveen, R. P. Waagepetersen (1998): 'Log Gaussian cox processes', *Scand. J. Statist.* **25**, pp. 451–482

43. Ogata, Y., K. Katsura (1988): 'Likelihood analysis of spatial inhomogeneity for marked point pattersn', *Ann. Inst. Statist. Math.* **40**, pp. 29–39
44. Ogata, Y., M. Tanemura (1985): 'Estimation of interaction potentials of marked spatial point patterns through the maximum likelihood method', *Biometrics* **41**, pp. 421–433
45. Ohser, J., D. Stoyan (1981): 'On the second–order and orientation analysis of planar stationary point processes', *Biom. J.* **23**, pp. 523–533
46. Onuora, L. I., P. A. Thomas (2000): 'The alignment of clusters using large–scale simulations', *MNRAS* **319**, pp. 614–618
47. Peebles, P. J. E. (1980): *The Large Scale Structure of the Universe* (Princeton University Press, Princeton, New Jersey)
48. Peebles, P. J. E. (1993): *Principles of Physical Cosmology* (Princeton University Press, Princeton, New Jersey)
49. Penttinen, A., D. Stoyan (1989): 'Statistical analysis for a class of line segment processes', *Scand. J. Statist.* **16**, pp. 153–168
50. Reichert, H., O. Klein, H. Dosch, M. Denk, V. Honkimäki, T. Lippmann, G. Reiter (2000): 'Observation of five-fold local symmetry in liquid lead', *Nature* **408**, p. 839
51. Schlather, M. (2001): 'On the second–order characteristics of marked point processes', *Bernoulli* **7(1)**, pp. 99–107
52. Schlather, M. (2002): 'Characterization of point processes with Gaussian marks independent of locations', *Math. Nachr.* Accepted
53. Sok, R. M., M. A. Knackstedt, A. P. Sheppard, W. V. Pinczewski, W. B. Lindquist, A. V. A, , L. Paterson (2000): 'Direct and stochastic generation of network models from tomographic images; effect of topology on two-phase flow properties'. In: *Proc. Upscaling Downunder*, ed. by L. Paterson, (Kluwer Academic, Dordrecht)
54. Spanne, P., J. Thovert, C. Jacquin, W. Lindquist, K. Jones, P. Adler (1994): 'Synchrotron computed microtomography of porous media: Topology and transport', *Phys. Rev. Lett.* **73**, pp. 2001–2004
55. Stoyan, D. (1984): 'On correlations of marked point processes', *Math. Nachr.* **116**, pp. 197–207
56. Stoyan, D. (2000): 'Basic ideas of spatial statistics'. In: *Statistical Physics and Spatial Statistics: The Art of Analyzing and Modeling Spatial Structures and Pattern Formation*, ed. by K. R. Mecke, D. Stoyan, Number 554 in Lecture Notes in Physics (Springer, Berlin)
57. Stoyan, D. (2000): 'Recent applications of point process methods in forestry statistics', *Statistical Sciences* **15**, pp. 61–78
58. Stoyan, D., W. S. Kendall, J. Mecke (1995): *Stochastic Geometry and its Applications* (John Wiley & Sons, Chichester), 2nd edition
59. Stoyan, D., H. Stoyan (1985): 'On one of matérn's hard-core point process models', *Biom. J.* **122**, p. 205
60. Stoyan, D., H. Stoyan (1994): *Fractals, Random Shapes and Point Fields* (John Wiley & Sons, Chichester)
61. Stoyan, D., O. Wälder (2000): 'On variograms in point process statistics, ii: Models for markings and ecological interpretation', *Biom. J.* **42**, pp. 171–187
62. Struble, M. F., P. Peebles (1985): 'Erratum: A new application of Binggeli's test for large–scale alignment of clusters of galaxies', *Astronomical Journal* **90**, pp. 582–589
63. Struble, M. F., P. Peebles (1986): 'A new application of binggeli's test for large–scale alignment of clusters of galaxies', *Astronomical Journal* **91**, p. 1474
64. Sylos Labini, F., M. Montuori, L. Pietronero (1998): 'Scale invariance of galaxy clustering', *Physics Rep.* **293**, pp. 61–226
65. Szapudi, I., G. B. Dalton, G. Efstathiou, A. S. Szalay (1995): 'Higher order statistics from the APM galaxy survey', *Astrophysical Journal* **444**, pp. 520–531

66. Szapudi, I., A. S. Szalay (1993): 'Higher order statistics of the galaxy distribution using generating functions', *Astrophysical Journal* **408**, pp. 43–56
67. Ulmer, M., S. L. W. McMillan, M. P. Kowalski (1989): 'Do the major axis of rich clusters of galaxies point toward their neighbors?', *Astrophysical Journal* **338**, pp. 711–717
68. van Lieshout, M. N. M., A. J. Baddeley (1996): 'A nonparametric measure of spatial interaction in point patterns', *Statist. Neerlandica* **50**, pp. 344–361
69. van Lieshout, M. N. M., A. J. Baddeley (1999): 'Indices of dependence between types in multivariate point patterns', *Scand. J. Statist.* **26**, pp. 511–532
70. Wälder, O., D. Stoyan (1996): 'On variograms and point process statistics', *Biom. J.* **38**, pp. 895–905
71. Wälder, O., D. Stoyan (1997): 'Models of markings and thinnings of poisson processes', *Statistics* **29**, pp. 179–202
72. Widom, B., J. Rowlinson (1970): 'New model for the study of liquid–vapor phase transitions', *J. Chem. Phys.* **52**, pp. 1670–1684
73. Willmer, C., L. N. da Costa, P. Pellegrini (March 1998): 'Southern sky redshift survey: Clustering of local galaxies', *Astronomical Journal* **115**, pp. 869–884

Spatial Jump Processes and Perfect Simulation

Kasper K. Berthelsen and Jesper Møller

Department of Mathematical Sciences, Aalborg University,
Fredrik Bajers Vej 7G DK-9220 Aalborg, Denmark

Abstract. Spatial birth-and-death processes, spatial birth-and-catastrophe processes, and more general types of spatial jump processes are studied in detail. Particularly, various kinds of coupling constructions are considered, leading to some known and some new perfect simulation procedures for the equilibrium distributions of different types of spatial jump processes. These equilibrium distributions include many classical Gibbs point process models and a new class of models for spatial point processes introduced in the text.

1 Introduction

Many dynamic systems of interacting "objects" evolving in time can be described by a stochastic process $X = \{X_t : t \geq 0\}$, where t denotes time. Often the objects can be viewed as points living in some appropriate space S, so that $X_t \subseteq S$ for all times t. For example, S could be a planar or spatial region; or a space for some geometric objects like line segments or discs. One class of models for such stochastic processes is that of spatial jump processes.

In this contribution we consider the case of finite spatial jump processes, i.e. when X_t is a finite subset of S. Recall that in general a jump process is a continuous-time Markov process $X = \{X_t : t \geq 0\}$ with piecewise constant sample path [7]. In the special case where each jump consists in adding/deleting exactly one point, we have a spatial birth-and-death process [25]. This is illustrated in Fig. 1 when $S = [0, 1]$ is a unit interval and in Fig. 2 when $S = [0, 1]^2$ is a unit square. For a general spatial jump process as considered later in the text, let x and y be finite subsets of S. Then conditionally on the current state $X_t = x$ and the previous history $\{X_s : s < t\}$, the waiting time τ_x to the next jump (i.e. X_s stays in x for $t \leq s < t + \tau_x$, and $X_{t+\tau_x} \neq x$) is exponentially distributed and depends only on x. Furthermore, conditionally on both $X_t = x$, $\{X_s : s < t\}$, and τ_x, the next state $X_{t+\tau_x} = y$ follows some probability measure P_x which only depends on x so that either $x \subset y$ or $x \supset y$.

Spatial jump processes are interesting for several reasons. They provide a large class of models for spatio-temporal processes, which e.g. may describe many interacting particle systems studied in physics. As we shall demonstrate, they also provide mathematical tractable models for spatio-temporal processes. Specificly, we consider different coupling techniques which allow us to study the ergodic behaviour and characterise the equilibrium distribution for a large class of models in the reversible case as well as in the irreversible case. Furthermore, due to certain thinning techniques and since the sample paths are piecewise constant, spatial jump processes can often easily be simulated on

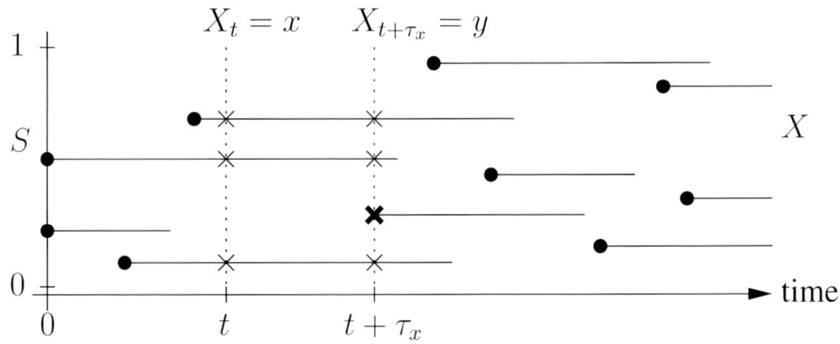

Fig. 1. Schematic description of a spatial birth-and-death process defined on $S = [0, 1]$. Each horizontal line represents the life span of a point. The point configuration $X_t = x$ at time t is shown (indicated by the crosses at time t). The waiting time τ_x to the next transition (here a birth) is indicated, and the new point configuration $X_{t+\tau_x} = y$ is shown (indicated by the crosses at time $t + \tau_x$, the bold cross being the new born point).

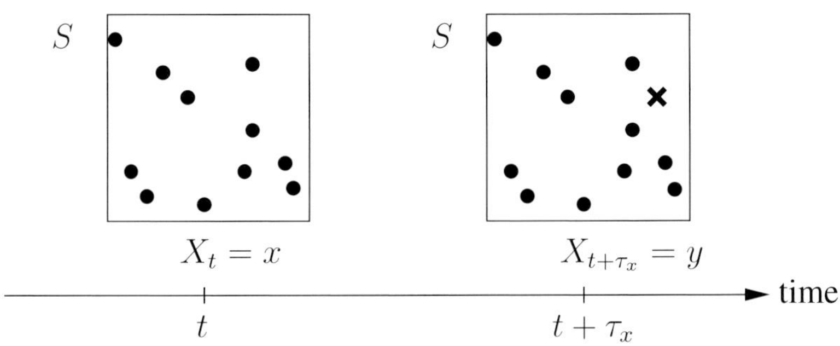

Fig. 2. Schematic description of a spatial birth-and-death process defined on $S = [0, 1]^2$. The point pattern $X_t = x$ at time t is shown, the waiting time τ_x to the next transition (here a birth) is indicated, and the new point configuration $X_{t+\tau_x} = y$ is shown (here the cross is the new born point).

a computer. Particularly, we extend the ideas of [26] and use the thinning techniques for various perfect (or exact) simulation algorithms based on dominated coupling from the past [14, 15] . As in [15], we prefer the term "perfect simulation" instead of "exact simulation", since random number generators always have defects and an algorithm may fail to deliver a simulation within practical constraints of time.

Currently perfect simulation is a hot research topic in statistics and applied probability, and as such the perfect simulation part (Sects. 4 and 5.2) is of its own interest. Perfect simulation techniques have proven to be particular useful in statistical physics, spatial statistics, and stochastic geometry. A primer on perfect simulation is provided in [32] and a recent review on perfect simulation in stochastic geometry is given in [22]. [9] and [11] deal with perfect simulation and some aspects of phase transition in the Widom-Rowlinson model [33] and related models. For a survey on the historic development of perfect simulation and a comprehensive list of references, see `http://www.dbwilson.com/exact`.

In this contribution we discuss various constructions of spatial jump processes, study the ergodicity properties of these processes, characterise their equilibrium distributions, and show how to start spatial jump processes in equilibrium by the use of perfect simulation techniques. Readers interested in a further discussion of the statistical and computational aspects (using perfect simulations) are referred to [3]. Furthermore, many of the ideas in the perfect simulation part of the text apply to Gibbs sampling (also known in statistical physics as the heat-bath algorithm) and other Metropolis-Hastings algorithms for spatial point processes [11, 15, 22, 23, 31].

The text is organised as follows. Section 2 provides some background material on simple jump processes (i.e. when the position of points is ignored), which becomes useful when we later couple simple jump processes with spatial jump processes. The main part of the text concerns spatial birth-and-death processes Sect. 3 is concerned with general properties of spatial birth-and-death processes, using certain coupling constructions. A description and comparison of the perfect simulation algorithms in [8] and [15] using spatial birth-and-death processes is given in Sect. 4. Section 5 concerns spatial birth-and-catastrophe processes, i.e. when a jump consists in either adding a new point or deleting all existing points. Section 6 comments on more general cases of spatial jump processes. Finally, the main notation is listed and briefly explained in the Appendix.

2 Simple Jump Processes

This section is a short diversion into simple jump processes as they will eventually control everything which goes on in this text. See for example [1] and [24] for more details.

A simple jump process $N = \{N_t : t \geq 0\}$ is a continuous time Markov process with state space $\mathbb{N}_0 = \{0, 1, 2, \ldots\}$ and piecewise constant right-continuous paths. It can formally be constructed as follows. Let $J_1 < J_2 < J_3 \ldots$ be the times at which N makes a jump, and set $J_0 = 0$. Further, for all $n \in \mathbb{N} = \{1, 2, \ldots\}$, let

$$H_n = \begin{cases} J_n - J_{n-1} & \text{if } J_n < \infty \\ \infty & \text{otherwise} \end{cases}$$

be the holding times, and set $M_n = N_{J_n}$. The discrete time Markov chain $M = \{M_n : n \in \mathbb{N}_0\}$ is called the jump chain (or the embedded Markov chain). For any $n \in \mathbb{N}$ and any $i \in \mathbb{N}_0$, the conditional distribution of H_n given $H_1, \ldots, H_{n-1}, M_0, \ldots, M_{n-1} = i$ is $\mathrm{Exp}(q_i)$, the exponential distribution with mean $1/q_i$, where $q_i > 0$ is a given parameter. Furthermore, for any $j \in \mathbb{N}_0 \setminus \{i\}$,

$$\mathbb{P}(M_n = j | H_1, \ldots, H_n, M_0, \ldots, M_{n-1} = i) = q_{ij}/q_i,$$

where the $q_{ij} \geq 0$ are given parameters so that $q_i = \sum_{j \neq i} q_{ij}$. The matrix $Q = \{q_{ij} : i, j \in \mathbb{N}_0\}$ with $q_{ii} = -q_i$ is called the generator of N.

In other words, conditionally on $N_t = i$, the waiting time τ_i to the next jump is independent of the previous history $\{N_s : s < t\}$, and $\tau_i \sim \mathrm{Exp}(q_i)$. Moreover, conditionally on both $N_t = i$, $\{N_s : s < t + \tau_i\}$, and τ_i, we have that $N_{t+\tau_i}$ jumps to j with probability q_{ij}/q_i for $j \neq i$.

In the sequel we impose the following technical conditions, which are commented below.

Conditions:
(i) $\pi = \{\pi_i : i \in \mathbb{N}_0\}$ is a given probability density function where the support $I = \{i : \pi_i > 0\}$ is given by either $I = \mathbb{N}_0$ or $I = \{0, \ldots, l\}$ for some $l \in \mathbb{N}_0$.
(ii) Q is irreducible, i.e. for all distinct $i, j \in I$ there exist $i_0, i_1, \ldots i_n \in I$ such that $i_0 = i$, $i_n = j$, and $q_{i_0 i_1} \ldots q_{i_{n-1} i_n} > 0$.
(iii) π is invariant for Q, i.e. $\pi Q = 0$.
(iv) Q is non-explosive, i.e. $\mathbb{P}(\sum_n H_n < \infty | N_0 = i) = 0$ for all $i \in I$; see Fig. 3.

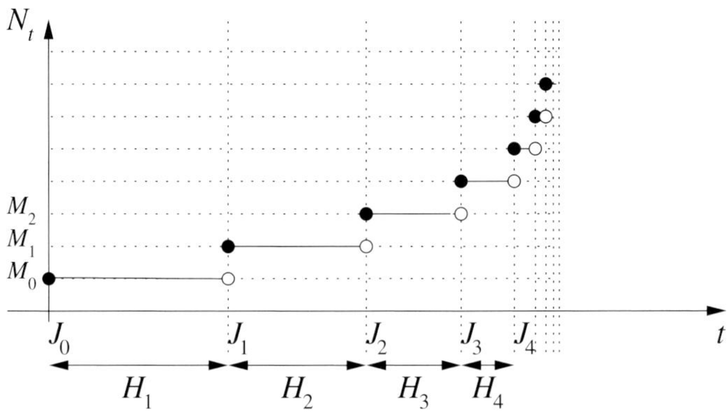

Fig. 3. Schematic description of a simple jump process which does not satisfy the non-explosive condition.

Remarks:
(a) The density π will play a key role in the sequel. That π is invariant for Q means that if N_s follows π, then N_t follows π for all $t \geq s$. Condition (i) is a kind of hereditary condition on π. It is equivalent to assuming that the support is of the form $I = \{0, 1, \ldots, l\}$ or $I = \mathbb{N}_0$.
(b) Conditions (ii) and (iii) imply that π is the unique invariant distribution for Q.
(c) Sometimes we assume that Q and π are in detailed balance, i.e.

$$\pi_i q_{ij} = \pi_j q_{ji} \text{ for all } i, j \in \mathbb{N}_0. \tag{1}$$

(d) Detailed balance implies that π is invariant for Q. Combining (ii), (iv), and (1) we obtain reversibility of N.
(e) Condition (iv) holds if and only if the only bounded solution to $Qk = k$ for column vectors $k = (k_0, k_1, \ldots)^{\mathrm{T}}$ is $k = (0, 0, \ldots)^{\mathrm{T}}$ (see e.g. [1]). A sufficient condition for this is $\sup_{i \in I} q_i < \infty$.
(f) Conditions (ii) and (iv) imply positive recurrence, i.e. for any $i \in I$, the time between two consecutive occurrences of i in N has finite mean.
(g) Conditions (i)–(iv) imply convergence towards π, see e.g. Theorem 3.6.2 in [24] or Theorem II.4.6 in [1].

(h) For later purposes we notice that the jump chain has invariant density $\pi'_i \propto q_i\pi_i$, and it is positive recurrent (by Theorems 3.4.1 and 3.5.3 in [24], the jump chain is recurrent; and recurrence implies positive recurrence as the support I is countable). Furthermore, let L be the number of jumps in a cycle of the jump chain, i.e. the number of jumps between two successive zeros in the jump chain. Then by Kac's theorem [20],

$$\mathbb{E}L = 1/\pi'_0. \tag{2}$$

3 Spatial Birth-and-Death Processes

This section considers finite spatial birth-and-death processes defined on some rather arbitrary space S. Section 3.1 concerns the simplest case where S consists of one element only; this case corresponds to a simple jump process with jumps of the type $i \to i \pm 1$. Section 3.2 introduces a general notation for both spatial birth-and-death processes and the spatial jump processes considered later in this text. The general case of a spatial birth-and-death process is described in Sect. 3.3, using a coupling to a dominating simple birth-and-death process. Section 3.4 discusses a detailed balance condition and an extension of the concept of local stability [29]. Section 3.5 considers how to construct in an easy way a spatial birth-and-death process by thinning from a simpler dominating spatial birth-and-death process, extending ideas in [14] and [15]. This coupling construction will also be used in Sect. 4 for perfect simulation purposes.

3.1 The Simple Case

A simple jump process with generator Q is a simple birth-and-death process if $q_{ij} = 0$ whenever $|i-j| > 1$. For convenience set $b_i = q_{i,i+1}$ and $d_{i+1} = q_{i+1,i}$ for $i \in \mathbb{N}_0$. Throughout this section we assume that

$$\pi_i b_i = \pi_{i+1} d_{i+1} > 0 \quad \text{whenever } \pi_{i+1} > 0. \tag{3}$$

This assumption is equivalent to the conditions (i)–(iii) and the detailed balance condition (1) in Sect. 2. Regarding the final condition (iv) in Sect. 2, by [27], a simple birth-and-death process is non-explosive if

$$\sum_{i\in I} \frac{1}{b_i} = \infty. \tag{4}$$

In the sequel we often refer to the following Examples 1–3.

Example 1: Consider a standard immigration birth-and-death process, i.e. $b_i \equiv \beta > 0$ and $d_{i+1} = i+1$ for $i \in \mathbb{N}_0$. Clearly, (3) and (4) are satisfied, and

$$\pi_i = e^{-\beta}\beta^i/i!, \tag{5}$$

i.e. π specifies a Poisson distribution with mean β. By (h) in Sect. 2, $\pi'_{i+1} \propto \pi_{i+1} + \pi_i$. Hence by (2), $\mathbb{E}L = 2e^\beta$.

Example 2: For $i \in \mathbb{N}_0$, let $b_i = p$ and $d_{i+1} = 1 - p$, where $0 < p < \frac{1}{2}$. Then (3) and (4) are satisfied, and

$$\pi_i = (1-q)q^i \tag{6}$$

specifies a geometric distribution with parameter $q = p/(1-p)$. The corresponding jump chain is a random walk with reflecting barrier at 0. Using (h) and (2), we obtain that $\mathbb{E}L = 2q/(1-q)$.

Example 3: In this example we start by specifying π and then determine the birth and death rates such that (3) and (4) hold. Assume that

$$\pi_i = \frac{\Gamma(i+\alpha)}{\Gamma(\alpha)i!}(1-q)^\alpha q^i \tag{7}$$

is negative binomial, where $\alpha > 0$ and $0 < q < 1$. Letting $q = p/(1-p)$, one solution to (3) is an extension of Example 2 such that

$$b_i = p(i+\alpha)/(i+1) \quad \text{and} \quad d_{i+1} = 1 - p.$$

Then (4) is seen to hold.

3.2 Notation for Spatial Jump Processes

We now consider some general notation which applies not only to spatial birth-and-death processes but all spatial jump processes considered in this text. We avoid topological and measure theoretical details; such details can be found in e.g. [4, 5, 16].

In the spatial case of jump processes $X = \{X_t : t \geq 0\}$, we consider this as a stochastic processes defined on a metric space S with metric $d(\cdot,\cdot)$; in many of our examples, $S = [0,1]^2$ is a unit square and $d(\xi,\eta) = \| \xi - \eta \|$ is the Euclidian distance. For simplicity we consider X_t to be a finite subset of S, though everything in the sequel easily extend to the case where X_t is allowed to have multiple points. For any finite $x \subseteq S$, we let $n(x)$ denote the number of points in x. Then the state space of X_t is $\Omega = \bigcup_{i=0}^\infty \Omega_i$, where $\Omega_i = \{x \subseteq S : n(x) = i\}$ is the set of finite point configurations of cardinality i; note that $\Omega_0 = \{\emptyset\}$ where $\emptyset$ denotes the empty point configuration. For $x \in \Omega$ and $\xi \in S$, we write $x\backslash\xi$ for $x\backslash\{\xi\}$, and $x \cup \xi$ for $x \cup \{\xi\}$.

Finally, we let λ denote an arbitrary diffuse probability measure on S (here "diffuse" means that $\lambda(\{\xi\}) = 0$ for all $\xi \in S$). For example, if $S \subset \mathbb{R}^d$ has finite volume, λ is typically taken to be the uniform distribution on S.

3.3 Description of Spatial Birth-and-Death Processes

Now, consider a spatial birth-and-death process X. For the following description of X the reader may find it illuminating to consider Fig. 1 and Fig. 2 in Sect. 1.

We specify the distribution of X by two functions $\beta, \delta : \Omega \times S \to [0,\infty)$. For any $x \in \Omega$, $\beta(x,\xi)$ is the birth rate at which a point $\xi \in S$ is added to x, and if $x \neq \emptyset$, $\delta(x\backslash\eta,\eta)$ is the death rate at which a point $\eta \in x$ is deleted from x. Roughly speaking, for an infinitesimal time interval $[t, t+dt]$ and an infinitesimal ball $d\xi$ with centre ξ,

$$\mathbb{P}(X_{t+dt} \setminus x \text{ is a point in } d\xi \,|X_t = x) \approx \beta(x,\xi)\lambda(d\xi)dt,$$
$$\mathbb{P}(X_{t+dt} = x\backslash\eta \,|X_t = x) \approx \delta(x\backslash\eta,\eta)dt,$$
$$\mathbb{P}(\text{more than one jump in } [t, t+dt]|X_t = x) \approx 0.$$

More precisely, define in a similar way as in Sect. 2, the jump times $J_1 < J_2 < \ldots$, the holding times $H_1, H_2, \ldots$, and the jump chain $Y = \{Y_n : n \in \mathbb{N}_0\}$ for X, where $Y_0 = X_0$. Further, for $x \in \Omega$, let $B(x) = \int_S \beta(x,\xi)\lambda(d\xi)$ and $D(x) = \sum_{\eta\in x} \delta(x\backslash\eta,\eta)$ be the total birth and death rates, assuming $B(x) < \infty$ and setting $D(\emptyset) = 0$. Then the conditional distribution of H_n given $H_1, \ldots, H_{n-1}, Y_0, \ldots, Y_{n-1} = x$ is $\text{Exp}\,(B(x) + D(x))$. If we also condition on H_n, the probability for a birth at time J_n is $B(x)/(B(x) + D(x))$. If we further condition on that a birth happens at time J_n, then $Y_n = x \cup \xi$, where ξ has density $\beta(x,\cdot)/B(x)$ with respect to λ. If we instead condition on that a death happens at time J_n, then $Y_n = x\backslash\eta$, where $\eta \in x$ is selected with probability $\delta(x\backslash\eta,\eta)/D(x)$.

It remains to clarify if such a spatial birth-and-death process is non-explosive. Given a generator Q for a non-explosive simple birth-and-death process N as in Sect. 3.1, we assume henceforth that

$$\beta(x,\xi) = b_{n(x)}\bar{b}(x,\xi) \quad \text{and} \quad \delta(x,\xi) = d_{n(x)+1}\bar{d}(x,\xi)/(n(x)+1) \tag{8}$$

where the functions $\bar{b}$ and $\bar{d}$ satisfy the following constraints:

$$0 \le \bar{b} \le 1 \quad \text{and} \quad \bar{d} \ge 1. \tag{9}$$

By (8) and (9),

$$B(x) \le b_{n(x)} \quad \text{and} \quad D(x) \ge d_{n(x)}. \tag{10}$$

It is then possible to couple N and X so that

$$N_t \ge n(X_t) \quad \text{for all } t \ge 0 \tag{11}$$

if $N_0 \ge n(X_0)$. Briefly, (X, N) is a jump process where X and N evolve independently of each other as long as $N_t > n(X_t)$, while if $N_t = n(x) = n$ and $X_t = x$, (X_t, N_t) jumps to (x', n') with a rate specified as follows:

$$\beta(x,\xi) \text{ for } (x', n') = (x \cup \xi, n+1),$$
$$b_n - B(x) \text{ for } (x', n') = (x, n+1),$$
$$d_n\delta(x\backslash\eta,\eta)/D(x) \text{ for } (x', n') = (x\backslash\eta, n-1),$$
$$(D(x) - d_n)\delta(x\backslash\eta,\eta)/D(x) \text{ for } (x', n') = (x\backslash\eta, n).$$

By (10) and (11), since Q is non-explosive, it is intuitively clear that X is non-explosive. A formal proof can be found in [25].

Combining (f) in Sect. 2, (11), and the renewal theorem (see e.g. Theorem V.1.2 in [1]), we obtain the following results. Recall that $\emptyset$ denotes the empty point configuration. Write X_{t-} for the state of X just before time t, let $\mathcal{T} = \inf\{t > 0 : X_{t-} \neq \emptyset, X_t = \emptyset\}$ be the return time to $\emptyset$ when $X_0 = \emptyset$, and set $\mu = \mathbb{E}(\mathcal{T}|X_0 = \emptyset)$. Then $\emptyset$ is an ergodic atom, i.e. $\mu < \infty$. Hence X jumps to $\emptyset$ infinite often, and at each such jump time X regenerates (i.e. X at such a time is independent of its past history). As $t \to \infty$, X_t converges in distribution towards an equilibrium distribution κ, say, independent of the initial state of X_0, i.e. for all bounded functions $f : \Omega \to \mathbb{R}$ which are continuous almost everywhere with respect to κ ,

$$\mathbb{E}[f(X_t)|X_0] \to \int f(x)\kappa(dx) \quad \text{as } t \to \infty. \tag{12}$$

Further,

$$\mu \int f(x)\kappa(dx) = \mathbb{E}\left[\int_0^{\mathcal{T}} f(X_t)dt \middle| X_0 = \emptyset \right], \tag{13}$$

and the support of κ is given by

$$\text{supp}(\kappa) = \Omega_0 \cup \Big\{ \{x_1, \ldots, x_n\} \in \Omega : n \in \mathbb{N} \text{ and} \\ \beta(\{x_1, \ldots, x_{i-1}\}, x_i) > 0 \text{ for } i = 1, \ldots, n \Big\}. \tag{14}$$

Finally, we often have geometrically fast convergence towards κ, cf. [21]. For instance, this is the case when b_i and d_i are specified as in Examples 1–3.

3.4 Detailed Balance and Local Stability Conditions

Let the situation be as in Sect. 3.3. In this section we try to get hold on the equilibrium distribution κ in (13). Particularly, we discuss a detailed balance condition which characterises κ in terms of an (unnormalised) density ϕ. Furthermore, we motivate an extension of the important concept of local stability [10, 15, 29].

In order to specify the invariant distribution κ, we assume that ϕ is an unnormalised density with respect to the probability measure on Ω given by

$$\nu = \sum_{i=0}^{\infty} \pi_i \lambda_i. \tag{15}$$

Here π is given by (3), λ_0 is the probability measure concentrated at the empty point configuration, and λ_i $(i \geq 1)$ denotes the distribution for a binomial point process given by i independent points in S with common distribution λ. Note that for a point process $Z \sim \phi$, conditionally on $n(Z) = i$, Z has unnormalised density ϕ with respect to λ_i, i.e. no matter the choice of π. In most applications π is a Poisson distribution with mean β. Then ν is simply a Poisson point process with intensity measure $\beta\lambda$. Moreover, if λ is the uniform distribution on a bounded set $S \subset \mathbb{R}^d$, then $-\ln \beta$ is the so-called chemical activity for a point process $Z \sim \phi$.

Now, as in [28], we obtain that the detailed balance condition,

$$\bar{b}(x, \xi)\phi(x) = \bar{d}(x, \xi)\phi(x \cup \xi) > 0 \quad \text{whenever } \phi(x \cup \xi) > 0 \tag{16}$$

ensures that κ has a density proportional to ϕ, and X is time reversible with respect to κ ([28] assumes π to be Poisson, but we do not need to make this restriction). Note that (9) and (16) always imply that ϕ is non-increasing,

$$\phi(x \cup \xi) \leq \phi(x) \tag{17}$$

which in turn implies integrability of ϕ with respect to ν no matter the choice of π. We refer to (17) as local stability of κ with respect to ν for the following reasons.

[10] and [15] consider the common case (5) where π is Poisson and

$$\phi(x \cup \xi) \leq K\phi(x), \quad \bar{d} \equiv 1, \quad \bar{b}(x,\xi) = \begin{cases} \phi(x \cup \xi)/\phi(x) & \text{if } \phi(x) \neq 0 \\ 0 & \text{otherwise} \end{cases} \tag{18}$$

for some constant K (independent of x and ξ). Then points are dying with uniform rate 1, and $\bar{b}$ is the Papangelou conditional intensity [5, 12]. Note that (18) implies (16), and the existence of the uniform upper bound K on $\bar{b}$ is local stability in the sense of [29]. As we can replace β in (5) by $\beta' = \beta K$, and ϕ in (18) by $\phi'(x) = \phi(x)/K^{n(x)}$, we can without loss of generality assume $K = 1$. Then (9) and (17) are satisfied.

On the other hand, if π is not Poisson, then in general we do not have Ruelle local stability. For instance, let π be given by (6), and let $\tilde{\nu} = \sum_i \tilde{\pi}_i \lambda_i$ where $\tilde{\pi}_i = e^{-1}/i!$ is Poisson. Then $\tilde{\phi}(x) = \phi(x) q^{n(x)} n(x)!$ is the corresponding unnormalised density with respect to $\tilde{\nu}$, but (17) does not ensure that

$$\tilde{\phi}(x \cup \xi)/\tilde{\phi}(x) = (\phi(x \cup \xi)/\phi(x))q(n(x)+1)$$

is bounded by a constant K.

We now consider two examples where (17) is satisfied. The first example is frequently used later in the text, and it shows the effect of the specification of π. The other demonstrates that the specification of π and ϕ sometimes requires a little thought.

Example 4: Suppose that

$$\phi(x) = \gamma^{s_R(x)} \tag{19}$$

where $s_R(x) = \sum_{\{\xi,\eta\} \subseteq x} \mathbb{1}[d(\xi,\eta) \leq R]$, and $0 \leq \gamma \leq 1$ and $R \geq 0$ are given parameters. In the case where π is Poisson, we obtain a usual Strauss process [13, 30]; if further $\gamma = 0$, we have a Gibbs hard core process (taking $0^0 = 1$ in (19); see Example 7 below). Clearly, ϕ is non-increasing.

Simulated realisations of different Strauss processes are shown in Fig. 4. For this the perfect simulation algorithm in Sect. 4.2 has been used. Due to the thinning procedure in that algorithm (as illustrated later in Fig. 6), the first point pattern in Fig. 4 contains the two others.

Figure 5 shows the mean number of points for different values of γ when π is either Poisson or geometric; the means are estimated by Monte Carlo methods using the perfect simulation algorithms in Sects. 4.2 and 5.2. In the geometric case, for γ close to 1, the mean is very sensitive to changes in γ, while for γ not close to 1, the number of points is very small.

Example 5: Assume that $S \subset \mathbb{R}^k$ has Lebesgue measure 1, λ is the uniform distribution on S, and $\pi_i \propto \beta^i/i!$ is Poisson. Consider an area interaction point process with unnormalised density

$$\phi(x) = \gamma^{-\lambda(U_x)}$$

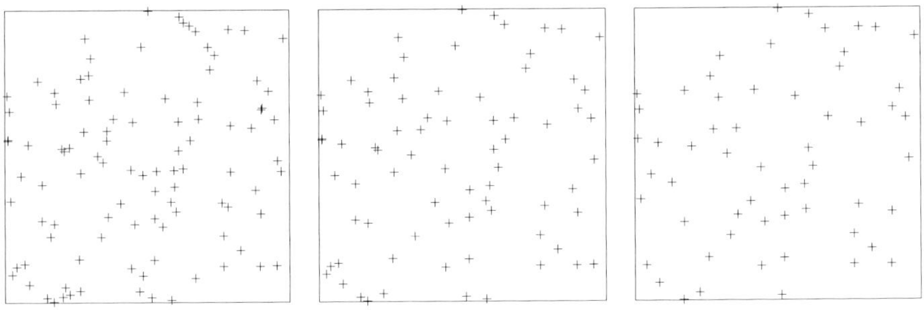

Fig. 4. Simulation of a usual Strauss process ($\pi_i \propto \beta^i/i!$) on $S = [0,1]^2$, when $\beta = 100$, $R = 0.05$, and $\gamma = 1, 0.5, 0$ (from left to right).

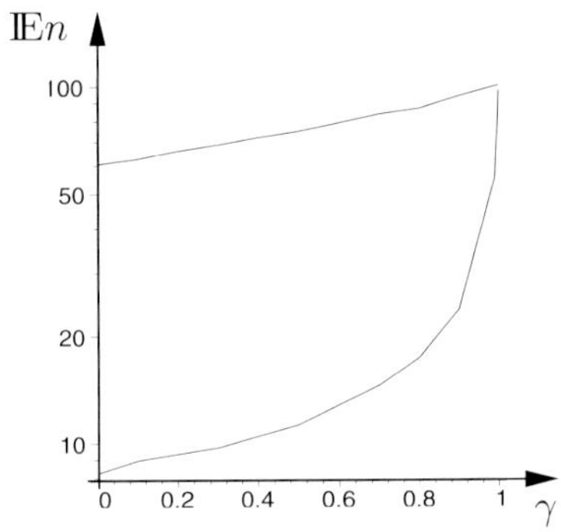

Fig. 5. Mean number of points versus γ in a usual Strauss process ($\pi_i \propto \beta^i/i!$; upper curve) and in a "geometric-Strauss" process ($\pi_i \propto q^i$; lower curve). In both models, $R = 0.05$ and π has mean 100 (i.e. $\beta = 100$ and $q = 100/101$). Each mean number of points is estimated from 500 i.i.d. samples.

with respect to ν [2, 11, 22, 33]. Here $U_x = \cup_{\xi \in x} \text{ball}(\xi, R)$ where $\text{ball}(\xi, R)$ denotes the closed ball centred in ξ with radius R, and $R > 0$ and $\gamma \geq 0$ are given parameters. The area interaction process and its generalisations are relevant in physics, see [19] and the references therein.

The area interaction process is said to be attractive for $\gamma > 1$ and repulsive for $\gamma < 1$, as

$$\phi(x \cup \xi)/\phi(x) = \gamma^{-\lambda(U_{x \cup \xi} \setminus U_x)} \tag{20}$$

is non-decreasing ($\gamma > 1$) or non-increasing ($\gamma < 1$) in x. It follows from (20) that (17) holds in the attractive case but not in the repulsive case. In the latter case we therefore redefine ν by taking $\pi_i \propto (\beta\gamma^{-B})^i/i!$ where B is the volume of a ball in $\mathbb{R}^k$ with radius R, and redefine ϕ by

$$\phi(x) = \gamma^{n(x)B - \lambda(U_x)}.$$

Then (17) is satisfied.

3.5 Coupling Construction

Consider again the situation in Sect. 3.3. This section presents a coupling constructing where X is obtained by thinning from a dominating spatial birth-and-death process D, which is easily constructed. The coupling construction extends that of [14] and [15], and it is advantageous for simulation of X as the computation of the integral $B(x) = \int_S \beta(x,\xi)\lambda(d\xi)$ is not needed. Furthermore, the coupling construction becomes later useful for perfect simulation.

As will be clarified below, we need to assume that

$$b_{i-1} \leq b_i \quad \text{and} \quad d_i/i \geq d_{i+1}/(i+1) \quad \text{for } i \in I\backslash\{0\}. \tag{21}$$

For instance, (21) is satisfied in Examples 1–3.

The dominating process D has birth and death rates

$$\beta^D(x,\xi) = b_{n(x)} \quad \text{and} \quad \delta^D(x,\xi) = d_{n(x)+1}/(n(x)+1).$$

This choice corresponds to a maximal birth rate and a minimal death rate in (8), cf. (9). By (16), D has invariant distribution ν. As the total birth and death rates satisfy

$$B^D(x) = b_{n(x)} \quad \text{and} \quad D^D(x) = d_{n(x)},$$

we can take $N = n(D)$. The process $D_t,\ t \geq 0$, can easily be generated: The simple birth-and-death process N is straightforwardly generated, and $D_t,\ t \geq 0$, is constructed forwards in time by

$$N_t = N_{t-} + 1 \Rightarrow D_t = D_{t-} \cup \xi_t,$$
$$N_t = N_{t-} - 1 \Rightarrow D_t = D_{t-}\backslash \eta_t,$$

where $\xi_t \sim \lambda$ and η_t is a uniformly selected point from D_{t-} (and where ξ_t and η_t are independent of the history of D before time t).

We wish to couple X to D so that D dominates X in the sense that

$$X_t \subseteq D_t \quad \text{for all } t \geq 0. \tag{22}$$

For simplicity, as in [14] and [15], we consider the case where $d_i \propto i$. This is fulfilled in Example 1. Further we assume that $\bar{d}$ and $\bar{b}$ are specified by (18) where ϕ satisfies (17). Then X_t can be constructed iteratively for $t \geq 0$ as follows. Each time $N_t = N_{t-} + 1$ we associate to the birth in D at time t a mark $R_t \sim$ Uniform$[0,1]$ which is independent of $\{D_s : 0 \leq s \leq t\}$ and previous marks $R_s,\ 0 \leq s < t$. Initially let $X_0 \subseteq D_0$ and then

$$N_t = N_{t-} \qquad \Rightarrow X_t = X_{t-} \tag{23}$$

$$N_t = N_{t-} + 1 \Rightarrow X_t = \begin{cases} X_{t-} \cup \xi_t & \text{if } R_t \leq \bar{b}(X_{t-},\xi_t) b_{n(X_{t-})}/b_{N_{t-}} \\ X_{t-} & \text{otherwise} \end{cases} \tag{24}$$

$$N_t = N_{t-} - 1 \Rightarrow X_t = X_{t-}\backslash \eta_t. \tag{25}$$

Clearly this construction satisfies (22) as illustrated in Fig. 6. It is straightforwardly verified that the birth and death rates of X are given by (8). Note that (22) and (24) imply the need of the first restriction in (21).

Example 3 (and hence also Example 2) does not satisfy $d_i \propto i$, but it is still to some extend possible to use the coupling construction above, letting now $d_i = (1-p)i$ and $b_i = (i+\alpha)p$. Then the detailed balance condition (3) is fulfilled, and Q is non-explosive as (4) is satisfied.

Now, consider the situation where we neither assume $d_i \propto i$ nor that $\bar{b}$ and $\bar{d}$ are specified by (18). In general we have that (22) together with (25) imply a lower bound on the death rate

$$\delta(x\backslash\eta,\eta) = d_{n(x)}\bar{d}(x\backslash\eta,\eta)/n(x) \geq d_{n(y)}/n(y) \quad \text{for } \eta \in x, y \in \Omega, x \subseteq y, x \neq \emptyset.$$

This explains why the second restriction in (21) is needed. By (8), for each $\eta \in X_{t-}$, we need a rate

$$d_{n(X_{t-})}\bar{d}(X_{t-}\backslash\eta,\eta)/n(X_{t-}) - d_{n(D_{t-})}/n(D_{t-}) \tag{26}$$

for deleting η from X but not from D. In the coupling construction below we show how this is possible. However, readers who prefer to skip these technical details should move on to Sect. 4 where only the construction in (23)–(25) is used.

The process X_t is iteratively generated forwards in time $t \geq 0$ as follows. Initially let $X_0 \subseteq D_0$, and let $X_t = X_{t-}$ whenever $N_t = N_{t-}$. Consider first the case $t = 0$. Set $x = X_t$, $i = n(x)$, and $j = n(D_t)$. Let $\tau = \min\{s > t : N_{s-} \neq N_s\}$ be the first jump time in N after time t. In accordance with (26), for each $\eta \in x$, define

$$a_\eta = d_i\bar{d}(x\backslash\eta,\eta)/i - d_j/j \quad \text{and} \quad A(x) = \sum_{\eta\in x} a_\eta, \tag{27}$$

setting $A(x) = 0$ if $x = \emptyset$. Then generate $\tau' \sim \text{Exp}(A(x))$ (independently of τ and the history of D and X before and including time t), setting $\tau' = \infty$ if $A(x) = 0$. If $\tau' < \tau - t$, then increase t by τ' and set $X_t = x\backslash\eta$, where $\eta \in x$ is chosen with probability $a_\eta/A(x)$ (independently of τ, τ', and the history of D and X before time t). Else set $t = \tau$ and let

$$N_t = N_{t-} + 1 \Rightarrow X_t = \begin{cases} x \cup \xi_t & \text{if } R_t \leq \bar{b}(x,\xi_t)b_i/b_j \\ x & \text{otherwise} \end{cases} \tag{28}$$

$$N_t = N_{t-} - 1 \Rightarrow X_t = x\backslash\eta_t, \tag{29}$$

where in (28) we associate to the birth in D a mark $R_t \sim \text{Uniform}[0,1]$ (independent of τ, τ', the history of D before and including time t, and the history of X before time t). Repeating this procedure it is straightforwardly verified that (22) is satisfied and the birth and death rates of X are given by (8).

4 Perfect Simulation

In this section we discuss perfect simulation procedures based on coupling from the past [26] for a point process with a distribution κ with respect to ν in (15). We assume that κ is the equilibrium distribution of a spatial birth-and-death process satisfying the

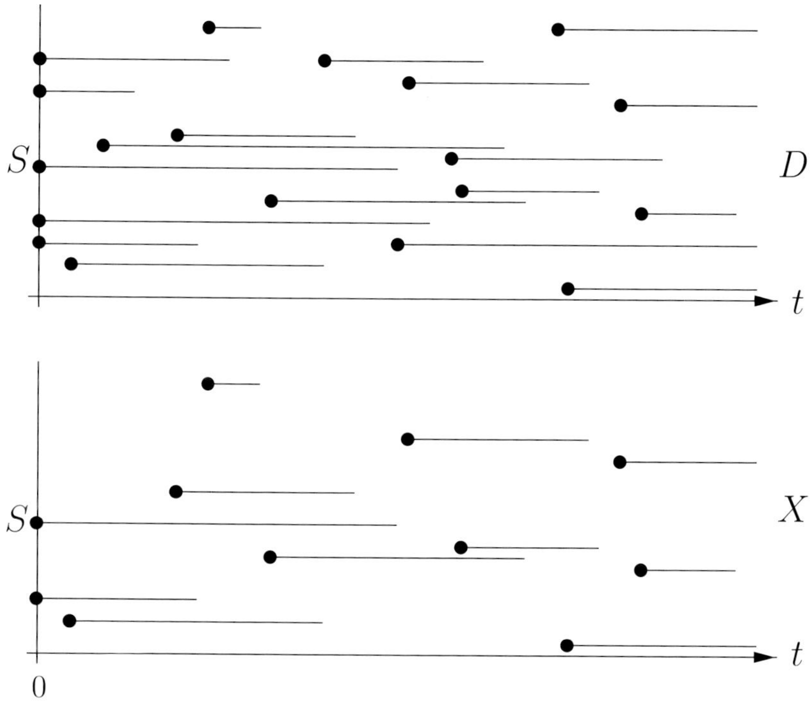

Fig. 6. Illustration of the coupling construction when $\bar{d} \equiv 1$, $X_0 \subseteq D_0$ and $S = [0, 1]$. Top: the dominating process D. Bottom: the target process X obtained by thinning from D as described in (24).

conditions in Sects. 3.3 and 3.5. We use the coupling construction in Sect. 3.5 in two ways, extending the methods of [15] in Sects. 4.1 and 4.2, and of [8] in Sect. 4.3. Finally, in Sect. 4.3 we compare the methods and report on some empirical findings.

Throughout this section we assume for simplicity that $d_i \propto i$, and $\bar{d}$ and $\bar{b}$ are specified by (18) where ϕ satisfies (17). However, it is possible to extend our perfect samplers to the case of the more complicated coupling construction described at the end of Sect. 3.5.

4.1 Dominated Coupling from the Past

In this section we describe the simplest version of dominated coupling from the past (dominated CFTP), also called horizontal CFTP, cf. the survey in [22]. For further details on dominated CFTP in a general setting, see [15]. Refined versions of dominated CFTP are given in Sects. 4.2 and 4.3.

The basic idea in dominated CFTP is as follows. Start the dominating process D from Sect. 3.5 in equilibrium at time 0, and extend it both forwards and backwards in time. This is generally easily done, since D has equilibrium distribution ν and D is time reversible. Further, for all forwards births $D_t \backslash D_{t-} \neq \emptyset$ with $-\infty < t < \infty$, generate independent marks $R_t \sim \text{Uniform}[0, 1]$, where the R_t's are independent of $\{D_t : -\infty < t < \infty\}$.

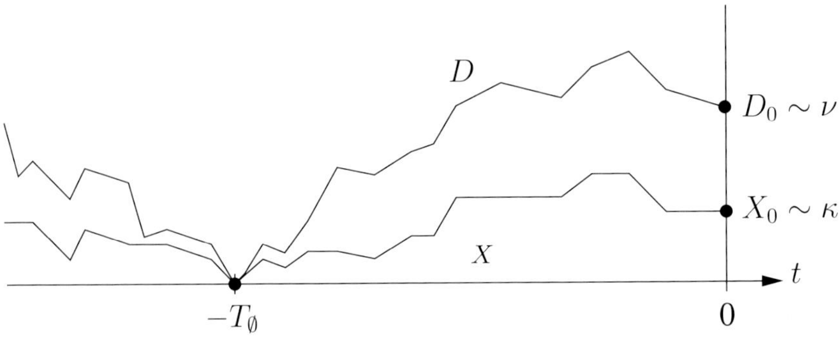

Fig. 7. Illustration of the idea behind dominated CFTP (coupling from the past).

Let τ_i, $i \in \mathbb{N}_0$, denote the times D enters $\emptyset$, i.e. $D_{\tau_i-} \neq \emptyset$ and $D_{\tau_i} = \emptyset$, and assume that $\tau_i < \tau_{i+1}$ for all $i \in \mathbb{N}_0$. For each $i \in \mathbb{N}_0$, construct a target process X_t on the time interval $[\tau_i, \tau_{i+1}]$ in exactly the same way as in Sect. 3.5 except that we are now starting with $X_{\tau_i} = \emptyset$. Then (D, X) regenerates each time D enters $\emptyset$, $X_t \subseteq D_t$ for all $-\infty < t < \infty$, and X is a spatial birth-and-death process with equilibrium distribution κ, where κ has unnormalised density ϕ. Because of time-stationarity of (D, X), for any two fixed times, e.g. 0 and -1, X_0 and X_{-1} follow the same distribution, which is then the invariant distribution κ. Hence $X_0 \sim \kappa$.

Let $-T_\emptyset \leq 0$ denote the first time that D enters $\emptyset$ when D is generated backwards from time 0. As $\emptyset$ is an ergodic atom, $T_\emptyset$ is finite almost surely. Note that for the generation of X_0, it suffices to consider the jump chains of D and X on the time interval $[-T_\emptyset, 0]$. Let $T'_\emptyset$ be the number of jumps D makes on $[-T_\emptyset, 0]$. As illustrated in Fig. 7, for the perfect simulation procedure described above, we need only to

(I) generate $D_0 \sim \nu$;
(II) generate the jump chain of D (together with the marks for forwards births), for $T'_\emptyset$ steps backwards in time from time 0;
(III) construct the jump chain of $\{X_t : -T_\emptyset \leq t \leq 0\}$ forwards in time;
(IV) return $X_0 \sim \kappa$.

However, for many applications $T'_\emptyset$ will be infeasibly large as shown in the following example.

Example 6: Consider the jump chain $\{\ldots, M_{-2}, M_{-1}, M_1, M_2, \ldots\}$ from Sect. 2 when this is in equilibrium and extended to all times $n \in \mathbb{Z}\setminus\{0\}$. Set $M_0 = M_{-1}$, $L_0 = T'_\emptyset + T''_\emptyset$ if $M_0 \neq 0$, and $L_0 = \inf\{n \in \mathbb{N} : M_n = 0\}$ if $M_0 = 0$, where $T'_\emptyset = \inf\{n \in \mathbb{N} : M_{-n} = 0\} - 1$ and $T''_\emptyset = \inf\{n \in \mathbb{N}_0 : M_n = 0\}$. By reversibility, $T'_\emptyset$ and $T''_\emptyset$ are identically distributed. Further, conditionally on $M_0 = 0$, we have that $L, T'_\emptyset, T''_\emptyset$ are identically distributed. Furthermore, by time-stationarity, $\mathbb{P}(L_0 = l) = \mathbb{P}(L = l)l/\mathbb{E}L$, hence $\mathbb{E}L_0 \geq \mathbb{E}L$, i.e. $\mathbb{E}T'_\emptyset \geq \mathbb{E}L/2$. Combining these properties with (2), it is straightforwardly derived that

$$\mathbb{E}T'_\emptyset \geq (1 + 1/\pi'_0)/2. \tag{30}$$

In the examples below we just use that $\mathbb{E}T'_\emptyset \geq 1/(2\pi'_0)$, since π'_0 is very close to 0 in cases which are of practical interest, and we let μ_π denote the mean of π.

Consider first the Poisson case $\pi_i \propto \beta^i/i!$. By (30), $\mathbb{E}T'_\emptyset \geq e^\beta$ is at least exponentially growing in β. For example, if $\beta = 100$ (i.e. $\mu_\pi = 100$),

$$\mathbb{E}T'_\emptyset \geq e^{100} \approx 2.7 \times 10^{43}. \tag{31}$$

Consider next the case where $\pi_i \propto q^i \Gamma(i+\alpha)/i!$ is negative binomial as in Example 3, where $\alpha > 0$ and $0 < q < 1$. Let $d_i = i$ and $b_i = (i+\alpha)q$ (as noticed in Sect. 3.5, Q is then non-explosive). By (30),

$$\mathbb{E}T'_\emptyset \geq (1-q)^{-(1+\alpha)}. \tag{32}$$

If $(1-q)^{1-\alpha} \leq \alpha q(1+q)$, a better bound is obtained as follows. The probability for a transition $i \to i+1$ in M is $q(i+\alpha)/(q(i+\alpha)+i)$. As i increases, $q(i+\alpha)/(q(i+\alpha)+i)$ decreases towards $p \equiv q/(1+q) < 1/2$. Hence it suffices to consider a random walk $\{S_n : n \in \mathbb{N}_0\}$ starting in $S_0 = N_0$ and with $\mathbb{P}(S_{n+1} - S_n = 1) = 1 - \mathbb{P}(S_{n+1} - S_n = -1) = p$, since we can couple the random walk to M so that $T''_\emptyset \geq \mathcal{T}_0$ where $\mathcal{T}_0 = \inf\{n \in \mathbb{N}_0 : S_n = 0\}$. By Wald's identity, $N_0 = (1-2p)\mathbb{E}(\mathcal{T}_0|N_0)$, and so $\mathbb{E}N_0 \leq (1-2p)\mathbb{E}T''_\emptyset$, i.e.

$$\mathbb{E}T'_\emptyset \geq \alpha q(1+q)/(1-q)^2. \tag{33}$$

For instance, in the case of a geometric distribution ($\alpha = 1$), this is a better bound than that in (32) when $q > (\sqrt{5}-1)/2 \approx 0.618$, and we obtain

$$\mathbb{E}T'_\emptyset \geq 20100 \tag{34}$$

if $q = 100/101$ and $\alpha = 1$ (corresponding to $\mu_\pi = 100$).

This indicates that the running time may be much smaller in the geometric case than in the Poisson case (at least when $\mu_\pi = 100$ in both cases). In the geometric case we expect the distribution of $T'_\emptyset$ to be heavy-tailed so that very long running times may occur. However, in the negative binomial case with $\alpha < 1$, a smaller running time is obtained using the algorithm in Sect. 5.2.

4.2 Upper and Lower Processes

A faster perfect simulation procedure can be obtained by constructing upper and lower processes $U^s = \{U^s_t : s \leq t \leq 0\}$ and $L^s = \{L^s_t : s \leq t \leq 0\}$ for $s \leq 0$. These are constructed forward in time as follows. Initially set $U^s_s = D_s$ and $L^s_s = \emptyset$. For $s < t \leq 0$, if $d = D_{t-}$, $u = U^s_{t-}$, and $l = L^s_{t-}$, let

$$\begin{aligned}
D_t = d \quad &\Rightarrow U^s_t = u \quad \text{and} \quad L^s_t = l, \\
D_t = d\setminus\eta_t \quad &\Rightarrow U^s_t = u\setminus\eta_t \quad \text{and} \quad L^s_t = l\setminus\eta_t, \\
D_t = d \cup \xi_t \quad &\Rightarrow U^s_t = \begin{cases} u \cup \xi_t & \text{if } R_t \leq \alpha_{\max}(d,u,l,\xi_t) \\ u & \text{otherwise} \end{cases}
\end{aligned} \tag{35}$$

$$\text{and } L^s_t = \begin{cases} l \cup \xi_t & \text{if } R_t \leq \alpha_{\min}(d,u,l,\xi_t) \\ l & \text{otherwise,} \end{cases} \tag{36}$$

where

$$\alpha_{\max}(d,u,l,\xi) = \max\{\bar{b}(x,\xi)b_{n(x)}/b_{n(d)} : l \subseteq x \subseteq u\}, \tag{37}$$
$$\alpha_{\min}(d,u,l,\xi) = \min\{\bar{b}(x,\xi)b_{n(x)}/b_{n(d)} : l \subseteq x \subseteq u\}. \tag{38}$$

Other choices of $\alpha_{\max}$ and $\alpha_{\min}$ are possible and may be convenient for computational reasons, cf. the discussion at the end of this section.

The construction in (35)–(38) ensures the following properties; see also Fig. 8. The sandwiching property,

$$L_t^s \subseteq X_t \subseteq U_t^s \subseteq D_t, \quad s \le t \le 0, \tag{39}$$

meaning that L^s and U^s are lower and upper bounds on X, and all processes are bounded by D. The funnelling property,

$$L_t^{s'} \subseteq L_t^s \subseteq U_t^s \subseteq U_t^{s'}, \quad s \le s' \le t \le 0, \tag{40}$$

meaning that a pair of lower and upper processes started at time $s < 0$ are bounded by any pair of upper and lower processes started at time $s' > s$. And the coalescence property,

$$L_t^s = U_t^s \quad \Rightarrow \quad L_{t'}^s = U_{t'}^s \quad \text{for } s \le t \le t' \le 0, \tag{41}$$

meaning that once a pair of upper and lower processes have coalesced, they stay in coalescence, and at time 0 they are equal to $X_0 \sim \kappa$, cf. (39).

Now the perfect simulation algorithm works as follows. Pick a sequence of times $\ldots s_2 < s_1 < 0$, where $\lim_{i\to\infty} s_i = -\infty$. Then start D in equilibrium at time 0 and generate it backwards until time s_1. For forwards birth times $t \in [s_1, 0]$ in D, generate i.i.d. marks $R_t \sim \text{Uniform}[0,1]$ independent of D. Then generate the upper and lower processes U^{s_1} and L^{s_1} as in (35)–(36). If U^{s_1} and L^{s_1} are in a common state at time 0 then stop. If not then extend D (together with the marks for forwards births) further backwards from $t = s_1$ to $t = s_2$, construct U^{s_2} and L^{s_2}, and check if $U_0^{s_2} = L_0^{s_2}$. This backwards-forwards step is repeated until the upper and lower chains have coalesced by time 0, i.e. at the coalescence time $-T_s = \sup\{s_i : U_0^{s_i} = L_0^{s_i}\}$. By (39), we have $U_0^{-T_s} = X_0$ and hence $U_0^{-T_s} \sim \kappa$. This procedure is illustrated in Fig. 8.

Notice that we need only to generate the jump chains of D and the upper and lower processes. Hence it is sensible to let $\{s_i\} \subseteq \{J_{-i}\}$, where $\ldots J_{-2} \le J_{-1} < 0$ are the jump times of D before time 0. As $-T_s \le -T \le J_{-n(D_0)}$, where $-T = \sup\{s \le 0 :$

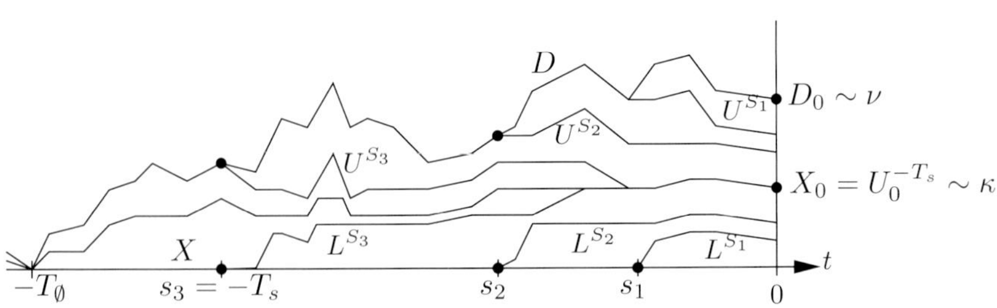

Fig. 8. Illustration of sandwiching, funnelling, and coalescence properties; see (39)–(41).

$U_0^s = L_0^s\}$ is the "true coalescence time", it is natural to let $s_1 \le J_{-n(D_0)}$. For efficiency reasons a doubling scheme is usually used [26], i.e.

$$s_1 = J_{-n}, s_2 = J_{-2n}, s_3 = J_{-4n}, s_4 = J_{-8n}, \ldots, \tag{42}$$

where $n \in \mathbb{N}$ is a user-specified parameter. Then we write T_n for T_s. Note that $T' \le T'_n$ for the coalescence times in the jump chains, i.e. T' and T'_s are defined by $-T = J_{-T'}$ and $-T_s = J_{-T'_s}$. The doubling scheme can be refined by letting

$$s_1 = J_{-T'_{\min}}, s_2 = J_{-T'_{\min}-n}, s_3 = J_{-T'_{\min}-2n}, s_4 = J_{-T'_{\min}-4n}, \ldots, \tag{43}$$

writing $T_{n,\min}$ for T_s, and where $T'_{\min} = \inf\{k \in \mathbb{N}_0 : D_{J_{-k}} \cap D_0 = \emptyset\}$ specifies the number of jumps D has to go through from the first point in D_0 is born until time zero. Often in applications, $T_n \ll T_\emptyset$, $T'_n \ll T'_\emptyset$ and $T'_{n,\min} \le T'_n$; this is illustrated later in Example 7.

The calculation of $\alpha_{\max}$ and $\alpha_{\min}$ is particular simple in the following cases. A point processes is attractive respectively repulsive if

$$\bar{b}(x,\xi) \le \bar{b}(y,\xi) \quad \text{whenever} \quad x \subseteq y, \xi \notin y, \tag{44}$$

$$\bar{b}(x,\xi) \ge \bar{b}(y,\xi) \quad \text{whenever} \quad x \subseteq y, \xi \notin y. \tag{45}$$

For example, for the point processes considered in Examples 4 and 5, either (44) or (45) is satisfied. In the attractive case (44) we obtain by (21) and (37)–(38),

$$\alpha_{\max}(d,u,l,\xi) = \bar{b}(u,\xi)b_{n(u)}/b_{n(d)},$$
$$\alpha_{\min}(d,u,l,\xi) = \bar{b}(l,\xi)b_{n(l)}/b_{n(d)}.$$

In the repulsive case (45), it is computational convenient to redefine $\alpha_{\max}$ and $\alpha_{\min}$ by

$$\alpha_{\max}(d,u,l,\xi) = \bar{b}(l,\xi)b_{n(u)}/b_{n(d)},$$
$$\alpha_{\min}(d,u,l,\xi) = \bar{b}(u,\xi)b_{n(l)}/b_{n(d)}.$$

Then the perfect simulation algorithm still works as (39)–(41) hold. Note that it is only in the attractive case that U^s and L^s are individual Markov chains. Observe also that in Fig. 4, the point pattern for $\gamma = 0$ is not contained in that for $\gamma = 0.5$. This is possible because the Strauss model is non-attractive.

4.3 Clan of Ancestors

An alternative perfect simulation algorithm is given by [8]. We assume that ϕ has finite range of interaction, i.e. there exists an $R < \infty$ such that for any $x \in \Omega$ and $\xi \in S\backslash x$, $\bar{b}(x,\xi) = \bar{b}(x \cap \text{ball}(\xi,R),\xi)$. This is fulfilled in Examples 4 and 5.

The algorithm is based on two central concepts defined as follows. Consider again the coupling construction in Sect. 3.5. When we have a birth $D_t = D_{t-} \cup \xi_t$, the *parents* of $\xi = \xi_t$ is the set $\text{pa}(\xi) = D_{t-} \cap \text{ball}(\xi,R)$ (we suppress in the notation that $\text{pa}(\xi)$ depends on D_{t-}). Whether $X_t = X_{t-} \cup \xi$ or $X_t = X_{t-}$ in (24) depends on D_{t-} only through $\text{pa}(\xi)$, or in fact only through those points in $\text{pa}(\xi)$ which have not been accepted

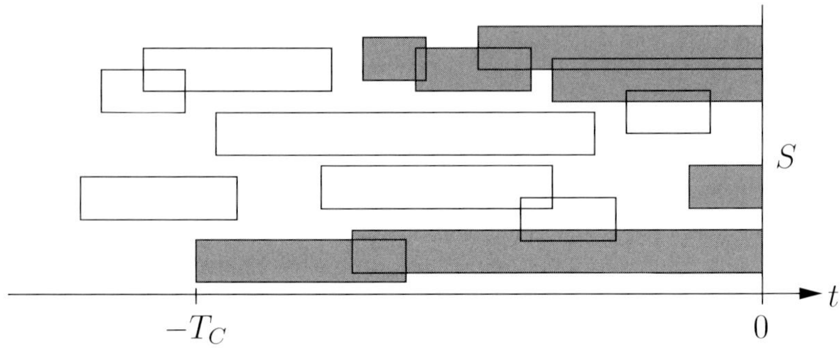

Fig. 9. Example of a clan of ancestors where shaded blocks represents members of the clan.

earlier in the thinning of D. In this sense whether ξ should be added to X depends not only on its parents, but also on the parents of its parents, and so on. Therefore we define recursively the ith generation ancestors of ξ by $\text{pa}^i(\xi) = \cup_{\eta \in \text{pa}^{i-1}(\xi)} \text{pa}(\eta)$, $i = 2, 3, \ldots$, where $\text{pa}^1(\xi) = \text{pa}(\xi)$ (again we suppress the dependence of D when defining $\text{pa}^i(\xi)$). Furthermore, let $\text{an}(\xi) = \cup_{i \in \mathbb{N}} \text{pa}^i(\xi)$ be the union of all generations of ancestors of ξ. If $\xi \in D_0$, then the points in $\text{an}(\xi)$ are the only points which can have any influence on whether ξ is in X_0 or not. So $C(D_0) = \cup_{\xi \in D_0} \text{an}(\xi) \cup D_0$ is the set of *all* points in D which can have any influence on the configuration of X_0. In other words, if $[-T_C, 0]$ is the time interval in which the points in $C(D_0)$ are living, then $T_C \leq T_\emptyset$, and X_0 is unaffected if we set $X_{-T_C} = \emptyset$ and generate X_t forwards in time $t \geq -T_C$ as usual. We refer to $C(D_0)$ as the *clan of ancestors*. An example of a clan is shown in Fig. 9.

Now, briefly the perfect simulation algorithm works as follows (noticing that we only need to generate the relevant jump chains). Let $D_0 \sim \nu$. As noticed above we only need to generate the dominating process D back until time $-T_C$, which may be determined as follows. Initially all points in D_0 are set to be in the clan. Then whenever a member of the clan dies in D (when D is considered backwards in time), e.g. $D_t = D_{t-} \cup \{\xi_t\}$, all points in D_{t-} within distance R of ξ_t are the parents of ξ_t and are therefore added to the clan. When all members of the clan have died, we have found the time T_C and the generation of D stops. Finally, we set $X_{-T_C} = \emptyset$, construct X_t, $t \geq -T_C$, as in Sect. 4.1, and return $X_0 \sim \kappa$.

As noticed, $T_C \leq T_\emptyset$, and it is not hard to see that $T \leq T_C$. Similarly, $T' \leq T'_C \leq T'_\emptyset$, where $T_C = -J_{-T'_C}$. Note that the running time of the algorithm depends only on b through R, and no monotonicity properties such as (44) and (45) are required.

Example 7: Consider a Gibbs hard core process as given by (19) with $\gamma = 0$ and hard core parameter $R \geq 0$, letting S be a unit square, and ν a homogeneous Poisson point process of rate $\beta = 100$. Such hard core processes are studied in the contributions by Schmid & Phuong and by Döge in this collection; see also [6, 17, 18]. Figure 10 shows perfect simulations of a hard core process for different values of R.

Figure 11 shows the mean coalescence times of T'_C, T'_n, and $T'_{n,\min}$ versus R, when $n = 1$ in the doubling schemes. For all values of R in Fig. 11, the mean coalescence times are much smaller than $\mathbb{E}T'_\emptyset \geq e^{100}$, cf. (31). For small values of R, the clan method and the refined doubling scheme are almost equal (in the sense that $\mathbb{E}(T'_C) \approx \mathbb{E}(T'_{n,\min})$) and slightly better than the (ordinary) doubling scheme (where the mean coalesce times

Fig. 10. Simulation of a Gibbs hard core process (see Example 7) when $R = 0.03, 0.07, 0.11$ (from left to right).

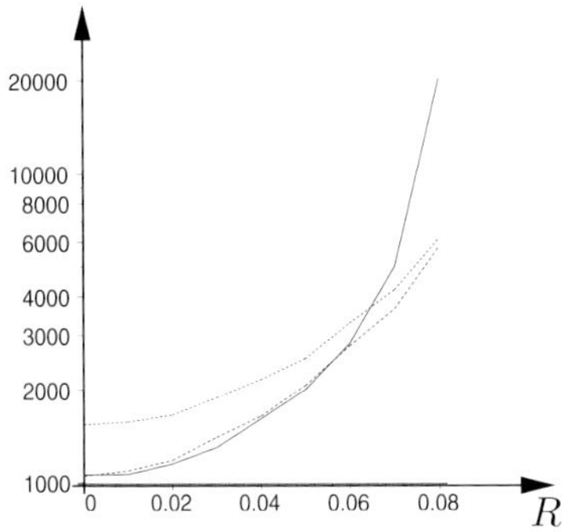

Fig. 11. Mean coalescence times versus the hard core parameter R for a hard core process. Full line: $\mathbb{E}(T'_C)$; upper dotted line: $\mathbb{E}(T'_n)$; lower dotted line: $\mathbb{E}(T'_{n,\min})$. Each mean value is estimated from 500 independent perfect simulations.

are about 2/3 smaller). As R increases, T'_n and $T'_{n,\min}$ do not increase as rapidly as T'_C, and the clan method is clearly slower than both doubling schemes for large values of R. This is not so surprising, as an increase in R implies an increase in the number of parents for each point, and hence a larger clan reaching further back in time. The specification of β is of course also crucial: T'_C tends to be smaller than T'_n for small values of β, while the opposite is the case for large values of β.

Further results on the running time for the algorithm based on upper and lower processes are given in [3].

5 Spatial Birth-and-Catastrophe Process

5.1 Description and Construction

5.1.1 The Simple Case. We start by specifying a simple birth-and-catastrophe process $N = \{N_t : t \geq 0\}$ with generator Q given by $q_{i+1,0} = d_{i+1}$ and $q_{i,i+1} = b_i$, while all other off-diagonal elements are zero. Hence $q_i = b_i + d_i$ for all $i \in \mathbb{N}_0$, setting $d_0 = 0$. Note that d_i now has another interpretation than in Sect. 3.

As in Sect. 2, (i)–(iv) are assumed to hold. For simplicity we assume that $I = \mathbb{N}_0$ and $d_{i+1} > 0$ and $b_i > 0$ for all $i \in \mathbb{N}_0$. Then (ii) is satisfied. Further (iii) is equivalent to that both

$$\pi_{i+1} = \pi_0 \frac{b_0}{q_1} \frac{b_1}{q_2} \cdots \frac{b_i}{q_{i+1}}, \quad i \in \mathbb{N}_0, \tag{46}$$

and

$$\lim_{n\to\infty} \frac{b_0 \cdots b_i}{q_1 \cdots q_i} = 0 \tag{47}$$

are satisfied. To see this, rewrite $\pi Q = 0$ as

$$\pi_0 b_0 = \sum_{i=1}^{\infty} \pi_i d_i \quad \text{and} \quad \pi_i b_i = \pi_{i+1} q_{i+1} \quad \text{for all } i \in \mathbb{N}_0. \tag{48}$$

From the latter equation, using induction, we obtain (46). Combining the equations in (48),

$$\pi_0 b_0 = \sum_{i=1}^{\infty} \pi_{i-1} b_{i-1} - \pi_i b_i = \pi_0 b_0 - \lim_{i\to\infty} \pi_i b_i,$$

whereby (47) is obtained.

5.1.2 The Spatial Case. We next construct a spatial birth-and-catastrophe process $X = \{X_t : t \geq 0\}$, proceeding along similar lines as in Sects. 3.3–3.5. The process is specified by two functions $\beta : \Omega \times S \to [0, \infty)$ and $\delta : \Omega \to [0, \infty)$, where for any $x \in \Omega$, $\beta(x, \xi)$ is the birth rate at which a point $\xi \in S$ is added to x, and if $x \neq \emptyset$, $\delta(x)$ is the catastrophe rate a which all points are deleted from x. As in Sects. 2 and 3.3, let $J_1 < J_2 < \ldots$ be the jump times, $H_1, H_2, \ldots$ the holding times, and $Y = \{Y_n : n \in \mathbb{N}_0\}$ the jump chain of X with $Y_0 = X_0$. The conditional distribution of H_n given $H_1, \ldots, H_{n-1}, Y_0, \ldots, Y_{n-1} = x$ is $\text{Exp}(B(x) + \delta(x))$, where $B(x) = \int_S \beta(x, \xi)\lambda(d\xi)$ is the total birth rate (assuming $B(x) < \infty$) and we set $\delta(\emptyset) = 0$. If we also condition on H_n, the probability for a birth at time J_n is $B(x)/(B(x) + \delta(x))$. If we further condition on that a birth happens at time J_n, then $Y_n = x \cup \xi$ where ξ has density $\beta(x, \cdot)/B(x)$. If instead a catastrophe happens at time J_n, then $Y_n = \emptyset$.

We assume that β and δ are related to the rates for N by

$$\beta(x, \xi) = b_{n(x)} \bar{b}(x, \xi) \quad \text{and} \quad \delta(x) = d_{n(x)} \bar{d}(x), \tag{49}$$

where the constraints $0 \leq \bar{b} \leq 1$ and $\bar{d} \geq 1$ are satisfied. Then, by (49), $B(x) \leq b_{n(x)}$ and $\delta(x) \geq d_{n(x)}$. Furthermore, we assume that $d_1 \geq d_2 \geq \ldots$. Then X and N can be coupled so that N dominates $n(X)$ as in (11). Briefly, (X, N) is a jump process where a transition from (x, n) to (x', n') happens with a rate given as follows, where $m = n(x) \leq n$:

b_n for $(x', n') = (x, n + 1)$ and $m < n$,

$\beta(x, \xi)$ for $(x', n') = (x \cup \xi, n)$ and $m < n$,

$\beta(x, \xi)$ for $(x', n') = (x \cup \xi, n + 1)$ and $m = n$,

$b_n - B(x)$ for $(x', n') = (x, n+1)$ and $m = n$,

d_n for $(x', n') = (\emptyset, 0)$ and $m \le n$,

$\delta(x) - d_n$ for $(x', n') = (\emptyset, n)$ and $m \le n$.

As in Sect. 3.3 we obtain the following: X is non-explosive; $\emptyset$ is an ergodic atom at which X regenerates; as in (12), X_t converges in distribution towards an equilibrium distribution κ, say, independent of the initial state of X_0 as $t \to \infty$; and κ has a support given by (14).

The equilibrium distribution is uniquely characterised by the equations

$$B(\emptyset)\kappa(\emptyset) = \int_{x \neq \emptyset} \delta(x)\kappa(dx) \tag{50}$$

and

$$\int_{F_i} [B(x) + \delta(x)]\kappa(dx) = \iint_{x \cup \xi \in F_i} \beta(x, \xi)\lambda(d\xi)\kappa(dx) \tag{51}$$

for all $i \in \mathbb{N}$ and events $F_i \subseteq \Omega_i$ (this follows from (56) in Sect. 6). If κ has an unnormalised density ϕ with respect to ν given by (15), then (50) and (51) become

$$b_0 \pi_0 \phi(\emptyset) \int \bar{b}(\emptyset, \xi)\lambda(d\xi) = \sum_{i=1}^{\infty} d_i \pi_i \int \bar{d}(x)\phi(x)\lambda_i(dx) \tag{52}$$

and

$$\int_{F_i} \pi_i \phi(x) \left(b_i \int \bar{b}(x, \xi)\lambda(d\xi) + d_i \bar{d}(x) \right) \lambda_i(dx) = \iint_{x \cup \xi \in F_i} \pi_{i-1} b_{i-1} \bar{b}(x, \xi)\phi(x)\lambda_{i-1}(dx). \tag{53}$$

Using (48), then (52) and (53) are easily seen to be satisfied if, for example, ϕ is non-increasing as in (17), $\bar{b}$ is the Papangelou intensity as in (18), and

$$\bar{d}(x) = 1 + \left(b_{n(x)} - B(x)\right) / d_{n(x)} \quad \text{for } x \neq \emptyset.$$

In the sequel we assume that $\bar{b}$ and $\bar{d}$ are defined in this way for a given unnormalised density ϕ which is non-increasing. These assumptions imply that the total rates in N and X agree in the sense that for any $x \in \Omega$, $B(x) + \delta(x) = q_{n(x)}$. In other words, if $X_t = x$, the waiting time τ to the next jump is Exp($q_{n(x)}$)-distributed, and we may generate $\xi \sim \lambda$ and return $X_{t+\tau} = x \cup \xi$ with probability $\beta(x, \xi)/q_{n(x)}$, and return $X_{t+\tau} = \emptyset$ otherwise.

5.2 Perfect Simulation

In this section we show that the simple perfect simulation procedure in Sect. 4.1 is feasible when the dominating simple jump process is of the irreversible type presented in Sect. 5.1. In addition to the assumptions in Sect. 5.1, we assume for simplicity that $q_i \equiv 1$ is constant for all $i \in \mathbb{N}$ and $b_0 \leq b_1$. Equivalently, by (48), $b_i = \pi_{i+1}/\pi_i \leq 1$ is non-decreasing for $i \in \mathbb{N}_0$, and $d_i = 1 - b_i$ for $i \in \mathbb{N}$. Hence the Poisson case $\pi_i \propto \beta^i/i!$ is excluded, since $\pi_{i+1}/\pi_i = \beta/(i+1)$ is decreasing in i. If $\pi_i \propto \Gamma(i+\alpha)q^i/i!$ is negative binomial (Example 3), $\pi_{i+1}/\pi_i = q(i+\alpha)/(i+1) \leq 1$ is non-decreasing if and only if $\alpha \leq 1$.

Now, a coupling construction is easily specified: Suppose N_t is generated for all times $t \geq 0$, and for all jump times t of N_t, there are generated mutually independent points $\xi_t \sim \lambda$ and numbers $R_t \sim \text{Uniform}[0,1]$ (independent of N). Let $n(X_0) \leq N_0$ and generate X_t forwards in time $t > 0$ as follows. Suppose that $X_{t-} = x$ and $N_{t-} = j$ with $n(x) = i \leq j$. If $N_t = j$ is unchanged, then $X_t = x$ is unchanged, while

$$N_t \neq j \Rightarrow X_t = \begin{cases} x \cup \xi_t & \text{if } N_t = j+1 \text{ and } R_t \leq \bar{b}(x, \xi_t) b_i / b_j \\ \emptyset & \text{otherwise.} \end{cases} \tag{54}$$

Since $\{b_j\}$ is non-decreasing, we obtain a spatial birth-and-catastrophe process with rates as required, and by induction, $N_t \geq n(X_t)$ for all $t \geq 0$.

For perfect simulation, we let $N_0 \sim \pi$ and imagine that N_t is generated backwards in time $t \leq 0$ until it reaches zero, i.e. at time $-\tilde{T}_\emptyset \equiv \sup\{t \leq 0 : N_t = 0\}$. The number of jumps in $[-\tilde{T}_\emptyset, 0]$ is simply given by N_0, since N_t can only move one step up when considered forwards in time between $-\tilde{T}_\emptyset$ and 0. Hence, in accordance with (54), we let $Y_0 = \emptyset$ and generate Y_i, $i = 1, \ldots, N_0$, by

$$Y_i = \begin{cases} Y_{i-1} \cup \xi_i & \text{if } R_i \leq \bar{b}(Y_{i-1}, \xi_i) b_{n(Y_{i-1})} / b_{i-1} \\ \emptyset & \text{otherwise,} \end{cases} \tag{55}$$

where $\xi_i \sim \lambda$ and $R_i \sim \text{Uniform}[0,1]$. Since $Y_0, \ldots, Y_{N_0}$ can be considered as the part of the jump chain of X_t ("started in the infinite past") where $t \in [-\tilde{T}_\emptyset, 0]$, we conclude that $Y_{N_0} \sim \kappa$ (a formal argument follows exactly the same lines as in Sect. 4.1).

So perfect simulation is very simple:

(I) generate $N_0 \sim \pi$;
(II) generate the jump chain $Y_0, \ldots, Y_{N_0}$;
(III) return $Y_{N_0} \sim \kappa$.

The number of jumps is given by $N_0 \sim \pi$. This should be compared to the case in Example 6: If $\pi_i \propto q^i$ is geometric, its mean $q/(1-q)$ is $(1+q)/(1-q)$ times smaller than the lower bound on $\mathbb{E}T'_\emptyset$ in (33). Hence it is much faster; for example, $(1+q)/(1-q) = 201$ if $\mu_\pi = 100$, cf. (34).

Unless $\bar{b}$ is very close to 1 (corresponding to a weak interaction), we might by (55) expect $n(Y_{N_0})$ to be small. This is in accordance with the observations in Example 4, and it explains why the perfect simulation algorithm is so fast. Furthermore, in contrast to the algorithms in Sect. 4, there is no need for storing any information: our perfect sampler is a simple example of a so called read-once algorithm [34].

6 The General Case

In a general setting, a spatial jump process $X = \{X_t : t \geq 0\}$ may be specified by functions $\beta, \delta : \Omega \times \Omega \to [0, \infty)$, where for any $x \in \Omega$, $\beta(x, y)$ is the rate at which a point configuration $y \in \Omega$ is added to x, and if $z = x \cup y \in \Omega$, $\delta(x, y)$ is the rate at which y is deleted from z; in order to ensure that the process really jumps, we set $\beta(x, \emptyset) = 0$ and $\delta(x, \emptyset) = 0$. Let further $B(x) = \int \beta(x, y)\nu(dy)$ and $D(z) = \sum_{x \subseteq z} \delta(x, z \backslash x)$, where ν is defined as in (15) with respect to a given probability density function $\pi = \{\pi_i\}$ on $\mathbb{N}_0$. Then X can be defined in terms of its jump times, holding times, and jump chain in the same way as in Sect. 3.3. Especially, for the jump chain $\{Y_n\}$, in case of a transition $Y_n = x \cup y$ given that $Y_{n-1} = x$, y has density $\beta(x, \cdot)/B(x)$ with respect to ν (assuming $B(x) < \infty$), and in case of a transition $Y_n = x$ given that $Y_{n-1} = z = x \cup y$, y has been selected with probability $\delta(x, y)/D(z)$.

Let us, as before, restrict attention to the non-explosive case. For instance, X is non-explosive if $B(x) + D(x)$ is uniformly bounded. By Proposition 8.1 in [25], a probability measure κ on Ω is invariant if and only if

$$\int_{F_i} [B(x) + D(x)]\, \kappa(dx) = \iint_{x \cup y \in F_i} \beta(x, y)\nu(dy)\kappa(dx) + \int \sum_{x \subseteq z : x \in F_i} \delta(x, z \backslash x)\kappa(dz) \tag{56}$$

for all $i \in \mathbb{N}_0$ and events $F_i \subseteq \Omega_i$. Here invariance means that $X_t \sim \kappa$ for all $t \geq 0$ if $X_0 \sim \kappa$. In particular, if X converges in distribution towards κ, then κ is the unique invariant distribution. For example, assume that Q and π are in balance, cf. (1), and

$$\beta(x, y) = q_{n(x), n(x)+n(y)} \bar{b}(x, y)$$

and

$$\delta(x, y) = q_{n(x)+n(y), n(x)} \bar{d}(x, y) \Big/ \binom{n(x) + n(y)}{n(x)}.$$

Then, if ϕ is an unnormalised density for κ with respect to ν, there is a detailed balance condition:

$$\phi(x)\bar{b}(x, y) = \phi(x \cup y)\bar{d}(x, y) \quad \text{for all } x, y \in \Omega. \tag{57}$$

It is straightforwardly verified that this implies both (56) and reversibility of X.

In Sects. 3 and 5.1 we demonstrated that a coupling of X to a simple jump process N is very useful. However, at the present level of generality it seems difficult to specify a coupling so that N dominates $n(X)$. Probably we need to consider this problem case by case; indeed the coupling constructions in Sects. 3 and 5.1 depend much on the choice of β and δ. For making perfect simulations from a given invariant distribution κ, we may simply try to use one of the algorithms in Sects. 4 and 5.2.

Example 8: Let $\pi_i \propto \beta^i / i!$ be Poisson, where $\beta > 0$,

$$q_{i,i+j} = p^i (\beta(1-p))^j / j!, \quad q_{i+j,i} = \tbinom{i+j}{i} p^i (1-p)^j,$$

where $0 < p < 1$, and

$$\bar{d} \equiv 1, \quad \bar{b}(x, y) = \phi(x \cup y)/\phi(y),$$

with $\phi > 0$ non-increasing. Then the detailed balance conditions (1) and (57) are satisfied, and X develops as follows. Suppose that $X_t = x$. Then

(I) with rate $p^{n(x)}e^{\beta(1-p)}$, first propose a birth $x \to x \cup y$, drawn from a Poisson process with intensity measure $\beta(1-p)\lambda$, next with probability $\phi(x \cup y)/\phi(x)$ accept $x \cup y$, and retain x otherwise;

(II) with rate 1 make an independent thinning of the points in x, where each point is retained with probability p.

Note that nothing may happen with rate $2p^{n(x)}$ (namely if $y = \emptyset$ in (I) or all points are retained in (II)). Observe also that y is independent of x in (I), and the retention probability p does not depend on x in (II).

Both N and X are non-explosive as $q_i = p^i e^{\beta(1-p)} + 1 - 2p^i$ and $B(x) + D(x) \leq q_{n(x)}$ are bounded. It is easily seen that $q_{i,i+j} \geq q_{i-k,i+j}$ for all integers $0 < k \leq i$ and $j \geq 1$ if and only if $p \geq \beta/(2+\beta)$. Hence, if $p \geq \beta/(2+\beta)$, we obtain that N dominates X by specifying a jump process (X, N) as follows. The rate for a jump $(x, n) \to (x', n')$ is as follows:

$$\beta(x,y) \text{ for } (x',n') = (x \cup y, n(x \cup y)) \text{ and } n < n',$$

$$q_{n,n'} - \int_{n(x \cup y)=n'} \beta(x,y)\nu(dy) \text{ for } x' = x \text{ and } n < n',$$

$$\beta(x,y) \text{ for } (x',n') = (x \cup y, n) \text{ and } n(x \cup y) \leq n,$$

$$q_{n,n'} \text{ for } x' = x \backslash y \text{ and } n' \leq n - n(y).$$

Because of this domination, we have the usual ergodicity properties: $\emptyset$ is an ergodic atom at which X regenerates, and X converges in distribution towards κ (independent of the initial state of X_0).

It is tempting to try to extend this coupling construction to a dominating spatial jump process $D \supseteq X$, but this idea fails because of the accept-rejection step (I). However, we can still simulate X_0 from its equilibrium distribution as any of the perfect simulation algorithms in Sect. 4 apply (for any $\beta > 0$ and $0 < p < 1$).

Acknowledgement

This research will be a part of KKB's PhD dissertation. JM was supported by the European Union's research network "Statistical and Computational Methods for the Analysis of Spatial Data, ERB-FMRX-CT96-0096", by the Centre for Mathematical Physics and Stochastics (MaPhySto), funded by a grant from the Danish National Research Foundation, and by the Danish Natural Science Research Council. We thank Bjarne Højgaard for helpful discussions.

Appendix: Notation Index

Symbol	Explanation	Page
S	The state space for "points"	391, 396
X (X_t)	In general: spatial jump process (at time t) In Sects. 3–4: spatial birth-and-death process In Sect. 5: spatial birth-and-catastrophe process	393
N (N_t)	Simple jump process (at time t)	393
Q (q_{ij},q_i)	Generator of N (with elements q_{ij}, where $q_i = -q_{ii}$)	393
π (π_i)	The invariant probability density function for N (with probabilities π_i for $i \in I$)	394
I	The support for π	394
b_i, d_i	In Sect. 3.1: birth and death rates for N In Sect. 5.1.1: birth and catastrophe rates for N	395, 409
β	If π is a Poisson distribution, β is the mean for π and $-\ln\beta$ is the chemical activity for $Z \sim \phi$	395, 398
q, α	Parameters for the geometric/negative binomial distribution defined in Example 2 and 3	396
Symbol	Explanation	Page
$\Omega = \cup_{i=0}^{\infty}\Omega_i$	The state space of X_t	396
Ω_i	The set of finite point configurations confined in S and of cardinality i	396
$n(x)$	The number of points in a finite point configuration $x \in \Omega$	396
$\emptyset$	The empty point configuration	396
λ	Probability measure on S	396
$\beta(\cdot,\cdot)$, $\delta(\cdot,\cdot)$	In Sect. 3.3: birth and death rates for X. In Sect. 5.1.2: birth and catastrophe rates for X In Sect. 6: rates for adding/deleting point configurations to/from X	396, 410, 413
$\bar{b}(\cdot,\cdot)$, $\bar{d}(\cdot,\cdot)$	Functions used for relating $\beta(\cdot,\cdot)$, $\delta(\cdot,\cdot)$ to b_i, d_i	397
κ	Invariant distribution for X	398
ϕ	Unnormalised density for κ w.r.t. ν	398
ν	Probability measure on Ω (used as an reference measure)	398

γ, R	In Example 4: the interaction parameter and interaction radius for a Strauss process In Example 5: the parameters for an area interaction process	399
$D, (D_t)$	The dominating process (at time t)	401
$T_\emptyset$	$-T_\emptyset$ is the first time before time 0 that D_t enters $\emptyset$	404
$T'_\emptyset$	The number of jumps D makes on $[-T_\emptyset, 0]$	404
μ_π	The mean for the distribution π	405
U^s (U^s_t), L^s (L^s_t)	Upper/lower processes started at time $s \leq 0$ (and considered at time t	405
T (T')	True coalescence time for upper and lower processes (or for their corresponding jump chains)	407
T_n (T'_n)	Coalescence time when a doubling scheme is used for upper and lower processes (or for their corresponding jump chains)	407
$T_{n,\min}$ ($T'_{n,\min}$)	Coalescence time when a refined doubling scheme is used for upper and lower processes (or for their corresponding jump chains)	407

References

1. Asmussen, S. (1987): *Applied Probability and Queues*. (Wiley, Chichester)
2. Baddeley, A. and van Lieshout, M. N. M. (1995): 'Area-interaction Point Processes'. *Ann. Inst. Statist. Math.* **46**, pp. 601–619.
3. Berthelsen, K. K. and Møller, J. (2001): 'Perfect simulation and inference for spatial point processes'. *Technical report*, R-01-2009, Department of Mathematical Sciences, Aalborg University. Submitted.
4. Berthelsen, K. K. and Møller, J. (2001): 'A primer on perfect simulation for spatial point processes'. *Technical report*, R-01-2026, Department of Mathematical Sciences, Aalborg University. Submitted.
5. Daley, D. J. and Vere-Jones, D. (1988): *An Introduction to Point Processes*. (Springer, Berlin)
6. Döge, G. (2000): 'Grand canonical simulations of hard-disk systems'. In: *Statistical Physics and Spatial Statistics*, ed. by K. R. Mecke and D. Stoyan, volume 554 of *Lecture Notes in Physics* (Springer, Berlin), pp. 373–393.
7. Feller, W. (1971): *An Introduction to Probability Theory and Its Applications*, volume 2 (John Wiley and Sons Inc., New York), 2nd edition.
8. Fernández, R., Ferrari, P. A. and Garcia, N. L. (1999): 'Perfect simulation for interacting point processes, loss networks and Ising models'. Unpublished.
9. Georgii, H.-O. (2000): 'Phase transitions and percolation in Gibbsian particle models'. In: *Statistical Physics and Spatial Statistics*, ed. by K. R. Mecke and D. Stoyan, volume 554 of *Lecture Notes in Physics* (Springer, Berlin), pp. 267–294.
10. Geyer, C. J. (1999): 'Likelihood inference for spatial processes'. In: *Stochastic Geometry, Likelihood and Computation*, ed. by O. E. Barndorff-Nielsen, W. Kendall and M. van Lieshout (Chapman & Hall), pp. 79–140.
11. Häggström, O., van Lieshout, M.-C. N. M. and Møller, J. (1999): 'Characterization results and Markov chain Monte Carlo algorithms including exact simulation for some spatial point processes'. *Bernoulli* **5**, pp. 641–658.

12. Kallenberg, O. (1984): 'An informal guide to the theory of conditioning in point processes'. *Int. Statist. Rev.* **52**, pp. 151–164.
13. Kelly, F. P. and Ripley, B. D. (1976): 'A note on Strauss's model for clustering'. *Biometrika* **63**, pp. 357–360.
14. Kendall, W. S. (1998): 'Perfect Simulation for the Area-Interaction point process'. In: *Probability Towards 2000*, ed. by L. Accardi and C. Heyde (Springer, New York), pp. 218–234.
15. Kendall, W. S. and Møller, J. (2000): 'Perfect simulation using dominating processes on ordered spaces, with application to locally stable point processes'. *Adv. Appl. Prob.* **32**, pp. 844–865.
16. Lieshout, M. N. M. van (2000): *Markov point processes and their applications*. Imperial College Press, London.
17. Löwen, H. (2000): 'Fun with hard spheres'. In: *Statistical Physics and Spatial Statistics*, ed. by K. R. Mecke and D. Stoyan, volume 554 of *Lecture Notes in Physics* (Springer, Berlin), pp. 295–331.
18. Mase, S., Møller, J., Stoyan, D., Waagepetersen, R. and Döge, G. (2001): 'Packing densities and simulated tempering for hard core Gibbs point processes', *Ann. Inst. Statis. Math.* **53**, pp. 661–680.
19. Mecke, K. (2000): 'Additivity, convexity and beyond: applications of Minkowski functionals in statistical physics'. In: *Statistical Physics and Spatial Statistics*, ed. by K. R. Mecke and D. Stoyan, volume 554 of *Lecture Notes in Physics* (Springer, Berlin), pp. 111–184.
20. Meyn, S. P. and Tweedie, R. L. (1994): *Markov Chains and Stochastic Stability*. (Springer-Verlag, London)
21. Møller, J. (1989): 'On the rate of convergence of spatial birth-and-death processes'. *Ann. Inst. Statist. Math.* **41**, pp. 565–581.
22. Møller, J. (2001): 'A review of perfect simulation in stochastic geometry'. In: *Selected Proceedings of the Symposium on Inference for Stochastic Processes*, ed. by I. V. Basawa, C. C. Heyde and R. L. Taylor, volume 37 of *IMS Lecture Notes & Monographs Series*, pp. 333–355.
23. Møller, J. and Schladitz, K. (1999): 'Extensions of Fill's algorithm for perfect simulation'. *J. Roy. Statist. Soc. Ser. B* **61**, pp. 1–15.
24. Norris, J. R. (1997): *Markov Chains*. (Cambridge University Press, New York)
25. Preston, C. (1977): 'Spatial birth-and-death processes'. *Bull. Int. Statist. Inst.* **46**, pp. 371–391.
26. Propp, J. G. and Wilson, D. B. (1996): 'Exact Sampling with Coupled Markov Chains and Applications to Statistical Mechanics'. *Random Structures and Algorithms* **9**, pp. 223–252.
27. Reuter, G. E. H. and Ledermann, W. (1953): 'On the differential equations for the transition probabilities of Markov processes with enumerably many states'. *Proc. Camb. Phil. Soc.* **49**, pp. 247–262.
28. Ripley, B. D. (1977): 'Modelling spatial patterns (with discussion)'. *J. Roy. Statist. Soc. Ser. B* **39**, pp. 172–212.
29. Ruelle, D. (1969): *Statistical Mechanis: Rigorous Results*. (W. A. Benjamin, Reading, Massachusetts)
30. Strauss, D. J. (1975): 'A model for clustering'. *Biometrika* **62**, pp. 467–475.
31. Thönnes, E. (1999): 'Perfect simulation of some point processes for the impatient user'. *Adv. Appl. Prob.* **31**, pp. 69–87.
32. Thönnes, E. (2000): 'A primer on perfect simulation'. In: *Statistical Physics and Spatial Statistics*, ed. by K. R. Mecke and D. Stoyan, volume 554 of *Lecture Notes in Physics* (Springer, Berlin), pp. 349–378.
33. Widom, B. and Rowlinson, J. S. (1970): 'A new model for the study of liquid-vapor phase transitions'. *J. Chem. Phys.* **52**, pp. 1670–1684.
34. Wilson, D. B. (2000): 'How to Couple from the Past Using a Read-Once Source of Randomness'. *Random Structures and Algorithms* **16**, pp. 85–113.

Statistics for Non-sparse Spatially Homogeneous Gibbs Point Processes

Gunter Döge and Dietrich Stoyan

Institute of Stochastics, Freiberg University of Mining and Technology,
Bernhard-von-Cotta-Str. 2, D-09596 Freiberg, Germany

Abstract. A successful approach in statistical analysis of point processes or systems of hard disks is to consider them as samples of Gibbs point processes with or without marks, where the disk radii are the marks. A practically important particular case are Gibbs processes with pair-wise interaction potentials. Such a pair potential may be a valuable statistical characteristic additionally to pair correlation functions or Mecke's morphological functions.

The main topic of the present text is statistical estimation of such pair potentials as well as chemical activities by the Takacs-Fiksel method. The particular problem faced here is that the patterns considered are by no means sparse, which was until now the case in most statistical analyses of point patterns analysed in spatial statistics by the Gibb process approach. The data come from a series of microscopic images showing water droplets on a naphtalin-brom surface at progressing stages of condensation.

1 Introduction

Many geometrical structures in statistical physics and material science are random systems of non-overlapping spheres or similar figures. Often these structures can be considered as finite parts of infinite spatially homogeneous structures the distributions of which are invariant with respect to translations and rotations (stationary and isotropic). A possible statistical model for such a structure is a grand canonical marked Gibbs point process, where the points are the sphere centres and the marks the radii. If this model is used for given empirical data, then it is necessary to estimate model parameters. These parameters determine usually the energy function; if the energy function is given by a pair potential, the model parameters are the chemical activity and pair potential parameters, collected in a parameter vector γ. Spatial statisticians use this approach also in situations in which physicists are sure that a Gibbs process is a wrong model. They nevertheless expect to obtain a model which can be easily simulated and they use estimated pair potentials as alternative descriptors of the variability of the patterns. This may help to get other information than that represented by the pair correlation function or Mecke's morphological functions.

In spatial statistics various statistical methods have been developed for solving the task of estimation of these parameters. A survey of the situation around 1994 is given in Diggle et al.[2]. Though, most of the methods described in that paper are designed for the non-marked case and finite patterns (canonical case). The methods used there are maximum likelihood and pseudo-likelihood. In particular, maximum likelihood estimation as described in [2] assumes sparse pattern since it uses Padé approximations of the partition function. Modern methods use instead Monte Carlo methods as described in [4]. Also the methods described in [5] and [6] assume sparse patterns. The case of marks is considered in [5] and [11]. For the case of samples from homogeneous systems, the Takacs-Fiksel method is frequently used; the pseudo-likelihood method can be considered as a particular case.

The present paper describes briefly the Takacs-Fiksel method for estimating the parameter vector γ and shows then how it can be applied to planar hard disk systems. If the disk radii are constant, then the only model parameter is the chemical activity α, i. e. $\gamma = \alpha$. The Takacs-Fiksel method yields reliable estimates of α if the size of the window of observation is large enough; for area fractions η of the corresponding disk systems larger than 0.6 this needs, however, very large windows, if not the functional relationship between α and η is used. As a second example, a pattern of water droplets on a naphtalin-brom surface is used. Here marks play a non-trivial role, and the estimation problem is more complicated. Now the parameter vector γ has chemical activities and pair potential parameters as components. Since only a pair potential model is used and because the pattern does not represent an equilibrium state, the estimated parameters lead only to a rough fit of the model to the geometrical data.

2 Fundamentals of the Theory of Marked Point Processes

2.1 Definition

A *marked point process* is a collection of randomly scattered marked points, denoted by $\Psi = \{[x_i, m_i]\}_{i=1,2,\ldots}$, with points $x_i \in \mathbb{R}^d$ and marks m_i in a mark space $\mathbb{M}$, see also [1] in this volume. Here we consider only *stationary* processes Ψ, i.e., the point process distribution is translation-invariant:

$$\{[x_1 + x, m_1], [x_2 + x, m_2], \ldots\} \stackrel{d}{=} \{[x_1, m_1], [x_2, m_2], \ldots\} \qquad \text{for every } x.$$

Let λ be the *intensity* of Ψ, the mean number of points per volume unit, and denote the volume or d-dimensional Lebesgue measure by ν. A *local specifications*

$$\Psi_{W \times L} = \Big\{[x, m] \in \Psi : \quad x \in W,\ m \in L\Big\}$$

for a subset W of $\mathbb{R}^d$ and mark set $L \subset \mathbb{M}$ is that subprocess of Ψ which consists of the points inside W with marks in L. $\Psi(W \times L)$ denotes the number of points of that subprocess. Especially for $W = \mathbb{R}^d$ we have the L-thinned process $\Psi_L \equiv \Psi_{\mathbb{R}^d \times L}$, which consists of all points of Ψ having marks in L is again stationary, with intensity λ_L. The *mark distribution* of Ψ is given by

$$M(L) = \frac{\lambda_L}{\lambda},$$

which can be also obtained by the ergodic limit

$$\lim_{W_n \uparrow \mathbb{R}^d} \frac{\Psi(W_n \times L)}{\Psi(W_n \times \mathbb{M})},$$

where $\{W_n\}$ is an increasing sequence of sets.

2.2 Stationary Marked Gibbs Point Process

For a subset W of $\mathbb{R}^d$ with finite volume the *conditional density* of $\Psi_{W\times\mathbb{M}}$ given a fixed configuration $\psi_{\overline{W\times\mathbb{M}}}$ outside $W\times\mathbb{M}$ is given by

$$f(\varphi \mid \psi_{\overline{W\times\mathbb{M}}}) = \frac{1}{\Xi_W(\psi_{\overline{W\times\mathbb{M}}})} \exp\left\{-\tilde{E}(\psi_{\overline{W\times\mathbb{M}}}, \varphi)\right\}, \tag{1}$$

for local configurations $\varphi = \psi_{W\times\mathbb{M}}$. Without the normalizing factor $\Xi_W(\psi_{\overline{W\times\mathbb{M}}})$ we write

$$f(\varphi \mid \psi_{\overline{W\times\mathbb{M}}}) \propto \exp\left\{-\tilde{E}(\psi_{\overline{W\times\mathbb{M}}}, \varphi)\right\}.$$

The *energy function* $\tilde{E}$ for an arbitrary fixed external configuration ψ and any finite configuration $\varphi = \{[x_i, m_i]\}_{i=1,\dots,n}$ is

$$\begin{aligned}\tilde{E}\,(\psi, \varphi) = E(\psi, [x_1, m_1]) + E(\psi \cup \{[x_1, m_1]\}, [x_2, m_2]) + \dots \\ + E\,(\psi \cup \{[x_i, m_i]\}_{i=1,\dots,n-1}, [x_n, m_n])\,.\end{aligned} \tag{2}$$

Here $E(\psi', [x, m])$ is the *local energy*, which can be interpreted as the energy required to add the marked point $[x, m]$ to the configuration ψ'. If the distribution is given by a *pair potential*, then the local energy depends only on pairwise interactions, i.e.

$$E(\psi, [x, m]) = \alpha(m) + \sum_{[x', m'] \in \psi} V(x - x', m, m'). \tag{3}$$

V is the *pair potential* and α the *chemical activity*, which depends on the mark of the added point. This leads to the following expression for the local energy:

$$\tilde{E}\,(\psi, \varphi) = \sum_{i=1}^{n} \left[\alpha(m_i) + \sum_{[x', m'] \in \psi} V(x_i - x', m_i, m') + \sum_{j=i+1}^{n} V(x_i - x_j, m_i, m_j)\right].$$

In the particular case of *free boundary*, where $\psi = \emptyset$, it is

$$\tilde{E}\,(\emptyset, \varphi) = \sum_{i=1}^{n} \alpha(m_i) + \sum_{1 \le i < j \le n} V(x_i - x_j, m_i, m_j).$$

In the case of an isotropic point process the vector difference $x_i - x_j$ can be replaced by the distance $r = \|x_i - x_j\|$:

$$V(x_i - x_j, m, m') \equiv V(r, m, m')$$

A stationary (and isotropic) Gibbs process is then defined by means of a stationary independently marked Poisson weight process Ψ_Q [12], p. 177. Assuming that the intensity of Ψ_Q is one, the intensity measure of Ψ_Q is

$$\Lambda_Q(W \times L) = \mathbf{E}\Psi_Q(W \times L) = \nu(W) M_Q(L)$$

with *mark weight distribution* M_Q on $\mathbb{M}$. That means, for the complete description of point process distribution the local densities $f(\varphi \mid \ldots)$ *and* the mark weight distribution M_Q are necessary. M_Q appears in simulation algorithms such as the Gibbs sampler as the distribution of newly generated marked points, which may or may not be accepted in the acceptance step of the algorithm. In most cases a uniform distribution M_Q is a good choice, i.e.

– for discrete marks $\mu_1,\ldots,\mu_k$:

$$M_Q = \frac{1}{k}\sum_{j=1}^{k} \delta_{\mu_j},$$

– for real-valued (continuous) marks in interval $[a,b]$:

$$M_Q(L) = \frac{1}{b-a}\nu(L \cap [a,b]).$$

2.3 The Nguyen-Zessin-Georgii Equation for Stationary Marked Gibbs Point Processes

The Takacs-Fiksel method is based on the famous Nguyen-Zessin-Georgii equation (NZG). It says that for any real-valued measurable function g it holds

$$\lambda \int_L \int_A g(\psi) P^!_m(d\psi) M(dm) = \int_L \int_A g(\psi) \exp\left\{-E(\psi,[0,m])\right\} P(d\psi) M_Q(dm). \tag{4}$$

The function g is sometimes called *test function*. (4) is a generalization of (5.5.18) in [12] to the case of a marked point process. $P^!_m$ is the reduced Palm distribution of Ψ, see [12] or [10] for a rigorous explanation. It can be interpreted as the conditional distribution of the marked point process $(\Psi\backslash\{[0,m]\})$, the point process without the marked point $[0,m]$, under the condition $[0,m] \in \Psi$. P is the stationary distribution of the marked Gibbs process. An explanation of the non-marked analogue of (4) is given in [12], and we refer to that book. Here we describe, in order to support understanding, how to estimate the left and right hand sides statistically, if a sample of the marked point process is given in a window W of area $\nu(W)$, and A is the whole space of point configurations. We assume that g is the particular function with $g(\psi) = 1$ if in the disk of radius r centred at 0 there is at least one point of ψ with mark l; otherwise $g(\psi) = 0$.
Left hand side: Take in turn marked points with marks in L as centres of disks of radius r and check whether in those disks there are points with mark l. Divide the total number of all these points by $\nu(W)$.
Right hand side: If we had not only one sample of Ψ in W we would look at the disk of radius r centred at 0. If in that disk would be a point with mark l we would calculate $\exp\{-E(\psi,[0,m])\}$ for all marks m in L and average with respect to M_Q. Otherwise the contribution of the sample would be zero. In the case of only one sample we use a lattice of points in W, which play the role of 0 relative to the same marked point pattern Ψ.

2.4 Marked Hard-Core Gibbs Point Processes

The hard sphere resp. hard disk model for the non-marked case (identical disks) is well known in statistical physics [8]. It can be easily extended for the marked case (variable radii) with the following pair potential:

$$V(r, m, m') = \begin{cases} \infty & \text{for } r \leq R(m) + R(m') \\ 0 & \text{else} \end{cases} .$$

Here $R(m)$ is a given radius function, which associates each mark to a hard-core radius (i.e. we assume that the radii are given by basic marks). Denote the corresponding chemical activities by $\alpha(m)$. They control the frequency of appearance of radii. It is natural to attach to each marked point configuration ψ the corresponding union of non-overlapping disks

$$\psi^* = \bigcup_{[x,m] \in \psi} b(x, R(m)).$$

Use in (NZG) for the test function g the indicator function

$$g(\psi) = \mathbf{1}\Big(b(0, R(\mu)) \cap \psi^* = \emptyset\Big), \tag{5}$$

$L = \{\mu\}$ and for A the whole set of all marked point sequences. Note that $g(\psi) = 0$ if the disk of radius $R(\mu)$ centred at 0 intersects the union set of disks ψ^*, otherwise it is $g(\psi) = 1$. Then we get

$$\lambda M(\{\mu\}) = e^{-\alpha(\mu)} \left(1 - H_s(R(\mu))\right) M_Q(\{\mu\}), \tag{6}$$

where H_s is the *spherical contact distribution function* of the random set Ψ^*, see [7]. $H_s(r)$ is the probability that a disk of radius r with randomly chosen centre x outside the disks of Ψ does not overlap any of those disks. Equation (6) gives a useful relation between the mark distribution M (which can be estimated statistically) and chemical activity $\alpha(m)$; M_Q can be chosen according to the statistician's taste, e.g. as uniform distribution.

3 Estimation of Gibbs Process Parameters by the Takacs-Fiksel Method

The Takacs-Fiksel method [14, 3] is based on (NZG), using a series of test functions $\{g_j\}$. The idea is to determine estimators of the left- and right-hand sides of (NZG) for the g_j. They depend on the parameter vector γ and are therefore denoted by $\hat{\mathcal{L}}_k(\gamma)$ and $\hat{\mathcal{R}}_j(\gamma)$. The estimator of γ is the solution of

$$\sum_j \left(\hat{\mathcal{L}}_j(\gamma) - \hat{\mathcal{R}}_j(\gamma)\right)^2 \underset{\gamma}{\rightarrow} \min! \tag{7}$$

In this section the following problem will be considered: Which test functions should be used for the Takacs-Fiksel method in the case of hard disk processes. [2] suggest choices of the g_i in the non-marked case.

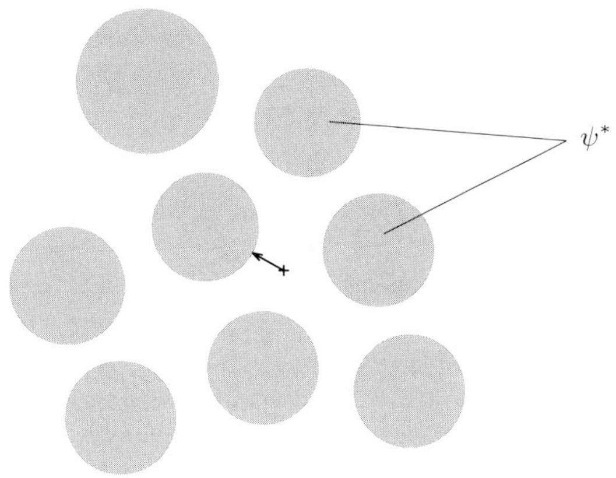

Fig. 1. H_s is the distribution function of the random distance from the randomly chosen point + outside of ψ^* to the random set ψ^* shown in grey.

Contact indicator functions. Consider the case of discrete marks $\mu_1,\ldots,\mu_k$ and uniform M_Q and use the test functions

$$g_j(\psi) = \mathbf{1}\Big(b(0, R(\mu_j)) \cap \psi^* = \emptyset\Big) \quad \text{for } j = 1, \ldots, k.$$

The explanation of g_j is the same as that of g in (5).

According to (6) it follows

$$\begin{aligned} \hat{\mathcal{L}}_j &= \hat{\lambda}\hat{M}(\{\mu_j\}) \\ \hat{\mathcal{R}}_j &= e^{-\hat{\alpha}_j}\Big(1 - \hat{H}_s(R(\mu_j))\Big)\frac{1}{k}. \end{aligned} \tag{8}$$

Here $\boldsymbol{\gamma}$ consists of the chemical activity values $\alpha_j = \alpha(\mu_j)$ for $j = 1, \ldots, k$ and possible parameters of the transition function $R(\cdot)$. The simplest case is $R(m) = m$. Then the optimal Takacs-Fiksel least square sum (7) can be made zero by solving the equations $\hat{\mathcal{L}}_j = \hat{\mathcal{R}}_j, j = 1, \ldots, k$, which yields the following estimators of the chemical activities:

$$\hat{\alpha}_j = \log \frac{1 - \hat{H}_s(R(\mu_j))}{\hat{\lambda}\, k \hat{M}(\{\mu_j\})}. \tag{9}$$

Here $\hat{H}_s(r)$ is an estimator of the spherical contact distribution function H_s of the random set ψ^*, see Sect. 2.4. Estimation formulas can be found in [12], p. 215. $\hat{M}$ is the *empirical* mark distribution; for discrete marks, $\hat{M}(\mu_j)$ is simply the relative frequency of mark μ_j in the point sample.

For the particular case of the non-marked hard-core Gibbs process (with identical diameters) the following table helps to estimate the chemical activity α based on estimated area fraction η. The values presented here in the first two lines result from simulations

η	0.2	0.3	0.4	0.5	0.55	0.6
α	-0.321	0.894	2.351	4.384	5.800	7.662
$\hat{\alpha}$	-0.334	0.887	2.356	4.36	5.74	7.54

described in [9]. There α was prescribed and η obtained statistically. The third line in the table shows estimates of α obtained by the Takacs-Fiksel method as described above, using the samples generated in [9].

What happens for ultra-dense packings? The spherical contact distribution then tends to one, especially for large $R(\mu_j)$. In fact, if there is no room for a further disk of radius $R(\mu_j)$ in a given disk union sample ψ^*, we get $\hat{H}_s(R(\mu_j)) = 1$ and estimation formula (9) is useless. (8) implies that the size of the observation window must grow exponentially in α to use estimation formula (9), in order to observe patterns with holes so large that the disk finds room. Consequently, for $\eta > 0.65$ we need samples which are too huge in size and computation time for present computer capabilities. Note that the main problem for large α is the behaviour of $\hat{H}_s$, not the simulation of the point samples; this can be done up to $\eta = 0.75$ with the Simulated Tempering algorithm [9].

Contact number functions. The test function

$$g_j(\psi) = \psi(b(0, r_j) \times M_j),$$

counts all points within distance r_j from the origin 0 the marks of which are in a subset M_j of $\mathbb{M}$. In the case of non-marked processes these test functions were proposed in [2] and [11] for repulsive pair potentials. For hard-core potentials with finite range, the radii r_j should be chosen between the hard-core distance h of the potential

$$h = \arg \sup_r \left\{ \inf_{m,m'} V(r, m, m') = \infty \right\}$$

and 1.5ϱ, where

$$\varrho = \arg \inf_r \left\{ \sup_{m,m'} V(r, m, m') = 0 \right\}.$$

Little is known about choosing M_j. For discrete marks $\mu_1, \ldots, \mu_k$ one may try singleton sets $M_j = \{\mu_j\}$

$$g_j(\psi) = \psi(b(0, r_{j_1}) \times \{\mu_{j_2}\}), \qquad j_2 = 1, \ldots, k,$$

for suitable radii r_{j_1} and marks μ_{j_2}. $g_j(\psi)$ is simply the number of points in $b(0, r_{j_1})$ having mark μ_{j_2}. In the case of such contact number functions the left-hand side of (NZG) is

$$\mathcal{L}_j = \lambda^2 K(r_j; M_j),$$

where $K(\cdot; \cdot)$ is related to Ripley's K-function $K(\cdot)$ (see [12] and [13] for explanation) as follows: $\lambda K(r)$ is the mean number of points within distance r of a typical point, while $\lambda K(r; M)$ is the mean number of points *with marks in* M within distance r of a typical point, the mark of which does not play a role.

For the estimation of the right-hand side some numerical integration has to be done:

$$\hat{\mathcal{R}}_j = \frac{1}{kn_y} \sum_{u=1}^{n_y} \sum_{l=1}^{k} \varphi\left(b(y_u, r_j) \times M_j\right) \cdot \exp\left\{ -\alpha_l - \sum_{i=1}^{n} V(x_i - y_u, m_i, \mu_l) \right\},$$

using a grid of points $\{y_j\}_{j=1,\ldots,n_y}$ on the observation window. By the way, this is an estimator without edge correction; it can be improved by edge correction principles; see [11] for details.

4 Water Droplets on Naphtalin-Brom Surface

The Takacs-Fiksel method described in the previous section is now applied on the following example. If water condenses on a plane surface of naphtalin-brom then characteristic patterns of water droplets appear. Figure 2 (a) shows an intermediate stage of this condensation process. We tried to analyse such patterns by the Gibbs process approach[1], where the points are the centres and the marks are the measured radii of the droplets on the photographic picture. In the given case, only three different radii were observed: 5, 6 and 7 μm. We considered the pattern as a sample of a marked Gibbs hard-core process (see Sect. 2.4) with $R(m) \equiv m$. For this simple model we got the estimates $\hat{\alpha}(5) = 3.93$, $\hat{\alpha}(6) = 2.44$ and $\hat{\alpha}(7) = 3.72$, obtained by (9). To check this model against our original data we simulated the stationary marked Gibbs hard-core process using the calculated activities $\hat{\alpha}(\mu)$. The result is disappointing: Fig. 2 (c) shows the pair correlation functions of the original pattern vs. the 95% confidence bands of simulated data (dashed lines).

A more careful look at the original picture shows that there is additional space between any pair of neighbouring disks, which is unusual for ordinary Gibbs hard-core processes. Also the peak positions of the pair correlation functions in Fig. 2 (c) may lead to this assumption. So one can expect that there are repulsive forces beyond the distance of the measured radii. For the marked Gibbs hard-core process that would mean $R(m) > m$ or, more likely, a more complex pair potential should be considered. The simplest case is the model $R(m) = a + m$ with an additional distance a to each radius, but this model turned out to be unable to fit the data. An alternative is linear expansion of disks, i.e. $R(m) = am$. The parameter vector is now $\gamma = (\alpha_5, \alpha_6, \alpha_7, a)$

- the chemical activity values $\alpha_j = \alpha(\mu_j)$ for the three observed marks $\mu_j = 5, 6$ and 7 and
- the expansion factor a, which is the only parameter of the pair potential V.

As described in Sect. 3 the three contact indicator functions are not sufficient to estimate additional parameters of the pair potential, so further test functions for the Takacs-Fiksel method are needed. We chose the following three contact indicator functions

[1] Klaus Mecke was so kind to inform us that physically a Gibbs process is not an appropriate model.

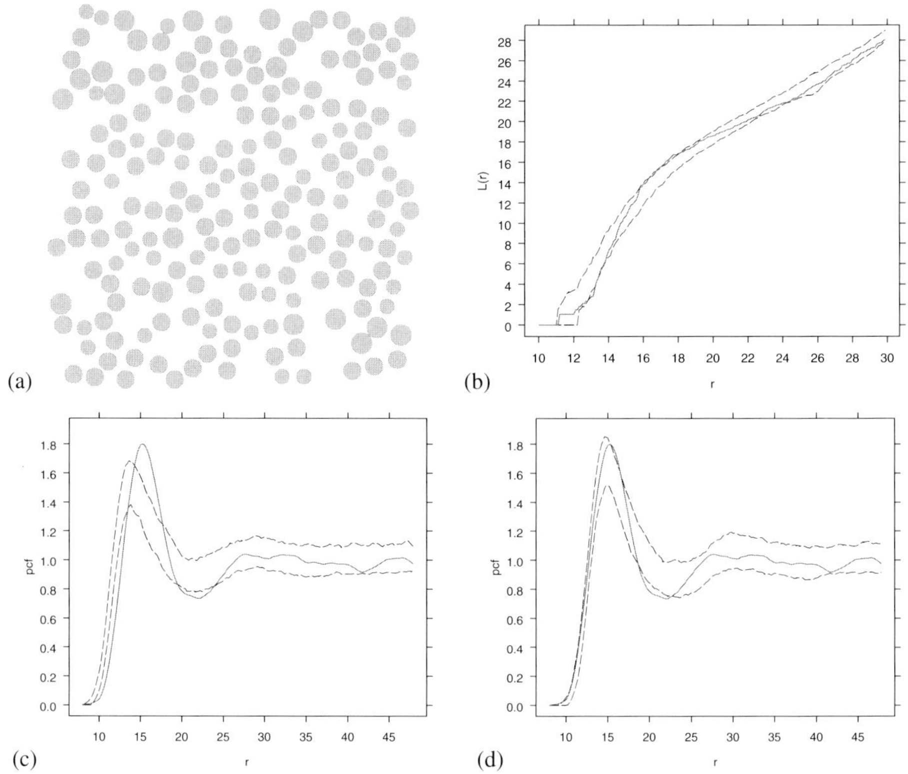

Fig. 2. (a) Water droplets on a surface of naphtalin-brom. Courtesy of K.-W. Stöckelhuber (b) 95% confidence bands of L-functions for $R(m) = 1.09m$ and empirical L-function; it is $L(r) = \sqrt{K(r)/\pi}$. (c) Pair correlation functions for $R(m) \equiv m$ (d) Pair correlation functions for $R(m) = 1.09m$

$$g_j(\psi) = \mathbf{1}\Big(b(0, a\mu_j) \cap \psi^* = \emptyset\Big), \qquad j = 1, 2, 3,$$

(as in the simple model $R(m) \equiv m$) and

$$g_j(\psi) = \psi(b(0, r_j) \times \mathbb{M}), \qquad j = 4, \ldots, 23$$

for 20 equidistant radii r_j between 5 and 20 as test functions. The estimation results are an expansion factor $\hat{a} = 1.09$ and chemical activities

μ	$R(\mu)$	$\hat{\alpha}(\mu)$
5	5.45	3.39
6	6.54	1.77
7	7.63	2.95 .

Figure 2 (d) shows 95% confidence bands of the pair correlation functions obtained by simulation. They are closer to their empirical counterparts than the pair correlation functions in Fig. 2 (c), but still the fit is not completely satisfying. Finally, Fig. 2 (b)

shows the related L-functions. Usually the L-function is used in goodness-of fit tests in point process statistics [11]. The fact that the empirical curve is nearly completely between the 95% confidence bands obtained by simulation of the model can be seen as a proof that the multiplicative model with $a = 1.09$ is close to be acceptable.

References

1. Beisbart, C., M. Kerscher, K. Mecke (2002): 'Mark correlations: relating physical properties to spatial distributions'. In: this volume.
2. Diggle, P.J., T. Fiksel, P. Grabarnik, Y. Ogata, D. Stoyan, M. Tanemura (1994): 'On Parameter Estimation for Pairwise Interaction Point Processes'. *Int. Stat. Rev.* **62**, 1, pp. 99 – 117.
3. Fiksel, T. (1988): 'Estimation of interaction potentials of Gibbsian point processes'. *statistics* **19**, pp. 77 – 86.
4. Geyer, C.J. (1999): 'Likelihood inference for spatial point processes'. In: *Stochastic Geometry: Likelihood and Computations*, ed. by O.E. Barndorff-Nielsen, W.S. Kendall and M.N.M. van Lieshout, (Chapman and Hall/CRC, London), pp. 79 – 140.
5. Goulard, M., A. Särkkä, P. Grabarnik (1996): 'Parameter estimation for marked Gibbs point processes through maximum pseudo-likelihood method'. *Scand. J. Statist.* **23**, pp. 365 – 379.
6. Gubner, J.A., W.-B. Chang, M.M. Hayat (2000): 'Performance analysis of hypothesis testing for sparse pairwise interaction point processes'. *IEEE Trans. Inform. Theory* **46**, no. 4, pp. 1357 – 1365.
7. Hug, D., G. Last, W. Weil (2002): 'A Survey on Contact Distributions'. In: this volume.
8. Löwen, H. (2000): 'Fun with Hard Spheres'. In: *Statistical Physics and Spatial Statistics*, ed. by K.R. Mecke, D. Stoyan (Lecture Notes in Physics 554, Springer), pp. 295 – 331.
9. Mase, S., J. Møller, D. Stoyan , R.P. Waagepetersen, G. Döge (2001): 'Packing Densities and Simulated Tempering for Hard Core Gibbs Point Processes'. *Ann. Inst. Statist. Math.*, **53**, 4, pp. 661 – 680.
10. Nagel, W. (2000): 'Stationary Models in Stochastic Geometry – Palm Distributions as Distributions of Typical Elements. An Approach Without Limits'. In: *Statistical Physics and Spatial Statistics*, ed. by K.R. Mecke, D. Stoyan (Lecture Notes in Physics 554, Springer), pp. 22 35.
11. Stoyan, D., H. Stoyan (1994): *Fractals, Random Shapes and Point Fields.* (John Wiley & Sons, Chichester)
12. Stoyan, D., W.S. Kendall, J. Mecke (1995): *Stochastic Geometry and its Applications*, 2nd edn. (Wiley & Sons, New York)
13. Stoyan, D. (2000): 'Basic Ideas of Spatial Statistics'. In: *Statistical Physics and Spatial Statistics*, ed. by K.R. Mecke, D. Stoyan (Lecture Notes in Physics 554, Springer), pp. 3 – 21.
14. Takacs, R. (1986): 'Estimator for the pair-potential of a Gibbsian point process'. *Math. Operationsf. Statist. ser. statistics* **17**, pp. 429 – 433.

Spatial Statistics of a Turbulent Random Multiplicative Branching Process

Martin Greiner

[1] Department of Physics, Duke University, Durham, NC 27708, USA;
[2] Max-Planck-Institut für Physik komplexer Systeme,
Nöthnitzer Str. 38, D–01187 Dresden, Germany

Abstract. Random multiplicative branching processes serve as empirical models for fully developed turbulence. It is necessary to discuss their spatial statistics in terms of a generating functional, so that a rigorous comparison with data can be performed. Some possible links to a Navier-Stokes dynamics are indicated.

All of us know about turbulence. It is a real-life phenomenon describing far-out-of-equilibrium situations: turbulent human relationships, turbulent ball games or turbulent stock markets, just to name a few. As those events come and go, no one is really able to predict, control and understand them. The latter is also exactly the case for the physics process coining the name: fully developed turbulence of fluid flows. Some even say that this process might represent the biggest challenge of classical statistical physics. In fact, this fortress has not been taken by cohorts of scientists besides many heroic attempts during the past century [1, 2]. How comes that turbulence is so tremendously difficult to understand?

To give a feeling for this, let us face the underlying equation of motion for incompressible fluid flows, the Navier-Stokes equation:

$$\partial_t \boldsymbol{v} + (\boldsymbol{v} \cdot \boldsymbol{\nabla})\boldsymbol{v} = -\boldsymbol{\nabla} p + \nu \boldsymbol{\nabla}^2 \boldsymbol{v} + \boldsymbol{f} \ . \tag{1}$$

It represents Newton's second law of classical mechanics: the left-hand side is the acceleration of a fluid particle moving with velocity $\boldsymbol{v}$ and the right-hand side are the forces acting on it. $\boldsymbol{f}$ stands for external forces representing in some respect the boundary conditions; this term introduces a large length-scale L. The friction force with the viscosity ν introduces another distinct length-scale, which is called the dissipation scale η and which is still macroscopic, but much smaller than L. In order to see what role the advection term $(\boldsymbol{v} \cdot \boldsymbol{\nabla})\boldsymbol{v}$ and the pressure term $-\boldsymbol{\nabla} p$ do play, we transform (1) into an energy balance equation by multiplication with $\boldsymbol{v}$, subsequent integration over a large volume and assuming stationarity:

$$\langle \boldsymbol{v} \cdot \boldsymbol{f} \rangle - \left\langle \frac{\nu}{2} \sum_{i,j=1}^{3} (\partial_i v_j + \partial_j v_i)^2 \right\rangle = 0 \ . \tag{2}$$

The first term with the external force feeds energy into the turbulent flow at the large scale L and dissipation takes it out at the small scale η as kinetic energy is then converted into heat. Somehow the energy has to be transported from L down to η. This is what the nonlinear local advection term and the nonlinear nonlocal pressure term are in charge for. Hence, turbulence is a multiscale process. The number of degrees of freedom are about $(L/\eta)^3$; since $L/\eta = \mathcal{O}(10^3 - 10^4)$ for realistic jet, wind-tunnel or atmospheric boundary layer turbulence this number is of the order 10^{10}. Because of the nonlinearities and nonlocalities involved in the energy cascade transport the interaction between these too-many degrees of freedom is strong. This is exactly the point why all analytical and numerical approaches starting with the Navier-Stokes equation have more or less failed to provide a complete understanding of turbulence.

The very large number of degrees of freedom justifies a statistical approach. Here two different routes have to be distinguished: a top-down approach, which again starts right away with the Navier-Stokes equation, but is again haunted by notorious difficulties. The other approach is bottom-up, aims at understanding the data phenomenology with simple empirical models first and only then tries to connect all the way back to the Navier-Stokes equation. It is the latter approach we are conveying.

Typically data is taken with an anemometer, which records components (typically only one) of the velocity field as a time series, as the turbulent flow is sweeping over it. Assuming fluctuations around the mean velocity to be small, such time-series can be interpreted as one-dimensional spatial series by the frozen flow hypothesis. These signals look very irregular, i.e. turbulent. The standard observables are velocity structure functions, which are moments of velocity increments $\Delta v_l = v(x+l) - v(x)$. They show an approximate scaling $\langle \Delta v_l^n \rangle \sim l^{\zeta_n}$ for distances $\eta \ll l \ll L$; scaling exponents are found to be $\zeta_n \approx n/3 - \tau_{n/3}$, where $\tau_{n/3}$ represent the intermittency corrections to the $n/3$-behavior of Kolmogorov's K41 picture [3] of fully developed turbulence. This scale-invariance is to be expected since, as we have already argued before, no distinct length scales are present in the Navier-Stokes equation other than the large and small scales L and η, respectively. However, the observed scaling is not perfect. Hence, legitimate questions to ask are: are velocity structure functions the ideal observational tool for scaling? Maybe other observables are more suited? Perhaps other fields than the velocity field should be looked at?

Picking up the last question and remembering our intuitive picture on the energy transport process through the scales, a kind of "energy" field appears to be more appealing. A good representative is the so-called energy dissipation field $\varepsilon(\boldsymbol{r})$, which is the quantity given within the angle bracket of the second term of (2). The surrogate field $\varepsilon(x) = 15\nu(\partial_x v_{\|})^2$ is easily constructed from the streamwise component of the measured velocity field and reveals how energy flux is taken out at the dissipation scale: spatial fluctuations $\varepsilon(x)/\langle\varepsilon\rangle$ around the average are anomalously strong and intermittent, values ranging between about 10^{-2} to 10^2 for large-Reynolds number turbulent flows. The spatial statistics of these fields now serve as guidance to learn about the details of the energy cascade process and to construct respective models.

A particular attractive and robust class of such data-driven models are random multiplicative branching processes. They introduce a nested hierarchy of scales $l_j = L/2^j$ ranging from the large scale L down to the small scale $\eta = L/2^J$. The synthetic field construction uses a cascade generator, which redistributes the energy flux density $\Pi_{k_1 \cdots k_j}$ associated to the parent interval with length l_j nonuniformly onto the left $(k_1 \cdots k_j 0)$ and the right $(k_1 \cdots k_j 1)$ offspring intervals with length l_{j+1}:

$$\begin{aligned} \Pi_{k_1\cdots k_j 0} &= q_{k_1\cdots k_j 0}\Pi_{k_1\cdots k_j} \ , \\ \Pi_{k_1\cdots k_j 1} &= q_{k_1\cdots k_j 1}\Pi_{k_1\cdots k_j} \ ; \end{aligned} \tag{3}$$

the random variables q are drawn from a scale-independent probabilistic splitting function $p_{\mathrm{sf}}(q_{k_1\cdots k_j 0}, q_{k_1\cdots k_j 1})$ and are independent from those belonging to other branchings. Starting at L the cascade generator is applied over and over again until η is reached. Then the energy flux amplitudes are set equal to the energy dissipation amplitudes:

$$\varepsilon_{k_1\cdots k_J} = \Pi_{k_1\cdots k_J} = q_{k_1}q_{k_1 k_2}\cdots q_{k_1\cdots k_J} \ . \tag{4}$$

They are a product of J iid random numbers with mean $\langle q\rangle = 1$, so that with some small probability the observed strong intermittent fluctuations around $\langle\varepsilon\rangle = 1$ are easily reproduced. A synthetic model field realization comes with support $[0, L]$ and can be expressed as

$$\varepsilon(x) = \sum_{k_1,\ldots,k_J=0}^{1} \varepsilon_{k_1\cdots k_J}\Theta\left(\frac{L}{\eta}\left(x - \sum_{j=1}^{J} k_j l_j\right)\right) \ , \tag{5}$$

where $\Theta(x) = 1$ for $0\leq x\leq L$ and 0 otherwise is an index function.

The qualitative reproduction of the observed intermittent fluctuations is only a first test of the model. A much more sophisticated and complete test would be to compare the spatial statistics. The latter is characterized by the multivariate probability distribution function

$$p_{\mathrm{field}}(\varepsilon(x)) = p_{\mathrm{field}}(\varepsilon_{0\cdots 0}, \ldots, \varepsilon_{1\cdots 1}) \ , \tag{6}$$

which gives the probability that a specific cascade field configuration with the 2^J amplitudes $\varepsilon_{k_1\cdots k_J}$ occurs. Another, but equivalent characterization of the field statistics is given by the Fourier transform of (6):

$$\begin{aligned} Z[\lambda_\kappa] &= \int d\varepsilon_{0\cdots 0}\cdots d\varepsilon_{1\cdots 1}\, p_{\mathrm{field}}(\varepsilon_{0\cdots 0}, \ldots, \varepsilon_{1,\ldots,1}) \exp\left(i\sum_{\kappa}\lambda_\kappa\varepsilon_\kappa\right) \\ &= \left\langle \exp\left(i\sum_\kappa \lambda_\kappa\varepsilon_\kappa\right)\right\rangle \\ &= 1 + i\sum_{\kappa_1}\lambda_{\kappa_1}\rho_{\kappa_1} + \frac{i^2}{2!}\sum_{\kappa_1}\sum_{\kappa_2}\lambda_{\kappa_1}\lambda_{\kappa_2}\rho_{\kappa_1\kappa_2} + \ldots \ ; \end{aligned} \tag{7}$$

here we have used the abbreviation $\kappa = (k_1\cdots k_J)$ for the binary address of the field amplitude ε_κ. $Z[\lambda_\kappa]$ contains the same information as $p_{\mathrm{field}}(\varepsilon(x))$. The second step of (7) is simply a shorthand notation of the first line, where the angle brackets symbolize an expectation value. The coefficients of the Taylor expansion of (7) are the n-point correlation densities

$$\rho_{\kappa_1\cdots\kappa_n} = \langle\varepsilon_{\kappa_1}\cdots\varepsilon_{\kappa_n}\rangle = \frac{1}{i^n}\left.\frac{\partial^n Z[\lambda_\kappa]}{\partial\lambda_{\kappa_1}\cdots\partial\lambda_{\kappa_n}}\right|_{\lambda=0} \ . \tag{8}$$

This equation is the reason why $Z[\lambda_\kappa]$ is called a generating or characteristic functional. All this demonstrates that

$$p_{\text{field}}(\varepsilon(x)) \quad \longleftrightarrow \quad Z[\lambda_\kappa] \quad \longleftrightarrow \quad \rho_{\kappa_1 \cdots \kappa_n} \tag{9}$$

all access the spatial statistics in a complete way and are thus equivalent to each other. Experimentalists like to analyse the n-point correlation densities, whereas mathematicians and theoretical physicists like to deal with the generating functional.

For the random multiplicative branching process of (3)–(5) an analytic expression for the generating functional can be given [4, 5]. The secret here is to use the logarithm of (4), i.e.

$$\ln \varepsilon_{k_1 \cdots k_J} = \sum_{j=1}^{J} \ln q_{k_1 \cdots k_j} \ , \tag{10}$$

and the modified generating functional

$$K[\lambda_\kappa] = \ln \tilde{Z}[\lambda_\kappa] = \ln \left\langle \exp \left(\sum_\kappa \lambda_\kappa \ln \varepsilon_\kappa \right) \right\rangle \ . \tag{11}$$

For demonstration we pick a branching process of only $J{=}2$ cascade steps; then (11) becomes:

$$\begin{aligned} & K[\lambda_{00}, \lambda_{01}, \lambda_{10}, \lambda_{11}] = \\ & = \ln \left\langle \varepsilon_{00}^{\lambda_{00}} \varepsilon_{01}^{\lambda_{01}} \varepsilon_{10}^{\lambda_{10}} \varepsilon_{11}^{\lambda_{11}} \right\rangle \\ & = \ln \left\langle (q_0 q_{00})^{\lambda_{00}} (q_0 q_{01})^{\lambda_{01}} (q_1 q_{10})^{\lambda_{10}} (q_1 q_{11})^{\lambda_{11}} \right\rangle \\ & = \ln \left\langle q_0^{\lambda_{00}+\lambda_{01}} q_1^{\lambda_{10}+\lambda_{11}} \right\rangle + \ln \left\langle q_{00}^{\lambda_{00}} q_{01}^{\lambda_{01}} \right\rangle + \ln \left\langle q_{10}^{\lambda_{10}} q_{11}^{\lambda_{11}} \right\rangle \\ & = Q[\lambda_{00}{+}\lambda_{01}, \lambda_{10}{+}\lambda_{11}] + Q[\lambda_{00}, \lambda_{01}] + Q[\lambda_{10}, \lambda_{11}] \ . \end{aligned} \tag{12}$$

In the third step we have used that the bituples of random multiplicative weights (q_0, q_1), (q_{00}, q_{01}) and (q_{10}, q_{11}) belong to different branchings and are thus independent from each other. The branching generating function

$$Q[\lambda_L, \lambda_R] = \ln \left(\int dq_L dq_R \, p_{\text{sf}}(q_L, q_R) q_L^{\lambda_L} q_R^{\lambda_R} \right) \tag{13}$$

is directly linked to the splitting function $p_{\text{sf}}(q_L, q_R)$ of the cascade generator and is the building block for the generating functional (12): one Q for the only branching of the first cascade step and two Qs for the two branchings of the second cascade step; see last line of (12). It is this structure, which directly generalizes to an arbitrary number J of cascade steps [4, 5]: the generating functional (11) is a sum of branching generating functions over all branchings.

From the analytic solution of the generating functional various types of n-point correlation densities can be calculated straightforwardly: n-point correlation densities between logarithmic field amplitudes,

$$\langle \ln \varepsilon_{\kappa_1} \cdots \ln \varepsilon_{\kappa_n} \rangle = \left. \frac{\partial^n \exp(K[\lambda_\kappa])}{\partial \lambda_{\kappa_1} \cdots \partial \lambda_{\kappa_n}} \right|_{\lambda=0} , \tag{14}$$

n-point cumulants between logarithmic field amplitudes,

$$\langle \ln \varepsilon_{\kappa_1} \cdots \ln \varepsilon_{\kappa_n} \rangle_c = \left. \frac{\partial^n K[\lambda_\kappa]}{\partial \lambda_{\kappa_1} \cdots \partial \lambda_{\kappa_n}} \right|_{\lambda=0} , \tag{15}$$

as well as n-point correlation densities between conventional field amplitudes,

$$\langle \varepsilon_{\kappa_1} \cdots \varepsilon_{\kappa_n} \rangle_c = \exp\left(K[\lambda_{\kappa=\kappa_i=1}, \lambda_{\kappa\neq\kappa_i=0}] \right) . \tag{16}$$

Their values depend on a set of mutual ultrametric distances $D(\kappa_1, \kappa_2)$ between two bins κ_1 and κ_2, which is a quantity inherent to the hierarchical binary tree-like structure of the random multiplicative branching process. An experimentalist is not able to observe it. Consider for example a two-point correlation density, which is sampled over point pairs with Euclidean two-point distance d. In order to realistically compare quantities like (14)–(16) to their experimental counterparts, we have to introduce an additional experimental-like nonrestrictive sampling using conditional probabilities like $p_{\mathrm{samp}}(D|d)$; see Refs. [6, 7] for more details. Then two-point correlators take on the form [7]

$$\frac{\langle \varepsilon^{n_1}(x+d)\varepsilon^{n_2}(x) \rangle}{\langle \varepsilon^{n_1}(x+d) \rangle \langle \varepsilon^{n_2}(x) \rangle} = a_{n_1 n_2} \left(\frac{L}{d} \right)^{\tau[n_1, n_2]} f_{n_1 n_2}(d/L) , \tag{17}$$

where $f_{n_1 n_2}(d \ll L) = 1$ represents a scaling function and $\tau[n_1, n_2] = \tau_{n_1+n_2} - \tau_{n_1} - \tau_{n_2}$ with $\tau_n = \log_2 \langle q^n \rangle$. Analogously, two-point cumulants become [6]

$$\left\langle \ln^{n-1} \varepsilon(x+d) \, \ln \varepsilon(x) \right\rangle_c = G(d, J) \langle \ln^n q \rangle_c , \tag{18}$$

where the geometric function G does not depend on the order n and is related to the first moment of $p_{\mathrm{samp}}(D|d)$. Observables like (17) and (18) help to test the validity of random multiplicative branching processes for fully developed turbulence and to learn about the best approximate cascade generator $p_{\mathrm{sf}}(q_L, q_R)$ via the extractable multifractal scaling exponents τ_n and the same-lineage cumulants $\langle \ln^n q \rangle_c$; consult [8] for more details.

There are also other observables which test the validity of random multiplicative branching processes and allow to bridge the latter to other data-driven models. Some of these non-standard observables are breakup coefficients [9, 10] and Markovian Kramers-Moyal coefficients [11].

Fundamentally the presented random multiplicative branching processes have the drawback of having no time degree of freedom. However, generalizations in this direction are possible [12] and might help to learn about dynamical properties of the turbulent energy cascade. – For the ultimate goal to understand the random multiplicative branching processes in terms of the Navier-Stokes equation still something can be learned from the geometric cascade models: a wavelet representation of the respective two-point correlation densities leads to a more-or-less complete diagonalisation [13, 14, 15, 16], revealing that wavelets $\Psi_{jm}(x)$ appear to be the correct degrees of freedom to describe turbulence. A very severe truncation

$$\boldsymbol{v}(\boldsymbol{r},t) \longrightarrow v(x,t) = \sum_{jm} v_{jm}(t)\Psi_{jm}(x) \; , \tag{19}$$

of the velocity field, once inserted into (1), motivates the dynamical toy-model amplitude equations

$$\partial_t v_1 = ik_1 \sum_{2,3} c(1;2,3) v_2 v_3 - \nu k_1^2 v_1 + f_1 \; , \tag{20}$$

known as the hierarchical shell model [17]. The last two terms only act at the small and large scales. Here it would be very interesting to find out which, if any, structure of the coupling $c(1{=}j_1m_1; 2{=}j_2m_2, 3{=}j_3m_3)$ between wavelet modes is required to reproduce the spatial statistics of the random multiplicative branching processes. This then would serve as a first link between the empirical random multiplicative cascade processes and the Navier-Stokes equation. Work and more dreams in this direction are presently in progress.

Acknowledgements

I thank J. Schmiegel, J. Cleve, T. Dziekan, H. Eggers, B. Jouault, J. Giesemann and P. Lipa for their cooperation over many years; without them these studies would not have come to such a mature stage.

References

1. Monin, A.S. and Yaglom, A.M. (1971): *Statistical Fluid Mechanics*, Vols. 1 and 2 (MIT Press, Cambridge).
2. Frisch, U. (1995): *Turbulence* (Cambridge University Press, Cambridge).
3. Kolmogorov, A.N. (1962): *J. Fluid Mech.* **13**, 82.
4. Greiner, M., H. Eggers, P. Lipa (1998): *Phys. Rev. Lett.* **80**, 5333.
5. Greiner, M., J. Schmiegel, F. Eickemeyer, P. Lipa, H. Eggers (1998): *Phys. Rev. E* **58**, 554.
6. Eggers, H.C., T. Dziekan, M. Greiner (2001): *Phys. Lett. A* **281**, 249.
7. Schmiegel, J., T. Dziekan, B. Jouault, J. Cleve, M. Greiner (2001): *preprint*.
8. Dziekan, T. (2001): *Zwei-Punkt Kumulanten in synthetischer Turbulenz* (diploma thesis, TU Dresden).
9. Jouault, B., P. Lipa, M. Greiner (1999): *Phys. Rev. E* **59**, 2451.
10. Jouault, B., M. Greiner, P. Lipa (2000): *Physica D* **136**, 125.
11. Cleve, J., M. Greiner (2000): *Phys. Lett. A* **273**, 104.
12. Schmiegel, J., H. Eggers, M. Greiner (2001): *preprint* cond-mat/0106347.
13. Greiner, M., P. Lipa, P. Carruthers (1995): *Phys. Rev. E* **51**, 1948.
14. Greiner, M., J. Giesemann, P. Lipa, P. Carruthers (1996): *Z. Phys. C* **69**, 305.
15. Giesemann, J., M. Greiner, P. Lipa (1997): *Physica A* **247**, 41.
16. Greiner, M., J. Giesemann (1998), In: *Wavelets in Physics*, ed. by L.Z. Fang, R.L. Thews (World Scientific, Singapore), pp. 89-131.
17. Benzi, R., L. Biferale, R. Tripiccione, E. Trovatore (1997): *Phys. Fluids* **9**, 2355.

Subject Index

Druck: betz-druck GmbH, D-64291 Darmstadt
Verarbeitung: Buchbinderei Schäffer, D-67269 Grünstadt

Lecture Notes in Physics

For information about Vols. 1–562
please contact your bookseller or Springer-Verlag

Vol. 563: F. C. Lázaro, M. J. Arévalo (Eds.), Binary Stars. Selected Topics on Observations and Physical Processes. 1999.IX, 327 pages. 2001.

Vol. 564: T. Pöschel, S. Luding (Eds.), Granular Gases. VIII, 457 pages. 2001.

Vol. 565: E. Beaurepaire, F. Scheurer, G. Krill, J.-P. Kappler (Eds.), Magnetism and Synchrotron Radiation. XIV, 388 pages. 2001.

Vol. 566: J. L. Lumley (Ed.), Fluid Mechanics and the Environment: Dynamical Approaches. VIII, 412 pages. 2001.

Vol. 567: D. Reguera, L. L. Bonilla, J. M. Rubí (Eds.), Coherent Structures in Complex Systems. IX, 465 pages. 2001.

Vol. 568: P. A. Vermeer, S. Diebels, W. Ehlers, H. J. Herrmann, S. Luding, E. Ramm (Eds.), Continuous and Discontinuous Modelling of Cohesive-Frictional Materials. XIV, 307 pages. 2001.

Vol. 569: M. Ziese, M. J. Thornton (Eds.), Spin Electronics. XVII, 493 pages. 2001.

Vol. 570: S. G. Karshenboim, F. S. Pavone, F. Bassani, M. Inguscio, T. W. Hänsch (Eds.), The Hydrogen Atom: Precision Physics of Simple Atomic Systems. XXIII, 293 pages. 2001.

Vol. 571: C. F. Barenghi, R. J. Donnelly, W. F. Vinen (Eds.), Quantized Vortex Dynamics and Superfluid Turbulence. XXII, 455 pages. 2001.

Vol. 572: H. Latal, W. Schweiger (Eds.), Methods of Quantization. XI, 224 pages. 2001.

Vol. 573: H. M. J. Boffin, D. Steeghs, J. Cuypers (Eds.), Astrotomography. XX, 434 pages. 2001.

Vol. 574: J. Bricmont, D. Dürr, M. C. Galavotti, G. Ghirardi, F. Petruccione, N. Zanghi (Eds.), Chance in Physics. XI, 288 pages. 2001.

Vol. 575: M. Orszag, J. C. Retamal (Eds.), Modern Challenges in Quantum Optics. XXIII, 405 pages. 2001.

Vol. 576: M. Lemoine, G. Sigl (Eds.), Physics and Astrophysics of Ultra-High-Energy Cosmic Rays. X, 327 pages. 2001.

Vol. 577: I. P. Williams, N. Thomas (Eds.), Solar and Extra-Solar Planetary Systems. XVIII, 255 pages. 2001.

Vol. 578: D. Blaschke, N. K. Glendenning, A. Sedrakian (Eds.), Physics of Neutron Star Interiors. XI, 509 pages. 2001.

Vol. 579: R. Haug, H. Schoeller (Eds.), Interacting Electrons in Nanostructures. X, 227 pages. 2001.

Vol. 580: K. Baberschke, M. Donath, W. Nolting (Eds.), Band-Ferromagnetism: Ground-State and Finite-Temperature Phenomena. IX, 394 pages. 2001.

Vol.581: J. M. Arias, M. Lozano (Eds.), An Advanced Course in Modern Nuclear Physics. XI, 346 pages. 2001.

Vol.582: N. J. Balmforth, A. Provenzale (Eds.), Geomorphological Fluid Mechanics. X, 579 pages. 2001.

Vol.583: W. Plessas, L. Mathelitsch (Eds.), Lectures on Quark Matter, XIII, 334 pages. 2002.

Vol.584: W. Köhler, S. Wiegand (Eds.), Thermal Nonequilibrium Phenomena in Fluid Mixtures. XVII, 470 pages. 2002.

Vol.585: M. Lässig, A. Valleriani (Eds.), Biological Evolution and Statistical Physics. XI, 337 pages. 2002.

Vol.586: Y. Auregan, A. Maurel, V. Pagneux, J.-F. Pinton (Eds.), Sound–Flow Interactions. XVI, 286 pages. 2002

Vol.587: D. Heiss (Ed.), Fundamentals of Quantum Information. Quantum Computation, Communication, Decoherence and All That. XIII, 265 pages. 2002.

Vol.588: Y. Watanabe, S. Heun, G. Salviati, N. Yamamoto (Eds.), Nanoscale Spectroscopy and Its Applications to Semiconductor Research. XV, 306 pages. 2002.

Vol.589: A. W. Guthmann, M. Georganopoulos, A. Marcowith, K. Manolakou (Eds.), Relativistic Flows in Astrophysics. XII, 241 pages. 2002

Vol.590: D. Benest, C. Froeschlé (Eds.), Singularities in Gravitational Systems. Applications to Chaotic Transport in the Solar System. XI, 215 pages. 2002

Vol.591: M. Beyer (Ed.), CP Violation in Particle, Nuclear and Astrophysics. XI, 334 pages. 2002

Vol.592: S. Cotsakis, L. Papantonopoulos (Eds.), Cosmological Crossroads. An Advanced Course in Mathematical, Physical and String Cosmology. XVI, 477 pages. 2002

Vol.593: D. Shi, B. Aktaş, L. Pust, F. Mikhailov (Eds.), Nanostructured Magnetic Materials and Their Applications. XII, 289 pages. 2002

Vol.594: S. Odenbach (Ed.),Ferrofluids. Magnetical Controllable Fluids and Their Applications. XI, 253 pages. 2002

Vol.595: C. Berthier, L. P. Lévy, G. Martinez (Eds.), High Magnetic Fields. Applications in Condensed Matter Physics and Spectroscopy. X, 493 pages. 2002

Vol.596: F. Scheck, H. Upmeier, W. Werner (Eds.), Noncommutative Geometry and the Standard Model of Elememtary Particle Physics. XII, 346 pages. 2002

Vol.597: P. Garbaczewski, R. Olkiewicz (Eds.), Dynamics of Dissipation. X, 512 pages. 2002

Vol.598: K. Weiler (Ed.), Supernovae and Gamma-Ray Bursters. X, 350 pages. 2002

Vol.600: K. Mecke, D. Stoyan (Eds.), Morphology of Condensed Matter. Physcis and Geometry of Spatial Complex Systems. XVIII, 439 pages. 2002

Vol.601: F. Mezei, C. Pappas, T. Gutberlet (Eds.), Neutron Spin Echo Spectroscopy. Basics, Trends and Applications. XVII, 351 pages. 2003

Vol.604: J. Frauendiener, H. Friedrich (Eds.), The Conformal Structure of Space-Time: Geometry, Analysis, Numerics. XIV 373pages. 2002

Vol.608: F. Courbin, D. Minniti (Eds.), Gravitational Lensing:An Astrophysical Tool. XI, 191 pages. 2002

Vol.611: A. Buchleitner, K. Hornberger (Eds.), Coherent Evolution in Noisy Environments. XV, 293 pages. 2002

Monographs

For information about Vols. 1–29 please contact your bookseller or Springer-Verlag

Vol. m 30: A. J. Greer, W. J. Kossler, Low Magnetic Fields in Anisotropic Superconductors. VII, 161 pages. 1995.

Vol. m 31 (Corr. Second Printing): P. Busch, M. Grabowski, P.J. Lahti, Operational Quantum Physics. XII, 230 pages. 1997.

Vol. m 32: L. de Broglie, Diverses questions de mécanique et de thermodynamique classiques et relativistes. XII, 198 pages. 1995.

Vol. m 33: R. Alkofer, H. Reinhardt, Chiral Quark Dynamics. VIII, 115 pages. 1995.

Vol. m 34: R. Jost, Das Märchen vom Elfenbeinernen Turm. VIII, 286 pages. 1995.

Vol. m 35: E. Elizalde, Ten Physical Applications of Spectral Zeta Functions. XIV, 224 pages. 1995.

Vol. m 36: G. Dunne, Self-Dual Chern-Simons Theories. X, 217 pages. 1995.

Vol. m 37: S. Childress, A.D. Gilbert, Stretch, Twist, Fold: The Fast Dynamo. XI, 406 pages. 1995.

Vol. m 38: J. González, M. A. Martín-Delgado, G. Sierra, A. H. Vozmediano, Quantum Electron Liquids and High-Tc Superconductivity. X, 299 pages. 1995.

Vol. m 39: L. Pittner, Algebraic Foundations of Non-Com-mutative Differential Geometry and Quantum Groups. XII, 469 pages. 1996.

Vol. m 40: H.-J. Borchers, Translation Group and Particle Representations in Quantum Field Theory. VII, 131 pages. 1996.

Vol. m 41: B. K. Chakrabarti, A. Dutta, P. Sen, Quantum Ising Phases and Transitions in Transverse Ising Models. X, 204 pages. 1996.

Vol. m 42: P. Bouwknegt, J. McCarthy, K. Pilch, The W3 Algebra. Modules, Semi-infinite Cohomology and BV Algebras. XI, 204 pages. 1996.

Vol. m 43: M. Schottenloher, A Mathematical Introduction to Conformal Field Theory. VIII, 142 pages. 1997.

Vol. m 44: A. Bach, Indistinguishable Classical Particles. VIII, 157 pages. 1997.

Vol. m 45: M. Ferrari, V. T. Granik, A. Imam, J. C. Nadeau (Eds.), Advances in Doublet Mechanics. XVI, 214 pages. 1997.

Vol. m 46: M. Camenzind, Les noyaux actifs de galaxies. XVIII, 218 pages. 1997.

Vol. m 47: L. M. Zubov, Nonlinear Theory of Dislocations and Disclinations in Elastic Body. VI, 205 pages. 1997.

Vol. m 48: P. Kopietz, Bosonization of Interacting Fermions in Arbitrary Dimensions. XII, 259 pages. 1997.

Vol. m 49: M. Zak, J. B. Zbilut, R. E. Meyers, From Instability to Intelligence. Complexity and Predictability in Nonlinear Dynamics. XIV, 552 pages. 1997.

Vol. m 50: J. Ambjørn, M. Carfora, A. Marzuoli, The Geometry of Dynamical Triangulations. VI, 197 pages. 1997.

Vol. m 51: G. Landi, An Introduction to Noncommutative Spaces and Their Geometries. XI, 200 pages. 1997.

Vol. m 52: M. Hénon, Generating Families in the Restricted Three-Body Problem. XI, 278 pages. 1997.

Vol. m 53: M. Gad-el-Hak, A. Pollard, J.-P. Bonnet (Eds.), Flow Control. Fundamentals and Practices. XII, 527 pages. 1998.

Vol. m 54: Y. Suzuki, K. Varga, Stochastic Variational Approach to Quantum-Mechanical Few-Body Problems. XIV, 324 pages. 1998.

Vol. m 55: F. Busse, S. C. Müller, Evolution of Spontaneous Structures in Dissipative Continuous Systems. X, 559 pages. 1998.

Vol. m 56: R. Haussmann, Self-consistent Quantum Field Theory and Bosonization for Strongly Correlated Electron Systems. VIII, 173 pages. 1999.

Vol. m 57: G. Cicogna, G. Gaeta, Symmetry and Perturbation Theory in Nonlinear Dynamics. XI, 208 pages. 1999.

Vol. m 58: J. Daillant, A. Gibaud (Eds.), X-Ray and Neutron Reflectivity: Principles and Applications. XVIII, 331 pages. 1999.

Vol. m 59: M. Kriele, Spacetime. Foundations of General Relativity and Differential Geometry. XV, 432 pages. 1999.

Vol. m 60: J. T. Londergan, J. P. Carini, D. P. Murdock, Binding and Scattering in Two-Dimensional Systems. Applications to Quantum Wires, Waveguides and Photonic Crystals. X, 222 pages. 1999.

Vol. m 61: V. Perlick, Ray Optics, Fermat's Principle, and Applications to General Relativity. X, 220 pages. 2000.

Vol. m 62: J. Berger, J. Rubinstein, Connectivity and Superconductivity. XI, 246 pages. 2000.

Vol. m 63: R. J. Szabo, Ray Optics, Equivariant Cohomology and Localization of Path Integrals. XII, 315 pages. 2000.

Vol. m 64: I. G. Avramidi, Heat Kernel and Quantum Gravity. X, 143 pages. 2000.

Vol. m 65: M. Hénon, Generating Families in the Restricted Three-Body Problem. Quantitative Study of Bifurcations. XII, 301 pages. 2001.

Vol. m 66: F. Calogero, Classical Many-Body Problems Amenable to Exact Treatments. XIX, 749 pages. 2001.

Vol. m 67: A. S. Holevo, Statistical Structure of Quantum Theory. IX, 159 pages. 2001.

Vol. m 68: N. Polonsky, Supersymmetry: Structure and Phenomena. Extensions of the Standard Model. XV, 169 pages. 2001.

Vol. m 69: W. Staude, Laser-Strophometry. High-Resolution Techniques for Velocity Gradient Measurements in Fluid Flows. XV, 178 pages. 2001.

Vol. m 70: P. T. Chruściel, J. Jezierski, J. Kijowski, Hamiltonian Field Theory in the Radiating Regime. VI, 172 pages. 2002.

Vol. m 71: S. Odenbach, Magnetoviscous Effects in Ferrofluids. X, 151 pages. 2002.

Vol. m 72: J. G. Muga, R. Sala Mayato, I. L. Egusquiza (Eds.), Time in Quantum Mechanics. XII, 419 pages. 2002.